W9-CKQ-594

Euclid Public Library
631 E. 222nd Street
Euclid, Ohio 44123
216-261-5300

GEM

GEM

THE DEFINITIVE VISUAL GUIDE

Penguin Random House

DK LONDON

Senior Editor	Anna Fischel
Senior Art Editors	Jane Ewart, Gadi Farfour
Project Editor	Hugo Wilkinson
Designers	Helen Spencer, Stephen Bere, Katie Cavanagh, Renata Latipova, Clare Joyce
Photographer	Ruth Jenkinson
US Senior Editor	Shannon Beatty
US Editor	Jane Perlmutter
Picture Research	Sarah Smithies, Sarah Hopper
DK Picture Library	Martin Copeland
Jacket Designer	Mark Cavanagh
Jacket Editor	Claire Gell
Jacket Design Development	Sophia M.T.T.
Producer	Mandy Inness
Pre-production Producer	Andy Hilliard
Managing Editor	Gareth Jones
Senior Managing Art Editor	Lee Griffiths
Art Director	Karen Self
Associate Publishing Director	Liz Wheeler
Publishing Director	Jonathan Metcalf

DK INDIA

Senior Managing Art Editor	Arunesh Talapatra
Senior Art Editor	Chhaya Sajwan
Art Editors	Priyansha Tuli, Roshni Kapur, Sudakshina Basu
Production Manager	Pankaj Sharma
Pre-production Manager	Balwant Singh
Senior DTP Designer	Neeraj Bhatia
DTP Designers	Jaypal Chauhan, Nityanand Kumar
Picture Research Manager	Taiyaba Khatoon
Picture Researcher	Sakshi Saluja

Smithsonian

Established in 1846, the Smithsonian—the world's largest museum and research complex—includes 19 museums and galleries and the National Zoological Park. The total number of artifacts, works of art, and specimens in the Smithsonian's collections is estimated at 138 million, much of which is contained in the National Museum of Natural History, which holds more than 126 million specimens and objects. The Smithsonian is a renowned research center, dedicated to public education, national service, and scholarship in the arts, sciences, and history.

SMITHSONIAN CURATOR

Dr. Jeffrey E. Post, Chairman, Department of Mineral Sciences and Curator, National Gem and Mineral Collection. National Museum of Natural History, Smithsonian

SMITHSONIAN ENTERPRISES

Product Development Manager	Kealy E. Gordon
Licensing Manager	Ellen Nanney
Vice President, Education and Consumer Products	Brigid Ferraro
Senior Vice President, Education and Consumer Products	Carol LeBlanc
President	Chris Liedel

First American Edition, 2016
Published in the United States by DK Publishing
345 Hudson Street, New York, New York 10014

Copyright © 2016 Dorling Kindersley Limited
DK, a Division of Penguin Random House LLC
Foreword copyright @ 2016 Aja Raden

16 17 18 19 20 10 9 8 7 6 5 4 3 2 1
001–282973–Oct/2016
All rights reserved

Without limiting the rights under copyright reserved above, no part of this publication may be reproduced, stored in, or introduced into a retrieval system, or transmitted, in any form, or by any means (electronic, mechanical, photocopying, recording, or otherwise), without the prior written permission of both the copyright owner and the above publisher of this book.

Published in Great Britain by Dorling Kindersley Limited.

A catalog record for this book is available from the Library of Congress.
ISBN 978-1-4654-5356-3

DK books are available at special discounts when purchased in bulk for sales promotions, premiums, fund-raising, or educational use. For details, contact: DK Publishing Special Markets, 345 Hudson Street, New York, New York 10014 or SpecialSales@dk.com.

Printed and bound in China

A WORLD OF IDEAS:
SEE ALL THERE IS TO KNOW

www.dk.com

Contents

GEMSTONES 64

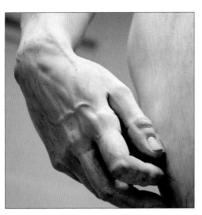

ORGANIC GEMS 290

ROCK GEMS AND ROCKS 320

Foreword

by Aja Raden

Whether they're stories about kings and queens, adventurers, wars, empires, or curses, the best stories are always about treasure. From sunken pirate ships to wicked diamonds to cities of gold, its sinister sparkle has a singular ability to bring out the very best and the very worst in each of us. History, mythology, or pure fantasy, the best stories all have that one thing in common: they all have something glittering at their heart.

But this book is not a story. A story has a beginning, a middle, and an end. This book is a map, a codex, a tool to help you understand all of those other stories. Throughout these pages you'll be introduced to hundreds of individual stones in their many forms; you'll learn about their singular qualities and unique beauties, and a little of the history that has defined them. You'll read about their objective qualities; their color, their clarity, their structural makeup, the geology of how they're formed, and how they're found. The hard science; the facts.

But the story of stones isn't wholly about facts. The story of stones is about beauty and desire, status and symbolism. It's also about value, which is never entirely objective or tangible. Gems were one of the first currencies, and to this day are the marker by which we judge monetary value. But what makes gemstones so valuable, so powerful? They are, after all, only stones…

And yet we've adored them and been adorned with them, imbued them with meaning and magic. Gems have been considered both poisons and medicines (and some actually are), barometers of everything from truth to danger. We've used them as signifiers of virtue, nobility, and even divinity. Conversely, they've often represented tyranny, greed, and death. Occasionally their power even strays from the symbolic to the literal, resulting in cults, relics, and legends of cursed gems. Wars have been fought, industries have been created, and empires have risen and fallen, all in pursuit of that which is beautiful, valuable, and rare.

Surely to desire, to covet, is human. As is the tendency to worship, to adorn, one might even say to treasure. But what catches the eye? What draws us to anything? A bright color? A flash of light? What is it about gems that so uniquely compels and possesses us? How have these glittering bits of the world had such power over our hearts and our minds throughout history? It's a deceptively simple question, but one that proves tricky to answer.

"Our need to see and possess glowing, shining jewels is as intrinsic to what we are, as who we are"

There are facts you can consider—but beware, facts aren't necessarily the same as truths. It's a fact that our motor cortex is stimulated when we behold something beautiful, meaning we're literally compelled to reach out for it. Free will be damned. It's another objective fact that there's a small cluster of neurons in our brain, the ventromedial prefrontal cortex (vmPFC), that both assigns an object its worth, and determines the depth of your emotional attachment to it. The vmPFC also plays a vital role in our assessment of morality. In other words, when confronted with great value or great beauty, of which jewels are the epitome, we're not only overcome with equally great emotion—our moral compass is set spinning and our very judgment is actually impaired.

But that singular human weakness for beauty and a collective obsession with gemstones goes deeper than emotional neurochemistry. Our need to see and possess glowing, shining jewels is as intrinsic to what we are, as to who we are. It's a surprising but clever trick of evolution. Another objective fact: we see certain colors more vividly and more clearly than we see anything else. Whether it's the abundance of green or the danger of red, we have actually evolved to look for colors, and to have an extreme emotional response to them. And even more so to shine, and above all to sparkle.

That's how we see water, after all: a shiny pool, ripples glittering across a surface, maybe just a glint in the distance. It was our first and still our most basic need—the need for water—that taught us to seek and to value that which glitters and shines. In a very real sense, sparkles equal life. And while a pretty bauble is not technically a matter of life and death, it's easy to see how our instincts can be confused. Whether it's the thrill we feel when we see a vivid color, a gleaming surface, or sparkles and shards of light—that need to see it, touch it, have it, goes all the way down to bone. It's the basis of what we are.

The story of stones must be chipped away at, rock by rock. It's impossible to tell the story of stones with a beginning, a middle, and an end, because it has no beginning, and it has no end. It is the very story of humanity. Everything we've done, from our earliest days to the very present, has been built on, immortalized in, and celebrated in, stone. From the crowns of the earliest chieftains, to modern diamond semiconductors, every age is the Stone Age. The story of stones is the story of us.

Introduction

Treasures of the Earth

The precious metals and gemstones that have been used for decoration and trade throughout human history have their origin in the rocks that surround us. Many of these began as mineral crystals that formed as a result of geological changes over millennia. The crystals are extracted, then cut, faceted, and polished to be used in jewelry and other decorative items. Organic gems are made of biologically derived matter, such as pearls produced by oysters, and amber, a form of fossilized tree resin. The financial worth and perceived value of these precious materials can vary from society to society—in some cultures, jade is more valuable than gold, for example. Rubies are among the most highly valued gemstones in the West—the 25.6-carat "Sunrise Ruby," mounted as a ring by Cartier, was sold for around $30 million in 2015.

Three rock types

Components of the Earth

There are three major classes of rock: igneous, sedimentary, and metamorphic. Igneous rocks are either formed from magma (molten rock) that has solidified underground, creating intrusive rocks such as granite, or has flowed onto the land or seabed, forming extrusive rocks such as basalt. Most sedimentary rocks, such as sandstone, are made of deposits laid down on the Earth's surface by wind, water, or ice. Metamorphic rocks are formed when the mineralogical composition of existing rocks is altered by heat or pressure. Quartzite, for example, is metamorphosed sandstone.

Igneous rock This example of intrusive igneous rock, granite, is formed inside the Earth when magma cools. Tiny crystals can be seen on its surface.

Sedimentary rock Sandstone usually contains quartz but other minerals can also be present. This example shows patches of iron oxide and flakes of mica.

Metamorphic rock The component minerals of gneiss—mainly quartz and feldspar—tend to separate out into distinct bands of different colors.

Beveled, tabular crystal, typical of the mineral wulfenite

Minerals

Most gems are cut crystals of minerals. A mineral is defined as a naturally occurring solid with a specific chemical composition and a distinctive crystal structure (see pp.14–15). Each mineral has a unique name based on these two criteria. If either of these changes, it becomes a different mineral with a different name.

Specimen of rough marble

Rocks

The Earth's rocks (see box, left) are made up of naturally occurring aggregates of one or more minerals, although there are a few rocks made from organic substances, such as decayed vegetation, which is the source of coal.

Gems in the Earth

The origins of rocks, minerals, and gemstones

Rocks and minerals are created in the rock cycle. All rocks begin as igneous, but over time they are altered by remelting, erosion, or metamorphosis—weathering and erosion lead to the formation of sedimentary rock, which can turn into metamorphic rock through temperature or pressure conditions.

GEMS IN NATURE

Mineral gemstones are directly mined from the rocks in which they were originally formed (see p.25): examples of these include diamond, tanzanite, ruby, kyanite, celestine, emerald, tourmaline, and aquamarine. Other types of gem, released from their original rock by weathering, can be mined from placer deposits found in stream gravels. Examples of these varieties include topaz, sapphire, chrysoberyl, garnet, zircon, and spinel.

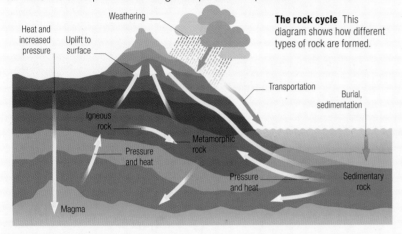

The rock cycle This diagram shows how different types of rock are formed.

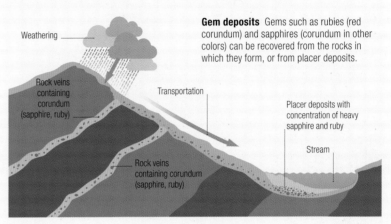

Gem deposits Gems such as rubies (red corundum) and sapphires (corundum in other colors) can be recovered from the rocks in which they form, or from placer deposits.

Crystals

A crystal is a solid, the component atoms of which are arranged in a particular, repeating, three-dimensional pattern. When these internal patterns produce a series of external, flat faces arranged geometrically, a crystal, such as the rhodochrosite above, is created.

Polished copper and silver in a rock matrix

Native elements

These are chemical elements that occur in nature uncombined with other elements (see also p.14), including gold, silver, and diamond (carbon). With the exception of platinum and gold, most metals are extracted from minerals that contain them.

Stalactitic amber

Organic gems

Created through organic processes, organic gems are not commonly crystalline. Jet is a form of coal, derived from plant matter; coral and shell are the secreted skeletons of marine animals, while pearls are formed by shelled mollusks; and amber and copal are tree resins, fossilized and semifossilized. Organic gems are generally softer than minerals and easier to work.

What is a mineral?

Minerals are the substances that make up the Earth's rocks. Each one has its own unique chemical composition and internal atomic structure—indeed, a mineral is defined by its chemical elements and by the atomic structure of its crystallization. Minerals are usually formed by inorganic processes, although there are organically produced substances such as the hydroxylapatite in teeth and bones that are also considered minerals. Certain substances, including opal and glass, resemble minerals in appearance,

chemistry, and occurrence, but do not have a regularly ordered internal arrangement and so do not exhibit crystallinity: these are known as mineraloids.

A few minerals occur as a single chemical element: these are known as "native elements" and include gold, silver, and diamond (see below). However, most minerals are chemical compounds, composed of two or more chemical elements. There are around 100 types of mineral that are considered common, out of more than 5,100 known minerals.

Mineral classification

Minerals are grouped according to their chemical composition. A mineral compound has positively and negatively charged atoms or groups of atoms: the element or elements that carry the negative electrical charge determine which chemical group a mineral is assigned to. The largest mineral group, the silicates, is further divided into six subgroups based on their different chemical structures.

Delicate blue-colored crystals

Blue azurite

Celestine cystals

Gold nugget

Fluorite crystals

Rough chrysocolla with azurite

Native elements

Minerals made up of atoms from a single element are known as native elements. The most common are metals such as copper, iron, silver, gold, and platinum, and nonmetals such as sulfur and carbon (as graphite and diamond). A few others occur in minute amounts, often alloyed with other native elements.

Halides

Halides consist of various metals combined with one of the common halogen elements: fluorine, chlorine, bromine, or iodine. There are three categories of halide: simple halides; halide complexes; and oxyhydroxy-halides. All halides are soft, and thus there are few gemstone varieties, except for fluorite.

Carbonates

A mineral in the carbonate group is characterized as having a carbon atom at the center of a triangle of oxygen atoms, which gives rise to trigonal symmetry (see pp.18–19). Examples of carbonates as gemstones include chrysocolla, calcite, smithsonite, and malachite.

Sulfates

These minerals have a crystal structure consisting of four oxygen atoms, with a sulfur atom in the center; this combines with one or more metals or semimetals. Examples of sulfates include baryte, celestine, and alabaster (a variety of the sulfate mineral gypsum).

Sulfides

Sulfide minerals are those in which sulfur is combined with one or more metals. Many of the sulfides are brilliantly colored, and most have low hardness and high specific gravity. Examples of sulfides include pyrite, marcasite, and sphalerite.

Oxides

Minerals of the oxide group consist of oxygen atoms combined with a metal or semimetal. An example of this is aluminum oxide, or corundum—ruby and sapphire. Other gemstone varieties include spinel—often mistaken for ruby—hematite, and rutile.

Phosphates

These minerals are grouped according to the similarity of their crystal structures—phosphate minerals contain phosphorus and oxygen combined in a 1:4 ratio. Some examples of phosphates that occur as gemstones include amblygonite, apatite, and turquoise.

Sphalerite rough

Deep blue coloring

Sapphire rough

Vitreous to pearly luster

Amblygonite rough

Violet color caused by iron and natural radiation

Amethyst rough

Silicates

All silicates consist of silicon and oxygen atoms, structured as a central silicon atom with oxygen atoms around it in various configurations. Silicates are divided into subgroups according to the varying structural configurations of their atoms; of these, inosilicates are subdivided into two further groups, as below. Silicates include many gemstones such as quartz and tourmaline.

Tectosilicates
Tectosilicates include lazurite (above), opal, quartz varieties such as amethyst, and more.

Phyllosilicates
This group includes chrysocolla, soapstone, and clay minerals, among others.

Single-chain inosilicates
Single-chain inosilicates include kunzite (above), enstatite, diposide, and others.

Double-chain inosilicates
Double-chain inosilicates include nephrite (above), edenite, and others.

Cyclosilicates
Emerald (above) is the best-known member of this group, which features tourmaline.

Sorosilicates
Sorosilicates include vesuvianite, zoisite, and other minerals.

Physical properties

When the need arises to identify or value a gem, it is sent to a professional gemologist who is certified by a professional gemological body. The gemologist will examine the gemstone for various physical and optical properties in order to make an identification and evaluate its quality.

An essential quality of a gemstone is durability. A gemstone's physical properties determine how durable it is, and how susceptible it is to wear, breakage, and deterioration, as well as the quality of its color. Note that gems with good cleavage (see right) can be very hard but still be susceptible to cracking.

Hardness

One of the main determinants of durability is the gem's hardness, or the relative ease or difficulty with which it can be scratched. Measured on the Mohs scale, hardness does not equate to strength, since very hard minerals can also be quite brittle. Gemstones below 5 on the scale are too soft for wear and even stones of 6 or 7 will scratch and abrade.

MOHS SCALE

Talc — Very easily scratched by a fingernail

Gypsum — Can be scratched by a fingernail

Calcite — Cannot be scratched by a fingernail; can be very easily scratched with a knife

Apatite — Scratched with a knife with difficulty

Fluorite — Easily scratched with a knife but not as easily as calcite

Orthoclase — Cannot be scratched with a knife; scratches glass with difficulty

Quartz — Scratches glass easily

Topaz — Scratches glass very easily

Corundum (sapphire) — Cuts glass

0   1000   2000   3000   4000   50

Specific gravity

Specific gravity (SG) is a measure of the density of a substance relative to that of water, and determines how dense a gemstone is. It is measured as the ratio of the mass of the substance and the mass of an equal volume of water— so a mineral with an SG of 2 is twice as heavy as water. Specific gravity can be determined using specialized balances or liquids that allow minerals below a given SG to float and those above it to sink. However, experts can often gauge the specific gravity of a gem purely by its heft.

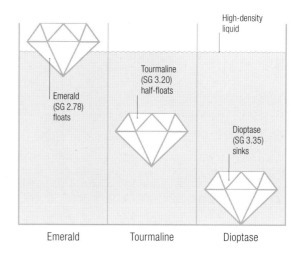

High-density liquid

Emerald (SG 2.78) floats

Tourmaline (SG 3.20) half-floats

Dioptase (SG 3.35) sinks

Emerald   Tourmaline   Dioptase

Cleavage

Cleavage is the property of a mineral that causes it to break along its atomic layers, where the forces bonding its atoms are the weakest. Some gems have cleavages in several directions, some of which may be very easy to trigger, meaning that the gem can be easily broken if sharply knocked. Because they follow the atomic planes, cleavage surfaces are often smooth.

Flat surface

Perfect cleavage is the breakage of a mineral along an atomic plane, where the bonds are weakest, and where the breakage leaves a flat surface.

Streak

Streak is the color of the powdered mineral, determined by rubbing a piece of the mineral across a piece of unglazed porcelain, leaving a streak of color. This can be used for identification, particularly with minerals that occur in different colors.

Streak can be more consistent in minerals than color—in fact, a mineral that occurs in different color varieties may have the same color streak. In this example, three color varieties of the mineral fluorite rough are shown. In a streak test, the purple, orange, and green fluorite specimens would all give a white streak.

Purple fluorite leaves white streak

Orange fluorite leaves white streak

Green fluorite leaves white streak

Fracture

Fracture is another way of describing how a mineral breaks. In fracture, however, breakage takes place across the mineral's atomic planes, rather than along them, as it does in cleavage, because there are no obvious planes of fracture. Distinctive fracture may help with identification.

Obsidian

Conchoidal fracture is where the breakage has a shell-like appearance. Quartz and glass gemstones—such as obsidian—show conchoidal fracture.

Chalcopyrite

Even fracture has a broken surface that is roughly textured, but flat.
Uneven fracture, as in the chalcopyrite above, has a rough and irregular surface.

Jagged edges

Gold

Hackly fracture shows an uneven surface with sharp edges and jagged points. It is characteristic of broken or torn metals and a few other minerals.

Aquamarine

Cuts glass

Diamond

6000   7000
KNOOP SCALE (kg/mm²)

Crystal systems and habits

A crystal is a solid in which the component atoms are arranged in a particular, repeating, three-dimensional pattern. When these internal patterns produce a series of external, flat faces arranged in geometric forms, this forms a crystal. These repeating structures are identical structural units of atoms or molecules, and are called unit cells. The unit cell is reproduced over and over in three dimensions, meaning that the shape of the crystal will resemble that of the individual unit cell. The crystals of different minerals can have unit cells that are the same shape but are made of different chemical elements. Because a crystal is built up of repeating geometric patterns, all crystals exhibit symmetry, depending on the basic geometry of their unit cells. These fall into seven main groups and are called crystal systems. The final external form a crystal takes is known as its habit, and the shape produced by a mass of numerous identical crystals is a growth habit (see opposite).

Minerals and crystal systems

Mineralogists and crystallographers have a complex set of criteria for determining which mineral belongs in which crystal system, based on symmetry. In practical terms, these systems can be understood as a group of three-dimensional cells starting with the basic cube (below). It should be noted that hexagonal and trigonal systems (right) are considered to be one system by some crystallographers.

Trigonal a=b=c The first hexagonal cell is further altered by squeezing two opposing short edges, so that all of the faces are lozenge shaped. The angle between a and b is 120 degrees. In the US, the trigonal is considered a division of the hexagonal.

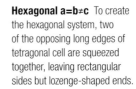

Hexagonal a=b≠c To create the hexagonal system, two of the opposing long edges of tetragonal cell are squeezed together, leaving rectangular sides but lozenge-shaped ends.

> When **two or more crystals of the same variety are intergrown symmetrically, they are referred to as twinned**

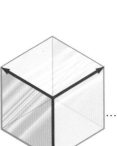

Cubic a=b=c The cubic unit cell has three right angles, and the lengths of its sides are all equal. Thus, relative to the length of its sides, a equals b equals c.

Tetragonal a=b≠c If the cubic cell is stretched vertically, there are still three right angles, but the length of the vertical side is longer than the other two. Now, a equals b does not equal c.

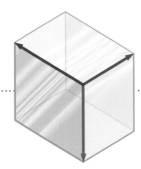

Orthorhombic a≠b≠c If the tetragonal cell is stretched horizontally, the three right angles remain, but now none of the sides are of equal length. Thus, a does not equal b does not equal c.

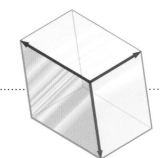

Monoclinic a≠b≠c The monoclinic system is created if the orthorhombic cell is skewed in one direction, only two right angles remain, and a does not equal b does not equal c.

Triclinic a≠b≠c To create the triclinic system, all of the faces are skewed so that no right angles remain, and none of the edges of the faces are equal.

GREATEST SYMMETRY

LEAST SYMMETRY

Pyramidal

Pyramidal crystal forms are, literally, in the shape of pyramids—pyramidal faces predominate in the crystal's shape. If pyramid faces occur in two directions with the pyramids base-to-base, the habit is dipyramidal.

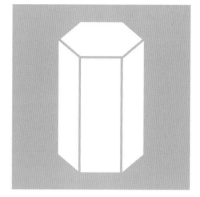

Pyramidal crystal form

Dipyramidal sapphire crystal

Prismatic

Prismatic crystals form long, pencillike shapes in which the length will be several times the diameter. Some prismatic crystals will have very regular, flat, rectangular faces, like aquamarine. In other minerals like tourmaline, the rectangular faces may be curved, to form a cross section like a triangle with curved edges.

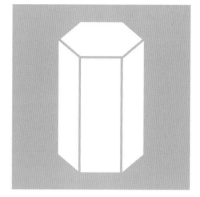

Prismatic crystal form

Prism-shaped aquamarine crystal

Acicular

Crystals are said to be acicular when they are thin and needlelike. In general there are few gemstones cut from acicular crystals, but when the needles are parallel and compact, they assume a fibrous form. Satin spar and tiger eye occur in such forms.

Acicular crystal form

Needlelike thomsonite crystals

Dendritic

In the dendritic crystal habit, aggregates of small crystals form in slender, divergent, somewhat plantlike branches. These are particularly common in copper, silver, and gold. Dendrites of iron and manganese oxide sometimes penetrate chalcedony to form dendritic agate.

Dendritic crystal form

Branchlike silver

Botryoidal

Botryoidal minerals form in globular aggregates, which often resemble bunches of grapes in shape. Hematite, chalcedony, and malachite are minerals that are found in botryoidal form. Malachite, in particular, is often cut and polished across the rounded masses, to reveal bull's-eye patterns.

Botryoidal crystal form

Bubblelike hemimorphite crystals

Massive

A mineral is said to be massive when it is a mass of tiny crystals that cannot be seen individually. Many gem minerals have their massive counterparts; other gem materials only occur in massive form. Massive minerals tend to be opaque, or at best translucent, and are cut as cabochons or used for carvings.

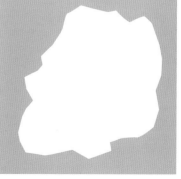

Massive crystal form

Massive sugilite (purple) on rock groundmass

What is a gem?

A gem is generally defined as any mineral that is highly prized for its beauty, durability, and rarity, is used for personal adornment, and has been enhanced in some manner by altering its shape—usually by cutting and polishing. A wider definition includes a few rocks, such as obsidian, and a few organic substances, such as amber (a fossilized resin). By far the majority of gems, however, are cut from the crystals of minerals. Precious metals are not considered to be gems, nor are items carved from minerals but not used for personal adornment, such as figurines, bowls, or vases.

Beauty

The first quality a gem must possess is that of beauty. This is subjective: some may prize a gem's interplay of light and color, while others may first be drawn to a gem's intricate cut. With an almost endless combination of color, shape, and fire (play of light), gemstones are capable of a range of aesthetic styles.

Blue sapphire, prized for its beauty

Durability

Hardness or toughness in a gem is a desirable quality, suggesting enduring value. Some gems require care to prolong their longevity. Certain gems resist chipping or scratching, but fade after long exposure to direct light; dry environments may cause some to crack, while others are susceptible to damage from acids.

Diamond, the hardest gemstone

Synthetic gems

Synthetic gems are identical to natural minerals physically, chemically, and optically, but are made in a laboratory. The two main ways to create them are from melt or solution. In production from melt, a powdered material, chemically equivalent to a natural mineral, is melted at high heat, then manipulated to solidify in a crystalline form. Production from solution involves dissolving one set of materials in a solution of different materials, again using high heat, then manipulating the solution so it precipitates into crystalline form. In both methods, crystals form on a seed crystal as the temperature is lowered.

Natural gemstones for comparison

Synthetic examples of gemstones

Large patches of color with snakeskin effect seen up close

Opal

Synthetic opal

Rarity

A gem may be rare for a number of reasons. The gem material itself may be rare, such as emerald, or a more common material may exhibit an unusual color or clarity. Some particularly soft or fragile stones are rare in cut form, since they require the work of highly skilled lapidaries.

Taaffeite, noted for its rarity

Other considerations

The desirability of a gem can depend on factors besides beauty, rarity, and durability. Gems may be symbolic of power, such as those mounted in crowns, or valued for their history or circumstances of origin. They may also be prized for their connection to astrology or mysticism, for their geological associations, or as fashion items.

Tanzanite, both rare and difficult to cut

Unusually flawless interior rarely found in nature

Diamond

Synthetic diamond

No visible internal flaws, unlike real emeralds

Emerald

Synthetic emerald

How qualities add up

Value, quality, and different varieties

Gems are defined by the very qualities that give them their value (see pp.30–31)—a gem would not be a gem without some level of material purity and craftsmanship. However, not all gems are equal.

COMPARATIVE GEM VALUES

The chart below compares a selection of popular gemstones according to their approximate monetary value (see p.27 for information on valuing diamonds, not included here). Varieties such as alexandrite, sapphire, and ruby are scarce and sought-after, and thus expensive; others, such as ruby, can vary hugely in cost, from modest to priceless. In general, the greater the clarity, size, and beauty of color, the more you will pay.

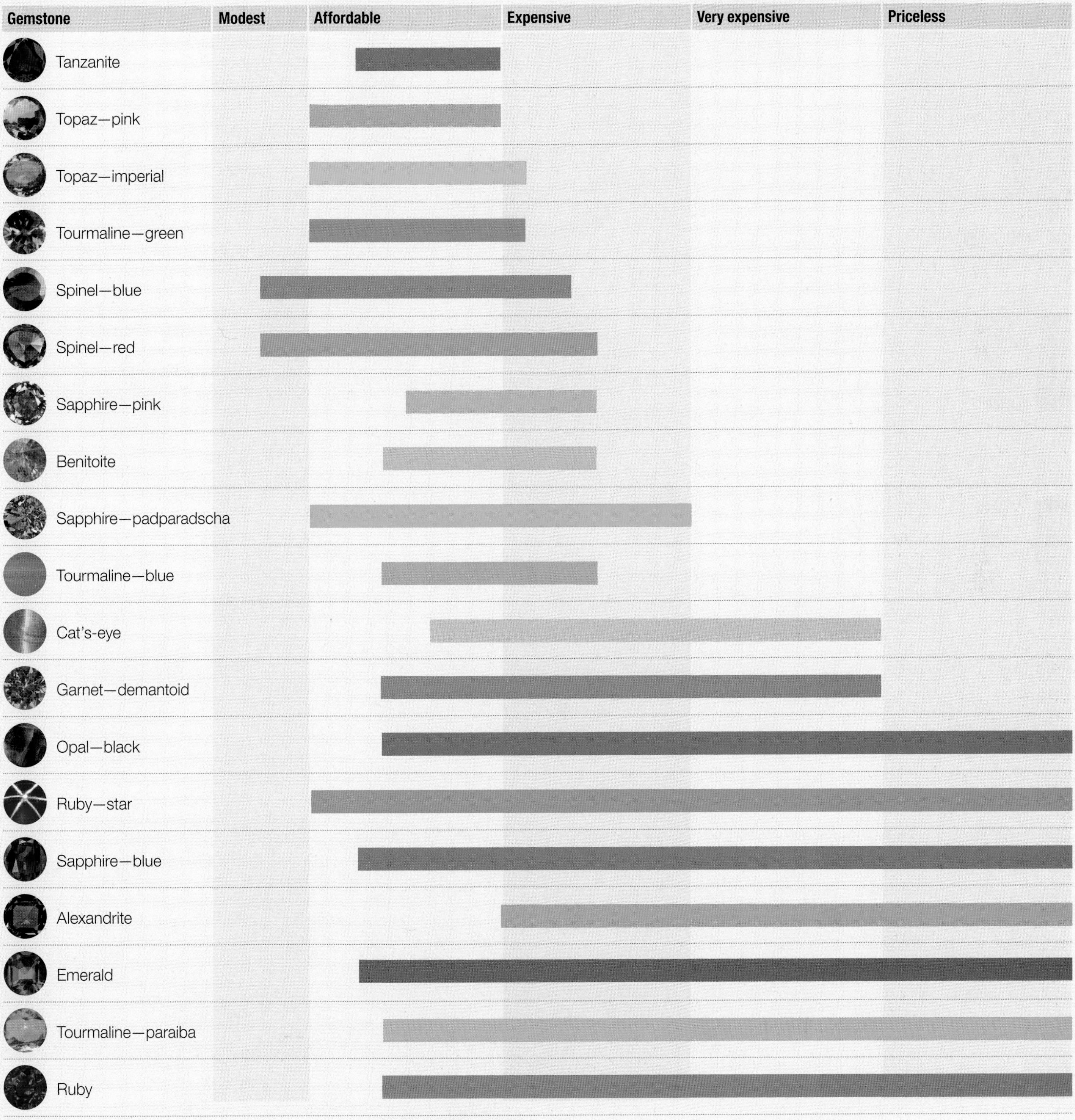

Gemstone	Modest	Affordable	Expensive	Very expensive	Priceless
Tanzanite		▬			
Topaz—pink		▬▬			
Topaz—imperial		▬▬▬			
Tourmaline—green		▬▬▬			
Spinel—blue		▬▬▬▬▬			
Spinel—red		▬▬▬▬▬			
Sapphire—pink		▬▬▬			
Benitoite		▬▬▬			
Sapphire—padparadscha		▬▬▬▬▬▬			
Tourmaline—blue		▬▬▬			
Cat's-eye		▬▬▬▬▬▬▬			
Garnet—demantoid		▬▬▬▬▬▬▬			
Opal—black		▬▬▬▬▬▬▬▬▬▬▬▬▬▬			
Ruby—star		▬▬▬▬▬▬▬▬▬▬▬▬▬			
Sapphire—blue		▬▬▬▬▬▬▬▬▬▬▬▬▬▬			
Alexandrite		▬▬▬▬▬▬▬▬▬			
Emerald		▬▬▬▬▬▬▬▬▬▬▬▬▬▬			
Tourmaline—paraiba		▬▬▬▬▬▬▬			
Ruby		▬▬▬▬▬▬▬			

ZIRCON SPINEL BARYTE TURQUOISE CAT'S-EYE

TANZANITE QUARTZ AMETHYST CHRYSOBERYL APATITE

Optical variations Each gem has a wide-ranging set of optical properties, both in terms of color and of clarity. During identification the gemologist has to take these into account, as well as the gem's refractive index (RI), its specific gravity (SG), hardness, luster, and dichroism.

CALCITE RUBY CHALCEDONY AGATE

AMMONITE JADEITE AMBER LAZURITE COPAL

Visual properties

How a gem interacts with light is the very essence of its nature. Light is the source of a gemstone's beauty, color, and sparkle; it is also a useful tool for the identification of gems, since each stone has its own particular set of optical properties. For example, there are a dozen or more red gemstones, and many of the red hues within each type of stone will have many different shades. All these properties are a way of identifying gemstones, although no single one is diagnostic in itself. Some categories such as luster are subjective observations; others, such as a mineral's refractive index, are objective. A gemologist identifying a stone will use a number of different methods and instruments to narrow the possibilities. Examination of one or all of the optical properties of the stone will show how it transmits, bends, and reflects light—just one of these may suffice for identification; in other instances, a complex combination of physical and optical properties may be needed.

Color

One of the most desirable qualities in a gemstone is beauty, and an important part of this is the stone's color. In gems, color is caused when light is absorbed within the crystal, or refracted—changing direction as it passes through the gem. White light is composed of many colors; when one or more of those colors is absorbed, the remaining light emerging from the gem is colored. This can be brought about by the presence of trace elements that cause certain wavelengths to be absorbed, or can be a part of the gem's chemical structure (see below).

Idiochromatic gems Idiochromatic gems are those sometimes described as "self-colored," as their color is inherent in their chemical makeup. Rhodochrosite is a manganese carbonate with a naturally pink to red color due to its manganese content, and peridot is an iron magnesium silicate, which is green as a result of its iron content.

Peridot

Allochromatic gems Allochromatic gems are those colored by trace elements in their structure. Amethyst and ruby are examples of these: amethyst is colorless quartz made purple by traces of iron, while ruby is corundum colored by traces of chromium.

Ruby

Particoloring Gems with different colors within the same stone are called particolored. Gems with two colors are called bicolored; those with three colors, tricolored. Rarely, a dozen or more colors can occur. Divisions between the colors can be abrupt or gradual. Parti-coloring is often caused by changes in the chemical medium in which the crystal has grown.

Watermelon tourmaline

Iolite seen from the top

Iolite seen from the side

Pleochroic gems As white light passes through a gemstone, colors are absorbed differently in different directions: as a consequence, a stone can be a different color when viewed from different angles. This effect is called pleochroism, and it can be an important aid to the identification of cut stones.

Luster

A gem's luster is the general appearance of its surface in reflected light. There are two basic types of luster: metallic and nonmetallic. Precious metals have metallic lusters, and gemstones nonmetallic, with the exception of a few like hematite and pyrite. Lusters that relate to gems include vitreous, waxy, pearly, silky, resinous, greasy, earthy, metallic, and adamantine.

Adamantine luster Gems that demonstrate an extraordinary brilliance and shine have an adamantine luster. It is a relatively uncommon luster, possessed by diamonds, some zircons, and a very few other gems.

Diamond

Refractive index

When light passes through a transparent gem, it changes speed and direction. This is called refraction. The change in the speed of light as it passes from the air into a gem is called the refractive index (RI). The change in direction, or bending, of the light can be used to calculate the gem's RI. Diamond's high RI results in flashes of light seen when the gem is moved— its "fire." The greater the dispersion of the white light, the greater the fire.

High refractive index (RI)   Low refractive index (RI)

Double refraction Gemstone minerals in the cubic system bend light equally in all directions. Other types of crystal system bend light in two directions, with the crystal structure causing light to bend at two different angles. This is called double refraction.

Spectroscopy

The study of the emission of light according to its wavelength is called spectroscopy. Devices known as spectroscopes are used to measure light waves as they pass through gemstones. The spectroscope has a small slit for light to pass through. When a gem is placed between a light source and the slit, a light spectrum is produced. Dark bands appear where certain wavelengths are absorbed by the stone. These bands are characteristic of various elements, enabling identification of the gem's chemical makeup. The three spectra on the right reveal much about the gems' composition.

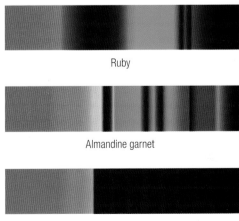

Ruby

Almandine garnet

Red glass

Spectra Ruby's simple composition shows only a few dark bands or lines. Almandine, meanwhile, shows numerous lines, due to its complex composition, while glass is made of only two elements, so displays only two absorption areas.

Where do gems come from?

Gems are found worldwide, but some areas are exceptionally rich sources: Myanmar contributes a huge amount of the world's ruby supply; Australia formerly dominated the precious opal market, although Ethiopia is now producing large quantities. The highest volume of fine emeralds comes from Colombia, while Madagascar is richest in sapphires. Diamond supply has been led by Botswana in recent years, although many stones are now coming out of Russia and Canada.

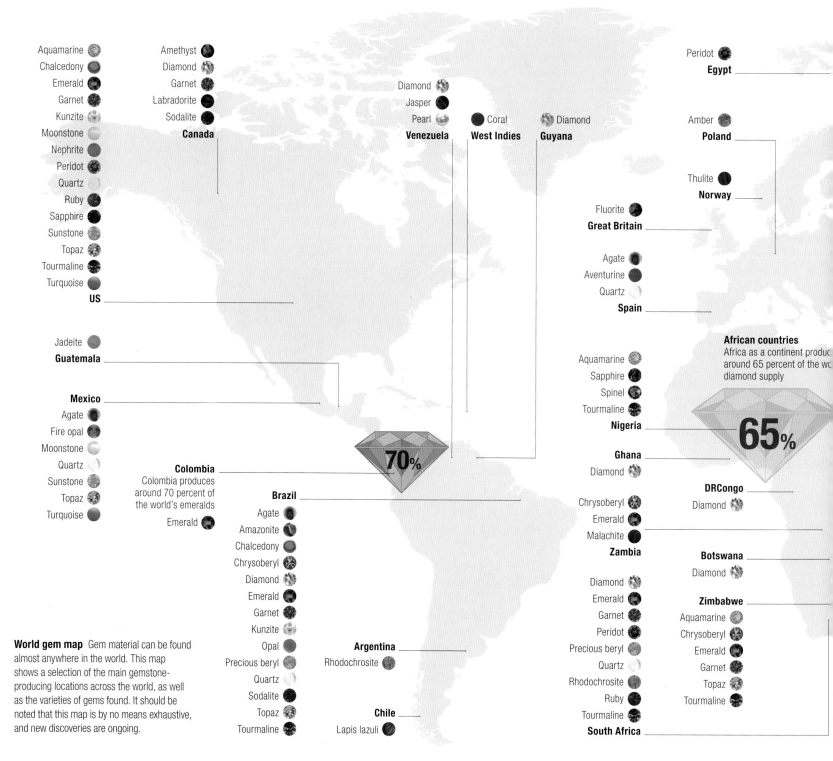

US
Aquamarine
Chalcedony
Emerald
Garnet
Kunzite
Moonstone
Nephrite
Peridot
Quartz
Ruby
Sapphire
Sunstone
Topaz
Tourmaline
Turquoise

Canada
Amethyst
Diamond
Garnet
Labradorite
Sodalite

Guatemala
Jadeite

Mexico
Agate
Fire opal
Moonstone
Quartz
Sunstone
Topaz
Turquoise

Colombia
Colombia produces around 70 percent of the world's emeralds
Emerald

70%

Venezuela
Diamond
Jasper
Pearl

West Indies
Coral

Guyana
Diamond

Brazil
Agate
Amazonite
Chalcedony
Chrysoberyl
Diamond
Emerald
Garnet
Kunzite
Opal
Precious beryl
Quartz
Sodalite
Topaz
Tourmaline

Argentina
Rhodochrosite

Chile
Lapis lazuli

Egypt
Peridot

Poland
Amber

Norway
Thulite

Great Britain
Fluorite

Spain
Agate
Aventurine
Quartz

Nigeria
Aquamarine
Sapphire
Spinel
Tourmaline

Ghana
Diamond

Zambia
Chrysoberyl
Emerald
Malachite

South Africa
Diamond
Emerald
Garnet
Peridot
Precious beryl
Quartz
Rhodochrosite
Ruby
Tourmaline

DRCongo
Diamond

Botswana
Diamond

Zimbabwe
Aquamarine
Chrysoberyl
Emerald
Garnet
Topaz
Tourmaline

African countries
Africa as a continent produces around 65 percent of the world's diamond supply

65%

World gem map Gem material can be found almost anywhere in the world. This map shows a selection of the main gemstone-producing locations across the world, as well as the varieties of gems found. It should be noted that this map is by no means exhaustive, and new discoveries are ongoing.

Mining for gems

Revealing the treasures of the Earth

Large-scale mining for precious materials is mainly reserved for "big names," such as gold, silver, and diamonds, as well as less precious material used in industry, such as copper. Some large-scale mines also produce fine gem material, such as malachite and turquoise, as by-products. Much gem mining is relatively small scale, however, and is often done using only hand tools.

The Superpit Located in Kalgoorlie in central Western Australia, the Superpit is Australia's largest open pit gold mine and is one of the largest open pit mines in the world.

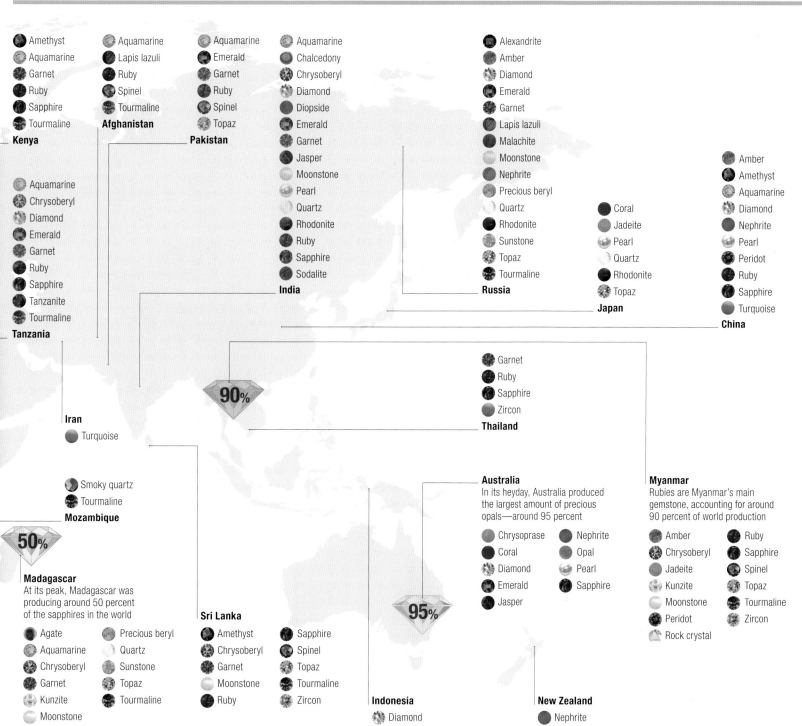

Kenya
- Amethyst
- Aquamarine
- Garnet
- Ruby
- Sapphire
- Tourmaline

Afghanistan
- Aquamarine
- Lapis lazuli
- Ruby
- Spinel
- Tourmaline

Pakistan
- Aquamarine
- Emerald
- Garnet
- Ruby
- Spinel
- Topaz

India
- Aquamarine
- Chalcedony
- Chrysoberyl
- Diamond
- Diopside
- Emerald
- Garnet
- Jasper
- Moonstone
- Pearl
- Quartz
- Rhodonite
- Ruby
- Sapphire
- Sodalite

Russia
- Alexandrite
- Amber
- Diamond
- Emerald
- Garnet
- Lapis lazuli
- Malachite
- Moonstone
- Nephrite
- Precious beryl
- Quartz
- Rhodonite
- Sunstone
- Topaz
- Tourmaline

Japan
- Coral
- Jadeite
- Pearl
- Quartz
- Rhodonite
- Topaz

China
- Amber
- Amethyst
- Aquamarine
- Diamond
- Nephrite
- Pearl
- Peridot
- Ruby
- Sapphire
- Turquoise

Tanzania
- Aquamarine
- Chrysoberyl
- Diamond
- Emerald
- Garnet
- Ruby
- Sapphire
- Tanzanite
- Tourmaline

Iran
- Turquoise

Mozambique
- Smoky quartz
- Tourmaline

Thailand
- Garnet
- Ruby
- Sapphire
- Zircon

90%

50%

95%

Madagascar
At its peak, Madagascar was producing around 50 percent of the sapphires in the world
- Agate
- Aquamarine
- Chrysoberyl
- Garnet
- Kunzite
- Moonstone
- Precious beryl
- Quartz
- Sunstone
- Topaz
- Tourmaline

Sri Lanka
- Amethyst
- Chrysoberyl
- Garnet
- Moonstone
- Ruby
- Sapphire
- Spinel
- Topaz
- Tourmaline
- Zircon

Indonesia
- Diamond

Australia
In its heyday, Australia produced the largest amount of precious opals—around 95 percent
- Chrysoprase
- Coral
- Diamond
- Emerald
- Jasper
- Nephrite
- Opal
- Pearl
- Sapphire

New Zealand
- Nephrite

Myanmar
Rubies are Myanmar's main gemstone, accounting for around 90 percent of world production
- Amber
- Chrysoberyl
- Jadeite
- Kunzite
- Moonstone
- Peridot
- Rock crystal
- Ruby
- Sapphire
- Spinel
- Topaz
- Tourmaline
- Zircon

Grading and evaluation

The grading and evaluation of gems can start even before they are removed from the ground. Within some gem deposits, certain areas are known to yield more or better-quality gem material, and are thus mined first. Only a small percentage of what is recovered is actually of gem quality, and this is further sorted to separate out the gem material. Any gem-quality roughs found are then carefully evaluated for color, clarity, and size. Even much of this selected material may remain uncut if it is too small, oddly shaped, or for some reason does not fit the current market demand. Cutting is expensive and time-consuming, and so meticulous grading at this early stage is essential. To be certain, the cutter will make his or her final decision about what is cut, but the greater the evaluation and grading before gems reach the cutter's workshop, the better.

Non-gem quality

The specimen shown here is of ordinary zoisite, and is typical of virtually all zoisite material. It is opaque, and even when well crystallized, it is not of gem quality. Until the discovery of tanzanite, zoisite's gem-quality variant (see p.253), only a tiny amount of gem-quality zoisite was known.

Non-gemmy material

Zoisite crystal This zoisite rough is not of gem quality and would be of value purely as a mineral sample.

> The jewel…
> is concentrated
> brilliancy, the
> quintessence
> of light
>
> Charles **Blanc**
> 19th-century author

Mid-gem quality

Because tanzanite is a relatively uncommon stone, even medium-quality gems are valuable enough to facet. In more common stones, such as amethyst, a mid-quality rough is rarely faceted because the cost of cutting it would exceed the final value of the finished gem.

Internal flaws and inclusions

Pale tanzanite rough This rough crystal is gem quality, but has a number of small flaws and inclusions.

Facets conceal imperfect interior

Cut pale tanzanite Faceting can be used to conceal or disguise internal imperfections in mid-quality gems, as seen in this example.

Superior quality

A relatively small portion of even a select gemstone rough is of superior quality. In this tanzanite example, both the rough and the cut stone are of visibly higher quality than the mid-quality grade (above).

Fine coloring

Superior-quality tanzanite rough Even as a rough, the quality of this tanzanite can be seen in its deep color and clear interior.

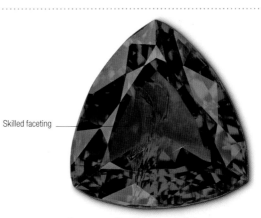

Skilled faceting

Cut tanzanite The best-quality gemstones combine superior-grade material with highly skilled craftsmanship, as in this stone.

Grading and valuing diamonds

The most precious gem of all

In general, intensely colored natural diamonds command very high prices. However, "colorless" diamonds, because they have a generally higher value than most "colored" stones, are graded by a more complex system. A single change in grade can result in a large difference in value. To avoid the large value changes between the grades that would occur if there were only a few grades, there are numerous grades based on each of the four "Cs" (see p.30), thus keeping the changes in value relatively small. The grades and their determinates below are those of the Gemological Institute of America (GIA).

Clarity grading scale The clarity scale runs in stages from visually flawless to stones with numerous visible inclusions.

IF (Internally flawless)	VVS1, VVS2 (Very, very slightly included 1 and 2)	VS1, VS2 (Very slightly included 1 and 2)	SI1, SI2 (Slightly included 1 and 2)	I1, I2, I3 (Included 1, 2, and 3)

Colour grading scale The most common color tint for "white" diamonds is yellow. This scale grades color according to the amount of yellow present, beginning at "D" for colorless, all the way to "Z" for light yellow (also brown or gray).

D	E	F	G	H	I	J	K	L	M	N	O	P	Q	R	S	T	U	V	W	X	Y	Z

Colorless	Near colorless	Faint yellow	Very light yellow	Light yellow

Type I gems

Gemstones of Type I level are usually "eye clean" as standard, with no visible inclusions. Stones in this category are usually of such high clarity that they will be free of even minor inclusions. For lapidaries, collectors, and jewelers, these stones represent the height of desirability.

Type II gems

Gems of the Type II category typically display some inclusions that are visible to the naked eye but do not detract from the desirability and overall beauty of the gemstone. Many such stones with visible inclusions are faceted for use in jewelry.

Type III gems

The Type III classification is mainly applied to gemstones that feature obvious inclusions or other imperfections. However, even stones with prominent inclusions are regularly cut for use in jewelry, and are considered beautiful objects in ther own right.

Aquamarine

Chrysoberyl, yellow

Chrysoberyl, green

Andalusite

Alexandrite

Garnet

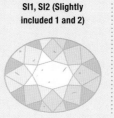
Emerald

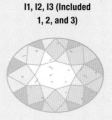
Red beryl

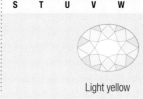
Tourmaline, watermelon (above), and red only

Helidore

Hiddenite

Kunzite

Iolite

Peridot

Quartz, amethyst

Only around 30 percent of diamonds mined worldwide are classified as being of gem quality

Morganite

Quartz, smoky

Tanzanite

Quartz, citrine

Quartz, ametrine

Ruby

Tourmaline, green

Zircon, blue

Diamond

Sapphire

Tourmaline: all except red, green, and watermelon

Zircon: all except blue

Gem cuts

Stones are reshaped to enhance their beauty and to increase their value. A finished gem can be many times the value of its rough, and it is also far more sellable. Gemstone rough may be sawn to remove poor areas, to separate valuable areas from within a larger stone, or to create a preliminary shape. Arriving at the final shape and form of a gem (its "cut") then involves various stages of grinding and polishing. The cut used on a particular piece of gem rough is determined by a number of factors in combination: the shape of the rough, the position of its flaws, its cleavage, the best orientation to display its color (in the case of starstones, for example, the cut is oriented so that the star is centered in the finished stone), and the most suitable cabochon shape if the stone is translucent or opaque.

Polished stones

If a stone is described as "polished," it can be anything from a slice of opaque gem material polished on the flat sides for use as a pendant to the intricate and detailed carving of a fine cameo.

The parts of a gem

Jewel terminology

Whatever the gem cut being used, there are a few fundamental parts: the crown, pavilion, and table facet. The other facets and the proportion between the crown and pavilion affect a gem's brilliance. Whether on a round, brilliant, or rectangular emerald cut, the facets are given the same names and fill the same relative positions on the gem. The angles at which the facets are cut are determined by the refractive index, and the cutter uses a set of tables to find the suitable face angles for each type of material. The angles shown here are for diamond. The usual ratio of the crown to the pavilion is 1:3, but this can vary depending on the angle of the crown mains.

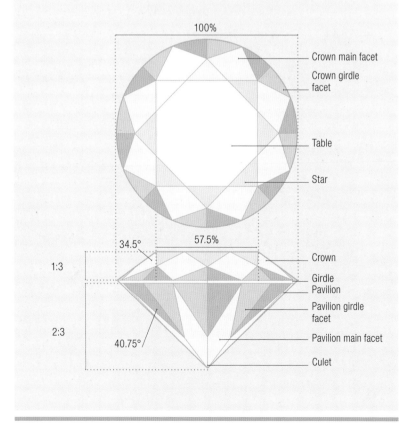

- 100%
- Crown main facet
- Crown girdle facet
- Table
- Star
- 34.5°
- 57.5%
- 1:3
- Crown
- Girdle
- Pavilion
- Pavilion girdle facet
- 2:3
- 40.75°
- Pavilion main facet
- Culet

Cutting techniques

The basic stages of sawing, grinding, and polishing are common to all lapidary processes, but each of the three requires skills and tools unique to that particular step. It is not unusual for a gem cutter to be able to do all three.

Choice of rough A gem rough is chosen for its color, size, clarity, shape, and freedom from flaws, fractures, and inclusions. If imperfections are present, the cut is oriented in such a way as to conceal them as much as possible.

Choice of cut The cut is chosen based on the shape of the rough, and the desired color and brilliance of the finished stone. The rough is sawn to the general shape of the stone or to provide the table facet on the top half of the stone, called the crown.

Faceting begins The best brilliance is achieved when facets are angled and positioned correctly. Large facets are cut first. On a brilliant cut, there are eight on the top and eight on the bottom. Usually the pavilion is faceted first, then the crown.

Further faceting Further smaller facets are added around the larger facets on the crown at the top and the pavilion at the bottom. On a brilliant cut there are 40 of these smaller ones. Each facet is cut to maximize its optical performance.

Finished off After all the final facets have been cut, the gem is then polished to remove any cutting scratches and to improve its luster. This stage can take place alongside the faceting process—as preferred by most cutters—or afterward.

Brilliant cut

Brilliant cuts maximize brilliance. They are also used on colored stones to deepen their color, conceal imperfections, and to even out patchy colour. Their facets, cut in a vertical direction from crown to pavilion, are roughly triangular or kite-shaped. The actual outline of the stone can vary from round to oval, to pendeloque, or even free form, so long as the facets are triangular.

Brilliant round demantoid garnet

Brilliant oval iolite

Brilliant round

Brilliant oval

Cabochon

Gems cut with a flat back and a domed top are called cabochons. The dome itself can be flat, or high in proportion to the outline dimensions of the stone. The high dome is used to emphasize the particular optical properties of certain stones, such as asterism, iridescence, or a cat's-eye effect. When the gem has color or pattern as part of its basic structure, the dome is usually shallower, in order to show off the color or pattern to best effect. In terms of their outlines, cabochons can be virtually any shape.

Cabochon

Sodalite oval cabochon

Step cut

Step cuts are used to enhance the color of a stone, although they generally produce less sparkle as a consequence. These cuts have rectangular facets in broad flat planes that resemble steps. The most widely used step cut is the emerald cut, which was originally designed to preserve valuable emerald rough. This is the preferred cut for brittle stones because it leaves them with no vulnerable sharp corners.

Square cut   Baguette   Emerald cut

Emerald-cut emerald

Mixed cut

Mixed cuts combine brilliant and step-cut facets, optimizing a stone's dimensions and visual properties to enhance the brilliance of a colored stone, while still emphasizing its color. The step-cut faces can be on the crown or pavilion. A mixed cut's outline shape can be virtually anything as long as the crown and pavilion display different cuts.

Mixed cut

Cushion cut

Cushion-cut heliodor

Mixed-cut topaz

Fancy cuts

Cuts with unusual outlines and facets are known as fancy cuts. Among these are hearts, free-form shapes with irregular outlines, kites, scissor, and pendeloque cuts, and standard shapes with unusual faceting—for example, where the facets form flat planes to create a checkerboard or zigzag pattern. Gems cut into elongated ovals with pointed ends exhibit the marquise cut.

Garnet heart

Scissor cut

Pendeloque

Marquise cut

Marquise-cut blue diamond

Carving

In general, carving means the shaping of a piece of gemstone rough into a three-dimensional figure. It is a very skilled lapidary art that can take several forms. These include intaglios, or relief carving, with the figure carved into the gem; cameos, with a figure or scene in relief on a contrasting colored background; or even full, three-dimensional figures, birds, and animals.

Cameo

Carved coral cameo

The value of a gem

Whether they are collectable specimens, used in everyday jewelry, or incorporated into beautiful works of art, gemstones are viewed as highly prestigious objects in many—if not most—cultures. It could be argued that their value is entirely man-made, but the fact remains that fine gems represent an apex of material quality, visual beauty, and fine craftsmanship. Diamonds are graded and valued in a slightly different way to other stones (see p.27)—the term "colored stones" refers to all non diamond gems, although diamonds can be colored—but the core principles remain the same.

Gem qualities

All gemstones are valued according to four "Cs": colour, clarity, cut, and carat. There is a final factor to add to these—their rarity. In general, larger stones are much more rare than smaller ones; for some stones, this means that an increase in weight can result in a disproportionately large price increase, so when a gemstone doubles in weight, its price may go up by four or five times.

Good clarity

Rock crystal set in a ring

Goshenite showing excellent clarity

Carat

A carat is a measurement of a gemstone's weight, equivalent to one-fifth of a gram. This should not be confused with karat, a measure of the proportion of gold in a gold alloy. 24-karat gold is pure gold; 18-karat gold is ¾ gold and ¼ another metal, often copper.

Clarity

Clarity refers to the lack of foreign matter—other minerals, hollows, or crystals—within the stone, known as inclusions. The resulting effect on the beauty of the stone determines value; although a lack of inclusions is valuable, certain types of inclusions are also desirable.

Rarity

The fifth characteristic

Rarity has a direct effect on value: a superb garnet will never command the same price as an equivalent ruby, simply because ruby is vastly rarer. Some stones usually occur only in small sizes due to their chemistry. In this case, larger stones are even more rare.

Excellent color

Ruby

Aquamarine in a round cut

Sherry topaz in a deep gold color

Cut

A gemstone's cut is graded on the basis of its technical perfection, and the brilliance it produces. Considerations might include whether the points of triangular facets meet without overlap, or if the sides of rectangular step-cut facets are parallel.

Color

The value attributed to a gemstone's color is usually determined by its purity and intensity. However, in some cases, it may be due to the rarity of a particular color. For example, natural red or blue diamonds command astronomical prices.

Enhancing gems

Many gemstones undergo treatment to enhance their natural characteristics in addition to the standard cutting and polishing processes that a lapidary will carry out. Gem sellers are expected to disclose any additional treatments a stone may have undergone. This is not only because a gem's value is based around its natural characteristics, but also because some treatments can also affect a gem's durability, or wear off. For example, the artificial coating of topaz gems (see right) can become scratched and wear off.

Tanzanite surface coated to deepen and improve color—before (left) and after (right)

Surface coating

Very thin coatings of gold, silver, and other metals can be applied to gems to alter their color or reflectivity. This can be seen in varieties such as "mystic" topaz and "aqua aura" quartz. While attractive, these coatings soon wear off.

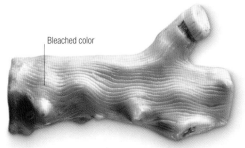

Bleached color

Coral bleached (above) then dyed (below)

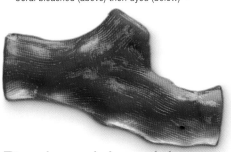

Enhanced color

Collection of irradiated gemstones in various colors

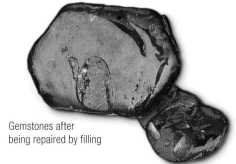

Gemstones after being repaired by filling

Dyeing, bleaching, and staining

The staining and dyeing of gems is widespread. Slices of agate are routinely dyed with vivid colors such as blue or red—often using ordinary household fabric dyes, although there are also dyes specifically for stones. If a dye comes off a gem onto the hands, it is a sign of poor quality.

Irradiation

A gemstone's color can be altered by irradiation—bombardment with neutrons, gamma rays, ultraviolet light, or electrons, which is often followed by heat treatment. Most blue topaz on the market is irradiated and heat-treated colorless topaz.

Filling, coating, and reconstruction

Some gemstones with cracks are subjected to fillers other than the oiling described for emerald. Fillers can be glass, resins, plastic, or waxes, and can be colored to match the gemstone. Heat, pressure, or solvents are used to fuse together small pieces.

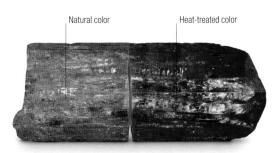

Natural color   Heat-treated color

Imperial topaz crystal sawn in half: the crystal on the right was heated. Both colors are highly desirable

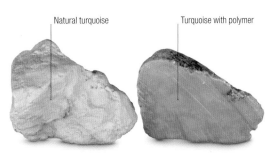

Natural turquoise   Turquoise with polymer

Porous turquoise impregnated with a wax or polymer substance to color and stabilize the material

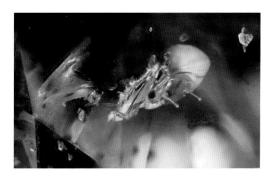

Laser-drilled table facet—the drill holes have created a cleavage crack, worsening the flaw

Heat treament

This is among the oldest forms of gemstone enhancement—zircons have been heat-treated to change their color for at least a millennium. Today, the process is used to change color or conceal inclusions.

Oiling

An old method of treating cracked emerald is simply to soak the stone in oil. This fills the cracks and makes the stone appear of a higher quality; however, it can give the gem an oily feel. Turquoise can also be soaked in polymer to improve color.

Laser drilling

Diamond is combustible, a property that allows infrared lasers to drill tiny holes into the material to reach flaws and inclusions. Once reached, inclusions can be dissolved, and flaws can be filled with epoxy resins.

What is a jewel?

A jewel is a precious stone, usually a single crystal or part of a hard, lustrous or translucent mineral that has been cut or shaped for decorative use, and typically set into a metal or other precious material—either as jewelry to be worn or as an ornamental object. Jewels and jewelry have been created since prehistoric times and rank among some of the earliest known artifacts. In some cases,

jewelry has originated in functional objects, such as brooches to fasten clothing. Some adornments had talismanic meaning, while in many cultures, jewelry and bejeweled objects were a means of storing wealth and indicating social standing. Below is a selection of jewelry and jeweled ornaments from prehistory to the modern world, covering a wide range of uses.

Mesolithic shell necklace
This necklace of snail shell beads found in Serbia is almost 10,000 years old, and is among early evidence of ornamental jewelry.

Lapis lazuli feathers

Egyptian falcon pectoral
Symbol of the sun god Horus, this gold, cornelian, and lapis lazuli pectoral is from the tomb of Amenemope c.1000 BCE.

Chinese deer pendant
This 1st-millennium BCE animal carving is of highly prized nephrite.

Cornelian insets

Babylonian gold pendant
One of a pair, this pendant from the 2nd millennium BCE represents the minor godddess Lama.

Griffin (half lion, half eagle)

Hippocamp (sea horse)

Greek fibula
Found in Crimea, this hollow gold fibula (brooch) dates from c.425–400 BCE, and depicts a mythical hippocamp and griffin.

Byzantine brooch
This 6th-century CE brooch in the form of a Greek cross displays the Byzantine love of gold and precious stones.

Eagle pendant
This Renaissance gold pendant from c.1620 is enameled and set with diamonds, rubies, and emeralds.

Art Nouveau hair comb
This 1904 ornamental poppy hair comb is made of horn, silver, enamel, and moonstones.

Elizabeth Taylor's charm bracelet
This circular link gold bracelet was given to Elizabeth Taylor by her husband Richard Burton in the mid-20th century.

Transylvanian crown
Given by the Ottomans to Prince Stephen Bocskay of Transylvania in 1605, this crown is inset with turquoise, rubies, and pearls.

Spanish caravel pendant
The design of this enamel, gold, and malachite pendant from the 1580s reflects Spain's seafaring glory.

French necklace
This 18th-century silver necklace is set with topazes and amethysts. The ribbons, flowers, and bows are outlined by round zircons.

Naga bracelet
This piece from 2011 in the shape of a mythical Cambodian ocean guardian is set with a 12.39-carat purple tourmaline, diamonds, sapphires, and more.

Native elements

Winged brooch | This gold Verdura winged brooch, made in 1939, is set with two large and rare pink topaz stones surrounded by diamonds. The actress Joan Fontaine owned it and wore it in Alfred Hitchcock's movie *Suspicion* (1941).

Gold wings

Diamonds

Pink topaz

Gold? Yellow, glittering, precious gold? This yellow slave will knit and break religions, bless the accursed...

William **Shakespeare**
Timon of Athens

Gold

△ **Crystalline gold**, made up of of octahedral gold crystals

ven before gold became the trading medium of the commercial world and the foundation for modern money, it was prized for its beauty and spiritual significance. To the ancient Egyptians, gold was the perfect material—it offered a glittering yellow surface soft enough to work, yet it was so durable that it would last essentially forever. In fact, of the three known metals stable enough to have been used for trade at that time, gold was the most suitable. It did not corrode and did not react with other substances. Unlike silver, it did not tarnish and, unlike copper with its high melting point, it could feasibly be melted into currency. Thus it became the most desired metal in the world, transcending geographical borders to become a universal symbol of power, both political and spiritual.

The color of gold

In its pure state, gold is always golden yellow, but it is too soft to make into jewelry: to increase its hardness, gold is alloyed with other metals. Adding silver, platinum, nickel, or zinc creates pale or white gold. Copper yields red or pink gold, and iron gives a blue tinge. The purity of alloyed gold is expressed in karats, which measures parts per 24: for example, 18-karat gold means there are 18 parts gold out of 24 in the alloy, while 24-karat gold is pure (and mostly too soft to be worn). Note that "karat" is distinct from "carat," a measurement of weight in gemstones.

Specification

Chemical name Gold | **Formula** Au | **Colors** Gold yellow
Structure Isometric | **Hardness** 2.5–3 | **SG** 19.3
RI n/a | **Luster** metallic

Locations
1 Canada 2 US 3 Brazil 4 South Africa 5 Russia
6 China 7 Australia (and many more)

Key pieces

Carved
scales

Ancient gold | Because of gold's chemically inert nature, ancient gold artifacts buried for thousands of years come out of the ground looking as bright as the day they were made. This Mycenean brooch is from c.1600–1100 BCE.

Roman gold | From the city of Pompeii buried by volcanic ash and lava in the 1st century CE, this gold armlet is in the form of a coiled snake and shows perfect preservation. The detail on the head and scales on the skin indicate high-quality craftmanship.

Cartier panther ring | This spectacular open-form, unisex ring in 18-karat yellow gold from the *Panthère de Cartier* collection features peridot eyes, an onyx nose, and lacquer accents. The wearer's finger goes through the open mouth.

Rough

Placer gold | This is a gold nugget recovered from a stream gravel deposit known as a placer (Spanish for alluvial sand). The nugget shows a typical battered and rounded form.

Gold in quartz | This specimen mined directly from one of the veins in the Earth is an example of the form in which it naturally occurs—as scattered grains in quartz.

Small grains

Grains of gold | Gold nuggets are relatively rare in stream gravels. Most stream gold is recovered in the form of small grains or flakes like these.

Gold nugget | The angular shape and rough texture of this nugget indicate that it has not moved far from the place where it weathered out (became exposed).

Settings

Roman gold | The dolphin was a common motif in ancient Roman art. These large-eyed, drop dolphin earrings date from around the 1st century CE.

Clasp secures pectoral

Cast animals

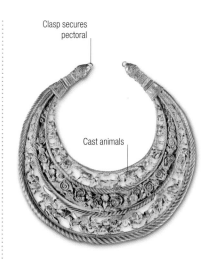

Scythian treasure | This elaborate gold pectoral was probably made by Greek goldsmiths in the 4th century BCE for a Scythian king in present-day Kazakhstan.

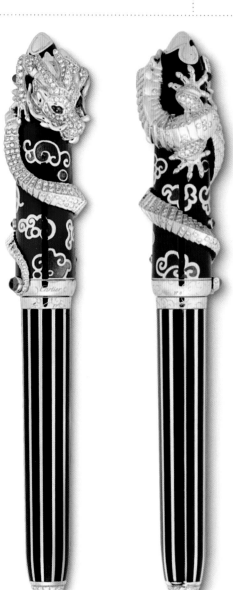

Cartier pen | Three views of this 2008 limited-edition Cartier gold pen show its dragon motif, set with 522 diamonds, six emeralds, and ruby eyes. The pen is finished in black lacquer.

Lacquer finish

The woman in gold

The *Mona Lisa* of Austria

With its unusual mix of naturalistic face and skin and Egyptian-influenced, jewel-like decorative detail, this painting is not just painted to look like gold but also *with* gold in a powdered pigment form. When the Nazis seized this work in 1940, they changed the title to *The Woman in Gold* so they could display it without reference to its Jewish sitter. Its restitution was the subject of a movie *Woman in Gold* in 2015.

Portrait of Adele Bloch-Bauer I
Gustav Klimt, 1907, 54 x 54 in (138 x 138 cm), oil and gold on canvas

When **gold** argues the case, eloquence is **impotent**

Publilius **Syrus**
Moral Sayings, 1st century BCE

Gold "petals"

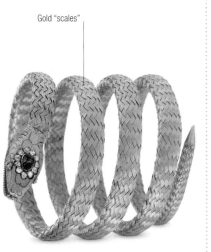

Gold "scales"

Gold, diamond, and sapphire bracelet | This antique bracelet from France takes the form of a serpent. Its head is set with a sapphire ringed with diamonds.

Wood-grain necklace | This American gold necklace has an unusual wood-grain texture. The clasp is cleverly concealed in one of the oversized links of the chain.

Sunflower | This delicately sculpted sunflower is made from yellow gold. Its petals have been etched with fine lines to create a textured appearance, and cut stones have been set in its stamens.

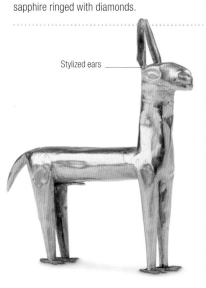

Stylized ears

Inca gold | Made in the 14th–15th centuries, this model llama from Peru is cast in high-karat gold. The animal's body has been pared down to simple geometric forms.

Engraved clasp

Charm bracelet | Bracelets like this one go in and out of fashion, but have the advantage of giving the wearer the chance to add personalized charms.

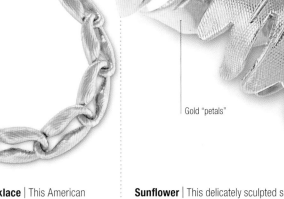

Onyx nose

Emerald eyes

Cartier panther 18-karat gold rings | The top ring is in yellow gold with green garnet eyes; the bottom one in white gold is set with 158 diamonds and emerald eyes.

Pink gold

White gold

Yellow gold

Bulgari triple-gold ring | Made by the house of Bulgari, this piece is composed of three rings of colored 18-karat gold in different alloys—yellow, pink, and white.

CAROLVS MA: IMP. EL
A°. 800: OB: A°. 814.

Crown of Charlemagne | c.960 CE | Eight hinged plates of 22-karat gold with 144 *en cabochon* gems

Crown of
Charlemagne

△ **Front view** of the crown

The Crown of Charlemagne is more than a jeweled medieval masterpiece—between the 10th and 19th centuries, it symbolized the might of the Holy Roman Empire, a vast European state with Germany at its heart. Later, the crown became such a powerful icon that the dictator Adolf Hitler used it in his campaign to create a new German-led empire in Europe in the 1930s.

Also called the Imperial Crown, or Crown of the Holy Roman Empire, it is named in honor of the first Holy Roman Emperor, Charlemagne I (Charles the Great), king of the Germanic Frankish tribes (c.747–814 CE). Although

Austrian 100-Euro coin depicting the Crown of Charlemagne

it is widely referred to as the Crown of Charlemagne, the surviving crown was probably made for the coronation of Otto the Great (912–73 CE), while it is thought that Charlemagne himself wore a simpler version for his coronation in 800 CE. Charlemagne was successful in conquering and unifying much of Western Europe and, after helping to put down a rebellion against Pope Leo III, was crowned Holy Roman Emperor by the grateful pope. The tradition of imperial rule continued until 1806, when the last Holy Roman Emperor, Franz II, dismantled the empire after military defeat by Napoleon. While Napoleon marched on Franz's base of Nuremberg, Franz moved the crown to Vienna for safekeeping.

Now preserved in the national treasury in the Hofburg Palace in Vienna as part of the Austrian crown jewels—where it is still on public display to visitors—the Crown of Charlemagne is octagonal and is made from eight hinged panels of 22-karat gold. The panels are set with a dazzling range of 144 precious gems, including sapphires, emeralds, amethysts, and more than 100 natural pearls. Typical of Byzantine jewelery, the stones are rough cut, because faceting techniques had not yet been developed. Four of its panels feature scenes from the Bible depicted in *cloisonné* enamel (a form of painting laced with silver or gold wire), a technique that was also characteristic of the Byzantine era.

Coronation of Charlemagne by Pope Leo III in 800, shown in the *Grandes Chroniques de France* (1375–79) featuring an earlier version of the crown

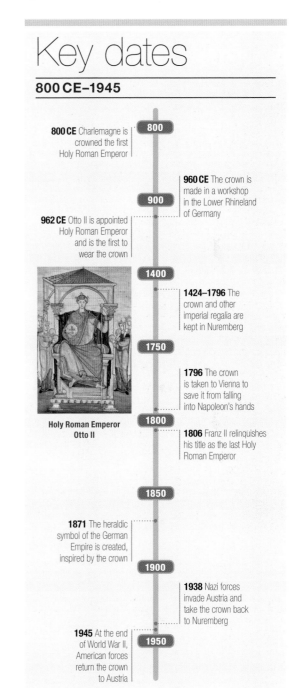

Key dates

800 CE–1945

800 CE Charlemagne is crowned the first Holy Roman Emperor

800

960 CE The crown is made in a workshop in the Lower Rhineland of Germany

900

962 CE Otto II is appointed Holy Roman Emperor and is the first to wear the crown

1400

1424–1796 The crown and other imperial regalia are kept in Nuremberg

1750

1796 The crown is taken to Vienna to save it from falling into Napoleon's hands

Holy Roman Emperor Otto II

1800

1806 Franz II relinquishes his title as the last Holy Roman Emperor

1850

1871 The heraldic symbol of the German Empire is created, inspired by the crown

1900

1938 Nazi forces invade Austria and take the crown back to Nuremberg

1945 At the end of World War II, American forces return the crown to Austria

1950

Silver

△ **English sterling-silver** christening mug of Edward VIII in Art-Deco style

Both gold and silver have been used as currency for thousands of years, and although gold has become synonymous with wealth, silver is increasing in value because of its scarcity. Since pure silver is easily damaged, jewelers work with sterling silver, which has copper added for strength. Silver in folklore is often related to the moon, a link made much of by silversmiths including legendary Danish designer Georg Jensen, who worked in the early 20th century. Since then, silver has grown in demand for use in both jewelry and industry.

Specification

Chemical name Silver | **Formula** Ag | **Color** Silver
Structure Cubic | **Hardness** 3.25 | **SG** 10.1–11.1
Streak Silver-white | **Locations** Mexico is the greatest single producer; also Peru, US, Canada, Norway, Australia, Russia, Kazakhstan

Rough

Wiry silver | Most silver in the ground is extracted from ores but, in its native state, silver can also appear as a coarse mass of tendrils. Here, it is growing in quartz.

Silver and copper melded together

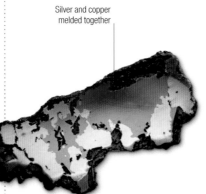

Polished silver and copper slice | Native copper and native silver can sometimes form together in a single specimen, commonly known in the US as a "half-breed."

Tarnished surface

Tarnished silver | The surface of silver is susceptible to tarnishing, which appears as a coating over the surface when exposed to oxygen or hydrogen sulphide, as here.

Settings

Moonlight brooch | Georg Jensen likened silver to the glow of the moon, and here he combines sterling silver with moonstones in an Art Nouveau design.

Silver in industry

Demand outstrips supply

Silver is naturally more abundant than gold in the Earth's crust, but over the past 20 years, stocks have dwindled dramatically due to the fact that it has become highly sought after in manufacturing for its exceptional ability to conduct both heat and electricity. Silver is a key component in photovoltaic panels for solar power, a growing sector that will only increase the demand for silver as an industrial commodity. It is also used in the production of almost all electronic devices, from circuit boards and TV screens to cell-phone batteries and computer chips. A cell-phone circuit board typically contains about 300 mg of silver.

Cell-phone circuit board Silver is used in circuits such as this.

Branchlike form

Silver dendrite | Some silver crystals grow in a dendritic (branching) formation, which has formed a naturally treelike structure in this specimen.

Amethyst cabochons depicting eyes and bubbles

Silver fish | This 1940s sterling-silver bracelet by Margot de Taxco is in the shape of a koi. De Taxco lived and worked in Mexico, the world's top silver-producing country.

Silver artifacts have been discovered that date back to around 4000 BCE

Silver maple seeds

Raised lifelike seed veins

Three-dimensional shaping of seeds

Deer horn

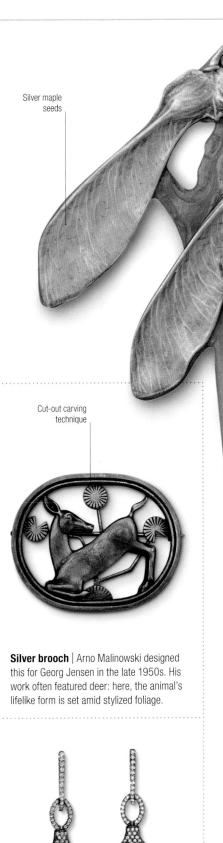

Diamonds totaling 1.55 carats

Cut-out carving technique

Aquamarine brooch | This 18th-century brooch, with a 4.80-carat aquamarine, follows the era's fashion for colorful stones set in silver with decorative elements.

Silver brooch | Arno Malinowski designed this for Georg Jensen in the late 1950s. His work often featured deer: here, the animal's lifelike form is set amid stylized foliage.

Silver Swiss pocket watch | From the 16th century onward, silver and gold replaced steel as the preferred metals for pocket watches. This one has a split-second dial.

Opal body

Plique-à-jour enamel wings

Cabochon sapphires

Dragonfly brooch | This brooch features a mixture of silver and gems. The dragonfly's articulated tail is made up of diamonds set in silver and sapphires set in gold.

Swing drop earrings | Featuring a pair of baroque pearls, these drop earrings also consist of blackened 14-karat gold, silver, and brilliant-cut diamonds.

Hair comb | Lucien Gaillard's comb from c.1902–06 is an example of the fashion for silver in the early 20th century. Silver was seen as both modern and functional.

Antique pansy brooch | Made in France, this colorful vintage brooch is made from silver with gold, amethyst, diamonds, and enamel.

Platinum

△ **Piece of platinum** in its natural state

When **Spanish conquistadors** in Columbia first found platinum in the 16th century, they called it platina, meaning "little silver." To them it was worthless, a distraction in their search for gold. Today, however, it is one of the most precious metals on Earth, both due to its scarcity and its properties as a catalyst, speeding up chemical reactions while remaining inactive itself. In addition to its use in fine jewelry, it is a vital component in converting crude oil into petroleum, and also plays a role in reducing pollution from cars as a form of filter (see box below).

Specification

Chemical name Platinum | **Formula** Pt | **Colors** White, silver gray, steel-gray | **Structure** Isometric (cubic) | **Hardness** 3.5 | **SG** 21.45 | **RI** 2.19 | **Luster** Metallic | **Streak** n/a **Locations** South Africa, Russia, Canada

Rough

Grains vary in size

Grains of platinum | Platinum grains occur naturally but usually include traces of other metals including iron, palladium, rhodium, and iridium.

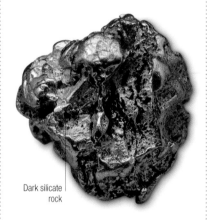

Dark silicate rock

Embedded in rock | When platinum is found as grains, flakes or thin layers in silicate rock, as in this example, it is typically mixed with other minerals and has to be separated out.

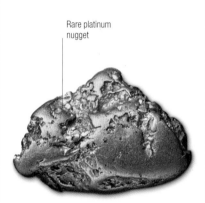

Rare platinum nugget

Platinum nugget | Although most platinum occurs naturally as grains, it is very rarely found in nugget form. Platinum nuggets such as this do not tarnish.

Settings

Channel-set diamonds

Eternity ring | Jewelry company De Beers invented the idea of the diamond eternity ring in the 1960s—the platinum version is the most valuable of their range.

Reducing gas

Catalytic converters

Platinum reduces pollutants from car engines by converting poisonous gases into less harmful substances. Since 1974, when the US introduced new laws on air quality, catalytic converters in vehicles have become a worldwide phenomenon. Catalytic convertors use platinum to minimize the emission of noxious gas from engines—the platinum catalyst rips apart the toxic nitrogen dioxide and allows the molecules to re-form in less toxic combinations.

Platinum at work This cross section of a catalytic converter shows the platinum grains it contains.

Angular crystal face

Sperrylite crystal | Sperrylite is a compound of platinum and arsenic, and is valued by collectors as a specimen rather than an ore of platinum.

Traces of platinum from around 1200 BCE have been found in ancient Egyptian tombs

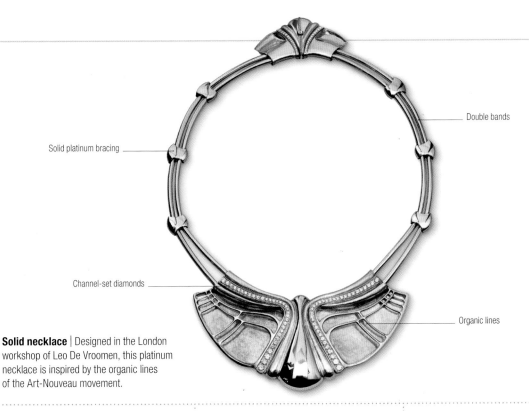

Solid platinum bracing

Channel-set diamonds

Double bands

Organic lines

Solid necklace | Designed in the London workshop of Leo De Vroomen, this platinum necklace is inspired by the organic lines of the Art-Nouveau movement.

Key pendant | The brilliant white color of platinum is evident in this Tiffany & Co. Quatra pendant, which also features white diamonds in a brilliant cut.

Knot set with brilliant-cut diamonds

2.9-carat diamond

Knot ring | This 1960s diamond and platinum ring with a knot-shaped setting is designed more for show than wearability, with a complex design and large center stone.

C-shaped tourbillon bridge

Cartier watch | Skeletonizing the movement of this Rotonde de Cartier platinum watch takes up to 200 hours by hand, with another 200 hours for assembly.

Prong setting

Solitaire ring | Cartier launched the solitaire in 1895. Since then, its platinum and diamond solitaires such as this example have become a benchmark for engagement rings.

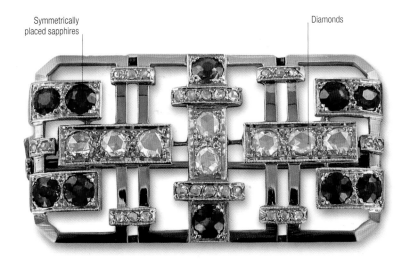

Symmetrically placed sapphires

Diamonds

Art-Deco brooch | Platinum jewelry was the metal of choice for fashionable women in the early 20th century. This geometric design is set with diamonds and sapphires.

Chandelier earrings | Made around 1915–20 by Marcus & Co., these earrings in an Art-Deco style feature round-cut diamonds in a platinum setting.

Old European-cut diamonds

Platinum setting

Openwork bracelet | This substantial openwork bracelet from France is inlaid with 411 diamonds set in platinum. It was made in around 1935.

Diamond pear-drop earrings | c.1770s–1780s | 14.25 and 20.34 carats | Made in France, probably by court jewelers Boehmer and Bassenge

Marie Antoinette's diamond earrings

△ **Marie Antoinette** pictured wearing the earrings

Reigning as the queen of France alongside King Louis XVI from 1774 to 1792, Marie Antoinette was the most glamorous woman in Europe in the 18th century, setting trends that were slavishly followed by fashionable ladies of the royal courts. Her extravagance provided fuel for the satirical newspapers of the day, and her love of fine clothes and jewelry earned her the nickname "Madame Déficit."

Among her indulgences were up to 300 dresses a year, countless pairs of perfumed gloves, and a hoard of sparkling jewelry. Some of it was made from paste (heavy flint glass) but much of it was real, including a favorite of the queen—a pair of diamond earrings with pear-shaped drops, one weighing 20.34 carats, the other weighing 14.25 carats. Thought to be a gift from Louis XVI, and commissioned from jewelers to the French court Boehmer and Bassenge, the earrings are believed to have passed down through the French royal family following Marie Antoinette's death by guillotine in 1793 during the French Revolution. They resurfaced some 60 years later as a wedding gift from Napoleon III to Empress Eugénie, who was fascinated by,

and modeled her style on, Marie Antoinette. After Eugénie was exiled to England in 1871, the earrings were sold to a Russian aristocrat, who in turn sold them to jeweler Pierre Cartier in 1928. In the same year, they were bought by American socialite Marjorie Merriweather Post, and in 1964 her daughter donated them to the Smithsonian Institution in the US, where they can still be seen in the Gem Gallery.

Portrait of Empress Eugénie, who received the earrings from Napoleon III as a wedding gift, and sold them after Napoleon's defeat in the Franco-Prussian War in 1871

Key dates
1755–1964

1725

1755 Marie Antoinette is born in Vienna, 15th child of the Empress of Austria, Maria Theresa

1750

1770 Marie travels to France to marry Louis Auguste, grandson of King Louis XV

1774 Louis Auguste takes the throne; Marie becomes queen

1775

1774–89 During their reign, Louis presents Marie with the pear-drop diamond earrings

1793 Marie Antoinette dies at the guillotine; her earrings remain in the French royal family

1800

1870–72 Empress Eugénie is thought to sell the earrings to Duchess Tatiana Yousupoff of Russia

1850

1853 Napoleon III gives the earrings to Empress Eugénie to celebrate their marriage

1900

1928 Pierre Cartier buys the earrings and sells them to Marjorie Merriweather Post

1959 The diamonds are mounted into replica settings by jewelers Harry Winston

1950

1964 Post's daughter, Eleanor Barzin, donates the earrings to the Smithsonian Institution

2000

Napoleon III

Every woman wanted to imitate the queen. Everyone rushed to get the same jewelry

Madame **Campan**
First lady-in-waiting to Marie Antoinette

Copper

△ **Sample of the mineral bornite**, a principal ore of copper

opper was the first metal to be used by humans—it naturally occurs in its pure form, and was used in early casting and decorative arts; it also forms a part of bronze, which was the first purposefully made alloy. Copper jewelry has been worn for millennia, and is popular in alternative medicine. It is also an extremely effective electrical conductor: this property, and the fact that it is extremely resistant to corrosion, have made copper the most widely used material for electrical wiring in the modern age.

Specification

Chemical name Copper | **Formula** Cu | **Colors** Orange red | **Crystal system** Cubic | **Hardness** 2.5–3.0
SG 8.89 | **RI** 2.43 | **Luster** Metallic | **Streak** n/a
Locations Chile, US, Indonesia

Rough

Branchlike growths

Dendritic copper | The most dramatic form of native copper, dendritic copper occurs when copper crystals form thin, branching, fernlike sheets such as this.

Copper sheets

Native copper | In this mixed rocky specimen, thin, leaflike sheets of native copper are interlayered with a piece of quartz groundmass.

Cast

Cast sphere | This bead of pure copper has been cast in the form of a sphere. Material such as this would be suitable for smelting or recasting.

Four-sided jar | This bronze, four-sided *fanghu*, or square jar, originates from China c.475–221 BCE. It was decorated with green malachite, most of which is now lost.

Copper jewelry is said to protect from health problems

Garnet "eye"

Copper alloy brooch | This Anglo-Saxon bird brooch made of a copper alloy was found at Bekesbourne Anglo-Saxon cemetery, UK. It originates c.5th–8th century CE.

Outsized horns

Suspension ring

Etruscan bronze | Made around 599–500 BCE, this cast Etruscan bronze amulet is in the form of two opposite-facing oxen with a common body.

Shell eyes

Bronze armor | This 3rd–2nd-century BCE armor is made from overlapping bronze plates, originally stitched or riveted to a leather jerkin.

Interlocking plates

Stylized features

Anatolian bronze | This gilt-bronze figurine from the first millennium BCE features a man riding a lion. It is probably based on an image from Anatolian mythology.

Loop for chain

Edo bronze | Originating from the Edo (Bini) people of Nigeria around 1520–80, this hollow-cast bronze head depicts a conquered king.

Roman brooch | This 1st-century CE Roman bronze brooch—one of a pair—was found in the UK and features elaborate swirling ornamentation.

Cast details

Chinese bronze | This classically ornamented cast bronze wine jar with a lid originates from the Western Zhou dynasty, which ended in 771 BCE.

Statue of Bast | The Egyptians revered cats as gods and had a temple dedicated to them at Bubastis on the Nile delta. This 22nd-dynasty bronze represents the goddess Bast.

Statue of Liberty

Icon in copper

Liberty Enlightening the World, known as the Statue of Liberty, stands on Liberty Island, New York, US. It was designed by French sculptor Frédéric Auguste Bartholdi, built by Gustave Eiffel, and dedicated in 1886. Its "skin" consists of around 200,000 lb (90,800 kg) of copper, $\frac{3}{32}$ in (2.3 mm) thick— at the time, the largest single use of copper in the world. Originally, the statue was a dull copper color, but later developed a green patina from oxidation. After investigation, it was decided that the patina should remain and actually helped to protect the exterior.

Statue of Liberty The iconic statue consists of copper plates attached to a rigid iron frame.

Artemision Bronze | 460–c.450 BCE | 6 ft 10 in (2.09 m) | Hollow cast bronze figure of Zeus or Poseidon | Severe style from the Classical period of ancient Greece

Artemision
Bronze

△ **Head of the statue** showing finely rendered hair

The Artemision Bronze is an ancient Greek sculpture found in a shipwreck off the north coast of the Greek island Euboea, in the Mediterranean. It represents either the Greek god Zeus, king of the gods, or Poseidon, god of the sea. Most ancient bronze statues have since been lost or melted down, making this one all the more precious.

Dating from 460–c.450 BCE, the nude bronze figure stands at 6 ft 10 in (2.09 m). It shows the realistic anatomy of the Greek Classical period, though the arms are disproportionately long, exaggerating the dramatic pose. The wide stance and extended arms—one poised to hurl a weapon and the other taking aim—suggest great power about to be unleashed. An

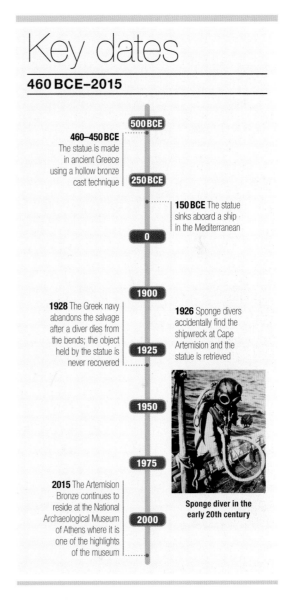

Front view of the bronze statue

object missing from the right hand, either a thunderbolt for Zeus or a trident for Poseidon, makes the god's identity debatable, although most scholars now think the sculpture represents Zeus. Ancient Greek pottery shows him holding his thunderbolt aloft in a similar stance, whereas Poseidon is normally depicted wielding his trident downward.

The shipwreck that contained the statue was first discovered by sponge divers in 1926, and a subsequent salvage operation by the Greek navy recovered the Artemision Bronze in two pieces, along with various other treasures. The exploration, however, was called off in 1928 after a diver died, and it was never resumed, despite the wreck lying only 130 ft (40 m) below the surface. It is thought that the ship may have been of Roman origin, bringing back looted treasures from Greece to Italy. Ironically, the statue was saved and preserved for future generations after it ended up on the bottom of the sea.

Key dates

460 BCE–2015

500 BCE

460–450 BCE The statue is made in ancient Greece using a hollow bronze cast technique

250 BCE

150 BCE The statue sinks aboard a ship in the Mediterranean

0

1900

1928 The Greek navy abandons the salvage after a diver dies from the bends; the object held by the statue is never recovered

1926 Sponge divers accidentally find the shipwreck at Cape Artemision and the statue is retrieved

1925

1950

1975

2015 The Artemision Bronze continues to reside at the National Archaeological Museum of Athens where it is one of the highlights of the museum

2000

Sponge diver in the early 20th century

Pottery figure of the Greek god Zeus, thought to be from around 500 BCE, depicted holding a spear or javelin in place of his customary thunderbolt

The bearded god once hurled a weapon held in his right hand, probably a thunderbolt, in which case he is Zeus

Fred S. **Kleiner**
Author

107.46-carat
"Graff Sunflower"

Diamond
surround

100-carat "Graff
Perfection"

Some early civilizations purposefully left diamonds uncut in the belief that this would preserve the stones' magical properties

"Royal Star of Paris" | This stunning brooch/pendant by Graff showcases a 107.46-carat yellow fancy diamond, and a 100-carat, class D, flawless pear-shaped diamond. In total, it contains diamonds weighing over 2,000 carats.

Diamond

△ **Platinum ring** set with yellow and white diamonds

With exceptional beauty, luster, and sparkle, diamond is the most iconic of all precious stones and highly prized in jewelry all over the world. However, this is only one of its uses. Industrial diamond is a vital component in oil drilling, specialized scalpels, tool manufacturing, and many other industries, all of which use the supreme hardness of diamonds for cutting tools and abrasive powders. There is no firm boundary between gem-grade and industrial-grade diamonds—around 80 percent of the diamonds mined each year are unsuitable for gemstone wear, and find other uses in industry. However, very small or lower-grade stones can be polished into gemstones rather than being used in industry.

Discovering diamonds

For over 2,000 years, diamonds were found only as crystals in river gravels, and, until 1725, India was the major source. As Indian production waned, diamonds were discovered in Brazil, and in 1867 they were found in gravels near the Orange River in the Kimberley region of South Africa. Further exploration there revealed volcanic pipes of a previously unknown rock type containing diamonds; this was named kimberlite and was recognized as the diamond source rock. Its discovery formed the basis of the modern diamond industry. Many similar pipes have since been found in other African countries, Siberia, Australia, and more recently in Canada, China, and the US.

Specification

Chemical name Carbon | **Formula** C | **Colors** All colors | **Structure** Cubic | **Hardness** 10 **SG** 3.4–3.5 | **RI** 2.42 | **Luster** Adamantine **Streak** None

Round brilliant   Oval brilliant   Pendeloque   Marquise

Baguette   Emerald   Mixed

Locations
1 Canada **2** US **3** Brazil **4** Ghana **5** Angola **6** Namibia **7** Botswana **8** South Africa **9** India **10** Russia **11** Borneo **12** Australia

Key pieces

Dresden Green | Probably from the Kollur Mine in India, this 41-carat natural green diamond is named after its home in Dresden, Germany (see pp.140–41). It is famous for its extraordinary green color and is set in a lavish hat ornament.

Cullinan I

Cullinan I | Originally part of a 3,106.75-carat rough stone from which several gems were cut (see p.54), the Cullinan I is part of the British crown jewels. At 530.1 carats, it is the largest polished white diamond in the world.

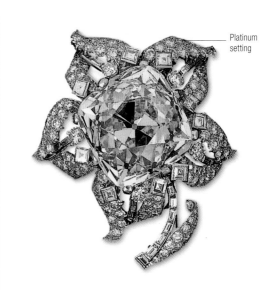

Platinum setting

Allnatt Diamond | Weighing 101.29 carats, this extraordinary stone is described as a fancy vivid yellow and is set in a platinum flower design. It is named after one of its former owners, businessman and art collector Alfred Ernest Allnatt.

Rough

Irregular surface

Rough diamonds | Diamond crystals can be found in a number of external forms, all related to their cubic structure. Here, they have formed rough cubes.

Thick carbon inclusions

Carbonado | This specimen consists of carbonado, a distinct cryptocrystalline variety of diamond originating from Brazil and Central Africa.

Tip of octahedron

Perfect octahedron | The majority of diamonds found are crystallized as octahedrons, as here. Originally, only the crystal faces were polished.

Flawless interior

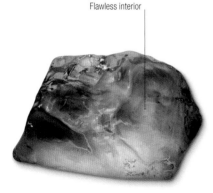

Cullinan rough model | This is a model of the largest diamond ever found—the Cullinan (see p.53). It weighed 3,106.75 carats and was the size of a potato.

> # We had with us salted birds… we placed the gem on them, they became animated and flew away

Reference to diamonds in the Babylonian Talmud

Colors

Natural blue coloring

Blue Heart Diamond | This natural blue diamond, cut into a heart shape, originated in the diamond fields of South Africa. It weighs 30.62 carats.

Facets visible through table

Brilliant cut | Looking downward through the table of this brilliant-cut diamond, the clarity of the stone and the reflectivity of its facets can be seen.

Crown facets

Brown diamond | Brown diamonds originate mainly from Australia. As seen in this pendaloque-cut gemstone, tend to lack the brilliance of diamonds in other colors.

Table facet

Deep green | This green diamond has been cut into a pendaloque. Its intense hue suggests it may have undergone color enhancement (see box, right).

Oppenheimer Diamond | One of the largest uncut diamonds in the world at 253.7 carats, this perfectly formed curved octahedron is a natural yellow stone.

Yellow coloring

Curved surfaces

Cut

Table facet

Girdle

Classic-cut diamond | Seen side-on, all of the facets of this champagne-colored diamond are visible, either directly or through the stone.

Extra pavilion facets

Fancy cut | This triangle-cut diamond is technically a fancy cut, in that it has a number of extra, nonstandard faces added to what is otherwise a brilliant cut.

Emerald cut facets

Blue emerald cut | The cutter of this diamond has chosen an emerald cut in order to enhance its blue tinge, but still retain its brilliance.

Tiny pavillion facets

Mixed cut | This white diamond has had a number of extra faces added to the pavilion of a scissors-cut crown to increase the sparkle under the table.

Steely coloring

Emerald cut | Even though the stone is small, the emerald cut of this gem has enhanced the steely sheen that this diamond possesses.

Finely faceted notch

Fancy heart | Hearts are the most difficult shape to facet because of the faces in the top notch. The cutter of this stone has achieved the heart shape with great skill.

Superb "fire"

Standard brilliant | This diamond has been cut in the "standard" 58-facet brilliant cut, specifically developed to maximize the brilliance, or "fire," of a diamond.

Diamond enhancement

Bringing out the best

"Fancy" (colored) diamonds demand high prices if the colors are definite and intense. Reds, violets, and blues bring the highest prices. They are not always what they seem. Today, a number of processes exist to change the color of white diamonds, from irradiation to flooding them with gases that are absorbed and produce color change. Other enhancements include laser drilling to remove inclusions and the application of sealants to fill cracks. Buyers should always purchase diamonds certified by a legitimate testing agency.

Marquise-cut blue Only a gemologist can tell if the color of this stone is natural.

Settings

Diamond-set chain

Victoria-Transvaal Diamond

Victoria-Transvaal Diamond | This 67.89-carat, pear-shaped, brownish-yellow stone was originally part of a 240-carat rough stone discovered in the Transvaal, South Africa. The Victoria-Transvaal Diamond Necklace has appeared in a Tarzan movie.

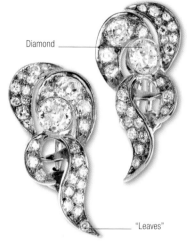

Diamond

"Leaves"

Diamond earrings | This pair of gold foliate diamond earrings features a large diamond as each flower head, and diamond-set leaves.

Aquamarine

Floral spray brooch | Set in 18-karat gold, this floral brooch has aquamarines for the flower centers, and diamonds for the leaves and petals.

Pavé-cut diamonds

Platinum owl brooch | This whimsical brooch is made out of platinum. The owl features yellow diamond eyes, a body set with pavé-cut diamonds, a black coral beak, and gold claws. The gem-encrusted bird rests on a black coral "branch".

Butterfly brooch | Made by jewelry artist Cindy Chao, this brooch is set with 2,138 gems, including large diamonds in the wings. These have been faceted on one side only.

Large, semi-faceted diamond "wings"

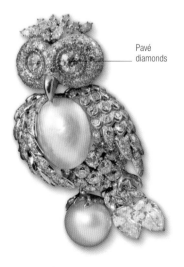

Pavé diamonds

Owl brooch | Made of platinum and yellow gold, this variation on the whimsical owl theme has an intricate pavé diamond head and body, and large pearl accents for chest and perch.

An ancient test for a diamond was to place the stone on an anvil and strike it with a hammer: if it broke, it wasn't a diamond

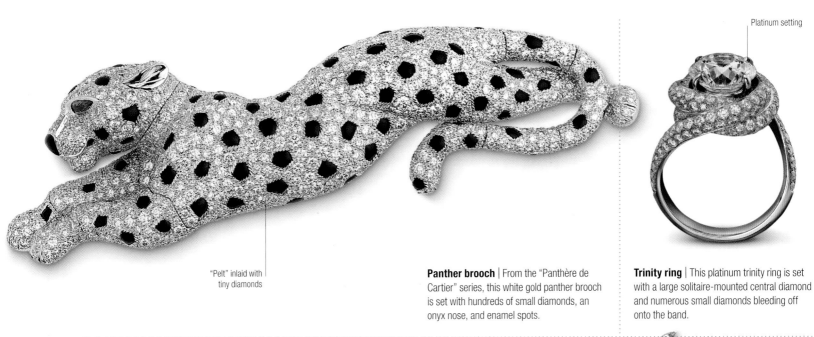

"Pelt" inlaid with tiny diamonds

Platinum setting

Panther brooch | From the "Panthère de Cartier" series, this white gold panther brooch is set with hundreds of small diamonds, an onyx nose, and enamel spots.

Trinity ring | This platinum trinity ring is set with a large solitaire-mounted central diamond and numerous small diamonds bleeding off onto the band.

18-karat white gold setting

Diamond centerpiece

Briolette diamond

"Buckle" ring | The central bezel of this ring has been designed to resemble a buckle. The band and bezel are both set with round and baguette-cut diamonds.

"Ribbon" brooch | Devised as a pair of ribbons passing through a circular buckle, this brooch is set with one large diamond and numerous smaller ones.

Briolette diamond brooch | Created for Van Cleef & Arpels, this phoenix brooch is set with diamonds and sapphires, with a 96.62-carat diamond in its beak.

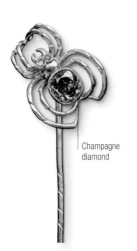

Champagne diamond

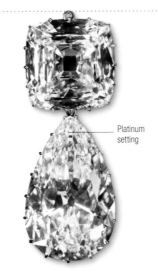

Platinum setting

The Orange

A natural rarity

Orange-colored fancy diamonds are noted of their rarity, and mostly occur only in smaller sizes. As a result, this gemstone caused a stir when it was put up for sale. At the time of its auction in 2013, The Orange was claimed to be the largest fancy vivid orange diamond in the world—estimated to be 14.82 carats. Its size, beauty, and rarity were reflected in its price, and it was sold by Christie's auctioneers in Geneva for over $35 million.

Bright, natural color The Orange is unusual for its combination of fine color and large size.

Spider stickpin | Made around 1900, this Art Nouveau stick pin is crafted in the form of a spider, and is set with a 0.80-carat champagne diamond.

Cullinan III and IV brooch | The diamonds in this platinum brooch are two of the smaller stones cut from the world's largest diamond, the 3,106-carat Cullinan (see pp.53 and 54).

Koh-i-noor | Oval brilliant-cut diamond | 105.6 carats | Seen here set in the center of the crown of Alexandra of Denmark (1844–1925), wife of King Edward VII

Koh-i-noor
diamond

△ **Replica of the Koh-i-noor** (center) in its original cut and setting

The Koh-i-noor ("Mountain of Light") diamond, like many famous gems, had a turbulent history. Mined in southern India, the stone was initially referred to in 1526 in the memoirs of Babur, the first Mogul king of India. It was a spoil of war, and it continued to change hands between kings over the course of several centuries, which may go some way toward explaining its reputation for being cursed, since whoever owned the huge diamond was a target for attack.

By the time five-year-old Duleep Singh came to power as the last ruler of the Punjab and Sikh Empire, the diamond belonged to him, since the previous four maharajas had been assassinated while still in possession of the stone. Just a few years later, the British dismantled Singh's kingdom, and the Koh-i-noor transferred to British ownership

Queen Elizabeth the Queen Mother, wearing the Koh-i-noor in a simple version of her crown

as part of the treaty that incorporated the Punjab into the British Empire. Presented to Queen Victoria in London in 1850, the diamond was apparently accompanied by a curse that read: "He who owns this diamond will own the world, but will also know all its misfortunes. Only God or Woman can wear it with impunity."

Far more controversial than the supposed curse was the criticism that the 186-carat gem looked dull due to poor cutting. Prince Albert decided to have the diamond recut in 1852, drastically reducing its size to 105.6 carats but eliminating several flaws in the process and creating an oval brilliant cut. Since then, the recut Koh-i-noor has been set in four different crowns, each worn by British queens, including Queen Alexandra, Queen Mary, and Queen Elizabeth the Queen Mother.

Koh-i-noor diamond

Koh-i-noor diamond (center) in Queen Elizabeth the Queen Mother's crown

Only **God** or **Woman** can wear it with impunity...

Curse said to have accompanied the Koh-i-noor

Key dates
1100–2015

Maharaja Duleep Singh

1100

1100–1300 The diamond is thought to have been mined in southern India

1500

1526 The diamond is first documented in the memoirs of the Mogul king, Babur

1600

1800

1850 Duleep Singh, last Maharaja of the Punjab and Sikh Empire, presents the Koh-i-noor to Queen Victoria

1851 On display at the Great Exhibition in London, the diamond is criticized for being dull

1852 The diamond is recut into an oval brilliant at Prince Albert's behest

1902 The diamond is set into the coronation crown of Queen Alexandra

1900

1937 Elizabeth (the late Queen Mother) wears the diamond at the coronation of her husband, George IV

1947 Newly independent India requests the diamond's return

1950

1976 Prime Minister Zulfikar Ali Bhutto lays claim to the diamond for his country, Pakistan

2015 A group of Indian investors launch legal proceedings for the diamond's return

2000

ANCIENT EGYPT

Clothing in ancient Egypt was drab by modern standards: it was typically off-white, the natural color of linen. It is not surprising, then, that the Egyptians adorned themselves with vividly-colored gems such as amber, turquoise, lapis lazuli, and carnelian, although a glasslike, glazed ceramic known as Egyptian faience was also used. Gems were showcased in wide semicircular collars worn by both men and women, with counterweights hanging down their backs to keep the jewelry in place. Wigs were popular, and richly ornamented headpieces and circlets kept them from slipping. Earrings, bracelets, and amulets were also worn by all classes, further enhancing a wearer's appearance.

Gems had great spiritual significance as well, and were often worn for protection, to ward off evil, or to attract the attention of good spirits. Red stones such as carnelian and red jasper were considered powerful because of their resemblance to the color of blood, which stood for life and longevity. Greenish-blue turquoise from the Sinai represented fertility, healing, and rebirth, while deep blue lapis lazuli from Afghanistan was especially significant, symbolizing the heavens, death, and the afterlife.

Gold, carnelian, and feldspar Egyptian necklace, c.1991–1786 BCE

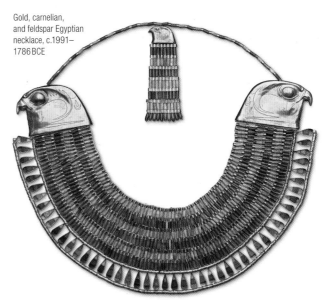

Egyptian guests at a banquet This wall painting from the tomb of Nebamun, an Egyptian offical who lived in the 18th dynasty, around 1350 BCE, depicts banquet guests with the women wearing elaborate gowns and wigs, and decorated with precious jewels.

Alternating square-cut diamonds

More diamonds surround central gem

Faceted girdle and extra facets on pavilion

Pin securing diamond setting

Hope Diamond | 1 x ¾ x ½ in (25.6 x 21.78 x 12 mm) | 45.52 carats, cushion antique brilliant, fancy dark grayish blue with whitish graining present

Hope Diamond

△ **Hope Diamond** in its necklace setting

The Hope Diamond is celebrated first and foremost for its stunning color and size. It weighs 45.52 carats and is the world's largest deep blue diamond to date. Its extraordinary color is caused by the mineral boron: most natural blue diamonds contain tiny particles of boron, averaging fewer than 0.5 parts per million (ppm), but areas of the Hope Diamond have as many as 8ppm. It also glows red under ultraviolet light.

King Louis XV of France, a former owner of the diamond

Adding to its aura of mystery, the Hope Diamond is said to carry a curse—various figures from its history have suffered ill fortune, including Marie Antoinette, guillotined in the French Revolution, and American heiress Evalyn Walsh McLean, who was struck by a catalogue of misfortunes. She bought the gem in 1911 and suffered bereavement, divorce, and bankruptcy. Its last private owner, jeweler

Harry Winston, posted it to the Smithsonian, its current owners, paying $155 in insurance, but even the mailman who delivered it attracted bad luck—he was allegedly hit by a truck.

For all the accursed tales surrounding it, the Hope Diamond has an illustrious, royal provenance. Discovered in an Indian mine in the 18th century, it was originally a larger stone weighing 115 carats. It was called the Tavernier Diamond after its first owner, Jean-Baptiste Tavernier. He sold it to Louis XIV, who had it cut. The larger part was a 67.12-carat heart-shaped diamond known as the French Blue. This gem, inherited by Louis VI and Marie Antoinette, was stolen during the French Revolution but surfaced in London in 1812 as a smaller, recut gem. The diamond was documented in 1839 owned by Henry Philip Hope, who gave it its name—then died in the same year.

Brilliant colors ... all are found enclosed in a morsel of pure carbon

Charles **Blanc**
Author

American heiress Evalyn Walsh McLean, pictured wearing the Hope Diamond in its necklace setting, and whose personal misfortune fueled rumors of a curse

Key dates

Mid-1600s–1958

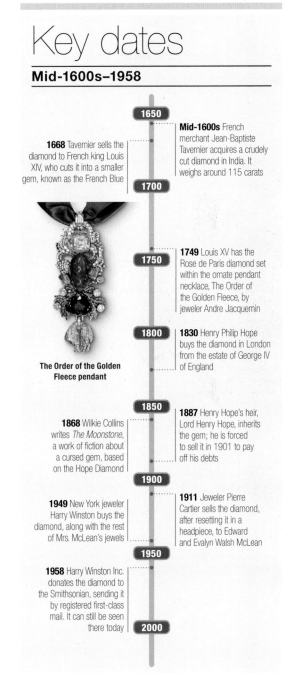

The Order of the Golden Fleece pendant

1650

1668 Tavernier sells the diamond to French king Louis XIV, who cuts it into a smaller gem, known as the French Blue

Mid-1600s French merchant Jean-Baptiste Tavernier acquires a crudely cut diamond in India. It weighs around 115 carats

1700

1749 Louis XV has the Rose de Paris diamond set within the ornate pendant necklace, The Order of the Golden Fleece, by jeweler Andre Jacquemin

1750

1800

1830 Henry Philip Hope buys the diamond in London from the estate of George IV of England

1850

1868 Wilkie Collins writes *The Moonstone*, a work of fiction about a cursed gem, based on the Hope Diamond

1887 Henry Hope's heir, Lord Henry Hope, inherits the gem; he is forced to sell it in 1901 to pay off his debts

1900

1911 Jeweler Pierre Cartier sells the diamond, after resetting it in a headpiece, to Edward and Evalyn Walsh McLean

1949 New York jeweler Harry Winston buys the diamond, along with the rest of Mrs. McLean's jewels

1950

1958 Harry Winston Inc. donates the diamond to the Smithsonian, sending it by registered first-class mail. It can still be seen there today

2000

Gemstones

Pyrite

△ **Pyrite crystals** in the form of pentagonal dodecahedrons, also called pyritohedrons

Known since antiquity, pyrite is better known by its informal title, "fool's gold." Its name is derived from the Greek word *pyr*, meaning "fire," because pyrite emits sparks when struck by iron. Nodules of pyrite have been discovered in prehistoric burial mounds: the sunlike color of pyrite probably assured its value. In later times, polished slices of its crystals were set edge to edge on wooden backing to make mirrors. Today, pyrite is polished as beads, and its bright crystals are themselves mounted as gemstones.

Specification

Chemical name Iron disulphide | **Formula** FeS_2
Colors Pale brass yellow | **Structure** Cubic | **Hardness** 6–6.5
SG 5.0–5.2 | **RI** 1.81 | **Luster** Metallic | **Streak** Greenish black to brownish black | **Locations** Spain, South America, US, Japan, Italy, Norway, Greece, Slovakia

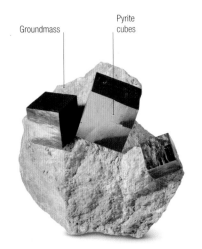

Groundmass
Pyrite cubes

Spanish pyrite | **Rough** | The source of this specimen—Almira, Spain—is famous for its abundance of pyrite. These well-formed cubes are in a lime-rich mudstone matrix.

Flat surfaces

Pyrite crystal | **Rough** | This dazzling, neatly cuboid pyrite crystal offers a good demonstration of how the mineral can form in its natural state.

Crystals with mixed shapes

Modified crystals | **Rough** | The pyrite crystals in this excellent specimen have developed into the form of cubes modified by octahedrons.

Pyrite necklace | **Set** | The spherical beads of this necklace are made of highly polished pyrite, finely crafted even though pyrite is brittle and difficult to work.

Octahedral pyrite

Quartz crystals

Pyrite and quartz | **Rough** | In this classic pyrite specimen, prismatic quartz crystals are growing on octahedral crystals of pyrite. The two often grow together.

Marcasite

Pyrite in disguise

Marcasite is a mineral that, in all likelihood, has never been used as a gemstone. However, the name is widely used to refer to both it and pyrite. So-called marcasite jewelry, popular in Victorian times, has always been made mainly from pyrite, since some genuine marcasite is chemically unstable and rapidly deteriorates in air. Although marcasite is chemically identical to pyrite, it has a different crystal structure.

Cut "marcasites" This mass of rose-cut "marcasites"—actually pyrite—are ready for setting.

Sphalerite

△ **Ruby blende**, a red variety of sphalerite

Sphalerite gemstones are rare. This is not because the stone itself is rare, but because it is possibly the most difficult of all gems to cut. The stone can easily shatter into small pieces during cutting: the ability to facet sphalerite is the mark of a master cutter. For this reason, stones are faceted only for collectors. Sphalerite takes its name from the Greek *sphaleros*, meaning "treacherous," referring to the fact that it occurs in a number of forms that can be mistaken for other minerals. Its usual color is greenish yellow, but it can also be ruby red.

Specification

Chemical name Zinc sulphide | **Formula** ZnS
Colors Yellow green, red, brown, black | **Structure** Cubic
Hardness 3.5–4 | **SG** 3.9–4.1 | **RI** 2.36–2.37 | **Luster** Resinous to adamantine | **Streak** Brownish to light yellow
Locations Russia, Spain, Mexico, Canada, US

Barite

Gem-quality sphalerite crystal

Internal variations

Facets of scissor cut

Barite on sphalerite | Rough | This specimen consists of a mass of elongated, platy barite crystals resting on a bed of sphalerite crystals.

Gem sphalerite | Rough | Here, a large, gem-quality sphalerite crystal can be seen embedded in smaller crystals of sphalerite and quartz.

Fine facets

Table facet

Faceted oval | Cut | The superb cut of this oval sphalerite gemstone brings out one of the mineral's more unusual colors, a deep red hue.

Emerald cut | Cut | Because of sphalerite's extreme brittleness, stones with corners of any kind are difficult to cut, so this emerald-cut gem displays exceptional craftsmanship.

Scissors cut | Cut | The cutter of this sphalerite gem has used the complexity of a modified scissors cut to help disguise the stone's internal color variations.

Octagonal rose-cut
St. Edward's Sapphire

Velvet cap with
ermine band

Fretted silver pavé set with
brilliant-cut diamonds

Stuart Sapphire

Stuart Sapphire | Approx 1½ x 1 in (3.8 x 2.5 cm) | 104-carat, oval-cut fine blue sapphire

Stuart
Sapphire

△ **Edward I of England**, who took the stone in 1296

Historians are not sure where the Stuart Sapphire came from, or of the identity of its first owner. What is certain, however, is that the gem has represented the might of Scotland and its royal family for hundreds of years. The 104-carat, oval-cut sapphire has a fine blue color and is drilled at one end, indicating that it has been worn in a pendant at some point in its history. It was named after the Stuart monarchs who united England and Scotland under their reign from 1603 to 1741. Before then, there is some evidence that it belonged to the first king of Scotland, Alexander II, and was set into his coronation crown in 1214.

Passed down through generations of Scottish royals, the sapphire is officially noted in the possession of the Stuart King James II when he ruled England and Scotland. Historians generally agree that when James II fled England in 1688, bound for France, he took the sapphire with him.

A century later, the sapphire was back on English soil in the possession of King George III.

By the time George's granddaughter Victoria came to the throne in 1837, the stone had become the centerpiece of the Imperial State Crown and was used for her coronation a year later. The sapphire took pride of place at the front of the crown until 1909 when it was moved to the back to make way for a stunning newcomer, the Cullinan II, cut from the largest diamond ever found. The Cullinan II and the Stuart Sapphire are now joined by a band studded with eight emeralds, eight sapphires, and two rows of pearls.

Medieval kings wore [sapphires] around their necks as a defense from harm

Beth **Bernstein**
Author

James I and VI, first King of England and Scotland, portrayed here c.1620 by Paul van Somer. King James owned the sapphire during his reign

Key dates

1214–1909

1200

1214 The sapphire is set into the crown of Scottish king, Alexander II

1296 Edward I of England takes the sapphire when he invades Scotland

1300

1371 David II's successor Robert II becomes the first Stuart king and now owns the stone

c.1360–70 Edward III returns it to his brother-in-law, David II of Scotland

1400

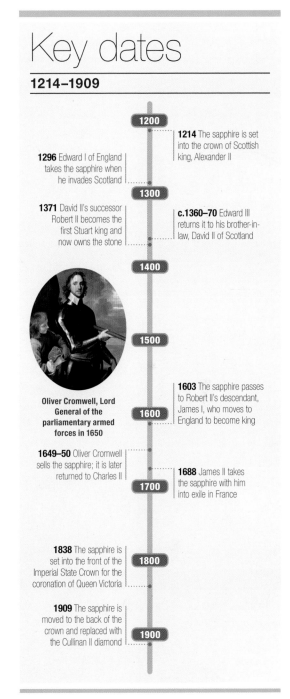

Oliver Cromwell, Lord General of the parliamentary armed forces in 1650

1500

1603 The sapphire passes to Robert II's descendant, James I, who moves to England to become king

1600

1649–50 Oliver Cromwell sells the sapphire; it is later returned to Charles II

1688 James II takes the sapphire with him into exile in France

1700

1838 The sapphire is set into the front of the Imperial State Crown for the coronation of Queen Victoria

1800

1909 The sapphire is moved to the back of the crown and replaced with the Cullinan II diamond

1900

Bismarck sapphire necklace | This necklace was created in 1959. The sapphire was originally set in a choker by Cartier in 1927. The stone hangs from a chain of baguette and round brilliant-cut diamonds.

Eight square-cut sapphires set off main stone

Diamonds

98.57-carat deep blue sapphire of exceptional clarity

Sapphires possess a beauty like that of the heavenly throne; they denote… those whose lives shine with their good deeds and virtue

Marbodius **of Rennes**
11th-century bishop and poet

Sapphire

△ **Rough sapphire** showing color gradation

Both ruby and sapphire are gem varieties of the same mineral, corundum, an aluminium oxide that is next to diamond in hardness. Although commonly thought of as blue, sapphire can also be colorless, green, yellow, orange, violet, and pink, among other hues. Before the end of the 19th century, when geologists realized that sapphires of all colors were the same mineral, terminology regarding the naming of the gem persisted from medieval times: green sapphire was called Oriental peridot and yellow sapphire was Oriental topaz. One of the oldest known stones clearly identified as sapphire is St. Edward's Sapphire: it is believed to date from the Anglo-Saxon king Edward the Confessor's coronet in 1042.

Fancy sapphires

With three exceptions, modern terminology simply uses the word "sapphire" preceded by the color of the stone—for instance, yellow sapphire or green sapphire. Two exceptions are the rare pink-orange stones that are called padparadscha (Sanskrit for "lotus blossom"), and sapphire that appears blue in daylight and reddish or violet in artificial light, which is called alexandrine or alexandrite sapphire. The third exception is blue sapphire, which is simply called "sapphire." Colors other than blue are often referred to as fancy sapphires. Many sapphires, whatever their color, have microscopic inclusions of rutile that produce a star when cut *en cabochon*.

Specification

Chemical name Aluminium oxide | **Formula** Al_2O_3
Colors Most colors | **Structure** Hexagonal, trigonal
Hardness 9 | **SG** 4.0–4.1 | **RI** 1.76–1.77 | **Luster**
Adamantine to vitreous | **Streak** Colorless

Round brilliant   Oval brilliant   Cameo

Step cut   Slab   Cabochon

Locations
1 Montana, North Carolina, US **2** Colombia **3** Brazil
4 Kenya **5** Malawi **6** Madagascar **7** Sri Lanka **8** India
9 Kashmir **10** Thailand **11** Vietnam **12** Australia

Key pieces

Pearl decoration

Gold and silver setting

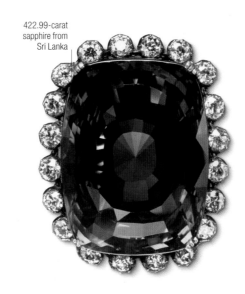

422.99-carat sapphire from Sri Lanka

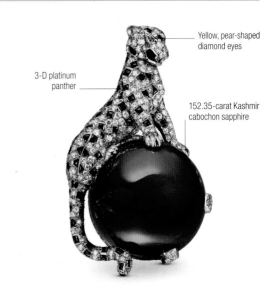

Yellow, pear-shaped diamond eyes

3-D platinum panther

152.35-carat Kashmir cabochon sapphire

Russian pectoral cross | Made in the Kremlin workshops in Moscow, Russia, during the second half of the 16th century, this cross was designed to be worn as a chest ornament. The central sapphire has been carved in the shape of Christ on the cross.

Logan sapphire | This flawless stone is the second largest known sapphire. The large table of the cushion cut shows the naturally perfect interior of this gem, which is set in a brooch surrounded by 20 round brilliant-cut diamonds.

Cartier clip brooch | The Duchess of Windsor revived the popularity of sapphires in the 1950s when she wore this clip. The sapphire is set in white gold and platinum, and smaller sapphire cabochons are used for the panther's spots.

Rough

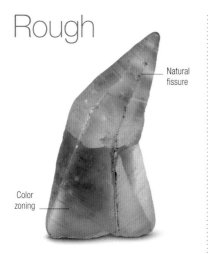

Natural fissure

Color zoning

Colorless rough | This crystal has the classic corundum prismatic shape, with a triangular termination. Colorless sapphire may also be described as white.

Fractured edge

Sapphire rough | This unusually large rough specimen of sapphire, around 1 in (22 mm) in length, displays a fine deep blue coloring. It also shows a number of imperfections.

Natural smoothing

Sapphire gravels | These uncut sapphires come from Philipsburg, Montana. For sheer volume, the US state of Montana is the world's most prolific producer of sapphires.

Striated bands of color

Padparadscha | The rare pink-orange variety of sapphire has its own name, as opposed to being described by its color. Most padparadscha comes from Sri Lanka.

Cut and color

Classic deep blue color

Oblong step | A large table brings out the color of sapphire. Blue is the most highly prized hue, and is given its color by traces of titanium and iron.

Large facets

Oval-cut green sapphire | Green is often found to be bands of yellow and blue sapphire, but skillful cutting of unevenly colored stones yields gems with a uniform appearance.

Table facet

Mixed-cut yellow sapphire | Skillful cutting reveals this stone's even yellow color, its transparency enhanced by the brilliant cut that accentuates highlights and shadows.

Crown facet

Colorless brilliant cut | Colorless sapphires are often cut with multiple facets to catch the light and make the most of their adamantine or vitreous (glassy) luster.

Jean Harlow

Sapphire on the screen

According to Hollywood legend, actress Jean Harlow accepted leading man William Powell's marriage proposal in 1936, but refused his offer of a diamond ring. The platinum-blonde bombshell felt that a large star sapphire would better suit her personal style, and Powell duly purchased one. The ring can be seen in the romantic comedy *Personal Property*, as Harlow wore it on set during her performance—the last of her tragically curtailed career.

Jean Harlow Harlow's sapphire engagement ring can be seen in this still from the movie *Personal Property*, 1937, the year she died.

Crown facet

Synthetic sapphire | This oblong stone features a brilliant cut. Once faceted, this variety of synthetic sapphire shows a full tonal range of pinks.

Asterism, or star pattern

"Star of Asia" | This Burmese gem display asterism, a starlike pattern seen in some stones when cut *en cabochon*. It is caused by tiny intersecting inclusions of rutile.

The sapphire shall be as blue as the great sea

Oscar **Wilde**

Settings

Gold setting

331 round brilliant-cut sapphires

Cluster ring | This dramatic cluster ring features a central, oval-cut sapphire displaying its classic blue coloring, surrounded by a cluster of diamonds.

Sapphire cabochon

Bands of color

Conchita sapphire butterfly | This versatile ornament can be worn as a brooch, pendant, or clasp. It shows the rainbow of sapphire colors, all found in Montana, US. Most Montana sapphires are heat treated to intensify the original colors.

Flower brooch | The petals of this distinctive gold brooch consist of pink sapphires, while the stem and center of the flower are set with diamonds.

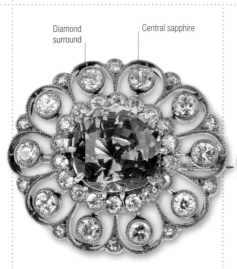

Diamond surround

Central sapphire

Sapphire and diamond brooch | This openwork brooch is based around a 9.32-carat central sapphire, framed by an old-cut diamond scalloped surround.

Multiple sapphires

Diamonds

Sapphire set ring | This twisted silver ring has multiple sapphires on one curve overlaying clustered diamonds set into the other curve, creating a contrast of color and shine.

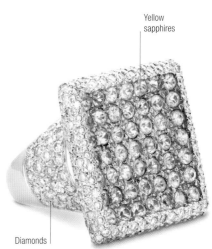

Yellow sapphires

Diamonds

Yellow sapphire ring | This striking dress ring is composed of 49 yellow sapphries set in a square grid pattern, ringed with smaller diamonds.

Danish ruby parure, worn by Queen Ingrid of Denmark c.1960s | 1804 | Rubies, diamonds, gold | Necklace, tiara, earrings, brooch, bracelet (ring later added by Crown Princess Mary)

Danish
ruby parure

△ **Ruby ring** from the parure in its current form

The Danish ruby parure is a breathtakingly beautiful set of jewelry, with a royal pedigree that stretches back over 200 years. Its story begins with the coronation of Napoleon I in 1804. To ensure that this was a spectacular occasion, he gave all his marshals funds to buy new jewelry for their ladies. Among them was Jean Bernadotte, who commissioned the parure for his wife, Désirée Clary. Both were commoners at this point, but Bernadotte was later elected heir to the Swedish throne, and Désirée became Queen Desideria (see pp.108–09). The parure passed into the Danish royal family in 1869, when Princess Louise received the jewels as a wedding present. Although Swedish herself, she was marrying the future King Frederik VIII of Denmark. The gift was deemed particularly appropriate since the diamonds and rubies echoed the colors of the Danish flag. The parure currently belongs to Crown Princess Mary of Denmark.

The showpiece of the parure is the stunning wreath tiara, composed of diamond-encrusted leaves and ruby "berries." The small rubies have been cleverly set in clusters, so that they appear

Napoleon I crowns his wife Josephine as empress in a painting by Jacques-Louis David, 1807

larger. Their coloring is light, closer to pink than blood red, but it has been claimed that this makes them more wearable, especially with blue or purple ensembles.

The precise makeup of the parure has varied over the years. Originally, the leaves were hair ornaments, and they were only modeled into a tiara in 1898. There have since been two major restylings, in 1947 and 2010. In these, the tiara has been made more compact, while the girandole earrings and necklace have been modified, to allow them to be worn with different accessories.

Key dates

1804–2010

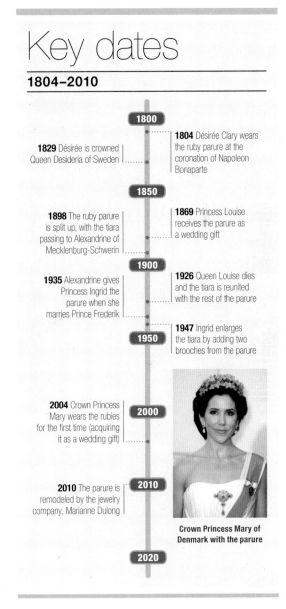

1800

1804 Désirée Clary wears the ruby parure at the coronation of Napoleon Bonaparte

1829 Désirée is crowned Queen Desideria of Sweden

1850

1869 Princess Louise receives the parure as a wedding gift

1898 The ruby parure is split up, with the tiara passing to Alexandrine of Mecklenburg-Schwerin

1900

1926 Queen Louise dies and the tiara is reunited with the rest of the parure

1935 Alexandrine gives Princess Ingrid the parure when she marries Prince Frederik

1947 Ingrid enlarges the tiara by adding two brooches from the parure

1950

2004 Crown Princess Mary wears the rubies for the first time (acquiring it as a wedding gift)

2000

2010 The parure is remodeled by the jewelry company, Marianne Dulong

2010

2020

Crown Princess Mary of Denmark with the parure

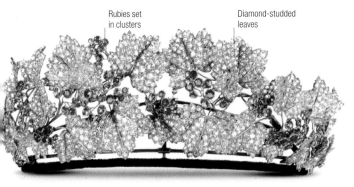

Rubies set in clusters

Diamond-studded leaves

Tiara from the parure featuring leaf shapes set with diamonds, interspersed with "berry" clusters composed of rubies

I see the **natural beauty of** the stones... I look at **colors, sharpening, brightness**

Per **Dirksen**
Goldsmith, on the remodeling of the tiara

Ruby

△ **Mid-20th century ruby** earrings set with 18-karat gold and diamonds

Ruby is the red variety of the mineral corundum, and its color seamlessly picks up where pink sapphire stops. Only darker stones are generally called rubies, but the distinction between ruby and pink sapphire can be a matter of opinion. Ruby is sometimes tinged with purple, and the most valued color is known as pigeon-blood red. It has been mined from the gravels of Sri Lanka since at least the 8th century BCE, the subject of speculation from its earliest days. Ancient Hindu and Burmese miners thought pale pink sapphires were unripe rubies.

Specification

Chemical name Aluminium oxide | **Formula** Al_2O_3
Color Red | **Structure** Trigonal | **Hardness** 9 | **SG** 4.0–4.1 | **RI** 1.76–1.78 | **Luster** Vitreous | **Streak** Colorless
Locations Myanmar, Sri Lanka, Nigeria, Thailand, Australia, Brazil, Kashmir, Cambodia, Kenya, Malawi, Colombia, US, and more

Rough

Termination of crystal

Raw ruby | This gem-quality crystal has the horizontal striations that indicate changes in its growing environment. Ruby ranges from deep cochineal like this to pale rose red.

Striations (parallel grooves)

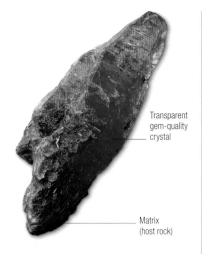

Transparent gem-quality crystal

Matrix (host rock)

Tapering crystal | This prismatic, gem-quality ruby crystal of good color retains a segment of the matrix in which it grew at the base.

Ruby crystals

Crystals in rock | Displaying the gemstone in its natural state, this specimen of rock groundmass is host to a number of prismatic Kashmir ruby crystals.

Cut

Brilliant cut | The faces of this round brilliant ruby illustrate one of the four Cs (color, clarity, cut, carats) that contribute to a gem's quality—in this case, a perfect cut.

Six-pointed asterism

Star ruby | The high-domed oval cabochon, good, intense color, and clearly defined star markings place this star ruby among the finest examples.

Ruby is the solitary and glowing eye which dragons and wyverns carry in the middle of their foreheads

Bishop **Marbodius**
(11th century)

Settings

Triangular-cut diamonds

Claw mount

Deep-red color often called "pigeon's blood"

Carmen Lúcia Ruby | At 23.10 carats, this is the largest faceted ruby in the United States National Gem Collection. It is also one of the finest large, faceted Burmese rubies known. High-quality Burmese rubies larger than 20 carats are exceedingly rare. The stone was mined from the fabled Mogok region of Myanmar in the 1930s.

52 square-cut rubies

46 round diamonds

Diamond trimming

Rubies

Sapphires

Floral brooch | This intricate gold brooch takes the form of a spray of flowers, set with rubies and sapphires (both varieties of corundum) and diamonds.

Diamonds

Rubies

Gold ring | The unusual hexagonal bezel on this ring is set with two rows of rubies, two rows of diamonds, and a large central diamond.

Platinum setting

1930s earrings | A total of 64 calibre-cut rubies are set in this ribbon and circle motif, with two central round-cut diamonds and 34 more baguette and round-cut diamonds.

Navette ring | Dating from around 1910, this Edwardian platinum and gold ring is topped with a diamond-set flower in a field of rubies framed with diamonds.

Diamonds in white gold

Ruby eye

Dragon pendant | In keeping with ancient traditions, this pendant takes the form of a dragon with a ruby for an eye. The setting consists of white gold and diamonds.

Brilliant-cut diamonds

"Pigeon's blood" coloring

Ruby and diamond ring | The pendeloque ruby set in this platinum ring, surrounded by 10 diamonds, has the slight purple tint called "pigeon's blood."

Red gemstones

What's in a name?

The name "ruby" has been applied to a number of red gemstones throughout history including garnets (see pp.258–63) and spinels (see pp.80–81)—Balas ruby was once another name for spinel. It was only through the development of chemistry and mineralogy in the 19th century that distinctions could be scientifically made. Today, many rubies are heat treated to improve their clarity or color. There are also synthetic rubies available, but their value is a tiny fraction of the real thing.

Stepped cut created for emeralds to show off color

Emerald-cut synthetic Despite the good color and clarity typical of synthetic rubies, their value is low.

Timur Ruby | 352.5 carats | Shown here in its necklace setting, in a photograph of Queen Elizabeth II by Cecil Beaton (1953)

Timur
Ruby

△ **Timur Ruby**, in the necklace made for Queen Victoria

Of the Royal Crown Jewels of Britain, there is one exceptional gem that is not what it seems, an impostor that passed through the hands of the rich and powerful for hundreds of years. Until 1851, it was the largest known ruby in the world, but at the time of its documented discovery in 1612, and for hundreds of years afterward, no one was aware that the Timur Ruby was not a ruby at all, but a spinel.

It was only in the latter half of the 19th century that gemologists began to differentiate between the two. The mistake, however, was understandable. Rubies and spinels look almost identical, and share a similar chemical composition and hardness. What separates them is the way in which they refract light—rubies are doubly refractive, while spinels are singly refractive. When light enters a ruby, it is split in two and each beam travels at a different speed. In contrast, when light enters a spinel only one beam of light is generated—an unusual characteristic that is also shared by diamonds and garnets.

Despite its changed classification, the Timur Ruby remains a highlight of the British Crown Jewels. It is named after the Mongol ruler who conquered Delhi in 1398, and who is believed to have taken the gem during the invasion. Returned to India in 1612, the gem was passed down through generations of Mughal emperors, each of whom inscribed their name on its surface. During Britain's annexation of India in the late 1840s, the Timur Ruby and other spinels were shipped to England and presented to Queen Victoria, who openly admired "the wonderful rubies."

Timur, one of the last great conquerors of the Eurasian steppe, pictured holding his crown

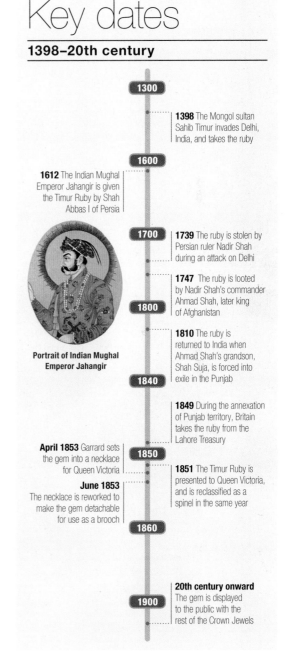

Key dates
1398–20th century

1300

1398 The Mongol sultan Sahib Timur invades Delhi, India, and takes the ruby

1600

1612 The Indian Mughal Emperor Jahangir is given the Timur Ruby by Shah Abbas I of Persia

1700

1739 The ruby is stolen by Persian ruler Nadir Shah during an attack on Delhi

1747 The ruby is looted by Nadir Shah's commander Ahmad Shah, later king of Afghanistan

1800

1810 The ruby is returned to India when Ahmad Shah's grandson, Shah Suja, is forced into exile in the Punjab

Portrait of Indian Mughal Emperor Jahangir

1840

1849 During the annexation of Punjab territory, Britain takes the ruby from the Lahore Treasury

April 1853 Garrard sets the gem into a necklace for Queen Victoria

1850

1851 The Timur Ruby is presented to Queen Victoria, and is reclassified as a spinel in the same year

June 1853 The necklace is reworked to make the gem detachable for use as a brooch

1860

20th century onward The gem is displayed to the public with the rest of the Crown Jewels

1900

[It] is the **largest in the world,** therefore even **more remarkable** than the **Koh-i-noor**

Queen **Victoria**
1851

Spinel

△ **Octagonal** mixed-cut ruby spinel

Gemstone spinel is a magnesium aluminium oxide, although the name is also given to a group of metal oxides, all of which have the same crystal structure. It is most familiar as a blue, purple, red, or pink gem, but it can occur in other colors; its blood-red variety is sometimes called "ruby spinel." Another variety, star spinels, are so named because they display stars created by natural light reflection within the gem. Most spinel is recovered from stream gravels; the earliest gem dates from 100 BCE and was discovered in a Buddhist tomb.

Specification

Chemical name Magnesium aluminium oxide | **Formula** $MgAl_2O_4$
Colors Red, yellow, orange, red, blue, green, black, colorless
Structure Cubic | **Hardness** 8 | **SG** 3.6 | **RI** 1.71–1.73 | **Luster** Vitreous | **Streak** White | **Locations** Myanmar, Sri Lanka, Vietnam, Madagascar, Afghanistan, Tajikistan, Pakistan, Australia, Tanzania

Rough

Distorted octahedron

Spinel group | These gem-quality spinels shows the mineral's color range. Some specimens are water rounded, while others have complete octahedral crystal forms.

Well-defined crystal face

Water-rounded surface

Single crystals

Octagonal crystals

Aggregate | This uncut stone is an aggregate of a number of small, gemmy, red spinel crystals that have naturally bonded together.

Magnetite | The distinctive, dark-colored iron oxide magnetite is one of the spinel group of minerals, seen here as a cluster of black octahedra.

The Black Prince's "ruby"

A case of mistaken identity

The superb spinel known as the Black Prince's Ruby was supposedly given to Edward, the Black Prince, the son of English King Edward III, by Peter the Cruel, king of Castille, after their joint victory at the Battle of Najera in 1367. Another English king, Henry V, wore it and nearly lost it at the Battle of Agincourt in 1415, and it was later set in the British Imperial State Crown. It was thought to be a ruby until the 19th century.

The "ruby" Shown here in its diamond-encrusted mount, the spinel is set with a small natural ruby near the top.

Cut

Fancy cut | This 7.27-carat flawless red spinel is faceted in the shape of a heart, one of the most challenging shapes for a gem cutter.

Crown facets

Crown facet

Brilliant cut | This fine purple spinel is faceted as a standard brilliant, displaying a total of 52 facets intended to provide a high level of light return.

Reflection of pavilion

Round brilliant | Varying the traditional brilliant cut, the cutter has doubled the number of facets on the pavilion of this mauve spinel to increase its brilliance.

Settings

Brilliant-cut diamonds

Baguette diamonds

Mixed-cut spinel

Spinel ring | A large, mixed-cut oval cushion red spinel graces this spectacular ring. The stone is surrounded by brilliant-cut diamonds, with baguette diamonds set into the shanks.

Brilliant-cut spinel

Rose-cut spinel

Spinel wonderland | Showcasing the variety of hues found in spinel gems, this gold ring is set with 14 spinels of different colors, cuts, and shapes, to dazzling effect.

Oval brilliant-cut spinel

Brilliant-cut diamond

Purple spinel ring | The vivid color of the brilliant-cut oval purple spinel in this gold ring is emphasized by the contrasting diamonds set on either side of it.

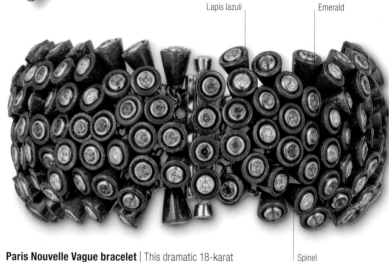

Lapis lazuli

Emerald

Spinel

Paris Nouvelle Vague bracelet | This dramatic 18-karat gold bracelet by Cartier features spinels, diamonds, pink and yellow sapphires, green garnets, amethysts, emeralds, and fire opals, all set into 252 cups carved from lapis lazuli.

The name "spinel" comes from the Latin **spinella,** meaning "little thorn," a reference to the **sharp points of its crystals**

Catherine II of Russia's "ruby" | c.14th century | Weight: 398.72 carats | Uniform color and excellent transparency; shown here in a portrait of Catherine the Great, c.1762

Catherine the Great's
spinel

△ **Portrait of Tsar Nicholas II on his coronation day**, wearing the Russian Imperial Crown with its spinel

Catherine the Great's "ruby" is the glittering showpiece of the Russian Imperial Crown. It is one of the "Seven Historic Stones," the rarest and most prized items in the royal collection of jewelry that was amassed by Peter the Great. This collection—now known as the Diamond Fund—was enlarged by later tsars, but always belonged to the state.

The "ruby" is actually a 398.72-carat red spinel, the second largest in the world. At the time, spinels were known as "balas rubies," taking their name from a famous mine in present-day Afghanistan. Russian envoy Nikolai Spafary acquired the gemstone in China while conducting trade negotiations with the emperor in 1676. He reportedly paid "a very pretty price" of 2,672 rubles for it.

The spinel already had a colorful history. According to legend, it was found in the 14th century by Chun Li, a Chinese mercenary in Turkic conqueror Timur's army. He found it in the mines of Badakhshan, where he had been exiled after stealing

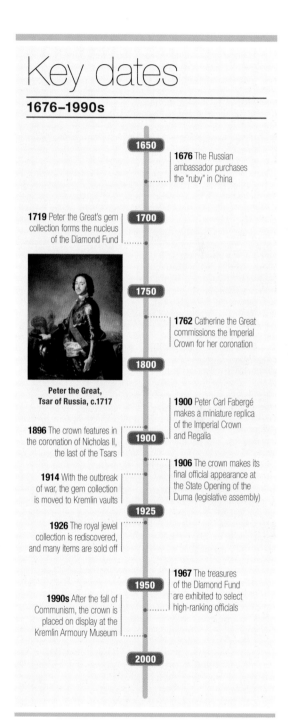

Catherine the Great's spinel

Russian Imperial Crown, displaying the spinel mounted on a central arch, accompanied by 9,936 diamonds and 74 pearls

gems in Samarkand. Chun Li tried to present it to the emperor in the hope of winning a pardon, but was murdered by a greedy palace guard who, in turn, was executed when his crime was discovered.

Catherine the Great commissioned the magnificent Imperial Crown for her 1762 coronation. It was fashioned by the court jeweler, Jérémie Pauzié, who removed the spinel from an earlier crown and added other jewels from the royal collection. Worn by Catherine's successors, the crown was concealed from view after the Revolution.

The spinel must have been an exceedingly unlucky stone

Emperor Paul I of Russia, Catherine's only son, wearing the crown c.1800

Diane **Morgan**
Author

Key dates

1676–1990s

1650

1676 The Russian ambassador purchases the "ruby" in China

1719 Peter the Great's gem collection forms the nucleus of the Diamond Fund

1700

Peter the Great, Tsar of Russia, c.1717

1750

1762 Catherine the Great commissions the Imperial Crown for her coronation

1800

1900 Peter Carl Fabergé makes a miniature replica of the Imperial Crown and Regalia

1896 The crown features in the coronation of Nicholas II, the last of the Tsars

1900

1906 The crown makes its final official appearance at the State Opening of the Duma (legislative assembly)

1914 With the outbreak of war, the gem collection is moved to Kremlin vaults

1925

1926 The royal jewel collection is rediscovered, and many items are sold off

1967 The treasures of the Diamond Fund are exhibited to select high-ranking officials

1950

1990s After the fall of Communism, the crown is placed on display at the Kremlin Armoury Museum

2000

Chrysoberyl

△ **Cat's-eye** chrysoberyl cabochon

Although crystals of chrysoberyl are not uncommon, the gemstone variety alexandrite is one of the most rare and most expensive gems in the world, with specimens seldom exceeding 10 carats. Alexandrite has the extraordinary visual property of appearing green in daylight but red under tungsten light; other forms of chrysoberyl occur in green, greenish yellow, and yellow. Alexandrite was discovered in the Ural Mountains in 1830, and was named after the Russian ruler Alexander II, on whose birthday it was supposedly found.

Specification

Chemical name Beryllium aluminum oxide | **Formula** $BeAl_2O_4$
Colors Green, yellow, brown | **Structure** Orthorhombic
Hardness 8.5 | **SG** 3.7 | **RI** 1.74–1.76 | **Luster** Vitreous
Streak White | **Locations** Russia, Myanmar, Zimbabwe, Tanzania, Madagascar, US, Brazil, Sri Lanka

Rough

Termination face

Center of twinning

Gemmy crystals | These large, wedge-shaped crystals are typical of chrysoberyl and are a good color. They display cyclic twinning (a group of crystals that have formed radially).

Multiple crystals

Alexandrite crystals | This specimen of alexandrite with mica was mined in Siberia, Russia, which remains a major source of high-quality material.

Twinning plane

Twinned crystals | These two crystals growing into a "V" formation from the same point at their base show classic chrysoberyl twinning.

Colors

Blue sheen

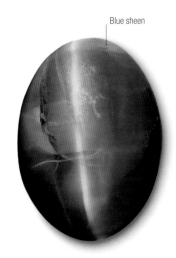

Cat's-eye chrysoberyl | This semitransparent oval cabochon, colored a typical deep yellow known as milk and honey, features a finely formed cat's-eye pattern.

Cat's-eye oval | This oval cabochon of yellow green cat's-eye chrysoberyl has an unusual bluish cast around the area of the "eye," and blemishes on its surface.

Cut

Pavilion facets visible

Cat's-eye cabochon | This transparent, cabochon shows not only the distinctive "eye" pattern but also the fibrous inclusions that created it, all in a hazy yellowish hue.

Green brilliant cut | The pale green color of this stone gives it a luminous shine, emphasized by its unusual 10 main facets, rather than the eight on a standard brilliant.

Settings

Orange sapphire

Pear-shaped
morganite

Chrysoberyl
beads

Cartier bracelet | This Cartier bracelet features strands of
small chrysoberyl beads supporting a 32.93-carat morganite,
an 8.16-carat orange sapphire, and four colored sapphires.
It is also set with brilliant-cut diamonds.

Gold detail

Moonstone

Chrysoberyl

Silver

Sapphire

Faceted stones

Cat's eye cluster ring | This yellow gold ring
from around 1900 features an 11.42-carat
central stone of cat's-eye chrysoberyl,
surrounded by diamonds.

Honey-yellow cat's-eyes | The 11 stones
set in this cross pendant are honey yellow;
along with greenish yellow, honey yellow is
the most desirable color of cat's-eye.

Arts-and-Crafts crescent brooch | This silver and gold
brooch by Dorrie Nossiter from around 1930 features varied
mixed-gem settings of moonstone, peridot, garnet, chrysoberyl,
ruby, sapphire, and green zircon.

Oval-cut
chrysoberyl

Vintage brooch | This Victorian brooch is set
with a selection of chrysoberyl gemstones in
oval cuts. The lines of the metal setting
suggest organic forms.

Alexandrite

Changing colors

The alexandrite variety of chrysoberyl
displays a color change, from greenish
to reddish, when seen in different
light conditions. Alexandrite appears
greenish in daylight, where a full
spectrum of light is present, but reddish
in incandescent light, because it
contains less of the green and blue
spectrum. The color change is due
to chromium atoms replacing the
aluminum in the chrysoberyl structure.
This causes intense absorption of light
over a narrow range of wavelengths.

Alexandrite in daylight This
cushion-cut alexandrite appears to be
green when seen in natural daylight.

Alexandrite in incandescent light
The same cushion-cut alexandrite takes
on a red hue in incandescent light.

Hematite

△ **Oval-cut** hematite

Hematite is an iron oxide and a relatively abundant mineral. Its name comes from the Greek word for "blood" and, although it can be various colors, it always produces a red streak. The reddish appearance of Mars is due to the presence of hematite on its surface and is the reason for its nickname, the "red planet." As a gemstone material it is dense with a high refractive index. In powdered form it provides the basis for many paints—pigment made from hematite has been found in cave paintings dating back 40,000 years.

Specification

Chemical name Iron oxide | **Formula** Fe_2O_3 | **Colors** Black, gray, silver, red, brown | **Structure** Trigonal | **Hardness** 5.5–6.5 **SG** 5.1–5.3 | **RI** 2.94–3.22 | **Luster** Metallic to earthy **Streak** Red to reddish brown | **Locations** China, Australia, Brazil, India, Russia, Ukraine, South Africa, Canada, Venezuela, US, UK

Group of well-formed hematite crystals | Rough |
This specimen of hematite has large, well-formed, trigonal crystals of fine cutting material and a bright metallic luster similar to polished silver.

Good crystal forms

Quartz crystal

Specular hematite | Rough | Specular hematite is hematite that has formed flattened, bright, and lustrous crystals, as in this example with quartz, from Cumbria, UK.

Metallic luster

Hexagonal specimen | Rough | This gem-quality hematite is well crystallized and exhibits good hexagonal form, with lustrous metallic faces.

Red coloring

Uncut hematite | Rough | This blocky specimen of hematite rough consists of a number of needlelike crystals, and shows a deep red color.

Carved eye

Textured finish

Hematite frog | Carved | Hematite is a popular, inexpensive carving material for objects such as this frog, and has been used in this way since the second millennium BCE.

Taaffeite

△ **Cushion-cut** taaffeite

When a mineralogist or gemologist hears of the discovery of a new and previously unknown gemstone, his or her thoughts go to a seam of rock in a distant mountain range, or the gravels of a stream or river flowing through some exotic jungle. Rarely do their thoughts go to a jewelers shop in Dublin, Ireland. Yet this is precisely where taaffeite was discovered by Richard Taaffe in 1945 among a number of faceted gems recovered from old jewelry. It is one of the rarest gemstones, and is cut exclusively for collectors.

Specification

Chemical name Beryllium, magnesium, and aluminum oxide
Formula $BeMg_3Al_8O_{16}$ | **Colors** Pale mauve, green, sapphire blue | **Structure** Hexagonal | **Hardness** 8–8.5 | **SG** 3.60–3.62 | **RI** 1.71–1.73 | **Luster** Vitreous | **Streak** White
Locations Sri Lanka, Tanzania, China

Cutter's window

Raw gem | Rough | This water-rounded crystal of taaffeite rough has a "window" polished into one end so the cutter can assess its clarity.

Brilliant-cut oval

Lavender color | Color variety | The pale, almost transparent, mauve color for which the gemstone is best known is displayed in this brilliant-cut oval cushion specimen.

Intense plum hue

Pentagonal taaffeite | Cut | This 8.5-carat taaffeite is exceptionally large and is faceted in a pentagonal step cut with a rich plum color.

Brilliant-cut crown

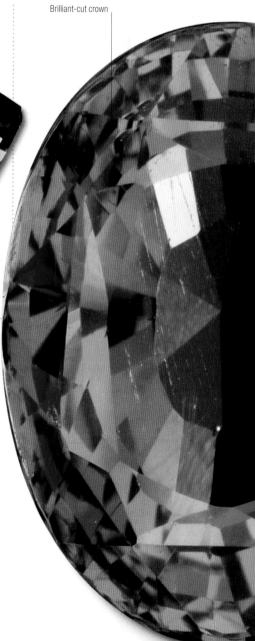

Taaffeite is the only gemstone in history to have been identified from a stone that had already been faceted

Double refraction

Simple cut

Cut taaffeite gemstone | Cut | Even a relatively simple brilliant cut has extra sparkle and "fire" due to the taaffeite's double refraction, as seen here.

Extraordinary color | Color variety | This 1.23-carat oval stone has a highly unusual bright purplish-red color and features a brilliant-cut crown.

Cassiterite

△ **Cassiterite rough**, also known as tin oxide

assiterite, a major source of tin, takes its name from *kassiteros*, the Greek word for the element. The vast majority of cassiterite is opaque black or brown, but occasionally transparent, reddish-brown crystals are found and can be faceted for collectors. Facet-grade crystals are sometimes recovered from rock, but most gemstone material is collected from stream gravels where it has weathered from the rocks in which it formed. Cassiterite continues to be mined for tin, especially in Malaysia, Thailand, Indonesia, and Bolivia.

Specification

Chemical name Tin oxide | **Formula** SnO_2 | **Colors** Medium to dark brown | **Structure** Tetragonal | **Hardness** 6–7 | **SG** 6.7–7.1 | **RI** 2.0–2.1 | **Luster** Adamantine to metallic | **Streak** White, grayish, brownish | **Locations** Portugal, Italy, France, Czech Republic, Brazil, Myanmar

Lustrous cassiterite faces

Muscovite groundmass

Cassiterite in matrix | **Rough** | Here, finely formed cassiterite crystals rest on a groundmass of muscovite. Their shiny luster is adamantine (diamondlike).

Transparent surface

Gemmy crystals | **Rough** | Transparent, reddish-brown, gem-quality crystals of cassiterite rest on this groundmass of massive cassiterite.

Well-crystallized shape

Cassiterite crystals | **Rough** | This cluster of sharp, well-defined, dark crystals with excellent luster grow outward from a lump of massive cassiterite.

Tin from cassiterite

Bronze Age to baked beans

Cassiterite-sourced tin has been traded across the Mediterranean world since the Bronze Age began in about 3000 BCE. Tin is the essential component (along with copper) of bronze, and the dark color of cassiterite makes it easy to see against the granite in which it typically forms. In modern times, the plating of steel with nontoxic and corrosion-resistant tin has revolutionized food storage with the creation of the tin can, sometimes just called a "tin."

Tin cans Tin derived from cassiterite has been essential in the long-term preservation of food.

Facets change color with light

Faceted oval | **Color variety** | This brilliant-cut oval shows the yellow-brown color typical of most faceted cassiterites. Its color flickers in the light like a diamond.

Inclusions within stone

Faceted round | **Color variety** | Here, the dark inclusions contrast with the colorless cassiterite, an unusual color considered rare enough to facet into a cut gemstone.

Cuprite

△ **Cuprite rough**, an uncommon copper mineral

The mineral cuprite is named from the Latin *cuprum*, meaning "copper," and it is sometimes known as ruby copper due to its distinctive carmine-red color. Almost every faceted stone larger than one carat has come from a single deposit in Namibia, which is now exhausted—and even these are rare. Faceted stones are too soft to wear, but their brilliance and garnet-red color is exceptional, making them highly desirable as collector's stones. Other localities that produce lesser amounts of smaller gem material are Australia, Bolivia, and Chile.

Specification

Chemical name Copper oxide | **Formula** Cu_2O | **Colors** Shades of red to nearly black | **Structure** Cubic | **Hardness** 3.5–4 | **SG** 6.1 | **RI** 2.85 | **Luster** Adamantine, submetallic **Streak** Brownish red | **Locations** Namibia originally; now also Australia, Bolivia, Chile

Tiny crystals

Small cuprite crystals | **Rough** | This group of tiny cuprite crystals provides much gem-quality material—the faces shine with adamantine luster.

Good-quality crystal face

Gem-quality crystals | **Rough** | These crisply formed cuprite crystals have good transparency, and each would cut a small, but fine, gem.

Scattered cuprite crystals

Flawless stone

Rectangular step | **Cut** | The superb clarity of this rare cuprite gem is emphasized by using a shallow step cut. It has intense color and a shiny luster.

Slightly cloudy surface

Oval brilliant | **Cut** | This cuprite gem has developed a slight metallic sheen on its surface, most likely as the result of a reaction with light over time.

Common cuprite | **Rough** | This cuprite specimen has a large number of minute crystals, and exhibits a rare form of fibrous cuprite, chalcotrichite.

Patiala necklace | 962.25 carats total (diamonds) | Diamonds, rubies, platinum

Maharaja's
Patiala necklace

△ **Patiala necklace** in its restored state

The Patiala necklace was a spectacular, five-tiered Art-Deco diamond necklace, which incorporated the famous De Beers diamond and was made for India's Maharaja of Patiala, Bhupinder Singh, by Cartier in 1928. It later disappeared for a time, but was rediscovered and restored.

The necklace's platinum base contained 2,930 diamonds, weighing a collective 962.25 carats. The world's seventh-largest diamond, the light yellow "De Beers," formed its centerpiece at 234.65 carats. Other highlights included two Burmese rubies and an 18-carat tobacco-colored diamond, one of the necklace's seven large diamonds ranging from 18 to 73 carats.

Yadavindra Singh inherited the necklace with his father's title in 1938. However, the Indian state came under financial pressure and was forced to sell off several of the necklace's stones. Its platinum base later

Watch with a portrait of Bhupinder Singh, c.1930

vanished from the royal treasury—presumably also sold—after India achieved independence in 1947. In 1998, the greatly diminished necklace resurfaced in a London antiques shop, where a Cartier representative recognized and purchased it. The largest of its stones were missing, including the famous De Beers diamond.

Cartier set about restoring the necklace, initially replacing missing diamonds with other natural stones such as white and yellow sapphires, white topaz, and garnets. However, these lacked the same brilliant effect as diamonds. The jewelers opted for a dazzling array of white cubic zirconias and other synthetic diamonds, as well as synthetic topaz, ruby, smoky quartz, and citrine stones.

In 2002, Cartier displayed the necklace in its New York boutique, attracting crowds of onlookers. Even without all of its original huge and valuable gems, the necklace was still a dazzling sight.

A wonder of natural beauty and supreme craftsmanship

Richard **Dorment**
Art critic

Key dates
1888–present

1888 The De Beers diamond is discovered in a South African De Beers mine

1889 The diamond is cut and displayed in Paris at the Exposition Universelle. Rajendra Singh, Maharaja of Patiala, purchases it

1880

1900

The De Beers diamond

1925 Bhupinder Singh visits Cartier in Paris with the De Beers diamond and numerous other unset stones to commission a necklace—Cartier's largest ever commission

1920

1928 Cartier completes the necklace and exhibits it before it is sent to India

1940

1947 The necklace disappears after several stones, including the De Beers, are sold separately

Poster for the Exposition Universelle of 1889

1960

1980

1998 Eric Nussbaum, a Swiss born gemologist working with Cartier, discovers the necklace in a London antique shop

1982 The De Beers diamond reappears at auction by Sotheby's in Geneva, where it sells for below the asking price

2000

2002–present Cartier exhibits the restored necklace internationally

2015

Bhupinder Singh's silver-gilt dinner service, made for a visit by future king Edward VIII in 1922

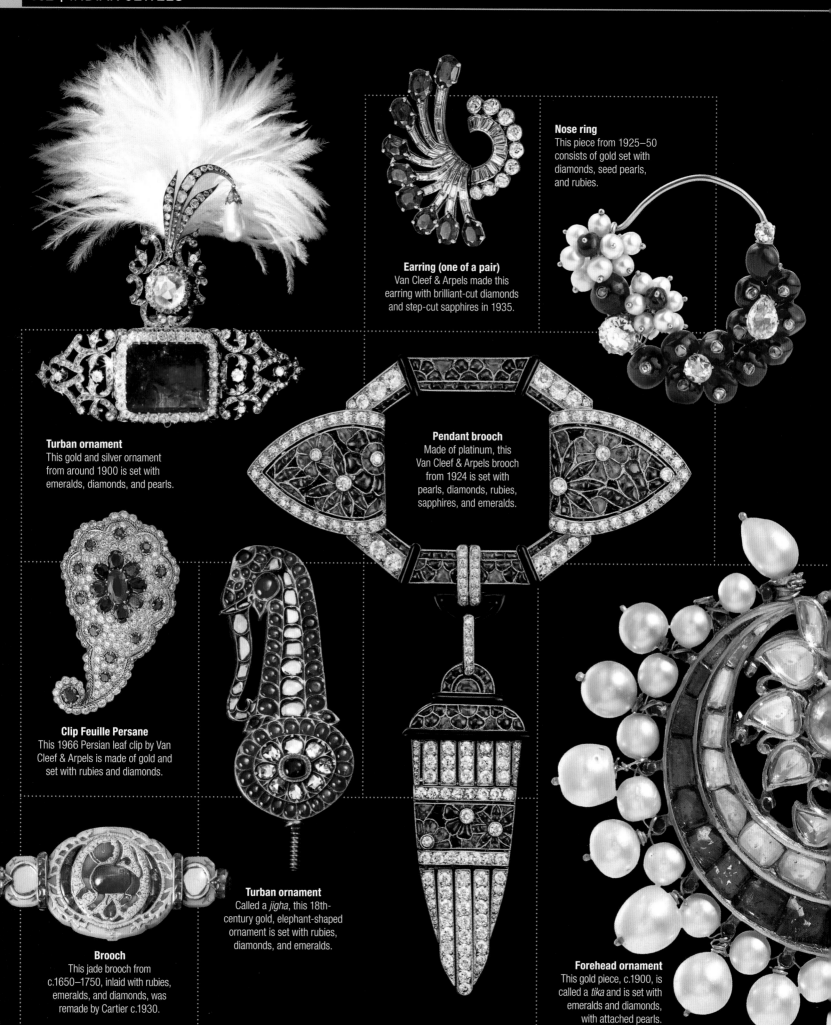

Nose ring
This piece from 1925–50 consists of gold set with diamonds, seed pearls, and rubies.

Earring (one of a pair)
Van Cleef & Arpels made this earring with brilliant-cut diamonds and step-cut sapphires in 1935.

Turban ornament
This gold and silver ornament from around 1900 is set with emeralds, diamonds, and pearls.

Pendant brooch
Made of platinum, this Van Cleef & Arpels brooch from 1924 is set with pearls, diamonds, rubies, sapphires, and emeralds.

Clip Feuille Persane
This 1966 Persian leaf clip by Van Cleef & Arpels is made of gold and set with rubies and diamonds.

Turban ornament
Called a *jigha*, this 18th-century gold, elephant-shaped ornament is set with rubies, diamonds, and emeralds.

Brooch
This jade brooch from c.1650–1750, inlaid with rubies, emeralds, and diamonds, was remade by Cartier c.1930.

Forehead ornament
This gold piece, c.1900, is called a *tika* and is set with emeralds and diamonds, with attached pearls.

Peacock brooch
This gold brooch from c.1905 is set with diamonds and enameled for color.

Turban jewel
Made of platinum and set with sapphire and diamonds in c.1920, this jewel was modified c.1925–35.

Indian
jewels

Sapphires, rubies, garnets—India has been famed for centuries as a source of some of the world's most precious gemstones, a treasure trove matched by the skills of its goldsmiths and jewelers. Cultural traditions and techniques flowing from India to Europe and the US and back have enhanced the jewelry of both East and West. Many of the pieces here come from the vast collection of Sheikh Hamad bin Abdullah Al Thani, which holds some of the world's rarest Indian jewels.

Rutile

△ **Cabochon** of rutile needles in quartz

Arguably, rutile is more important as a mineral that imparts desirable characteristics to other stones than it is as a gem in its own right. Rutile commonly forms microscopic, oriented inclusions in other minerals, and produces the asterism shown by star rubies and sapphires. It is also familiar as the golden, needlelike crystals trapped inside crystals of rutilated quartz, a material that has been used as an ornament since ancient times. Some reddish rutile crystals are darkly transparent and these are sometimes faceted for collectors.

Specification

Chemical name Titanium dioxide | **Formula** TiO_2 | **Colors** Reddish brown to red, golden | **Structure** Tetragonal | **Hardness** 6–6.5 | **SG** 4.2–4.3 | **RI** 2.62–2.90 | **Luster** Sub-adamantine to sub-metallic | **Streak** Pale brown to yellowish | **Locations** Sweden, Italy, France, Austria, Brazil, US

Reddish brown rutile crystals

Quartz crystal with rutile | **Rough** | This example of natural, clear quartz crystal is shot through in all directions with multiple acicular (needlelike) crystals of rutile.

Multiple inclusions

Golden rutile | **Color variety** | Rutile often takes on a bright golden color when enclosed in quartz, as can be seen in this example with sheaflike needles.

Rutile needles

Cabochon | **Cut** | This polished quartz cabochon shows a large number of inclusions of rutile needles forming a dense pattern beneath its surface.

Scent bottle | **Carved** | This carved scent bottle is made from rutilated quartz, with fine golden rutile needles throughout. It is also adorned with gold and onyx.

Titanium from rutile

Everyday uses

Although it may not be a household name, rutile is a vital part of modern life. Titanium is derived from rutile and, because it does not react with organic tissues, it is used to make artificial joints for hip and knee replacements and other prosthetic devices. It is also extensively used in aircraft and other applications that require high strength, low density, and excellent corrosion resistance. Titanium dioxide, meanwhile, is the main white pigment in paint, plastics, and white enamel.

Guggenheim Museum Architect Frank Gehry used titanium as the surface for his curving building in Bilbao, Spain.

Reverse intaglio | **Carved** | Rutile needles are visible in the stopper of this scent bottle. The body is carved from crystal quartz with a reverse intaglio carving.

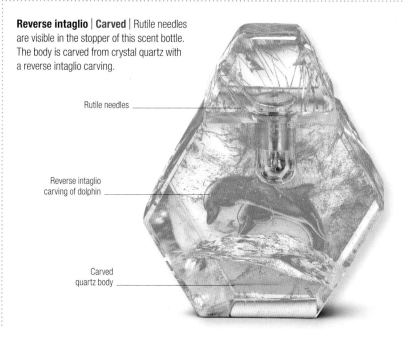

Rutile needles

Reverse intaglio carving of dolphin

Carved quartz body

Diaspore

△ **Emerald-cut** diaspore gemstone

While its scientific name comes from the Greek word *diaspora*, meaning "scattering", and refers to the way diaspore crackles under high heat, this gem was marketed under the trade name Zultanite, which is more suggestive of its pleochroic (color-changing) beauty (the name was later replaced by Czarite). Diaspore displays light-green tints in sunlight and raspberry purplish pinks in candlelight, and it gleams a champagne color under indoor lighting. In mixed lighting, variations on all of these colors can occur.

Specification

Chemical name Hydrous aluminum oxide | **Formula** AlO(OH) | **Colors** White, yellow, lilac, pink | **Structure** Orthorhombic | **Hardness** 6.5–7 | **SG** 3.3–3.4 | **RI** 1.70–1.75 | **Luster** Vitreous | **Streak** White | **Locations** Turkey, Russia, US

Emery groundmass

Diaspore crystals | Rough | In this specimen, a number of purplish diaspore crystals rest on a groundmass of emery— a granular form of corundum.

Good transparency

Gemmy crystal | Rough | This almost colorless, gemmy crystal of diaspore contains a number of striations parallel to the stone's long faces.

Visible striations

Fracture plane

Crystal of diaspore | Rough | The material comprising this facet-grade gem crystal of diaspore exhibits some transparency, but this is barely visible beneath the many characteristic striations that have developed on its surface.

Multiple color flashes

Zultanite gem | Color variety | Cut in a square cushion, this superb natural gem shows typical coloring, with flashes of green, blue, red, and other hues.

Slim facets

Fine gem | Cut | The clarity and brilliance that diaspore gems can achieve are illustrated by the stunning facets and light refraction in this square, scissors-cut stone.

Table facet

Brilliant-cut gem | Cut | This skillfully faceted diaspore gemstone features an oval, modified brilliant cut with an unusually large number of faces.

Diaspore was first described in 1801 after it was found in the Ural mountains of Russia

Fluorite

△ **Finely crystallized**, multiple grouping of interpenetrating green fluorite cubes

Representing one of the widest color ranges of any mineral, fluorite's hues tend to be vibrant, with violet, green, and yellow the most common. Its colors are commonly found as zones of different colors within a single crystal, and these zones typically follow the contour of the crystal faces. Fluorite is easily cleaved (broken along its atomic planes) and so is faceted for collectors only. For this, the stone is carefully oriented to avoid its four cleavage planes, and then cut and polished slowly to avoid heat or vibration.

Specification

Chemical name Calcium fluoride | **Formula** CaF_2
Colors Colorless, blue, green, purple, orange | **Structure** Cubic
Hardness 4 | **SG** 3.0–3.3 | **RI** 1.43–1.44 | **Luster** Vitreous
Streak White | **Locations** Canada, USA, Mexico, South Africa, China, Mongolia, Thailand, Peru, Europe, UK

Rough

Cubic crystals

Fluorite cubes | A group of purple fluorite cubes cap a block of massive white fluorite. The purple cubes in this specimen are capped with partially dissolved brown calcite.

Massive fluorite

Color layering

Blue John | This uncut piece of Blue John shows the layering of purple and yellow colors characteristic of this fluorite variety (see box, opposite).

Colors vary from pale to dark

Shadows of enclosed crystal

Fluorite slice | This thin, semitransparent, slice of fluorite, around ¼ in (5 mm) in thickness, displays a variety of colors, as well as visible layering.

Color zoning

Color

Twinned crystals

Interpenetrating crystals | These three striking orange fluorite crystals show a cubic structure, and demonstrate a typical pattern of twinning for this type of mineral.

Blue fluoresence

Fluorescent fluorite | The majority of fluorite fluoresces under ultraviolet light, as here, the color of the fluorescence depending on the trace elements present.

Opalescent Art Deco glass by René Lalique and Louis Tiffany was produced using fluorite in the glass blend

Cut

Numerous facets in cushion cut

Cushion cut | The multifaceted, round-cornered cut brings out the intense, deep green coloring of this large, 9.24-carat fluorite gemstone, found in the UK.

Multiple facet reflection

Brilliant cut | This brilliant cut demonstrates a high level of skill on the part of the faceter: multiple facet reflections are visible coming from the bottom half of the stone.

Pavilion facets

Cushion cut | The superbly executed cut of this oval, cushion-cut fluorite emphasizes reflections and hints of blue. Blue is a common color in fluorite.

Settings

Color zoning

Fluorite bead necklace | This necklace is composed of numerous circular, biconvex beads of green and purple fluorite, many showing typical color layering. Although fluorite is rarely used in jewelry because it fractures easily, it can be worn as beads.

Fluorite carvings

Use of massive fluorite

Fluorite with a massive habit— numerous intergrown crystals, rather than single ones—has been carved since ancient times. The ancient Egyptians used it for statues and scarabs, and Chinese artisans have carved fluorite for more than 300 years, recently making "New Age" items such as spheres and obelisks. Blue John, a massive English fluorite with layers of yellow and purple, has been carved since Roman times.

Fluorite bowl The craftsmanship of this finely carved, layered bowl is evident in the thinness of its walls.

Calcite

△ **Calcite crystals** displaying a vivid purple coloring

Although calcite forms spectacular crystals of varied shapes and in virtually all colors, most calcite occurs in the form of limestone, marble, or travertine, all of which are used as ornamental and carving stones. Travertine is a dense, banded rock formed by the evaporation of river and spring waters, depositing colored layers of calcite. Sliced travertine and marble were used extensively as a facing stone for buildings in ancient Greece and Rome, and many ancient Egyptian "alabaster" carvings are actually calcite.

Specifications

Chemical name Calcium carbonate | **Formula** CaCO$_3$
Colors Colorless, white, various | **Structure** Trigonal
Hardness 3 | **SG** 2.7 | **RI** 1.48–1.66 | **Luster** Vitreous | **Streak** White | **Locations** Iceland, US, Germany, Czech Republic, Mexico

Apatite crystal

Calcite and apatite | Rough | Calcite commonly occurs with other gemstones, usually providing a later infilling, as with these apatite crystals.

Scalenohedron spars

Stunning crystals | Rough | Calcite has the largest number of different crystal structures of any mineral, including these perfectly formed scalenohedrons.

Internal fracture

Scalenohedron | Rough | The form of calcite scalenohedrons is essentially a high-angle, hexagonal pyramid, as seen in this single example.

Oval brilliant cut

Vibrant color

Faceted stone | Cut | This superb stone from Tanzania was faceted by a master cutter. Calcite is remarkably difficult to facet because it is soft and breaks easily.

The Viking sunstone

Crystal compass

Seafaring Vikings relied on the sun for navigation, but cloudy days posed a problem. Norse sagas talk of a "sunstone" that could help find its position on any day, gray or bright. Some scientists think this may have been calcite: a calcite crystal polarizes light into two beams, which can be lined up to find the location of the sun. Modern calcite "detectors" achieve this within 1 percent accuracy.

Vikings at sea This rock painting at Alta in Norway depicts two Vikings fishing—one casts a net. They may have been guided by a sunstone.

Calcite alabaster | Carved | Ancient Egyptians used calcite for items such as this canopic stopper from Tutankhamun's tomb.

Headdress of carved calcite

Eyes picked out in black paint

Aragonite

△ **Crystals of aragonite** in a rock groundmass

ragonite occurs in rocks in the same way as other minerals, but it is also produced by certain biological processes—the shells of many marine mollusks, as well as corals and pearls, are composed mainly of aragonite. Like all carbonates, it is soft, fragile, and very difficult to facet, with transparent crystals only rarely faceted for collectors. Facet-quality crystals come from the Czech Republic, and there are superb cave formations in Mexico, but its type locality is Molina de Aragón in Spain after which it is named, following its discovery there in 1797.

Specifications

Chemical name Calcium carbonate | **Formula** CaCO₃
Colors Colorless, white, gray, yellowish, reddish, green
Structure Orthorhombic | **Hardness** 3.5–4 | **SG** 2.9
RI 1.53—1.68 | **Luster** Vitreous to resinous | **Streak** White | **Locations** Spain, Italy, China

Pseudohexagonal crystal

Spanish aragonite | Rough | The purple aragonite from the classic Spanish locality forms pseudohexagonal prisms, a cluster of which is shown here.

Lively coral color

Radiating petallike crystals

Aragonite sputnik | Rough | Mineral collectors sometimes refer to these radiating groups of orange pseudohexagonal aragonite crystals as "sputniks" or star clusters. The clusters are formed from multiple twin crystals.

Rock groundmass

Flos Ferri | Rough | Aragonite sometimes forms in treelike crystal groups. Known as flos ferri, or "popcorn" aragonite, it is brittle and extremely fragile.

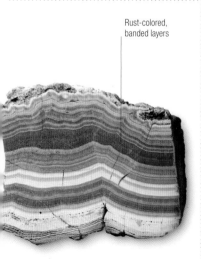

Rust-colored, banded layers

Aragonite slice | Cut | Layered aragonite forms in caves as stalactites. It is polished as cabochons, used as jewelry mounting or, if large enough, as wall paneling.

Bright orange color

Orange aragonite | Color variety | This naturally smoothed aragonite pebble displays a particularly bright orange coloring and faint pale layering.

Iron oxide layer

Turquoise cabochon | Cut | Aragonite layered and banded together in mixed colors is sometimes cut *en cabochon*, as seen in this turquoise, pear-shaped stone.

Brown aragonite

Monkey ornament | Carved | Aragonite is soft and easily carved, so when it is compact enough, ornaments such as this brown monkey can be produced.

Rhodochrosite

△ **Fine-quality, transparent** brilliant-cut rhodochrosite

Rhodochrosite's classic color is rose pink. It occurs both as transparent crystals and banded stalactitic rock, and is soft and very fragile; faceted clear crystals are rare and sought after. Vibrant, facet-quality, cherry-red crystals are found in Colorado, US, and at Hotazel in South Africa. Banded "Inca rose" rhodochrosite comes from Argentina: this is the main source of banded material, which is cut *en cabochon*, into beads, and used in carvings. Concentrically banded slices of rhodochrosite stalactites are polished and mounted in silver as pendants.

Specification

Chemical name Manganese carbonate | **Formula** $MnCO_3$
Colors Rose pink, cherry red | **Structure** Hexagonal or trigonal
Hardness 3.5–4 | **SG** 3.6 | **RI** 1.6–1.8 **Luster** Vitreous to pearly | **Streak** White | **Locations** US, South Africa, Romania, Gabon, Mexico, Russia, Japan

Rhombohedral crystal

Rhodochrosite on quartz | **Rough** | In this magnificent specimen, facet-grade rhombohedral crystals of rhodochrosite rest on a group of quartz crystals.

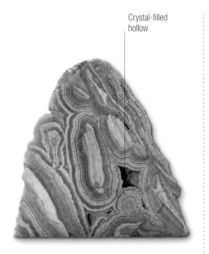

Crystal-filled hollow

"Inca rose" | **Cut** | The distinctive swirls and patterns of this slice of massive rhodochrosite are characteristic of the Argentinian variety called "Inca rose."

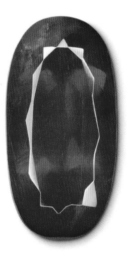

Brilliant oval | **Cut** | One of the most difficult gems to facet because of its softness, this brilliant oval has been cut by a highly skilled lapidary.

Cabochon | **Cut** | The swirls and layering typical of massive rhodochrosite are clearly seen in this irregular-shaped rhodochrosite cabochon.

Rhodochrosite drop

Necklace | **Set** | Created by Tony Duquette (1914–99), this glorious necklace contains pearl, amber, rose quartz, amethyst, garnet, and a large rhodochrosite.

Color layering

Rhodochrosite bowl | **Carved** | The lapidary has shaped this bowl to emphasize the superb natural pattern of the soft and easy to carve rhodochrosite.

Rhodochrosite parrot | **Carved** | Made by a master lapidary in Idar-Oberstein, Germany, in the late 20th century this parrot has a black agate beak, and stands on a base of quartz.

Fine carving

Cerussite

△ **Group of cerussite crystals** from Cumbria, UK

Cerussite has been known since antiquity and is named after the Latin *cerussa*, or white lead pigment. A lead carbonate, cerussite is the most common ore of lead after galena. It is generally colorless, but may be blue to green when copper impurities are present. Its refractive index is nearly as high as that of diamond, making its faceted stones especially brilliant. Unfortunately, such stones are rare; difficult to facet due to the gem's softness, brittleness, and tendency to break, they are also too soft to be worn.

Specification

Chemical name Lead carbonate | **Formula** $PbCO_3$ | **Colors** White, blue to green | **Structure** Orthorhombic | **Hardness** 3–3.5 | **SG** 6.5 | **RI** 1.8–2.1 | **Luster** Adamantine to vitreous | **Streak** Colorless | **Locations** Namibia, Morocco, Australia, US

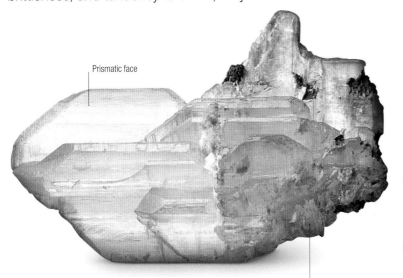

Prismatic face

Fractured surface

Prismatic gem crystal | Rough | This transparent, well-formed, prismatic single crystal of white cerussite consists of fine gem-quality material.

Large crystal

Cumbrian crystals | Rough | The county of Cumbria, in northern England, has produced cerussite since Roman times. This group of crystals is a fine example

Twinning centre

Twinned crystals | Rough | Cerussite is one of only a few minerals that produces star or cross-shaped twin crystals, such as in this excellent specimen.

Unusually long crystals of cerussite were found in the Pentire Glaze mine in Cornwall, UK

Crown facets

Faceted cerussite | Cut | Similar to rhodochrosite, cerussite is soft and extremely difficult to facet. Gems like this brilliant are rare and cut only for collectors.

Cerussite in cosmetics

The deadly beauty product

From around the 16th century, cerussite was widely used in cosmetic products designed to lighten the skin—a popular variety was known as "Venetian ceruse." However, due to the the mineral's lead content, such beauty treatments were also poisonous to the user. Symptoms included swelling of the eyes, changes to skin texture, and hair loss—perhaps related to this, by the 18th century it had become fashionable to shave the top of the hairline. In severe cases, lead poisoning could cause death.

Elizabeth I The English queen was rumored to be a user of the "Venetian ceruse" cosmetic.

BYZANTINE JEWELS

Just like the Romans had done before them, the people of the Byzantine empire wore jewelry for decoration and to indicate status, and gave it as diplomatic gifts. Between the 4th and 15th centuries, the huge wealth of the empire and its expanded trading network meant that Byzantine jewelers had unprecedented access to vast quantities of gold and a variety of gemstones, especially pearls and garnets. As a consequence, the Byzantine era is noted for its abundance of lavish jewelry.

Byzantine jewelry often featured polished cabochons prominently set in gold. Showy, colorful pieces were the most popular, and rings, bracelets, and necklaces often featured stones in alternating colors. The empire's extensive gold mines supplied the jewelers, and the metal was intricately worked into detailed open patterns known as *opus interrasile*, always with the intention of showing off the brightly-colored stones. Religion played an important role in jewelry design: crucifix neck pendants, earrings, and rings engraved with images of Christ, angels, and the saints were thought to provide spiritual protection and express devotion, as well as displaying the wearer's wealth.

[They wore] collars of gold and translucent necklaces of sparkling gems and precious pearls

Niketas **Choniates**
Greek Byzantine civil servant

Empress Theodora, Byzantine mosaic, San Vitale, Ravenna, c.6th century
This mosaic depicts the empress and her attendants adorned with gems. She wears a diadem studded with sapphires, emeralds, and red stones, strands of pearls, and a square pendant set with emeralds, pearls, and sapphires.

Variscite

△ **Sawn variscite** nodule showing its internal pattern

F ound as fine-grained masses, in veins, crusts, or in nodules, and occasionally as crystals, variscite is valued as a semiprecious gemstone for cabochons, carvings, and as an ornamental material. Black webbing sometimes occurs in the matrix of the variscite local to Nevada, US, and this form of the mineral is often confused with green turquoise; cabochons originating in Nevada that look like turquoise but are in fact variscite may be sold as "variquoise." Variscite is porous and can discolor if worn next to the skin.

Specification

Chemical name Aluminium phosphate | **Formula** $AlPO_4.2H_2O$
Colors Pale to apple green | **Structure** Orthorhombic | **Hardness** 3.5–4.5 | **SG** 2.5–2.6 | **RI** 1.55–1.59 | **Luster** Vitreous to waxy | **Streak** White | **Locations** Austria, Czech Republic, Australia, Venezuela, US, especially Utah

Crystalline variscite | Rough | In this specimen, a crust of crystalline variscite has formed on top of a large piece of massive (lacking a definite shape) variscite.

Massive form

Crystalline form

Polished end

Raw gem | Rough | An end of this piece of variscite gem rough has been polished to reveal its color and solidity. It has a waxy, semimatte luster.

Veining

Base color

Tumbled variscite | Cut | Variscite unsuitable for cabochons is often tumble polished to disclose interesting swirls and patterns, and sold for ornamental use.

Variscite is named for Variscia, the old name for the German district of Voightland, where it was first discovered in 1837

High dome

Oval cabochon | Cut | Variscite with a consistent color and density can be cut into attractive cabochons like this one, which has been polished to reveal its vitreous luster.

Smithsonite

△ **Rectangular cabochon** of smithsonite

Smithsonite can be various colors including yellow and pink, but blue-green is the most prized of all. Crystals are found occasionally—spectacular examples come from Tsumeb, Namibia—and sometimes faceted for collectors. Most gemstone material is cut *en cabochon* or carved into ornaments, but it is too soft for general wear as jewelry. Aside from its use as a gem, it is mined as a major source of zinc; it is thought that it may have provided the zinc component of brass in ancient metallurgy. One of the main sources is the Kelly Mine, New Mexico, USA.

Specification

Chemical name Zinc carbonate | **Formula** ZnCO$_3$ | **Colors** White, blue, green, yellow, brown, pink, lilac, colorless | **Structure** Hexagonal | **Hardness** 4–4.5 | **SG** 4.3–4.5 | **RI** 1.62–1.85 | **Luster** Vitreous to pearly | **Streak** White | **Locations** Namibia, Zambia, Australia, Mexico, Germany, Italy, USA

Iron oxides

Rounded masses showing botryoidal habit

Smithsonite on rock groundmass | **Rough** | This smithsonite specimen exhibits layers of intense color set in a groundmass of iron oxides. Its botryoidal habit (like bunches of grapes) is visible.

Smithsonite layer

Groundmass hosting smithsonite growth

Greek specimen | **Rough** | This smithsonite from Avron, Attica, Greece, shows a yellow hue rather than the more commonly seen blue green shades.

Rare strength of color

Cabochon | **Cut** | Smithsonite is brittle, soft, and easily abraded or chipped, as at the base of this example, but this cabochon is desirable for its intense blue color.

Translucent surface

Kelly specimen | **Cut** | This rare, faceted smithsonite, is an example of the superb-quality material that came from the Kelly Mine, New Mexico, US.

Uniform color

Oval cabochon | **Cut** | The solid translucent material of this oval cabochon brings out the blue-green hue, the color traditionally associated with smithsonite.

Changing names

Calamine to Smithson

Smithsonite was originally called by the umbrella name calamine—as in the anti-itch lotion used to treat skin problems, which contains the powdered mineral. The English chemist and mineralogist John Smithson discovered that calamine was in fact three different minerals, and smithsonite was named in his honor in 1832. The other two minerals were called hemimorphite and hydrozincite. Smithson also made the bequest that led to the establishment of the Smithsonian Institution.

James Smithson (1765–1829)
Smithsonite is named after Smithson, the mineralogist who discovered it.

Azurite

△ **Sphere** of radiating azurite crystals

Azurite is thought to have been used in blue glaze in ancient Egypt, and it was used as a blue pigment in Renaissance European art. It takes its name from the Persian *lazhuward*, meaning "blue." Azurite is cut *en cabochon* and, in rare cases, is faceted for collectors. Spheres of radiating azurite crystals more than 1 in (2.5 cm) in diameter are sometimes worn as jewelry, and slices of these are mounted in silver frames as pendants. Banded azurite and malachite used for ornamental purposes is called chessylite, after Chessy, France, where it was found.

Specification

Chemical name Copper carbonate | **Formula** $Cu_3(CO_3)_2(OH)_2$ | **Colors** Azure to dark blue | **Structure** Monoclinic | **Hardness** 3.5–4 | **SG** 3.7–3.9 | **RI** 1.72–1.85 | **Luster** Vitreous to dull to earthy | **Streak** Blue | **Locations** France, Mexico, Australia, Chile, Russia, Morocco, Namibia, China

Large crystals

Large crystals | Rough | This stunning mixed specimen features a cluster of unusually large and finely crystallized azurites that have developed on a groundmass of goethite, a form of iron hydroxide.

Chrysocolla

Goethite groundmass

Mixed minerals | Rough | Azurite often occurs with other copper minerals, as in this specimen of azurite and chrysocolla in a rock groundmass.

Fine crystals

Australian azurite | Rough | Australia is a mineral-rich country, with various locations yielding azurite sources as well as extensive copper deposits.

Blue pigment

An alternative source of color

Renaissance painters conventionally used lapis lazuli as a blue pigment. However, it was expensive, and azurite made a cheap and plentiful alternative. Unfortunately, powdered azurite is unstable in open air: moisture causes a replacement of some of the carbon dioxide in azurite with water, converting it to green malachite. This is why some of the blue in old paintings has turned green over time, including Giotto's fresco cycle in Padua, Italy, dating back to the early 14th century.

Giotto's Lamentation of Christ
Patches of azurite have deteriorated from their original blue color.

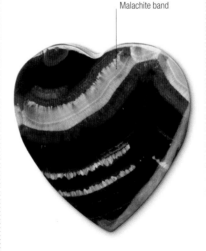

Malachite band

Azurite heart | Cut | Azurite and malachite are commonly intermixed, and can make spectacular cabochons, as in this carved, heart-shaped stone.

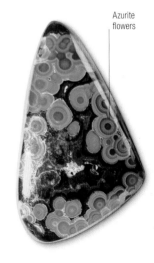

Azurite flowers

Mixed cabochon | Cut | When cut in the proper direction, the patterns produced by mixtures of malachite and azurite can be spectacular, as demonstrated here.

Malachite

△ **Piece** of fibrous malachite

Malachite powder was used as eyeshadow, pigment for wall painting, and in glazes and colored glass in ancient Egypt around 5,000 years ago. It was also a major source of copper, as it still is. The ancient Greeks used it in children's amulets, the Romans to ward off the evil eye, and the Chinese to decorate vases. In the 19th century, huge quantities were mined in the Ural Mountains, Russia, and an entire cathedral was decorated with it. Today, it is an important gemstone and ornamental mineral, used in cabochons, polished slabs, and carvings.

Specification

Chemical name Copper carbonate | **Formula** $Cu_2CO_3(OH)_2$
Colors Bright green | **Structure** Monoclinic | **Hardness**
3.5–4 | **SG** 3.2–4.1 | **RI** 1.65–1.91 | **Luster** Adamantine
to silky | **Streak** Pale green | **Locations** DR Congo,
Australia, Morocco, US, France

Botryoidal ("bubbly") surface

Gem malachite | Rough | When sliced across the "bubbles," this malachite rough will produce fantastic "bulls-eye" patterns, as in this example.

Malachite crystals

Peruvian malachite | Rough | This example from the Atacama desert of Peru features malachite crystals that have developed on a bed of atacamite.

Rough edges

Stalactitic habit of malachite | Rough | It is reasonably common for malachite to occur in stalactitic form, sometimes giving rise to irregular shapes such as this.

Linear texture

... a field of ripe cabbages with their prevailing hue of malachite green...

Walt **Whitman**
Author

"Bull's-eye" patterns

Spectacular patterns | Cut | When sliced across the "bubbly" structure of malachite, a stunning pattern is revealed, as shown in this polished slice.

Pendant support

Malachite pendant | Set | This polished piece of malachite is cut to show a cross section of its layers in a sweeping, linear arrangement, in contrast to the popular "bubble"-style patterns. It is set in an unusual pendant with silver and diamonds.

Diamond setting

Portrait of Queen Desideria of Sweden and Norway, 19th century | Owner of the malachite parure

Queen Desideria's
malachite parure

△ **One of the malachite stones** featuring carvings of classical scenes

Unlike many treasures held by Europe's royal families, the 19th-century parure owned by Queen Desideria of Sweden and Norway is made from something quite common—an inexpensive green stone called malachite.

Despite its lowly status as a gemstone, malachite became the height of fashion in the early 1800s, due in part to the era's obsession with new geological discoveries. It was often set into jewelry, and was even used to inlay entire rooms. Malachite is not transparent and is not cut into facets like diamonds; rather, it is formed into smooth cabochons or carved into detailed shapes.

Queen Desideria's parure is something of a mystery. Although it appears on an official list of the queen's jewels, there is no record of her wearing it. There are some clues, however, as to its origin. The back of the tiara bears the initials "SP" and the French assay mark 1819–39, almost certainly indicating it was made by high-society Parisian jeweler Simon Petiteau,

Malachite Hall, Winter Palace, St. Petersburg, Russia, designed in the 1830s by the architect Alexander Briullov and named after its malachite columns and fireplaces

probably during the 1820s and 1830s when the queen was living in Paris. Decades earlier, Désirée Clary—as she was known—the daughter of a wealthy French merchant, was engaged to Napoloen Bonaparte, who abruptly left her to wed Josephine de Beauharnais. Two years later, Désirée married General Jean-Baptiste Bernadotte, who, possibly at Napoleon's suggestion, was elected Crown Prince of Sweden. While Bernadotte spent much of his career on military campaigns, the queen lived mostly in Paris, where she is thought to have acquired the parure. Adding to the intrigue, Napoleon's wife Josephine also owned a malachite parure set with carved cameos.

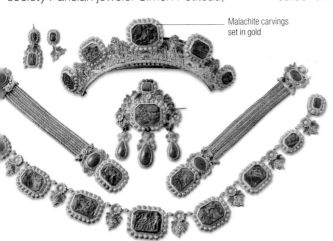

Malachite carvings set in gold

Queen Desideria's parure, with malachite carvings featuring classical scenes from antiquity

Key dates
1777–1954

1750

1777 Désirée Clary is born in Marseille, France, to wealthy merchant François Clary and his second wife

1795 Napoleon Bonaparte becomes engaged to Désirée, but soon breaks off the engagement

1810 Jean-Baptiste Bernadotte is elected Crown Prince of Sweden; Désirée becomes Crown Princess

1800

1798 Désirée marries Jean-Baptiste Bernadotte, Napoleon's most accomplished general

1820–30 Parisian jeweler Simon Petiteau creates the malachite parure for the queen

1829 Désirée is crowned Queen Desideria of Sweden and Norway

1850

1860 On Desideria's death, the parure is inherited by Josephine, wife of Desideria's only son, Oscar I

1871 When Josephine dies, the necklace goes to Sofia, wife of Oscar II

1900

1913 After death of Sofia, her family donates the necklace to the Nordic Museum in Stockholm

1950

1954 Jean Simmons plays Désirée Clary opposite Marlon Brando in the movie biopic *Désirée*

2000

The movie *Désirée*, based on the best-selling novel by Annemarie Selinko

It was **my destiny to be attractive** to heroes

Queen **Desideria**

Turquoise

△ **Cabochon** incorporating typical spiderweb markings

Beads made from turquoise dating back to c.5000 BCE have been found in Mesopotamia (present-day Iraq), making it one of the first gems to be mined and cut. It is relatively soft and easy to work and can be polished, made into beads, carved, and used for cameos. For most gem uses, however, turquoise is cut *en cabochon*. It varies in color from sky blue to green, depending on the amount of iron and copper it contains. Turquoise is porous and its color may deteriorate if worn frequently close to the skin.

Specification

Chemical name Copper aluminium phosphate | **Formula** $CuAl_6(PO_4)_4(OH)_8 \cdot 4H_2O$ | **Colors** Blue, green | **Structure** Triclinic | **Hardness** 5–6 | **SG** 2.4–2.9 | **RI** 1.61–1.65 | **Luster** Waxy | **Streak** White green | **Locations** Iran, China, US, Mexico, Chile, Africa, Australia, Siberia, England, Belgium, France, Poland

Rough

Spiderwebbing from rock fragments

Bisbee rough | This fine example from Bisbee, Arizona, US, shows the "spiderwebbing" (dark veining) typical of blue turquoise from that locality.

Prized pure color

Matrix (host rock)　Thin layering

Turquoise rough | A piece of massive turquoise like this, with thin layers of the mineral interlayered with matrix, needs skillful cutting to extract the gem.

Cut

Spiderwebbing

Pendeloque cabochon | The high-domed, teardrop shaped cabochon cut of this classic Bisbee turquoise complements the gem's attractive black spiderwebbing.

Persian blue

Top of the range

Turquoise from Nishapur, Iran (formerly Persia), is considered by many to be the finest quality and has been mined for centuries. This turquoise, usually referred to as "Persian," tends to be harder and of a more even color than North American turquoise, and it is always sky blue, never green. Turquoise has embellished thrones, sword hilts, horse trappings, daggers, bowls, cups, and other ornaments over the centuries, as well as being used extensively in jewelry.

Persian ornament This ornament is engraved and inlaid with gold in the highest expression of the lapidary art.

Iron oxides

Bisbee sample | This oval stone, naturally patterned with iron oxides, comes from the same type of turquoise as the Bisbee rough (above left).

Color looks "muddy"

Imitation stone | This oval cabochon is cut from synthetic turquoise. It lacks the color and texture of natural turquoise, though its uniform hue can be an advantage.

Settings

Vintage earrings | This pair of gold-mounted turquoise earrings comes from the British Arts and Crafts movement of the early 20th century.

Gold suspends turquoise drop

Twisted gold bezel

Gold ring | The simplicity of this ring highlights the cushion-shaped Persian turquoise cabochon. The twisted bezel mount is a backdrop to the smooth, opaque stone.

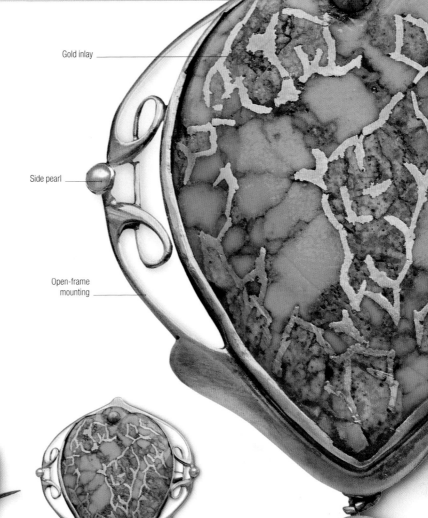

Gold inlay

Side pearl

Open-frame mounting

Pearl drop

Size-matched cabochons

Silver mounting for 88 turquoise cabochons

Navajo bracelet | This large silver bracelet set with turquoise from Morenci, Arizona, USA, is typical of the bold pieces beloved of the Navajo people.

Art-Nouveau pendant | Made at the end of the 19th century, this rare gold and Arizona turquoise pendant/brooch has two side pearls and a suspended oval pearl, and is inlayed with gold.

Biconical gold spacer

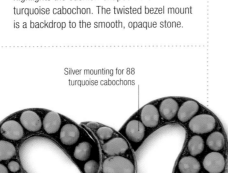

Turquoise and gold necklace | Featuring irregular, polished turquoise nuggets, this necklace is interspersed with spherical and biconical gold spacers.

Persian turquoise beads

Late 19th-century memorial brooch | This silver and turquoise brooch, designed as a bow suspending a heart and a cross, has a hair panel to the reverse of the heart.

Turquoise first came to Europe via Turkey, probably accounting for its name, which is French for "Turkish"

Marie-Louise's diadem | c.1810 | Originally set with emeralds, subsequently set with turquoise cabochons | Shown in a portrait by Giovanni Battista Borghesi—the artist has picked out the stones in red

Marie-Louise's **diadem**

△ **Marie-Louise's diadem**, reset with turquoise cabochons

Few pieces of historic jewelry have undergone such a dramatic makeover as the Marie-Louise Diadem. Dating from 1810, the tiara was originally studded with 79 deep green Colombian emeralds totaling 700 carats. It is named after Empress Marie-Louise of France, who received the headpiece from her husband Napoleon I to mark their wedding in 1810. Made in Paris by Francois-Regnault Nitot of Etienne Nitot et Fils (which later became the jeweler Chaumet), it was part of an emerald and diamond parure that included a necklace, comb, belt buckle, and earrings.

When Napoleon's empire crumbled, Marie-Louise fled to Austria and, on her death, left the parure to her aunt, Archduchess Elise. However, in the 1950s, jewelry maker Van Cleef & Arpels acquired the diadem from a descendent of Elise, and removed and sold the emeralds at auction.

Replica of the parure's emerald necklace

The company advertised them with the catchphrase, "An emerald for you from the historic Napoleonic Tiara…" The gems consisted of a central emerald weighing 12 carats, along with 21 other large emeralds and 57 smaller ones.

Van Cleef & Arpels replaced the emeralds with 79 turquoise cabochons, a change that horrified some, but appealed to others. One such admirer was American socialite and breakfast cereal heiress Marjorie Merriweather Post. She purchased the tiara in 1971, adding it to her extraordinary collection of jewelry, which had included a 263-carat diamond necklace commissioned by Napoleon as a gift for Marie-Louise when she gave birth to their son in 1811 (see pp.284–85). After wearing the reworked turquoise diadem a few times, Merriweather Post donated it to the Smithsonian Institution in the US, where it can still be seen today.

Key dates
1810–1971

1810 Napoleon commissions the diadem as a gift for his second wife, Marie-Louise

1800

1814 Marie-Louise returns to the family home in Vienna, taking her jewelry

1825

1847 The diadem is bequeathed by Marie-Louise to her aunt, Archduchess Elise

1850

French Emperor Napoleon I

1875

1900

1925

1953–56 Van Cleef & Arpels buys the diadem; the emeralds are removed and sold

1950

Marjorie Merriweather Post

1956–62 The emeralds are replaced with turquoises

1962 The parure is displayed at the Louvre, Paris

1971 The diadem is bought by heiress Marjorie Merriweather Post for the Smithsonian Institution

1975

1967 The largest emerald from the diadem is set in a brooch for American philanthropist Sybil Harrington

A turquoise given by a loving hand carries with it happiness and good fortune

Arabic proverb

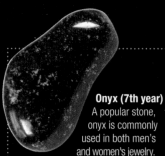

Onyx (7th year)
A popular stone, onyx is commonly used in both men's and women's jewelry.

Tourmaline (8th year)
Its extraordinary range of colors makes tourmaline a wonderful anniversary stone.

Lapis Lazuli (9th year)
Prized for its stunning blue color, lapis lazuli is the gem commonly used to mark the seventh year.

Diamond (10th year)
The diamond's link with the 10th anniversary is a modern idea promoted by jewelers.

Turquoise (11th year)
Few gems have a longer pedigree than turquoise, which dates back to ancient Egypt.

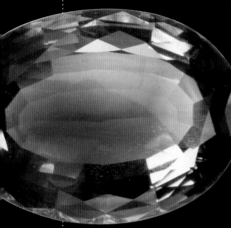

Amethyst (6th year)
In more frugal times, the traditional gift for a sixth anniversary was iron. Amethyst is now the popular choice.

Gems for
anniversaries

Stones have been assigned to planets, days of the week, and, most enduringly in modern times, to anniversaries. Many of the well-known gems and precious metals linked to five-yearly landmarks in longevity, such as 50 (gold) and 60 years (diamond) are also associated with anniversaries in the first 20 years. Like most lists of gems, definitions and interpretations can vary from one country to another, and likewise from dealer to dealer.

Sapphire (5th year)
In addition to the fifth anniversary, sapphires are also an alternative choice to mark the 23rd year.

Blue topaz (4th year)
Blue topaz was once rare, but with modern processes like irradiation, this is no longer the case.

Pearl (3rd year)
Once associated with the 30th anniversary, pearls are now the principal choice for the third year.

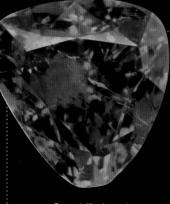

Garnet (2nd year)
A popular modern choice for the second year, garnet replaced the traditional gift of paper.

Gold (1st year)
Gold is also common associated with the 50th anniversary.

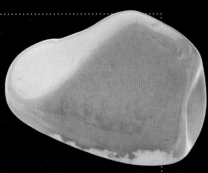

Jade (12th year)
Jade is a flexible gem
that can be carved into
a vast array of objects.

Gold (50th year)
In the Holy Roman Empire, wives
were crowned with gold wreaths
on their 50th anniversary.

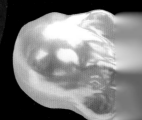

Ruby (40th year)
Like gold and diamond,
ruby is one of the more
traditional gem gifts.

Pearl (30th year)
Prized for their rarity, pearls
have been the traditional gift
for a 30-year anniversary.

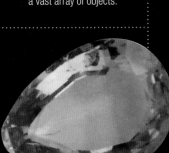

Citrine (13th year)
Named after the French word for
"lemon," citrine is often heat treated
to produce a more golden color.

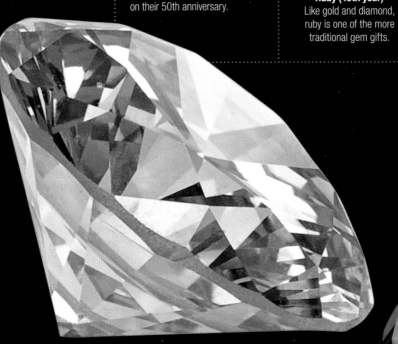

**Silver
(25th year)**
In medieval
Germany, silver
wreaths were
given to mark 25
years of marriage.

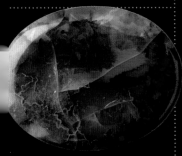

Opal (14th year)
Ivory was once the choice for
the 14th year, but for conservation
reasons, opals are now given.

Diamond (60th year)
Its origin in the Greek word *adamas*,
meaning enduring, makes diamond
a fitting gift to mark this anniversary.

**Emerald
(20th year)**
Replacing china,
emerald is now
the standard
gift for a 20th
anniversary.

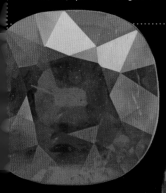

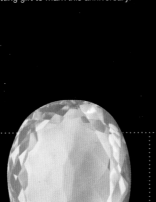

Peridot (16th year)
Peridots have rather exotic
associations, having sometimes
been found in meteorites.

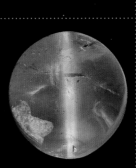

Ruby (15th year)
Originally used to mark 40
years of marriage, rubies are now
the gift for a 15th anniversary.

Quartz (17th year)
Watches were once the
traditional 17th-year gift:
the modern choice of quartz
may be a link to a watch's
quartz movement.

Chrysoberyl (18th year)
The attractive cat's eye
chrysoberyl is the official
gemstone for 18th
anniversaries.

Aquamarine (19th year)
Aquamarine, with its glorious
sea-blue coloring, makes for a
wonderful 19th anniversary gift.

Brazilianite

△ **Pale yellow** brazilianite rough

Brazilianite was only discovered and named—for the South American country of its discovery—in 1945, making it a relatively "new" gemstone. Most brazilianite is chartreuse yellow to pale yellow in color, and it is relatively hard for a phosphate mineral; it is also brittle. It is scarce, and very little gem-grade material is found each year. Because of these factors, it is faceted only for collectors. Since brazilianite was recognized, small amounts of the gem have also been found in Maine and New Hampshire in the US.

Specification

Chemical name Sodium aluminium phosphate
Formula $NaAl_3(PO_4)_2(OH)_4$ | **Colors** Yellow, green | **Structure** Monoclinic | **Hardness** 5.5 | **SG** 3.0 | **RI** 1.60–1.62
Luster Vitreous | **Streak** Colorless | **Locations** Brazil; Maine and New Hampshire, USA

Prismatic crystals | Rough | This group of finely formed, prismatic brazilianite crystals with accessory apatite orginates from Minas Gerais, Brazil.

Brazilianite crystals

Apatite accessory

Good transparency

Crystal | Rough | This strikingly colored lime green brazilianite crystal comes from the Brazilian state of Minas Gerais. It is part of the Smithsonian's gem and mineral collection.

Facets on back of stone are visible

Emerald cut | Cut | For this faceted gem, the cutter has chosen a step-cut variety—the emerald cut—to emphasize the stone's fine color and transparency.

Cutting brazilianite

Handled with care

Vivid yellow brazilianite's beautiful appearance would make it a popular gem, were it not for two factors—its fragility and its brittle texture. To cut it, the lapidary (gem-cutter) usually sticks the stone to its holder with easily removable adhesive, and must take great care not to knock it against anything. Likewise, any vibration in the grinding and polishing stages will shatter the brittle stone. The high level of skill required means that faceted stones are comparatively rare.

Gem cutting Brazilianite requires extreme care in cutting, or its brittleness will cause it to shatter.

Star facet

Fancy cut | Cut | Featuring a classic example of a fancy cut, this yellow brazilianite gemstone has been faceted in a traditional triangular step cut.

Table facet

Brilliant cut | Cut | This greenish brazilianite is faceted in a standard brilliant cut with 52 facets, as opposed to the 58-facet version, which is more common.

Amblygonite

△ **Transparent, colored** amblygonite rough

The Greek words *amblus* (**"blunt"**) and *gouia* ("angle"), are the origin of amblygonite's name—an allusion to the shape of its crystals. Most amblygonite is found as large, white, translucent masses, and it is often used as a rich source of lithium. Gem-quality amblygonite is less common and tends to be transparent, with a yellow, greenish-yellow, or lilac color. Although it can be faceted and used as a gemstone, it is vulnerable to breakage and abrasion from general wear when set into jewelry, and so is cut principally for collectors.

Specification

Chemical name Lithium, Sodium Alumino-Phosphate
Formula (Li,Na)AlPO$_4$(F,OH) | **Colors** White, yellow, lilac
Structure Triclinic | **Hardness** 5.5–6 | **SG** 3.0–3.1
RI 1.57–1.64 | **Luster** Vitreous to greasy or pearly
Streak White | **Locations** France, Brazil; California, US

Irregular surface

Amblygonite rough | **Rough** | The transparency of this piece of amblygonite rough is obscured by the reflections from its uneven surface.

Amblygonite coating

Wavellite body

Amblygonite with wavellite | **Rough** | In this example, the raw amblygonite has coated another phosphate mineral, wavellite, with a translucent layer.

Stone shows no flaws

Brilliant oval | **Cut** | The clarity and flawlessness of this colorless amblygonite is brought out by a simple yet effective, oval brilliant cut.

Angles catch light

The largest documented single crystal of amblygonite measured 530 ft³ (15 m³)

Background facets are visible

Emerald cut | **Cut** | The use of a classic emerald-step cut emphasizes the extremely rare blue-green coloring of this amblygonite stone.

Additional facets

Yellow-green transparent | **Coloration** | The cutter of this oval amblygonite has added a number of extra facets to the brilliant cut, to maximize the play of light and therefore bring out the stone's subtle coloring.

Apatite

△ **Fine, medium-blue, step-cut** oval apatite

The name apatite is derived from the Greek *apate*, meaning "deceit," as it often resembles the crystals of other minerals such as aquamarine, amethyst, and peridot. It can be intensely colored, occurring in vivid greens, blues, violet blues, purples, and rose reds. As a relatively soft crystal, it is not widely used as a gemstone, and although transparent apatite is sometimes faceted and mounted in jewelry, care must be taken when wearing it since it can scratch. Some of the largest apatite crystals have been found in Canada, weighing up to 485 lb (200 kg).

Specification

Chemical name Fluorapatite, chlorapatite, hydroxyapatite
Formula $Ca_5(PO4)_3(F,OH,Cl)$ | **Colors** Various | **Structure** Hexagonal or monoclinic | **Hardness** 5 | **SG** 3.1–3.2
RI 1.63–1.64 | **Luster** Vitreous, waxy | **Streak** White
Locations Madagascar, Brazil, Myanmar, Mexico

Apatite and muscovite | Rough | Fine green apatite crystals can be seen growing within the white tabular muscovite, or mica, crystals in this rough specimen.

Muscovite or common mica crystals

Apatite in calcite | Rough | Apatite is found in a number of geological environments, seen here as green crystals within a calcite groundmass.

Prismatic crystal | Rough | Apatite can look deceptively similar to other crystals. To an amateur, the blue crystals here could be mistaken for aquamarine.

Mexican apatite | Rough | Yellow apatite crystals, such as this Mexican example with a pyramidal end and a hexagonal prism, is popular with collectors and jewelry-makers.

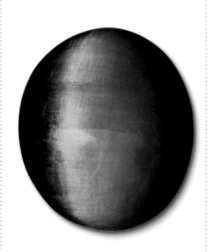

Cabochon of apatite | Cut | An attractive dark blue specimen, this apatite has been cut and polished *en cabochon* and shows a cat's eye effect.

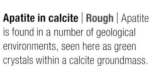

Step-cut apatite | Cut | Yellow apatite crystals from Durango, Mexico, were often faceted into fine gems, such as this rectangular step-cut.

Oval brilliant | Color variety | Apatite gems displaying a fine blue color, such as this 6.16-carat blue oval brilliant, are among the most popular varieties.

Lazulite

△ **Single, dipyramidal crystal** of lazulite from Afghanistan

aking its name from the old German word *lazurstein*, meaning "blue stone," lazulite is usually azure blue, sky blue, bluish white, or blue green. Rare faceting material is sometimes found, which can appear blue or white, depending on the angle it is viewed from. Granular lazulite is cut *en cabochon*, and can be tumble polished; it is sometimes fashioned into beads and carved into artifacts. Its appearance can be similar to lapis lazuli (see pp.174–77), and it is sometimes confused with lazurite (the main component of lapis lazuli) or azurite.

Specification

Chemical name Magnesium aluminium phosphate | **Formula** $(Mg,Fe)Al_2(PO_4)_2(OH)_2$ | **Colors** Various shades of blue
Structure Monoclinic | **Hardness** 5–6 | **SG** 3.1 | **RI** 1.61–1.64 | **Luster** Vitreous | **Streak** White | **Locations** Sweden, Austria, Switzerland, Canada, US, Afghanistan

Mottled coloring

Lazulite on muscovite | Rough | This specimen displays intensely colored lazulite crystals encrusting muscovite mica, accompanied by pink feldspar.

Dazzling blue crystals

Lazulite in quartz mass

Crystals in matrix | Rough | Afghanistan has produced some of the finest crystals of lazulite ever found, such as these specimens in a matrix of quartz.

A single crystal | Rough | This finely-formed single crystal of lazulite in quartz from Afghanistan features a perfect dipyramidal form.

Lazurite | Rough | The intense blue color displayed in this example of lazurite rough from Chile reveals how easily it can be confused with lazulite.

Cabochon | Cut | A typically mottled blue appearance is enhanced by this low-domed cut *en cabochon,* which also shows off the stone's vitreous luster.

Baryte

△ **Baryte rough** on sphalerite

The name "baryte" originates from the Greek *barys*, meaning "heavy," a reference to the mineral's high specific gravity. It is very soft and breaks readily in a number of directions—it is faceted with difficulty, purely as a collector's gem. Golden baryte from Colorado, US, is the most prized gemstone color; blue baryte is also faceted for collectors. Baryte is found in the form of stalactites, too, and round, banded sections of these are sometimes polished and mounted in silver frames as pendants. It is the most important single source of barium (see box, below).

Specification

Chemical name Barium sulfate | **Formula** $BaSO_4$
Colors Colorless, golden, bluish, greenish, beige | **Structure** Orthorhombic | **Hardness** 3–3.5 | **SG** 4.5 | **RI** 1.62–1.64
Luster Vitreous, resinous, pearly | **Streak** White | **Locations** England, Italy, Czech Republic, Germany, Romania, US

Golden coloring

Prismatic crystals | **Rough** | This group of baryte crystals exhibits a prismatic form and the resinous luster characteristic of the mineral.

Tabular crystals

Golden baryte crystals | **Rough** | The golden baryte material that is mined near Canyon City, Colorado, US, is world famous for its crystal forms and colors, as can be seen in this brightly colored specimen.

Flakelike form

Baryte crystals | **Rough** | Originating from Wet Grooves Mine, Yorkshire, UK, this specimen consists of a large group of tabular baryte crystals.

Medicine and industry

Baryte and barium

Baryte powder is used in medicine for a "barium X-ray," an image of the stomach and intestines. It is also an important mineral in oil and gas production, used as drilling mud to seal boreholes and prevent oil or gas blowout in oil- and gas-wells, which accounts for about 70 percent of its industrial output. It is also used as a filler in paper- and cloth-making, and as an inert body in colored paints.

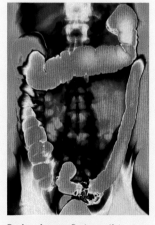

Barium image Barium sulfate is visible in X-rays when ingested by the patient.

Baryte crystal | **Rough** | This tabular, double-ended crystal shows some damage on its left side. Growth zones can be seen in pale banded sections at its base.

Rectangular table facet

Mixed-cut gemstone | **Cut** | Baryte is one of the most difficult of all collector's gems to facet. The cutter has done a fantastic job with this stone.

Celestine

△ **Crystallized** blue celestine

Celestine often forms beautiful, transparent, light to medium-blue crystals—if it were harder and more durable, it might be one of the world's favorite gemstones. It takes its name from the Latin *coelestis*, meaning "heavenly," an allusion to its "heavenly" sky-blue crystals. Because celestine is soft and easily broken, it is faceted only for collectors and museums by skilled lapidaries. Single crystals are sometimes sold as pendants, but they are too fragile for general wear. Facet-grade material is found in Namibia and Madagascar.

Specification

Chemical name Strontium sulfate | **Formula** SrSO$_4$
Colors Colorless, red, green, blue | **Structure**
Orthorhombic | **Hardness** 3–3.5 | **SG** 4.0 | **RI**
1.62–1.63 | **Luster** Vitreous, pearly on cleavage
Streak White | **Locations** US, Namibia, Madagascar

Sulfur groundmass

Crystals on sulfur groundmass | **Rough** | This group of very light blue celestine crystals is growing at all angles off a sulfur groundmass.

Iron oxide

Fine crystals | **Rough** | These small but perfectly crystallized celestine crystals have formed on a sheet of the iron-oxide mineral limonite.

Prism face

"Heavenly" crystals | **Rough** | These stunning dark blue crystals from Madagascar live up fully to the name celestine, derived from the Latin for "heavenly."

Color fades to clear

Double-terminated crystal | **Rough** | This unusual bicolored celestine crystal is double terminated, meaning it has termination faces on both ends.

Numerous small facets

Mixed-cut gem | **Cut** | Celestine is another of the extremely hard-to-cut collector's gems; this mixed cut shows the very high skill of the cutter.

Prism face

Celstine crystals

Calcite banding

Banded barian celestine | **Rough** | This variety, barian celestine, is rich in barium (see box, left). Here crystals have grown alongside sphalerite and calcite.

Alabaster

△ **Specimen** of rough gypsum alabaster

F ine-grained masses of gypsum are known as alabaster. The origin of the word is probably Middle English, but is ultimately derived from the Greek *alabastos*, a term referring to a vase made of alabaster. Alabaster vessels called *a-labaste* were widely used by worshippers of the goddess Bast in ancient Egypt, which may also reflect the origin of the name. Alabaster has been carved into ornaments, containers, and even utensils, for thousands of years, and it is sometimes made to look more like marble by heat treating it to reduce translucency.

Specification

Chemical name Calcium sulfate hydrate | **Formula** $CaSO_4{\cdot}2H_2O$ | **Colors** Colorless, white, yellow, light brown | **Structure** Monoclinic | **Hardness** 2 | **SG** 2.3 **RI** 1.52–1.53 | **Luster** Subvitreous to pearly | **Streak** White | **Locations** Egypt, Italy

Italian alabaster | **Rough** | While Italian marble is famous worldwide, the country's fine alabaster is less well known. These pieces of Italian rough have an almost waxlike appearance.

Waxy surface texture

Color variations

Calcite alabaster jar | **Carved** | Calcite alabaster was widely used in ancient Egypt (see box). This canopic jar of Psamtikpadineith is from the 26th dynasty, c.600 BCE.

Alabaster bust | **Carved** | This Italian gilt alabaster bust dates back to around 1900, and is modeled in the style of a Renaissance maiden.

Calcite alabaster

Alabaster in Tutankhamun's tomb

Even today, ancient items produced from banded calcite are still referred to as being made from "calcite alabaster." A famous ancient source of this was Hattsub, Egypt, and it is likely that some of the alabaster objects in Tutankhamun's tomb were carved in stone from this source, particularly the vases and canopic jars that held his internal organs. It was also used to make buildings, bowls, the inlaid eyes of statues, and more.

Egyptian carving An elaborately decorated ancient Egyptian casket carved from calcite alabaster

Ancient vase | **Carved** | This early vase was carved from alabaster in the ancient city of Ur (now in Iraq) some time in the 2nd millennium BCE.

Alabaster bust | **Carved** | This bust in Italian alabaster by the 18th-century artist Giovanni Battista Cipriani shows the delicate tones and fine detail possible in alabaster carvings.

Gypsum

△ **Example** of rough satin spar gypsum

Transparent crystals of gypsum are called selenite, named after the Greek moon goddess, Selene. The name may originate from the ancient belief that certain transparent crystals waxed and waned with the moon, and the crystal is still popular with enthusiasts of spiritual gems. A fibrous variety of selenite with a silky luster is known as "satin spar"; when cut *en cabochon*, fibrous gypsum produces a cat's-eye effect, but it is too soft for general wear.

Specification

Chemical name Calcium sulfate hydrate | **Formula** $CaSO_4 \cdot 2H_2O$ | **Colors** Colorless, white | **Structure** Monoclinic | **Hardness** 2 | **SG** 2.3 | **RI** 1.52–1.53 | **Luster** Subvitreous to pearly | **Streak** White | **Locations** Mexico, US

Fishtail twin | Rough | Twin crystals of gypsum growing in mirror image along a center line such as this are referred to as fishtail twins.

Gypsum crystals

Rock groundmass

Gypsum crystals | Rough | These gypsum crystals growing from a rock groundmass have their faces highlighted by an iron-oxide coating.

Desert rose | Rough | Spherical aggregates of gypsum crystals, such as this, that form in some relatively dry climates are called "desert roses," a reference to the flowerlike appearance of their crystals.

Bladed crystals

Internal imperfections

Selenite crystal | Rough | Transparent or highly translucent gypsum is called selenite, and has several crystal forms, one of which is shown here.

Satin spar | Cut | Gypsum sometimes forms masses of long, parallel crystals known as satin spar, and shows an "eye" when cut *en cabochon*, as here.

SACRED STONES

The Bible is full of references to precious gems—sapphires, diamonds, rubies, and pearls in particular. In both the Old and New Testaments, jewels are used as a metaphor to express how beautiful heaven will be. As a consequence, early medieval churches often used gems in their regalia, to decorate altars, and on the special vessels and vestments used in church services and processions. Some larger European monasteries had their own goldsmiths, and secular goldsmiths were also commissioned to make sacred, jewel-studded treasures.

Jewels also played a role in the Christian tradition of holy relics—the remains of a holy person, or objects they had touched—which were considered to be a bridge between heaven and earth and were the church's most valuable possessions. Skeletons believed to be saintly relics were draped in jewelry of gold, silver, and precious stones, while smaller relics were housed in ornate reliquaries (see pp.144–45). These containers made from precious metals and gems were often donated by pious worshippers and pilgrims. Such artifacts were intended to be physical manifestations of the spiritual treasures of the afterlife.

I will make your battlements of rubies, your gates of carbuncles… all your walls of precious stones

The **Bible**, Isaiah 54:11-12

The *Adoration of the Mystic Lamb* by Hubert and Jan van Eyck
Painted in 1432 by the van Eyck brothers for the altar of St. Bavo Cathedral, Ghent, in the Netherlands, this oil painting shows luminous jewels set into the regalia of popes, bishops, and deacons.

Scheelite

△ **Specimen of scheelite** in a rock groundmass

Crystals of scheelite can be opaque or transparent. The latter are sometimes cut as gemstones for collectors, and these exhibit almost as much dispersion (fire) as diamond. For this reason, synthetic, colorless scheelite is sometimes used as a diamond simulant, although it is too soft to wear well. Synthetic scheelite is also colored by trace elements to simulate other gemstones. Opaque crystals can grow very large—some weighing up to 15 lb (7 kg) come from Arizona, US. Most scheelite crystals fluoresce under ultraviolet light.

Specification

Chemical name Calcium tungstate | **Formula** $CaWO_4$
Colors Yellow, white, pale green, orange | **Structure** Tetragonal
Hardness 4.5–5 | **SG** 5.9–6.3 | **RI** 1.92–1.94 | **Luster** Vitreous to greasy | **Streak** White | **Locations** Austria, Italy, Brazil, Rwanda, US, UK, China

Muscovite

Scheelite crystal

Scheelite on muscovite | **Rough** | In this specimen originating from China, a large crystal is seated on a groundmass of muscovite, a common variety of mica.

Bipyramidal crystal

Crystals in matrix | **Rough** | This specimen consists of scheelite crystals resting on a groundmass of magnetite. The crystals are bipyramidal in shape.

Good transparency

Gem crystal | **Rough** | This finely formed, gemmy scheelite crystal originates from a major scheelite deposit location in China. Stones such as this can be cut into gems.

Tungsten

Turning up the heat

Scheelite is a major source of tungsten, which has the highest melting point of all elements and is a vital part of modern industry. Electric light filaments are made from pure tungsten, while tungsten carbide is used in drill bits, dies, and tools for shearing metal; cobalt-chromium-tungsten alloys are used in the surfaces of wear-resistant valves, bearings, propeller shafts, and cutting tools. Tungsten steel is used in high-temperature hardware such as rocket nozzles.

Rocket nozzle High-temperature alloys such as this rocket nozzle depend on tungsten derived from scheelite.

Transparent core

Orange crystal | **Color variety** | This large, finely transparent crystal is notable for its vivid orange coloring. A stone such as this would flouresce under ultraviolet light.

Crown facets

Brilliant cut | **Cut** | The cut of this fine, brilliant-cut, yellow-brown scheelite gem enhances the high reflectivity associated with this type of mineral.

Howlite

△ **Several nodules** of howlite on a rock matrix

Popular with collectors, howlite is generally found in nodules, in which it is usually white with veins of other minerals running through it. It is relatively porous and absorbs dye well, in particular blue dye: when altered in this way, it resembles, and is sometimes erroneously sold as, turquoise. Fortunately it is easily distinguished from turquoise since it is much softer, although it can still be polished. Howlite is found in quantity in Death Valley, California, US. It was named after the Canadian chemist, Henry How, who discovered it in 1868.

Specification

Chemical name Calcium borosilicate | **Formula** $Ca_2B_5SiO_9(OH)_5$ | **Colors** White | **Structure** Monoclinic | **Hardness** 3.5 | **SG** 2.6 | **RI** 1.58–1.61 | **Luster** Subvitreous | **Streak** White | **Locations** US, Canada, Mexico, Germany, Russia, Turkey

Rough natural surface

Howlite nodule | **Rough** | Howlite is sometimes found in cauliflower-like nodules like this one, which can be dyed to resemble turquoise nuggets (see far right).

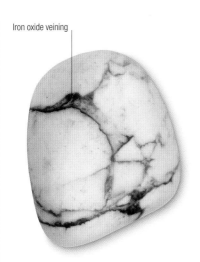

Iron oxide veining

Polished pebble | **Cut** | A tumble-polished example of natural, uncolored howlite showing the veining common in many specimens.

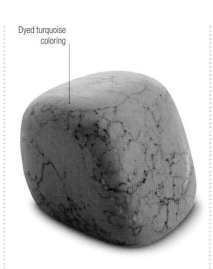

Dyed turquoise coloring

Dyed and tumbled | **Cut** | A popular form of howlite for collectors, many specimens are tumbled and dyed various shades of blue green to resemble turquoise.

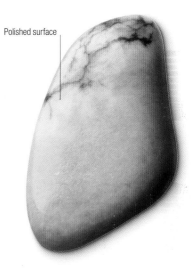

Polished surface

Dyed howlite | **Cut** | This tumble-polished specimen of howlite has also been dyed to look like turquoise, with a different shade and a more highly polished surface.

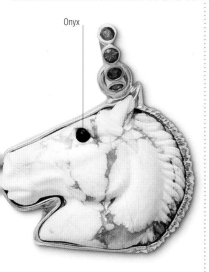

Onyx

Howlite pendant | **Carved** | This finely sculpted, veined howlite horse-head carving featuring onyx eyes is set in a frame fashioned from 18-karat gold.

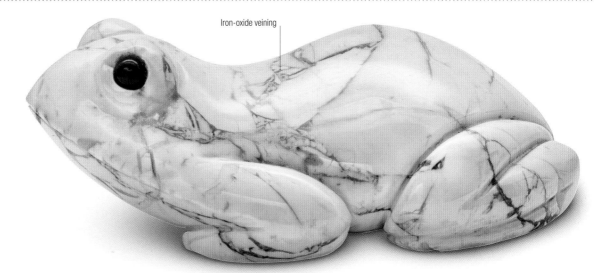

Iron-oxide veining

Carved frog | **Carved** | Howlite is soft but tough, which means it makes an excellent material for carving. This fanciful, veined howlite frog carving has a smooth surface and is set with onyx cabochons for eyes.

Enamel pelican on square-cut ruby | Shown here hanging from a gold-set diamond amid pearls in a portrait of Queen Elizabeth I attributed to Nicholas Hilliard, c.1573–75

Queen Elizabeth's
pelican brooch

△ **Enamel pelican**, Elizabeth I's signature emblem

Van Cleef & Arpels clip
featuring the pelican motif

While much is known about the symbolism of Queen Elizabeth I's pelican brooch, information on the piece itself is scant. The last known image of it is the *Pelican Portrait of Queen Elizabeth I*, attributed to artist Nicholas Hilliard. The painting was produced about halfway through Elizabeth's reign, when she was around 40 years old, a time when religious iconography became more important to her public portrayal. In the painting, the brooch is fastened to Elizabeth's richly adorned dress and depicts a pelican with a bloodied breast and her young around her. The enamel pelican hangs from a square-cut diamond set in gold and rests atop a square-cut ruby.

Elizabeth is known to have favored two symbols—the phoenix and the pelican. While the former signified her endurance and her long reign, the pelican symbolized her devotion to her subjects. In ancient legend, it was thought that a mother pelican, in times of scarcity, will peck at her body in order to feed her young with her own blood, perhaps based on the way pelicans press their bills against their breasts to fully empty food from their throat pouches. This legend, which predates Christianity, was adopted by early Christians to represent Christ's sacrifice and the gift of his body and blood in the Eucharist—he was sometimes referred to as "the Pelican."

Elizabeth adopted the symbol, wanting to be seen as the selfless mother of her people, placing her subjects' needs before her own. Her courtiers, aware of her personal adoption of these emblems, gave her gifts of pelican and phoenix jewelry, such as appear in this portrait and in Hilliard's matching *Phoenix Portrait*.

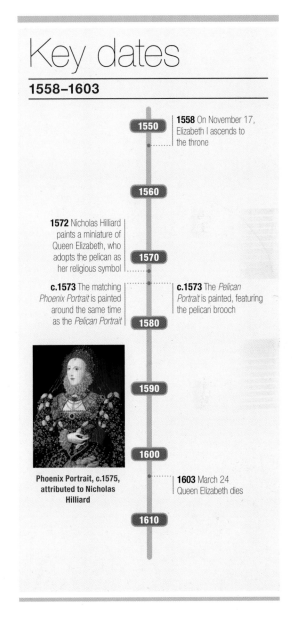

Key dates

1558–1603

1558 On November 17, Elizabeth I ascends to the throne

1572 Nicholas Hilliard paints a miniature of Queen Elizabeth, who adopts the pelican as her religious symbol

c.1573 The matching *Phoenix Portrait* is painted around the same time as the *Pelican Portrait*

c.1573 The *Pelican Portrait* is painted, featuring the pelican brooch

Phoenix Portrait, c.1575, attributed to Nicholas Hilliard

1603 March 24 Queen Elizabeth dies

Artist Nicholas Hilliard (1547–1619), famed for his portraits of Elizabeth I

... that good Pelican that to feed her people spareth not to rend her own person

John **Lyly**
English writer, c.1553–1606, on Queen Elizabeth I

Mysticism and
medicine

Since ancient times, gemstones have been regarded as talismans, warding off the evil eye and protecting the wearer from illness. Medieval alchemists ascribed curative properties to gems, and rich patients were given powdered stones as medicine. New Age practitioners today believe that crystals have healing powers when placed against the body—although there is no medical evidence for this.

Diamond
The Greeks believed that diamonds were the tears of the gods, the Romans that they were splinters of fallen stars.

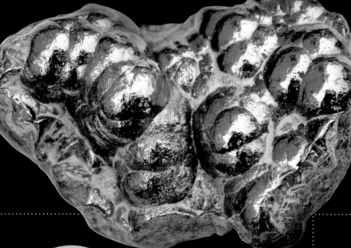

Hematite
A popular ancient belief stated that hematite formed on battlefields where soldiers' blood had been spilled.

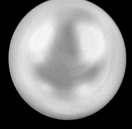

Ruby
In ancient lore, rubies were thought to be petrified drops of the blood of dragons.

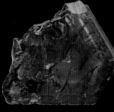

Pearl
Pearls are still used in medicine, ground to a powder and used as pharmaceutical calcium.

Blue sapphire
Ancient Egyptians used blue sapphire as an antidote to poison and to treat eye problems.

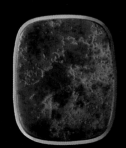

Bloodstone
Medieval legend holds that bloodstone was formed when drops of Christ's blood fell on the ground.

Yellow sapphire
Yellow sapphire is said to energize relationships and strengthen the wearer's inner will.

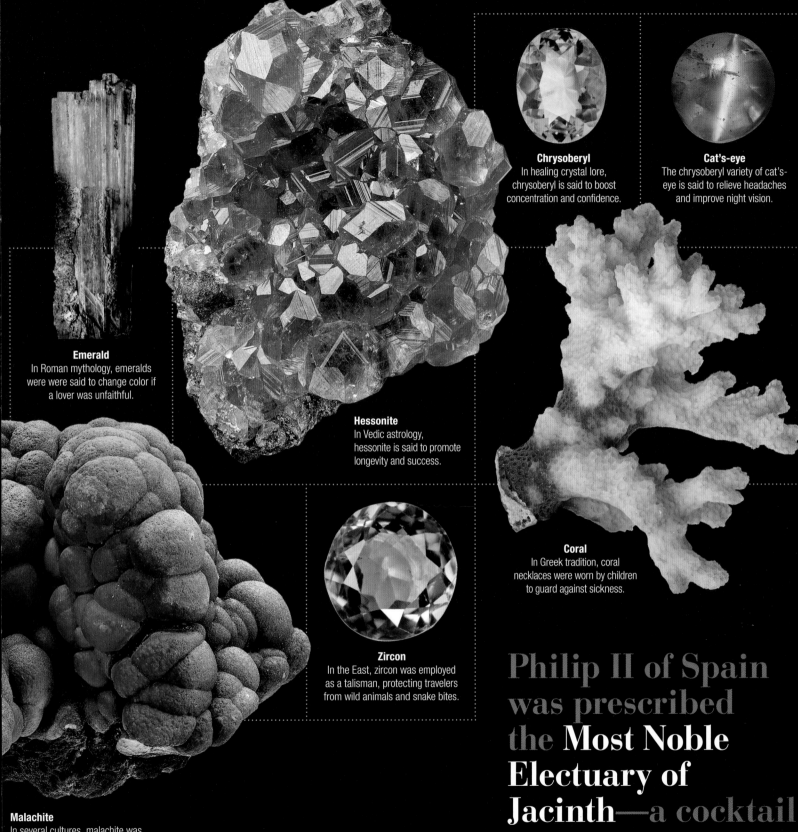

Chrysoberyl
In healing crystal lore, chrysoberyl is said to boost concentration and confidence.

Cat's-eye
The chrysoberyl variety of cat's-eye is said to relieve headaches and improve night vision.

Emerald
In Roman mythology, emeralds were were said to change color if a lover was unfaithful.

Hessonite
In Vedic astrology, hessonite is said to promote longevity and success.

Coral
In Greek tradition, coral necklaces were worn by children to guard against sickness.

Zircon
In the East, zircon was employed as a talisman, protecting travelers from wild animals and snake bites.

Malachite
In several cultures, malachite was said to ward off the evil eye, as well as safeguarding the wearer during pregnancy and childbirth.

Philip II of Spain was prescribed the **Most Noble Electuary of Jacinth**—a cocktail of **gems** including zircon (jacinth). **He died** two days later

Edwardian amethyst brooch | This stunning Brazilian amethyst and diamond brooch features a 96-carat, heart-shaped amethyst surrounded by diamonds, set in gold and platinum.

Diamond surround

Amethyst gemstone

Amethyst dissipates evil thoughts and quickens the intelligence

Leonardo **Da Vinci**
Artist and inventor

Quartz

△ **Brazilian amethyst crystal**, with broken base

Quartz is the third most common mineral on the Earth's crust, after ice and feldspar. Of all minerals, it has the greatest number of gem varieties, including prized gems such as amethyst, chalcedony, and agate. Quartz comes in two basic forms: crystalline (as distinct crystals), and cryptocrystalline (formed of microscopic crystals). The optical and electrical properties of colorless, transparent quartz have led to its extensive use in lenses and prisms, and as oscillators for electronic devices such as watches.

Stone of wonder

The word "quartz" comes from Old German and first appears in the writings of Georgius Agricola in 1530. However, long before this, Roman naturalist Pliny the Elder (23–79 CE) believed quartz to be ice that had been permanently frozen after great lengths of time, his evidence being that quartz is found near glaciers in the Alps, but not on volcanic mountains. Egg-sized, white quartz pebbles are found in Bronze-Age tombs in Europe, and in early Christian churches and chapels in Ireland and the North of England. Even now, rock crystal—transparent, colorless quartz—is commonly regarded in shamanistic practice as a "light stone," an instrument of clairvoyance between the visible and invisible. Australian Aborigines used rock crystal both as a talisman and to produce visions, while the Navajo believed that it first caused the Sun to cast its light upon the world.

Specification

Chemical name Silicon dioxide | **Formula** SiO_2 | **Colors** All colors | **Structure** Trigonal | **Hardness** 7 | **SG** 2.65 **RI** 1.54–1.55 | **Luster** Vitreous | **Streak** White

Cabochon   Mixed   Step

Pendeloque   Cameo

Locations
1 Brazil **2** Scotland **3** Spain **4** France **5** Swiss Alps
6 Russia **7** Sri Lanka **8** Madagascar

Settings

Amuletic pendant | **Set** | This ancient amulet from the Egyptian New Kingdom takes the form of a lion's head carved from amethyst, set on a gold base featuring baboon figures. It dates from around 700 BCE.

Engraved decorations

Crystal ewer | **Carved** | This stunningly crafted ewer was carved and hollowed from a single piece of rock crystal in the Fatimid Period (969–1161 CE) in Egypt, the source of a number of rock crystal artifacts.

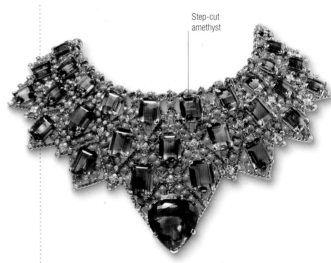

Step-cut amethyst

Duchess of Windsor's Cartier necklace | **Set** | Incorporating 29 smaller, step-cut amethysts and a large, central heart-shaped amethyst gem, this bib necklace is also set with turquoise and diamonds.

Rough

Rock crystal layer

Rock crystal on agate | This cross-sectional mineral slice reveals quartz that has grown in two different forms—as a base layer of cryptocrystalline agate, and as the upper layer of rock crystal upon which the former rests.

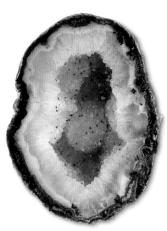

Agate geode | This geode—a mineral infilling an air bubble left in lava after cooling—was lined with agate, then later overgrown by tiny quartz crystals.

"Pyramid" terminations

Amethyst crystals | These crystals show pyramid-shaped terminations. They are a section of a massive, amethyst-lined geode, some of which can be several feet across.

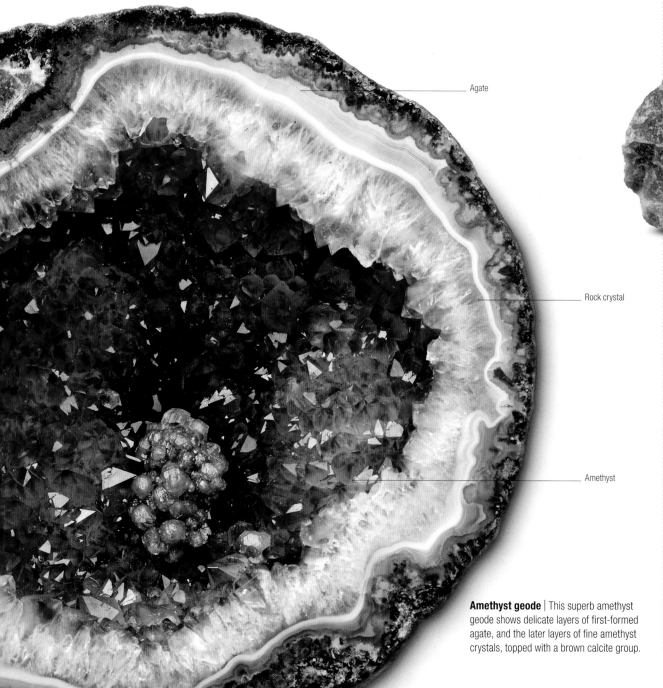

Agate

Rock crystal

Amethyst

Amethyst geode | This superb amethyst geode shows delicate layers of first-formed agate, and the later layers of fine amethyst crystals, topped with a brown calcite group.

Mica spangles

Aventurine rough | Dotted with tiny scales of mica or hematite, the quartz variety crystalline aventurine comes in various colors, and is popular for cabochons or tumbled stones.

Double-terminated quartz

Smoky quartz | This beautifully crystallized, double-terminated smoky quartz rests in a groundmass of milky quartz. Black quartz is sometimes known as morion.

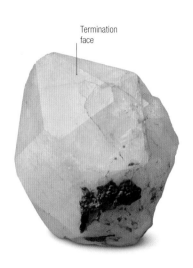

Termination face

Milky quartz | Long disregarded for its lack of clarity, translucent to opaque milky quartz has, in recent years, become highly valued by gem cutters and New Age collectors.

Namibian quartz | Quartz specimens from different localities all bear the same characteristic internal structure. This piece originates from Namibia.

Prism face

Small crystals

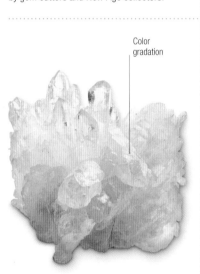

Color gradation

Rose quartz crystals | Crystals of rose quartz are exceedingly rare. Here, a group of crystals up to ½in (1 cm) in length surmount massive rose quartz.

Parallel inclusions

Cat's-eye quartz rough | The parallel needles of another mineral that create the "eye" when cut *en cabochon* can be seen in this piece of cat's-eye quartz rough.

Rock crystal | This stunning group of perfectly formed quartz rock crystals originates from the state of Arkansas, US. It is large at 13 cm (5 in) in height and also features a base of tiny quartz crystals.

Natural citrine | Much of the "citrine" quartz variety in the marketplace today is heat-treated amethyst. This Brazilian crystal is totally natural, and shows some water wear.

Hawk's-eye rough | Hawk's-eye is the mineral crocidolite saturated with quartz. In this variety the crocidolite is not oxidized; the oxidized variety is known as tiger's-eye.

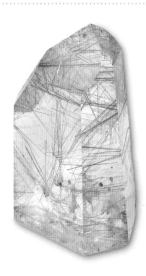

Rutilated quartz | This quartz crystal is shot through with needlelike crystals of the titanium mineral rutile. Quartz can also have black-to-green needles of tourmaline.

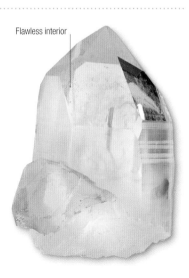

Flawless interior

Rock crystal rough | This well-formed crystal is internally flawless, and is suitable for cutting into gems, carving, or being sliced into oscillator plates for electronics.

Cut and color

Amethyst mixed cut | This hexagonal gemstone has an unusual mixed cut consisting of a faceted pavilion and a cabochon-domed crown.

Rounded corner

Step cut | The faceter of this square, step-cut, internally flawless amethyst has cleverly given a slight rounding to the corners to prevent chipping.

Mixed cut | This mixed-cut oval amethyst, with a brilliant cut crown and a step-cut pavilion, was cut with many tiny faces to disguise any color variation within.

Milky quartz | Milky quartz is rarely faceted, but this intricate, brilliant oval cut adds an air of haunting mystery to the cloudy interior of the stone.

Fancy free-form amethyst | Cuts such as this 40.3-carat amethyst are known as "free-form" cuts: that is, the placement of the facets does not follow standard patterns.

Girdle facet

"Steely" interior

Rock crystal | Faceted rock crystals were the original "rhinestones"—from quartz found in the Rhine. This brilliant cushion cut shows the "steeliness" of some stones.

Traingular facets

Rose quartz | Facet-grade rose quartz is fairly uncommon; the detail of this briolette cut enhances the exceptional quality of the material, which is usually cloudy.

Amethyst in myth

Origins and superstitions

According to myth, amethyst was created by Bacchus, the Greek god of intoxication, wine, and grapes. He was pursuing a woman, Amethyste, who refused his affections and prayed to the gods to remain chaste. The goddess Diana responded, transforming Amethyste into a white stone. In shame, Bacchus poured a goblet of wine over the stone as an offering, dyeing its crystals purple. In Greek legend, amethyst is also said to ward off drunkenness, while medieval European soldiers wore it into battle, believing it had healing properties. The "blasted" Heron-Allen amethyst (right), meanwhile, is famously cursed.

Heron-Allen amethyst
This stone was said to bring misfortune to all who touched it.

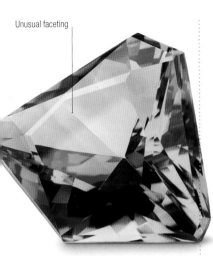

Free-form citrine | The cut on this spectacular 60.29-carat citrine is classifed as free form rather than fancy, in that it has facets placed at all angles and positions.

Unusual faceting

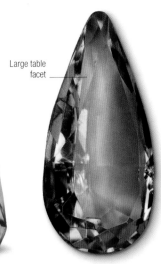

Large table facet

Pendaloque cut | The cutter of this splendidly colored citrine pendaloque has created a large table facet to expose more of the interior color.

Mixed faces

Cushion cut | This fancy cut cleverly combines a mixture of small and large faces to refract light and lighten the interior of this smoky quartz.

Drill hole

Brown smoky quartz | This gem is one of a pair of briolette-cut gems intended as earring drops. It is a perfect example of this type of gem cut.

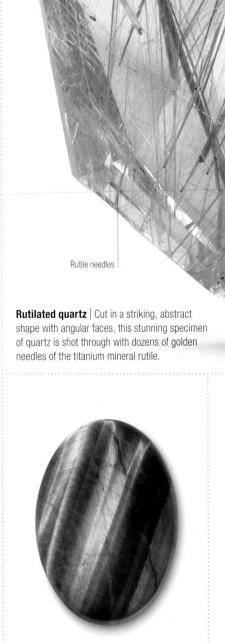

Rutile needles

Rutilated quartz | Cut in a striking, abstract shape with angular faces, this stunning specimen of quartz is shot through with dozens of golden needles of the titanium mineral rutile.

Cloudy interior

Rose quartz | Facet-grade rose quartz is somewhat cloudy: this 16.34-carat rectangular cushion cut uses large facets to emphasize color rather than brilliance.

Numerous "eyes"

Hawk's-eye quartz | The rich blue color in hawk's-eye quartz comes from numerous parallel fibers of the mineral crocidolite enclosed within it.

Tiger's-eye quartz | Tiger's-eye quartz contains crocidolite fibers exactly as in blue hawk's-eye, but in tiger's-eye they are oxidized to a golden color.

Cat's-eye quartz | In cat's-eye quartz the "eye" isn't as sharp as in other minerals, but when cut in a high-domed cabochon, as here, it is still prominent.

Settings

Ametrine ultramodern | This cut ametrine typifies a new trend in gem cutting, combining faceting and carving to create optical illusions and unusual forms.

Roman cameo | In ancient Rome, the Hellenistic fashion of cameos continued. This 1st–2nd-century CE amethyst cameo features the head of a Roman empress.

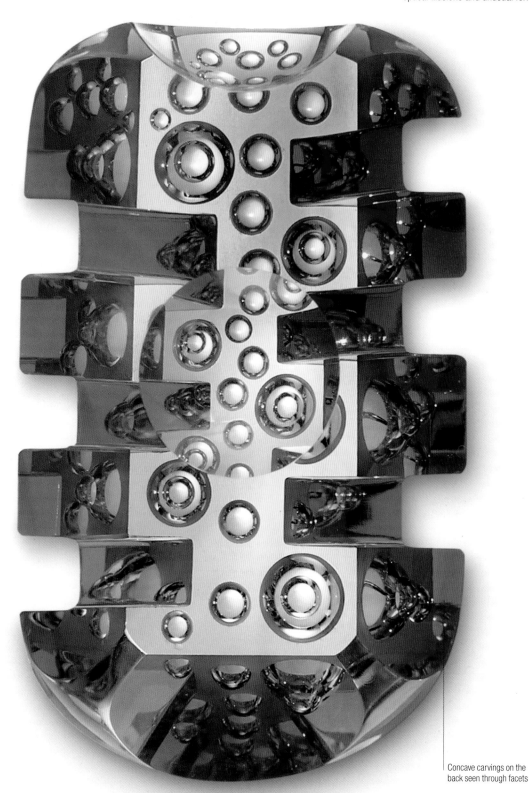

Concave carvings on the back seen through facets

Art-Nouveau brooch | Made by German craftsman Theodor Fahrner around 1910, this brooch consists of silver leaf patterns framing an emerald-cut amethyst.

Step-cut stones

Amethyst and seed pearl brooch | Set in gold and with a flower-and-leaf motif, this brooch's "petals" are faceted amethysts with seed pearl centers.

Rose-bloom fell on her hands together prest, And on her silver cross soft amethyst

John **Keats**
Romantic Poet

Rock crystal pendant | This rock crystal heart-shaped pendant, suspended from a gold and blue enamel bow, originates from the 19th century.

Rock crystal brooch | In the 19th century, gold brooches such as this with rock crystal "windows" were commonly used to store a lock of a loved one's hair.

Protective mounting

Crown facets

White gold setting

Faceted egg | This full-size rock crystal egg has been faceted with hundreds of perfectly meeting facets in a display of the highest gem-cutting skill.

Rock crystal brooch | Made in 1972 in Birmingham, UK, this 18-karat gold, sapphire, and rock crystal brooch features an unusual abstract design.

Rock crystal ring | In this white gold ring, a dazzling lozenge-shaped, fancy-cut rock crystal gemstone has been cleverly set to protect its sharp points from chipping.

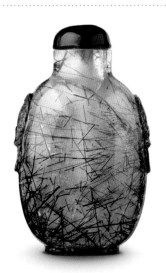

Tourmalated quartz | The material from which this extremely fine Chinese snuff bottle is carved is rock crystal with hundreds of tourmaline needles within it.

Citrine and amethyst | The oval-cut citrine gemstone in the center of this ornate brooch is surrounded by gold leaves and brilliant-cut amethysts.

Crystal skulls

Ancient mystery or hoax?

In modern culture, skulls cut from quartz have become popular for their alleged mystic qualities. Supposedly ancient Mesoamerican artifacts, none of the examples submitted for scientific testing have so far been authenticated as pre-Columbian. All have shown the marks of 19th-century tools; all are made from a type of quartz found in Madagascar; and many of them also seem to have passed through the hands of the same 19th-century antiques dealer, Eugène Boban.

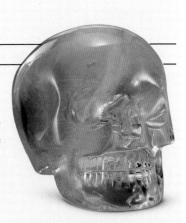

Carved skull This artifact is made from rock crystal and measures 10 in (25 cm) in height.

The Royal Household at Delhi on the Occasion of the Birthday of the Grand Mogul Aurangzeb | c.1701–08 | 22 x 56 x 45 in (58 x 142 x 114 cm) | Gold, silver, enamel, various gems, and lacquer

Treasure chambers of
Augustus II

△ **Order of the Polish White Eagle** from the collection

The Grünes Gewölbe (Green Vault) in the Residenzschloss in Dresden, Germany, is home to the largest collection of treasures in Europe. Founded in 1730 by Augustus II (1670–1733) to hold the royal jewel collection, it contains over 3,000 unique treasures. Augustus the Strong, as he was known, opened the doors of the Baroque rooms to the public, thus creating Europe's first public museum.

Among the featured exhibits is the extraordinary tableau of "The Royal Household at Delhi on the Occasion of the Birthday of the Grand Mogul Aurangzeb." The miniature model features the seventh Mughal Emperor of Hindustan, Emperor Aurangzeb (1658–1707), seated under a canopy and surrounded by 137 enameled figures of men, animals, and objects of gold, ivory, silver, and jewels. The model originally comprised 5,223 diamonds, 189 rubies, 175 emeralds, 53 pearls, two

Elephant with howdah showing the fine detail from the Aurangzeb model

cameos, and a sapphire—today, 391 precious stones and pearls are missing. The piece was created by one of Europe's greatest goldsmiths, Johann Melchior Dinglinger, c.1701–08. He was not commissioned to create the piece, but Augustus, who was delighted with the fabulous details, eventually paid him more than he spent constructing his castle at Moritzburg.

The piece embodies European society's fascination with Indian palaces and their riches. The wealth and power of the Mughal Empire reached its zenith under Aurangzeb; in the model he is depicted receiving 32 birthday gifts from the Empire's most powerful princes. These reference some of Dinglinger's other works, along with ancient Egyptian, Chinese, Greek, and Germanic objects and symbols, the significance of which was detailed in an accompanying treatise.

Death drops the curtain even on Emperors

Emperor **Aurangzeb**

Photograph showing the devastation of Dresden in 1945. The treasures had been moved on the brink of World War II, thus surviving the Allied bombing of the city.

Key dates

1658–1959

1650

1658 Aurangzeb becomes the seventh Mughal emperor

1694 Known as "the Strong," "Saxon Hercules," and "Iron Hand," Augustus II begins his reign

1698 Augustus II appoints Johann Melchior Dinglinger as court jeweler

1700

1701–08 Johann Dinglinger and brothers create the Aurangzeb tableau

1723 The Grünes Gewölbe in Dresden is founded by Augustus II

1725

1723–30 Augustus II creates a Baroque chamber for the treasures at Grünes Gewölbe, including the Aurangzeb model

1750

Augustus II, King of Saxony

1900

1930s Artworks from the Grünes Gewölbe are evacuated to Konigstein Fortress outside Dresden

1925

1945 The surviving treasures are looted by the Soviet Red Army

February 13, 1945 Along with most of the city, the Grünes Gewölbe is almost completely destroyed in the firebombing of Dresden

1950

1958 The Soviet government returns the pieces to Dresden

1959 Parts of the collection are put on display in the Albertinum modern art museum in Dresden

1975

Surface
luster

M ost gems have a vitreous luster—their surface reflects light like glass. A few are metallic or adamantine (like a diamond). More rarely, gems may glimmer like silk or look matte not shiny. Some feel greasy or waxy to the touch, while organic gems may be resinous or pearly. Luster is subjective—judged by sight or touch rather than scientific criteria.

Diamond
Polished stones have an adamantine luster, while rough stones may exhibit a greasy luster.

Gold
Gold has a metallic luster, which is opaque and reflective. It does not tarnish or discolor.

Jadeite
Jadeite has a greasy or oily luster, which can occur if a mineral has a huge number of microscopic inclusions.

Tsavorite garnet
An extremely rare gem, tsavorite has a vitreous luster, bordering on adamantine.

Citrine
A form of crystalline quartz, citrine displays a classic vitreous luster.

Amethyst
In common with most other silicates, amethyst has a vitreous luster.

Howlite
Howlite has a subvitreous luster—not quite or only partly vitreous.

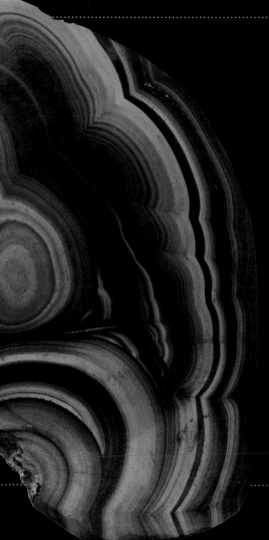

Malachite
Malachite material features a luster that is defined as adamantine to silky.

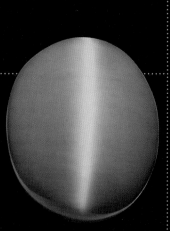

Satin spar
Satin spar normally displays a silky luster, which is caused by microscopic inclusions.

Mother-of-pearl
Not surprisingly, the luster of mother-of-pearl is defined as pearly. This material is also known as nacre.

Turquoise
The surface of turquoise generally has a waxy luster, though this can range to subvitreous.

Amber
Amber has a resinous luster, resembling the smooth surface of plastic. It is an organic gem, formed from tree resin.

Kaolinite
This clay mineral's luster is earthy. It is not gem-quality.

All that glisters is not gold

William **Shakespeare**
The Merchant of Venice

Opal eye

Pearl plumes on horse and helmet

Chalcedony horse

Emerald- and ruby-studded dragon

Sapphire and white enamel oblique fusils on Bavarian arms

Gold and silver-gilt pedestal

St. George statuette | 1586–97 | 20 in (50 cm) tall | Gold, silver gilt, diamonds, rubies, emeralds, opals, agate, chalcedony, rock crystal and other precious stones, pearls, and enamel

St. George
statuette

△ **St. George** slaying the dragon in an early 16th-century painting by Raphael

This dazzling reliquary features a statuette of England's patron saint, St. George, astride his horse, trampling underfoot the dragon he is famous for slaying. The horse is chalcedony, clad with a jewel-encrusted enamel caparison and crested with rubies and pearls. The dragon, scaled with emeralds and ruby studded, has a white enamel belly. St. George's armor is minutely detailed and the helmet's visor may even be lifted to reveal his face, which resembles the work's commissioner, Duke Wilhelm V of Bavaria. The gold pedestal that supports knight, horse, and dragon is richly adorned with diamonds, rubies, emeralds, pearls, agate, opals, and other gems. A drawer decorated with the Bavarian coat of arms in sapphire and enamel contains the relics of St. George.

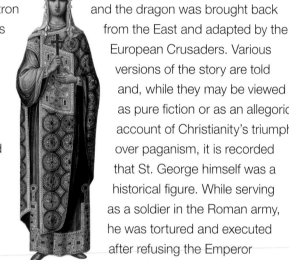

St. Alexandra martyr, Roman empress, and wife of Diocletian

The original legend of St. George and the dragon was brought back from the East and adapted by the European Crusaders. Various versions of the story are told and, while they may be viewed as pure fiction or as an allegorical account of Christianity's triumph over paganism, it is recorded that St. George himself was a historical figure. While serving as a soldier in the Roman army, he was tortured and executed after refusing the Emperor Diocletian's personal request to renounce his Christianity. Impressed by St. George's faith, Diocletian's wife, the Empress Alexandra, converted to Christianity and was also executed. It is possible she is the origin of the "princess" the knight rescues in the legend. St. George's tomb is located in Lydda, Israel, and relics from his remains are preserved in holy sites all around the world.

It is **not an impure idol,** it is a pious memorial

Bernard of **Angers**
11th-century chronicler

A band of armed Crusaders embarking for the Holy Land, where the legend of St. George originated, in an illuminated manuscript from the *Statutes of the Order of St. Esprit*

Key dates
275 CE–2000s

200 CE	**c.275–285 CE** St. George is born to a Greek noble family in Lydda (modern-day Israel). He later joins the Emperor Diocletian's army at Nicomedia
300 CE	
February 24, 303 CE Diocletian issues an edict declaring all Christian soldiers should be arrested. St. George publicly rejects the emperor's edict	**April 23, 303 CE** St. George is tortured on a wheel of swords and beheaded. He is soon after regarded as a martyr
400 CE	
494 CE St. George is canonized by Pope Gelasius **500 CE**	

St. George on the torture wheel

1586 Archbishop Ernst of Cologne (1554–1612) sends relics of St. George to his brother Duke Wilhelm V of Bavaria (1548–1626), known as "the Pious" **1500**	
1600	**1590** Duke Wilhelm commissions the statuette. Court goldsmith Hans von Schwanenburg of Utrecht designs it; Hans Schleich of Munich creates it
1600s The statuette is displayed on important feast days on the altar of the Rich Chapel in the Munich Residence, Germany	
1800	
2000	**2000s** The reliquary remains in the Treasury of the Munich Residence

Snake bracelet | This stunning bracelet in 18-karat gold features flexible segments with blue enamel scales, ruby eyes, and a diamond-encrusted head with a chrysoprase cabochon.

Enamel scales

Chrysoprase cabochon

Ruby eye

Chalcedony drives away phantoms and visions of the night

Josephi **Gonnelli**
18th-century physician

Chalcedony

△ **Leaf-shaped** carnelian cabochon

Chalcedony is a compact variety of quartz, composed of crystals that are microscopic (microcrystalline) or too small even for a standard optical microscope (cryptocrystalline). It forms in cavities, cracks, and by replacement when low-temperature, silica-rich waters percolate through preexisting rocks, in particular volcanic rocks. It is relatively porous, and much chalcedony on the commercial market has been dyed to enhance or color it artificially. Chalcedonies of all kinds have been used as gems, beads, carvings, and in seals for thousands of years. The earliest stone tools were generally made of some form of chalcedony.

Varieties

Pure chalcedony is white. However, when trace elements or microscopic inclusions of other minerals occur, it can yield a range of colors. Many of these have their own variety names: chalcedony that shows distinct banding is called agate; blood-red to reddish-orange translucent chalcedony colored by inclusions of iron oxide is know as carnelian; bloodstone is dark, opaque green colored by traces of iron silicates and with patches of bright red jasper. Chrysoprase is a translucent apple-green variety, colored by nickel; sard is light to dark brown chalcedony, while sardonyx is color-banded sard; jasper, chert, and flint are opaque, fine-grained or dense, impure varieties of cryptocrystalline quartz.

Specification

Chemical name Silicon dioxide | **Formula** SiO_2 | **Colors** All colors | **Structure** Hexagonal/trigonal | **Hardness** 7 | **SG** 2.65 | **RI** 1.54–1.55 | **Luster** Vitreous

Cabochon   Cameo   Slab

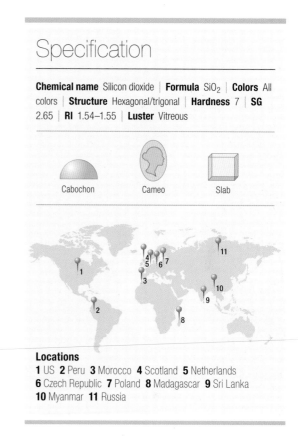

Locations
1 US **2** Peru **3** Morocco **4** Scotland **5** Netherlands
6 Czech Republic **7** Poland **8** Madagascar **9** Sri Lanka
10 Myanmar **11** Russia

Key pieces

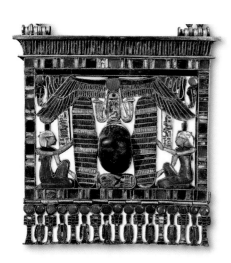

Ancient Egyptian gold pectoral | The lavish gold funerary pectoral of Pharaoh Psusennes I (3rd Pharaoh, 21st Dynasty), from around 1040–996 BCE, is decorated with inlays of red jasper and lapis lazuli.

Fine-grained flint blade

Turquoise mosaic pieces

Aztec knife | The blade of this highly decorated Aztec sacrificial knife is chipped from fine flint, with a mosaic handle of turquoise, coral, and jet. It is throught to originate from around the 15th to 16th centuries.

Gold with enamelling

Gray carved chalcedony

Chalcedony cup | Combining superb lapidary and enameling work, this antique cup has been delicately carved from waxy gray chalcedony material and set with gold trimming. The enamel decoration is particularly fine on the handles.

Rough

Red jasper | This variety of red jasper from Arizona, US, shows color mottling that will make colorful and interesting cabochons.

Color variations

Inclusion

Fractured edges

Good translucency

Coloring caused by iron oxide

Gem rough | This fine piece of massive gem carnelian rough has good color all the way through, and excellent translucency. It is naturally dyed with iron oxide.

Banded jasper | The dramatic color banding in this example of jasper rough clearly shows why it has been a popular carving and gem material since antiquity.

Deep green coloring

Cross-sectional view of interior

Heliotrope rough | This specimen consists of heliotrope, a green variety of chalcedony shot through with red patches of iron oxide, for which it is also known as bloodstone.

Water agate | Originating from a single locality in Brazil, agates of this variety are actually hollow geodes filled with preserved water, revealed by grinding a "window" into the inside.

Water-filled interior

Chrysoprase | This gem chrysoprase specimen displays a beautiful deep green color. Chrysoprase remains one of the most sought-after varieties of chalcedony.

Settings

Chalcedony ring | Unpatterned blue chalcedony is relatively uncommon, but comprises this subtle, pastel-blue cabochon set in a 14-karat yellow gold ring.

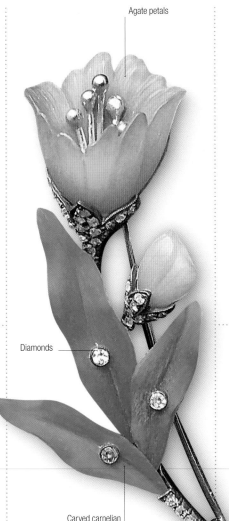

Agate petals

Diamonds

Carved carnelian

Chalcedony's name is thought to originate from the ancient port of Kalkedon in Asia Minor (modern Turkey)

Russian urn | Carved from a single piece of Ural Mountains jasper, this spectacular multicolored urn demonstrates the finest in 19th-century lapidary skill and craftsmanship.

Pendant brooch | Bearing the signature E. Paltscho, this beautiful gold-mounted pendant brooch features carnelian leaves and an agate flower, highlighted by diamonds.

Giardinetto brooch | This fabulous giardinetto brooch features diamond and emerald accented flower heads carved from amethyst, chrysoprase, carnelian, turquoise, and coral.

Bloodstone watch case | The back of this open-face pocket-watch case is carved from bloodstone, inlaid with 18-karat gold, and surrounded by seed pearls.

Chalcedony cabochon

Silver pin | Based around four smoky chalcedony cabochons, this silver pin was designed by the influential Danish silversmith Georg Jensen.

Van Cleef & Arpels pendant | Adorned with cabochons of chrysoprase and lapis lazuli, this gold pendant is also set with diamonds in a rope motif.

Chrysoprase cabochon

Garnets

Enamel decoration

Silver-gilt kovsh | This stunning kovsh (ceremonial drinking vessel) from Tsarist Russia features the finest in cloisonné enamel, and is highlighted by cabochons of chrysoprase and garnet. A kovsh of this quality would have been a royal gift.

Top view

Side view

Bottom view

Chrysoprase

Gems tinted by foil
backgrounds

Varicolored gold details

Snuffbox | c.1765 | Carved chrysoprase, gold, gems, and foiled diamonds

Frederick the Great's
snuffbox

△ **Detail showing tinted diamonds** in varicolored gold setting

Frederick II "the Great" of Prussia (r.1740–86) loved snuffboxes, and his collection supposedly included one for each day of the year. He was also known for his liking of the green gemstone chrysoprase, and he commissioned eight snuffboxes made from the mineral. London-trained designer Jean Guillaume George Kruger is thought to have made this example in about 1765, and Frederick later presented it to his brother Augustus Wilhelm, Prince of Prussia, as a gift.

The oval box and its cover are both made from single pieces of chrysoprase, a green variety of chalcedony (see pp.146–49). Diamonds and other gems are mounted on varicolored gold in the forms of scrolls, vines, and sprays of flowers; and pale pink, green, and lemon-yellow foil has been placed behind the diamonds to tint them subtly. The

Jeweled mother-of-pearl snuffbox, commissioned by Frederick the Great

interior of the lid is bordered by gold and engraved with more flowers and scrolls.

Frederick was renowned for his military achievements, but also for his patronage of the arts. He was fond of fine materials, and among these he so favored chrysoprase that toward the end of his life he had pieces of chrysoprase set out for him to look at alongside his boxes and jewels. His interest in elegant boxes was influenced by his mother Sophia-Dorothea's collection, and he kept a snuffbox on his person at all times. This proved fortuitous in 1759 at the Battle of Kunersdorf during the Seven Years' War, when he was hit by a Russian bullet—it was deflected by the snuffbox in his pocket, saving his life.

His approach to aesthetic quality... was robust

Tim **Blanning**
Author, on Frederick the Great

Fredrick the Great of Prussia pictured at the Battle of Kunersdorf in 1759, where his snuffbox deflected a Russian bullet, and so saved his life

Key dates

1712–1786

1700

1712 Frederick II is born

1753 Jean Guillaume George Kruger, a London-trained designer, moves to Berlin and designs a series of snuffboxes for the Prussian royal collection

1750

1740 Frederick II takes the throne and conquers Silesia (now mostly in Poland), where chrysoprase is mined, in the First Silesian War

1755

1759 During the Battle of Kunersdorf in the Seven Years' War, a snuffbox in Frederick's pocket saves him from a Russian bullet

1756 Frederick enters the Seven Years' War, which involves most of the world's major powers of the period

1760

1763 The Seven Years' War ends, leaving Prussia a major power

c.1765 The chrysoprase snuffbox is created, probably by Kruger; Frederick presents it to his brother, Augustus Wilhelm

1770

1780

1786 Frederick the Great dies

Frederick's brother, Augustus Wilhelm

Agate

△ **Fire agate** with unusually fine coloring

he **microcrystalline, compact variety** of quartz, agate is a common, semiprecious chalcedony. Agate is mostly characterized by color bands in a concentric form, and less often by mosslike inclusions, when it is called moss agate. Other names, such as fire, or Brazilian, often precede the word agate and can describe the locality where they are found, or denote a particular appearance or coloration. Agates are almost always cut *en cabochon*, carved, or used as beads or ornaments.

Specification

Chemical name Silicon dioxide | **Formula** SiO_2 **Colors** All
Structure Trigonal | **Hardness** 7 | **SG** 2.6 | **RI** 1.53–1.54
Luster Vitreous | **Streak** White | **Locations** Worldwide, notably Brazil, Botswana, South Africa, Egypt, China, Mexico, and Scotland; fire agate only in northern Mexico and southwestern US

Rough

Agate slice | Seen here in cross section, the bright and varied circles of color displayed in this agate slice indicate that it began life as a nodule of layered carnelian, and was then overlain by multiple layers of varicolored chalcedonies.

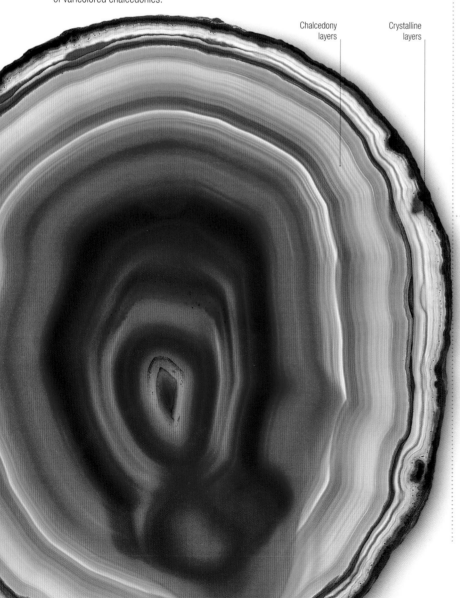

Chalcedony layers

Crystalline layers

Carnelian center

Color diffraction

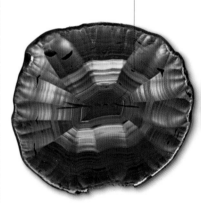

Brazilian agate | This slice shows stages of the agate's formation: the lava cavity was lined with crystallized quartz, then the inner hollow was layered with varicolored chalcedony.

Iris agate slice | The chalcedony layers in this agate are extremely thin, and act as diffraction gratings to produce rainbow colors according to their thickness.

Classic bubbly botryoidal surface

Agate rough | The surface of this specimen is botryoidal—it appears as a mass of globular forms, resembling a bunch of grapes. The layering typical of many types of agate is also visible.

Carnelian interlayered with chalcedony

Cut

Pale carving stands out

Composite cameo | The layering of agate lends itself to cameo work. This example has a carved figure applied to a background of mossy agate.

Mosslike mineral inclusions

Chlorite

Moss agate cabochon | The cabochon cut shows off the mineral inclusions, often chlorite, as here, which give the appearance of moss growing within the agate stone.

Settings

Mineral inclusions resemble moss

Agate bowl | The polished surface of this shallow bowl carved from moss agate shows the intricate appearance of the stone. The irregular form of the piece complements the natural, random pattern of the gem.

Color

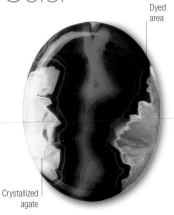

Dyed area

Crystallized agate

Dyed agate | Agate is relatively porous, and so it can easily be dyed. Blue, as here, is typical; red or purple are also common. It can be difficult to tell dyed agate from natural.

Iron oxide inclusions

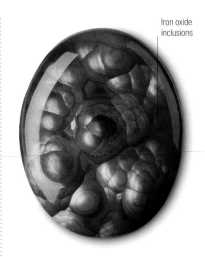

Fire agate cabochon | The cabochon cut of this fire agate stone brings out its natural iridescence. Its bubbling, oily appearance is caused by iron oxide inclusions.

Rose quartz

Rock crystal

Agate

Multicolored necklace | This necklace combines spherical beads of a number of quartz minerals, including agate, rose quartz, and rock crystal.

Naturally colored banding

Silver-mounted brooch | Typical of Celtic-style pieces produced from the 19th century, the agate used for this brooch was recovered from the beaches of northern Scotland.

In the Middle Ages, wearing agate was thought to cure insomnia and ensure sweet dreams

Manganese staining

Moss agate | In this cabochon, the "moss" is in fact staining from one of the iron or manganese oxides that penetrated the agate after it was formed.

Types of agate

Lace and fire

Fire agate is an unusual variety that has iridescent rainbow colors, with brown to honey-colored base material. Cutting is a meticulous process, removing only enough stone to reveal the "fire." Fortification agate is a general term for banded agate with angularly arranged bands. Brazilian agate is a fortification agate with banding in angled concentric circles, and Mexican lace agate—called "crazy-lace"—is a multicolored fortification agate with convoluted layering.

Lace agate rough This uncut piece of Mexican lace agate shows the intricate swirls and folds typical of its patterning.

Onyx

△ **Carnelian onyx cabochon** showing multiple layers of banding

Onyx is a striped, semiprecious variety of chalcedony quartz with alternating bands of black and white. Its varieties include carnelian onyx, which has white and red bands, and sardonyx, with white and brown bands; the name "onyx" is only properly applied to the black-and-white variety, but is also informally used for all varieties. Onyx is a popular material for cameos and intaglios because its layers can be cut to create a color contrast. A quantity of the modern onyx on the market is produced artificially by dyeing pale, layered chalcedony.

Specification

Chemical name Silicon dioxide | **Formula** SiO_2 | **Colors** White color banded | **Structure** Trigonal | **Hardness** 7 **SG** 2.65 | **RI** 1.54–1.55 | **Luster** Vitreous | **Streak** White **Locations** India, South America (onyx); Sri Lanka, India, Brazil, Uruguay (sard and sardonyx)

Rough

Multiple layering

Onyx rough | This example is a fine-quality piece of onyx rough displaying multiple banding in several different colors, principally white, gray, brown, and purple. Color layering such as this is highly desirable in onyx carving, and lends the characteristic layered appearance to cameos (see below).

Colored layers

Polished slab | This excellent quality slab of onyx features dramatic, characteristic color banding, and could be shaped to create a superb cabochon.

Cut

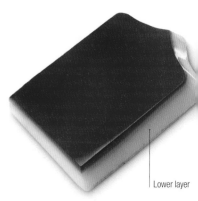

Lower layer

Shield shape | This shield-shaped onyx cabochon could be carved into a cameo: the top layer would form the subject, with the bottom layer as the background.

Roman cameos

Layers and contrast

The ancient Roman world produced some of the finest cameos ever made, and sardonyx in various colors and shades was the preferred medium. The intricacies of Roman carving and their use of the color contrasts of the various layers remain unsurpassed. One unusually fine group of cameos, often referred to as the "State Cameos," were of the Emperor Augustus, and show him with various divine attributes. One of the most stunning is the Blancas Cameo, now in the British Museum.

Roman cameo Onyx was a favorite medium for Roman cameo carvers—this one depicts an empress.

Pale layering

Dark central band

Slab with light banding | Onyx does not always need bold contrasts to be desirable. The light-colored banding seen here makes it excellent carving material.

Flat top

White layer

Cabochon | The strong banding of onyx offers the lapidary a number of choices when cutting, as with this flat-topped cabochon showing clear contrast.

Settings

Platinum setting

Onyx cap

Brilliant-cut diamond

Onyx brooch | This stunning brooch combines a circle of black onyx set with diamonds with two platinum and pink coral side bars set with diamonds.

Onyx dial

Dragon Mystérieux watch | Designed as a stylized dragon, this Cartier watch features 18-karat white gold, fire opal, diamonds, coral, and emeralds framing an onyx dial.

Faceted onyx   Central diamond

Onyx ring | Here a platinum ring is set with triangular stones of black onyx, with a large central diamond and crossbars set with numerous small diamonds.

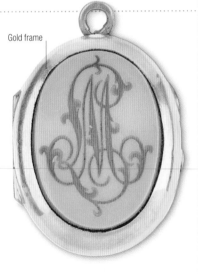

Gold frame

Gold and onyx pendant | This gold pendant is set with a multilayered onyx engraving with a monogram cut into the second layer. The red background gives the top layer a pink hue.

Sardonyx cameo | This cameo of a classical figure cut into multilayered sardonyx is a masterpiece of the carver's art. The color banding creates shading on the figure.

Onyx faceted in unusual shape

Georgian seal | The handle of this Georgian seal is intricately carved in onyx with the color banding showing high contrast: it is a fine example of Georgian lapidary work.

Onyx and diamond pendant | This gold pendant claw mounts an unusual black onyx stone, surrounded by a lavish setting of 21 diamonds.

The onyx, if worn on the neck, was said to cool the ardors of love

George Frederick **Kunz**
Mineralogist

GOLD AND POWER

The goldsmiths and jewelers of Italy were innovators during the Renaissance, elevating their craft to that of art under the patronage of the powerful de' Medici family, a banking and political dynasty that was established by Cosimo the Elder and effectively ruled Florence from the 15th to the early 18th centuries. The goldsmith's trade incorporated painting and sculpture, and many of the great Renaissance artists emerged from the Medici workshops, including Filippo Brunelleschi, Sandro Botticelli, and Benvenuto Cellini. Francesco I de' Medici, son of Cosimo I and Grand Duke of Tuscany, was especially interested in metalwork and jewelry. He established a workshop in the Uffizi Palace to broaden and develop jewelry-making techniques and the artistry of its practitioners.

Not suprisingly, the Medici crown jewels were renowned throughout Europe. Included in the trousseau of Catherine de' Medici for her marriage to King Henry II of France were pear-shaped pearls among the largest in Europe and a casket inset with engraved rock crystal by gem cutter Valerio Belli, whose patron was Giovanni de' Medici, the Renaissance Pope Leo X.

> **Gold is a treasure, and he who possesses it does all he wishes to in this world**
>
> Christopher **Columbus**
> 15th-century explorer

The Goldsmiths' Workshop, Alessandro (Il Barbiere) Fei, 1572
Grand Duke of Tuscany and patron of the arts Francesco I de' Medici (far left) inspects his father's crown and other items in a Florentine jewelry workshop.

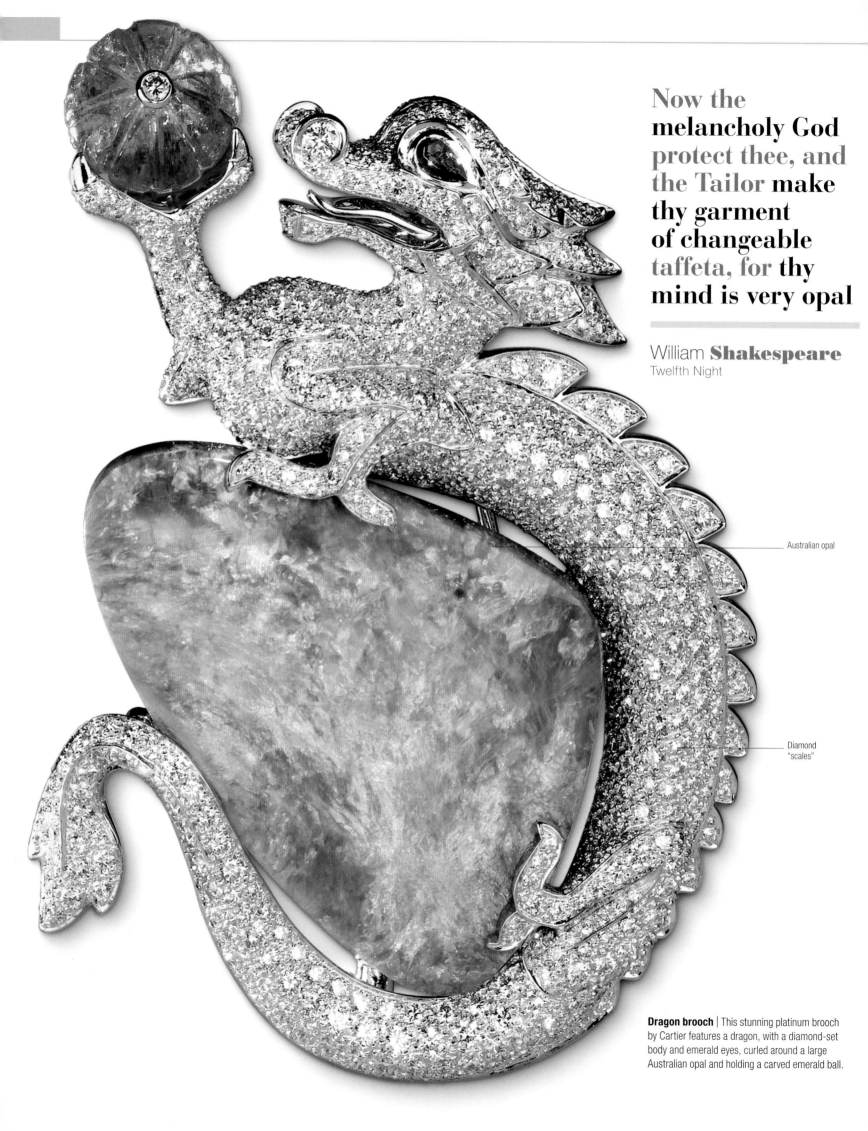

Now the melancholy God protect thee, and the Tailor make thy garment of changeable taffeta, for thy mind is very opal

William **Shakespeare**
Twelfth Night

Australian opal

Diamond "scales"

Dragon brooch | This stunning platinum brooch by Cartier features a dragon, with a diamond-set body and emerald eyes, curled around a large Australian opal and holding a carved emerald ball.

Opal

△ **Ethiopian opal** displaying a light-base, full-spectrum play-of-color

Opal falls into two categories: precious and common. The former displays highly prized rainbow iridescence with a white to dark body color, while the latter has a strong, attractive body color and no iridescence. Both kinds consist of hardened silica gel, and usually contain 5–10 percent water in submicroscopic pores. Precious opal consists of a regular arrangement of tiny, transparent, silica spheres, and its color play occurs when the spheres are regularly arranged and of the correct size, causing the diffraction of light and its consequent breakup into the colors of the spectrum: the actual colors that appear depend on the size of the spheres. Opal is deposited at low temperatures from silica-bearing waters, usually in sedimentary rocks. In ancient times, the primary source was present-day Slovakia; more recently, Australia was the main producer, and is also the source of fossil bones and seashells that have been replaced by precious opal. Ethiopia is now the main source of gem opal.

Common opal

Mineralogically, common opal refers to fire opals, which are transparent to translucent and do not usually show a play of color; it can also refer to opals with no color or transparency and no gemstone value. Fire opals, sometimes called jelly opals, are prized for their rich colors: yellow, orange, orange yellow, or red. Transparent fire opals tend to be faceted, and are often set into moderately expensive silver jewelry.

Specification

Chemical name Hydrous silicon dioxide | **Formula** $SiO_2.nH_2O$
Colors Colorless, white, yellow, orange, rose red, black, dark blue | **Structure** Amorphous | **Hardness** 5–6 | **SG** 1.9–2.5
RI 1.37–1.52 | **Luster** Vitreous | **Streak** White

Round brilliant   Cabochon

Locations
1 US **2** Mexico **3** Honduras **4** India **5** Australia
6 New Zealand **7** Ethiopia

Key pieces

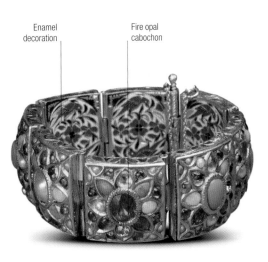

Enamel decoration   Fire opal cabochon

19th-century bracelet | Originating from Jaipur in India, this intricate gold bracelet is inlaid with fire opals, turquoises, and other precious and semiprecious stones. It also features enameled decorative panels.

Baroque pearl

Peacock brooch | Designed by French jeweler Georges Fouquet, this gold brooch from around 1900 is decorated with opals, garnets, pearls, and enamel. Its delicate, almost organic form is typically Art Nouveau in style.

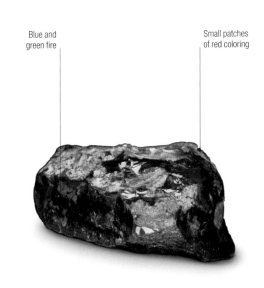

Blue and green fire   Small patches of red coloring

The Roebling Opal | Found in Nevada, US, this 2,585-carat black opal exhibits vivid blue and green flashes of color. Originally from the collection of Washington A. Roebling, it was donated to the Smithsonian in 1926 by his son, John A. Roebling II.

Rough

Common opal | The majority of opal is "common" opal—material that has neither transparency or fire. Here, pink common opal is in a rock groundmass.

Opal material

Opal nodule | Much Australian precious opal is found in nodules, occasionally replacing fossils. This white-base nodule is from Coober Pede, Australia.

Opal in ironstone | Some Australian precious opal has formed simultaneously with ironstone and is interlayered with it, as in this specimen.

Cut

Boulder opal | This cabochon has been cut from small layers of opal intermixed with ironstone matrix. This mixture is known as "boulder" opal.

Conchoidal fracture

Fire opal | Opal that is transparent or translucent red or orange is called "fire" opal. This rough specimen has a fine deep coloring.

Yellow common opal

Precious opal

Ironstone matrix | This specimen represents a colorful example of Australian opal. It has developed in an ironstone matrix and consists of a mixture of precious opal and yellow common opal, or "potch" opal.

Faceted opal | Some fire opal (sometimes called "jelly" opal) is transparent enough to facet. This orange brilliant cushion is cut from Mexican fire opal.

"Island Sunset" opal | This stunning black opal weighs 28.10 carats and was found at Lightning Ridge, Australia (see pp.162–63). It has been cut into a wide drop shape.

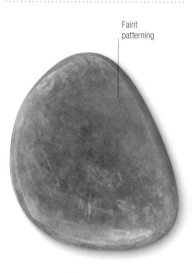

Faint patterning

Ethiopian opal | In recent years, a new discovery in Ethiopia has produced notable amounts of precious opal, as seen in this free-form cabochon.

Settings

Known since antiquity, opal derives its name from the Latin word "opalus," meaning "precious stone"

Pink opal bracelet | Created by Van Cleef & Arpels, this delicate bracelet is set with 29 cabochons of pink opal from the Wello region of Ethiopia.

Opal and garnet ring | Based around a 4.18-carat opal cabochon, this intricate ring is also set with various garnets as beads or carvings, and brilliant-cut diamonds.

Pink opal ring | This playful Cartier ring features pink opal set into 18-karat pink gold, with a diamond. Its motif reoccurs throughout the Amulette de Cartier range.

Louis Comfort Tiffany opal | Some opal is so valuable, it is cut in a shape to match the rough to avoid waste. Such is the case with this black opal with its irregularly shaped cut.

Fire opals

Opal earrings | In these gold earrings, a pair of double-domed fire opals are surrounded by foliate wreaths set with diamonds, with diamond-set suspensors.

Arts and Crafts necklace | With blue opal cabochons, emeralds, and pink tourmalines, this necklace was made by Georgie and Arthur Gaskin in the early 20th century.

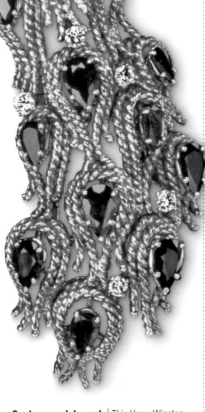

Opal peacock brooch | This Harry Winston brooch features a 32-carat black opal from Lightning Ridge, Australia, set with sapphires, rubies, emeralds, and diamonds.

Doublet earrings | These modern earrings consist of four opals surrounded by 1.82 carats of diamonds, set in 18-karat white gold.

Halley's Comet opal | Discovered 1986 | 1,982.5 carats

Halley's Comet
opal

△ **Full view** of the Halley's Comet opal

According to the *Guinness Book of World Records*, this impressive rock, which is roughly the size of a man's clenched fist, is the largest uncut black opal in the world. It was found in November 1986 by a group of five Australian miners known as "the Lunatic Hill Syndicate." They named it after the comet that was passing through the southern skies at the time of their discovery, which is only visible from Earth every 75 years.

This nodule, or "nobby," was found at an open-cut mine near the outback town of Lightning Ridge in New South Wales, which boasts the largest deposits of black opal on the planet. The syndicate consisted of two brothers and a small company, which provided financial backing as well as the earth-moving equipment. They operated at the Leaning Tree Claim on Lunatic Hill. The hill's curious nickname dates back to the early days of mining on the site. Most experienced prospectors

Halley's Comet opal
at auction, Bonhams,
Los Angeles, US

worked on the shallow flats below it, making their finds just a few feet below the surface. Only a madman would start at the top of the hill, they joked, since he would need to dig a very long way before discovering anything. Nevertheless, a lone miner tried it and his claim proved to be the most successful of all. The syndicate's efforts certainly vindicated their adoption of this painstaking approach, since the Halley's Comet opal was eventually located at a depth of 66 ft (20 m) below the surface.

Opals hold a special place in the Australian psyche. It is the country's national gemstone and it features in many legends that predate the arrival of European settlers. Aboriginal mythology relates how, in the Dreamtime, the Creator came down to earth on a rainbow, bearing his message of peace. Then, at the point where his foot touched the ground, the stones turned to opals, sparkling with all the colors of the rainbow.

Key dates

1705–2013

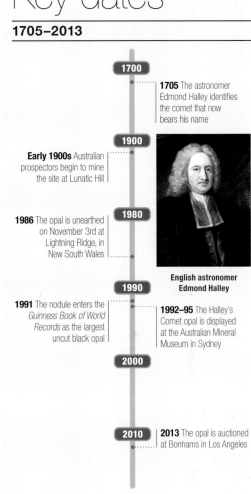

1700

1705 The astronomer Edmond Halley identifies the comet that now bears his name

1900

Early 1900s Australian prospectors begin to mine the site at Lunatic Hill

1980

1986 The opal is unearthed on November 3rd at Lightning Ridge, in New South Wales

**English astronomer
Edmond Halley**

1990

1991 The nodule enters the *Guinness Book of World Records* as the largest uncut black opal

1992–95 The Halley's Comet opal is displayed at the Australian Mineral Museum in Sydney

2000

2010

2013 The opal is auctioned at Bonhams in Los Angeles

Opal mine in rural Australia, showing the typical open-cut technique used to mine opals, whereby material is extracted from the surface rather than from deeper ground using tunnels

... opal resembles a fraction of the rainbow softened by a milky cloud

Charles **Blanc**
Author

Moonstone

△ **Cameo-carved** portrait with distinct blue sheen

Moonstone, an opalescent variety of anorthoclase and other feldspars, has been used in jewelry for centuries. Ancient Romans believed it came from solidified rays of the moon, and linked it to their lunar deities. Typically made up of layers of sodium- and potassium-rich feldspars, moonstone has a blue or white sheen—the result of the scattering or reflection of light by tiny intergrowths of minerals. The gem marketed as "rainbow moonstone" is more properly classified as a colorless labradorite (see p.169).

Specification

Chemical name Sodium potassium silicate (Anorthoclase)
Formula $(Na,K)AlSi_3O_8$ | **Color** Colorless, white
Structure Triclinic | **Hardness** 6–6.5 | **SG** 2.6
RI 1.50 | **Luster** Vitreous | **Streak** White | **Locations** India, Sri Lanka, Tanzania, Kenya, New Zealand, Australia, Norway

Rough

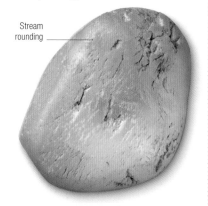

Stream rounding

Delicate shine caused by interference between light rays

Rounded pebble | The pitted surface of this water-worn moonstone pebble has the appearance of frosted glass. The best gem material is often found in this form.

Shimmering stone | In this specimen, each layer of feldspar reflects light, which causes a soft sheen or bright iridescence to shimmer with an ethereal glow.

Cut

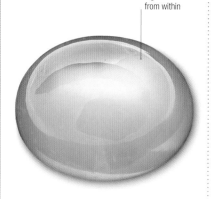

Light reflected from within

Detailed carving

Oval cabochon | With a high dome and subtle iridescence, this moonstone is excellent quality. Moonstone is mainly cut *en cabochon* to bring out the sheen.

Cameo carving | The iridescence of moonstone adds shadows and highlight to carvings, giving an impression of depth, as in this example.

Natural fractures

Moonstone rough | This large piece of uncut moonstone shows typical iridescence caused by minute interlayerings of rock affecting the path of light.

Color

Variation of color

Unusual hue | This cabochon has a dark honey tint to the base material combined with typical light iridescence like the shine of the moon after which moonstone is named.

Blue transparency | This unusually fine moonstone cabochon cut from a nearly transparent rough has a bluish iridescence that makes the stone glow.

Settings

Green diamond

Pink spinel

Moonstone cabochon

Innovative ring | This high-set ring features a number of round gemstones set within the sides, topped with a high-domed cabochon of moonstone.

Moonstone

Cartier Paris Nouvelle Vague ring | This gold ring is set with a large moonstone as well as sapphires, diamonds, chalcedony, turquoise, lapis lazuli, and aquamarine.

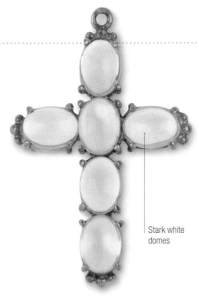

Stark white domes

Gold cross | This cross is set with six moonstone cabochons selected for their uniform whiteness, which stands out against the rich gold setting.

Gold rim with brilliant-cut diamonds

Lotus flowers in relief

Moonstone gives a shimmery background

Cartier watch and brooch | This detachable brooch is clipped over a hanging garden of lotus flowers and fish enameled over a moonstone studded with gems.

Moonstone is the state gem for Florida, from where the Moon landings took off—but it does not naturally occur there

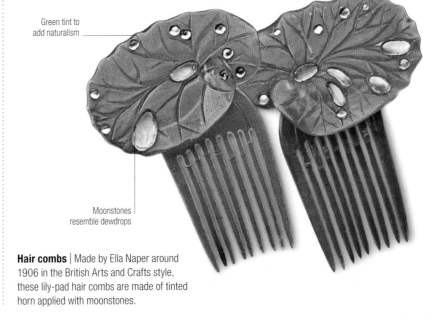

Green tint to add naturalism

Moonstones resemble dewdrops

Hair combs | Made by Ella Naper around 1906 in the British Arts and Crafts style, these lily-pad hair combs are made of tinted horn applied with moonstones.

Shimmering
color

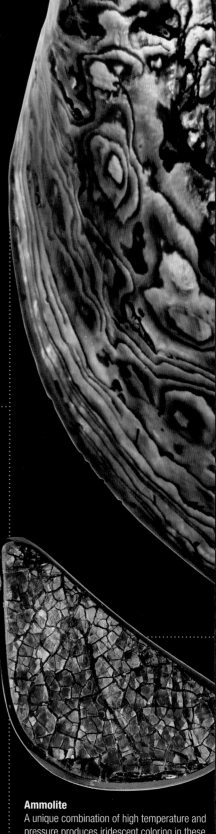

Play of color is caused by the interference of light interacting with a gemstone. Depending on the internal structure of the gem, light waves bounce off different layers, creating interference. The result is a shimmer of surface or internal color.

Precious opal
An opal's amazing play of color is produced by tiny spheres of silica gel, which reflect and diffract light.

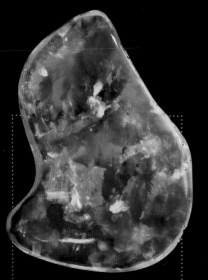

Aquamarine
This light blue aquamarine has been faceted to maximize the interplay of light in its transparent interior.

Mystic topaz
This type of gem is a recent innovation, dating back to the late 1990s. The color effects are artificial, produced by adding a thin, chemical coating to a white topaz; the colors change when the gem is tilted.

Ammolite
A unique combination of high temperature and pressure produces iridescent coloring in these rare organic gemstone fossils.

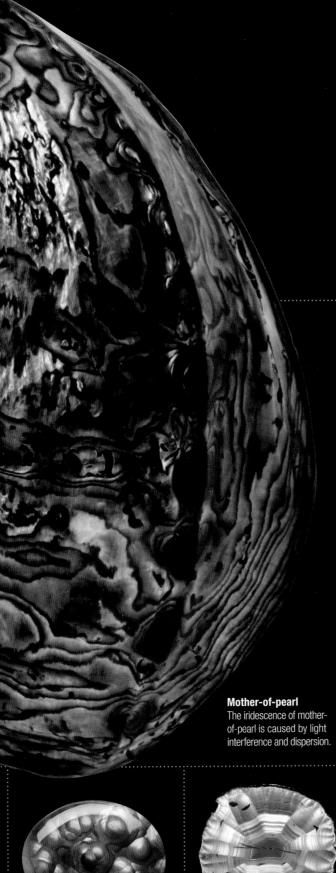

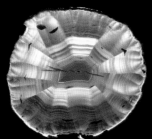

Common opal
Common opal is less sought-after than precious opal (far left), but can still be beautifully colored, as here.

Sunstone
This stone's glittering finish, called aventurescence, is caused by tiny inclusions of red copper or hematite.

Moonstone
Moonstone displays an effect called adularescence, caused by the unusual, layered structure of the mineral, which refracts and scatters light rays.

Mother-of-pearl
The iridescence of mother-of-pearl is caused by light interference and dispersion.

Labradorite
When viewed from certain angles, this feldspar mineral displays an iridescent luster known generally as the schiller effect (from the German word for "twinkle").

Fire agate
The layering of limonite or iron oxide and silica within this mineral produces its vibrant coloring.

Iris agate
The "iris" or rainbow effect can be achieved with backlighting on agate with fine color banding.

Sunstone

△ **Marquise-cut sunstone** sparkling with hematite inclusions

Sunstone is a gem that takes its name from its appearance, rather than as a result of the specific mineral it is made from. All types are characterized by minute, platelike inclusions of iron oxide or copper oriented parallel to one another, which give the stones a spangled appearance and often a reddish glow. The mineral classification of sunstones can be either oligoclase (a plagioclase feldspar) or orthoclase (an alkali feldspar). Other feldspars also produce sunstone in small quantities. Oligoclase sunstone is the most common type.

Specification

Chemical name Sodium, calcium aluminosilicate (oligoclase)
Formula $(Na,Ca)Al_2Si_2O_8$ | **Colors** Gray, white, orange brown, yellow | **Structure** Triclinic | **Hardness** 6.0–6.5
SG 2.62–2.65 | **RI** 1.53–1.55 | **Luster** Vitreous | **Streak** White | **Locations** US, Norway, India, Canada, Russia

Vitreous luster

Hematite flakes

Oligoclase | Rough | This specimen of uncut oligoclase sunstone shows the platy inclusions of hematite in parallel lines that give the gem its typical warm glow.

Brilliant-cut facet

Fancy cut | Cut | This Oregon sunstone in the US national collection was faceted in a triangle cut by award-winning cutters Darryl Alexander and Aivan Pham.

Symmetrical facets

Multiple faces catch light and maximize glow

Virtuoso piece | Cut | The versatility of sunstone in the hands of a master cutter is demonstrated by this piece titled "Snowflake," cut by jewelry artist Darryl Alexander.

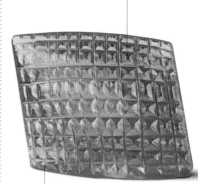

Multiple tiny facets

Play of light and shade gives appearance of depth

Fancy cut | Cut | Oregon, US, is the source of much gem-quality oligoclase. This Oregon sunstone with copper inclusions was faceted by the renowned cutter Larry Winn.

Sunstone | Cut | This faceted oval shows the spangled appearance created by the numerous inclusions that raise the mineral to gem-quality grade.

Densely clustered inclusions

Transparent surface between inclusions

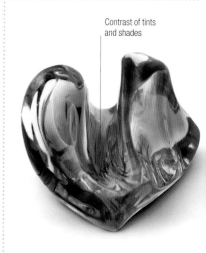

Contrast of tints and shades

Heart shape | Carved | This piece of sunstone from Oregon was carved into a heart by American Naomi Sarna, who specializes in sculptural gem cutting.

Labradorite

△ **Labradorite rough** in typical gemstone-grade base color

A **type of feldspar**, labradorite is named for the Canadian province of Labrador, where it was first identified in 1770. Gemstone labradorite is commonly characterized by its rich play of iridescent colors, principally blue, on broken surfaces. Crystals that display this effect are cut *en cabochon* or used in carvings. Nearly transparent material with a beautiful iridescence comes from southern India. Fully transparent labradorite is found from time to time, and can be yellow, orange, red, or green.

Specification

Chemical name Sodium, calcium aluminosilicate | **Formula** $NaAlSi_3O_8$—$CaAl_2Si_2O_8$ | **Colors** Blue, gray, white
Structure Triclinic | **Hardness** 6–6.5 | **SG** 2.65–2.75
RI 1.56–1.57 | **Luster** Vitreous | **Streak** White | **Locations** Madagascar, Finland, Russia, Mexico, US; Labrador, Canada

Labradorite

Labradorite in combination | **Rough** | This piece of labradorite rough shows gem-quality blue material interlayered with another feldspar.

Multicolored play of light and color

Square cabochon | **Cut** | This labradorite cabochon has fine blue, gold, and green schiller. Material like this is found in Mexico and the US.

Subtle schiller

Deep carving

Animal carving | **Carved** | The schiller in labradorite, if properly oriented, adds depth and life to carvings, as here. It combines with the vitreous luster of the mineral to give a glowing, greenish surface sheen reminiscent of the slimy skin of a frog.

Orthoclase layer

Cameo head | **Carved** | Skillful carving through the layers of labradorite brings out flashes of blue, green, yellow, and red as the stone is turned.

Diamonds

Labradorite cabochon

Pair of earrings | **Set** | The irregular rounded shapes of these iridescent earrings are set with rows of tiny diamonds around the borders.

Schiller

Lit from within

The iridescence in labradorite is technically called schiller. It is caused by the scattering of light from thin layers of a second type of feldspar that develops through internal chemical separation during the cooling of what was originally a single feldspar. These layers act as diffraction gratings, separating light into its component colors. The color that results is determined by the thickness of the layers, although the base color of labradorite is generally blue, dark gray, colorless, or white. High-quality labradorite from Finland is sometimes called spectrolite.

Dramatic coloration This labradorite specimen has superb schiller.

Orthoclase

△ **Rare, 250-carat**, yellow orthoclase gem, unusual for its size and clarity

The pink crystals of orthoclase give common granite its characteristic pink color. It is also an important rock-forming mineral that yields gemstones. Yellow and colorless orthoclase is faceted for collectors when transparent, and it sometimes produces gems called sunstone (see p.168). A cat's-eye effect results when some yellow and white specimens are cut *en cabochon*. A variety of orthoclase exhibiting adularescence is called moonstone. This adularescence results from the interlayering of orthoclase with albite (see p.172).

Specification

Chemical name Potassium aluminosilicate | **Formula** $KAlSi_3O_8$
Colors Colorless, white, cream, yellow, pink, brown red
Structure Monoclinic | **Hardness** 6–6.5 | **SG** 2.5–2.6
RI 1.51–1.53 | **Luster** Vitreous | **Streak** White
Locations Myanmar, Sri Lanka, India, Brazil, Tanzania, US, Mexico

Blocklike surface lines

Orthoclase crystal | Rough | This orthoclase crystal provides a good illustration of the mineral's classic blocky shape in its natural state.

Water wear

Gem orthoclase | Rough | In this water-worn piece of yellowish orthoclase, a high degree of transparency is readily apparent.

Textured surface

Cabochon | Color variety | Pink orthoclase that is translucent rather than transparent is commonly cut into attractive cabochons, such as this stone.

Smoky texture

Table facet

Moonstone | Cut | This gem is a cushion, brilliant-cut moonstone (see box, left), exhibiting a characteristic silvery-white texture, emphasized by the cut's many facets.

Sacred moonstone

Legends and beliefs

Orthoclase is one of several feldspars that show a white or silvery adularescence when cut *en cabochon* and are called moonstone. Other moonstones are anorthoclase, sanidine, albite, and oligoclase (see p.164, p.172, p.168). Moonstone was sacred in India, where it was said to inflame passions, and that if lovers placed it in their mouths at full moon, their futures would be revealed. In 11th-century Europe, moonstone was believed to reconcile lovers, and in 16th-century England, a moonstone dedicated to King Edward VI was said to wax and wane with the moon.

Moonstone cabochon This polished cabochon shows characteristic adularescence.

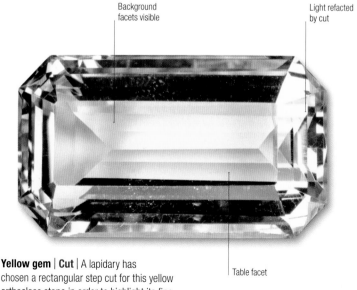

Background facets visible

Light refacted by cut

Table facet

Yellow gem | Cut | A lapidary has chosen a rectangular step cut for this yellow orthoclase stone in order to highlight its fine color and transparency.

Microcline

△ **Rough specimen** of amazonite

Microcline is one of the most common potassium aluminosilicate feldspar minerals; the other is orthoclase. Blue-green to green specimens of microcline are called amazonstone or amazonite. Although deep blue green is the most sought-after color, it varies from yellow green to blue green and may exhibit white streaks. Gem material is usually opaque and is cut *en cabochon*; it is rarely used for carvings or beads, being relatively brittle. Gem-quality amazonite is found in Minas Gerais in Brazil, Colorado in the US, and the Ural Mountains in Russia.

Specification

Chemical name Potassium aluminosilicate | **Formula** $KAlSi_3O_8$ | **Colors** White, pale yellow, green, blue green
Structure Triclinic | **Hardness** 6–6.5 | **SG** 2.6
RI 1.52–1.53 | **Luster** Vitreous | **Streak** White
Locations Russia, US, Brazil

Rock groundmass

Microcline crystals | **Rough** | In this specimen, a cluster of light-colored, blocky microcline crystals is set in a groundmass of rock.

Pink microcline

Amazonite crystal | **Rough** | This superbly formed, blue-green rough amazonite crystal is intergrown with contrasting layers of pink microcline.

> The name "microcline" originates from the Greek for "small slope"

Fine blue-green coloring

Internal structures

Amazonite slice | **Color variety** | The deep blue-green hue of this rough slice of amazonite is widely considered to be its most desirable color for use in gemstones.

Cabochon | **Color variety** | This polished cabochon of amazonite demonstrates the texture and fine turquoise color of excellent, gem-quality material.

Aquamarine

Crystal shows fine transparency

Crystal group | **Rough** | This group of three minerals shows the classic pegmatite assemblage—blue aquamarine and quartz perched on a microcline crystal.

Quartz

Microcline

Albite

△ **Nest of gemmy albite crystals** surmounted by brookite

Albite is mainly significant as a rock-forming mineral, but it also has some use as a gemstone. It is found as well-formed, glassy, and brittle crystals, and these are often of transparent, gem quality. However, since it is relatively soft and brittle, albite is faceted exclusively for collectors. Indeed, the variety known as peristerite—a mixture of albite and oligoclase (see p.168)—produces a pleasing bluish, moonstonelike sheen when cut *en cabochon*. The mineral mostly occurs as colorless material, but it can also be yellowish, pink, or green.

Specification

Chemical name Sodium aluminosilicate | **Formula** $NaAlSi_3O_8$ | **Colors** White, colorless, yellow, green **Structure** Triclinic | **Hardness** 6–6.5 | **SG** 2.6–2.7 **RI** 1.53–1.54 | **Luster** Vitreous to pearly | **Streak** White **Locations** Canada, Brazil, Norway

Elbaite tourmaline, quartz, and albite | **Rough** | In this mixed-mineral specimen, albite is host to impressive crystals of pink-purple tourmaline and clear quartz.

Tourmaline crystals

Quartz crystals

Albite

Prismatic tourmaline

Albite and tourmaline | **Rough** | This spectacular specimen features prismatic crystals of elbaite tourmaline resting on albite and quartz.

Large topaz

Albite and topaz | **Rough** | In this striking specimen from Afghanistan, snowy white, gemmy albite is the groundmass for a topaz weighing around 1 lb (0.5 kg).

Twinned crystal

Pearly luster

Albite group | **Rough** | The white albite crystals in this dramatic group have the characteristic blocky crystal form of albite, and many show twinning.

Brilliant-cut crown

Mixed-cut albite | **Cut** | This flawless, oval, bluish albite gemstone is faceted in a mixed cut with a brilliant-cut crown and step-cut pavilion.

Bytownite

△ **Marquise-cut** bytownite gemstone

Bytownite is the rarest member of the plagioclase feldspar group; the other members of the group that have gem varieties include labradorite, albite, and oligoclase. Bytownite is seldom found in well-developed crystals, but these can be gemmy when found. Its gemstones are usually faceted, with the transparent gems varying in color from a pale, straw yellow to a light brown. A variety from Mexico is marketed under the name Golden Sunstone, but is different from the various other feldspar sunstones.

Specification

Chemical name Sodium, calcium aluminosilicate | **Formula** $NaAlSi_3O_8$—$CaAl_2Si_2O_8$ | **Colors** White, gray, yellow, brown | **Structure** Triclinic | **Hardness** 6–6.5 | **SG** 2.7 | **RI** 1.56–1.57 | **Luster** Vitreous to pearly | **Streak** White | **Locations** Mexico, Oregon, Scotland, Greenland, US, Canada

Small gemmy areas

Bytownite in rock groundmass | Rough | Bytownite rarely forms distinct crystals, but is more often found intergrown with other plagioclases, as here.

Striations

Plagioclase | Rough | This plagioclase specimen shows surface striations, which are the prime characteristic of all plagioclases, including bytownite.

Basalt   Bytownite

Polished bytownite | Cut | This tumble-polished specimen of bytownite in basalt is one of the more unusual occurrences, with the local name of "Lakelandite."

Pavilion facets visible through table

Fine bytownite | Cut | The bytownite material comprising this stone is unusually flawless, and has been faceted in a step-cut cushion.

Bytownite specimen | Rough | This rough bytownite crystal originates from Ottawa, Canada, and exhibits a remarkable level of clarity and transparency.

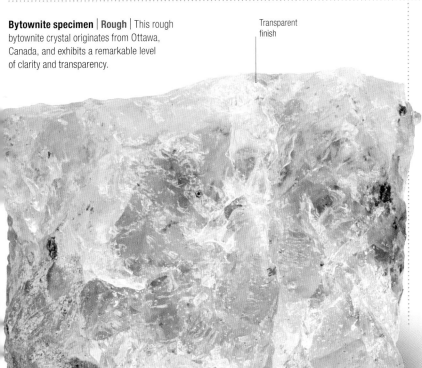

Transparent finish

Table facet

Step-cut stone | Cut | The faceter of this emerald-cut bytownite has found an unusually long piece of rough to work from, yielding a striking gem.

Bytownite is one of the minerals known to occur in stony meteorites

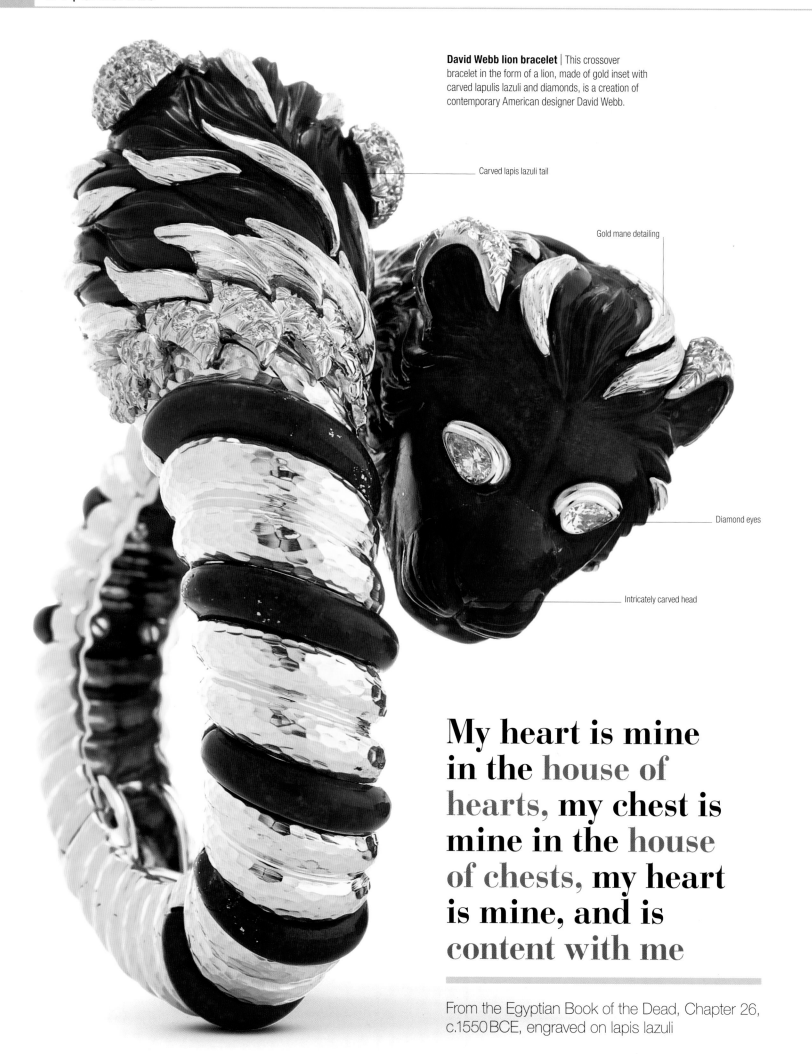

David Webb lion bracelet | This crossover bracelet in the form of a lion, made of gold inset with carved lapulis lazuli and diamonds, is a creation of contemporary American designer David Webb.

Carved lapis lazuli tail

Gold mane detailing

Diamond eyes

Intricately carved head

My heart is mine in the house of hearts, my chest is mine in the house of chests, my heart is mine, and is content with me

From the Egyptian Book of the Dead, Chapter 26, c.1550 BCE, engraved on lapis lazuli

△ **Piece of fine, rich blue** lapis lazuli rough

Lapis lazuli

For over **6,000 years**, people have been drawn to the intense blue of lapis lazuli, often flecked with golden glints like stars in the night sky. It is relatively rare, and commonly forms in crystalline limestones as a product of heat and pressure; the strong blue color is mostly caused by the mineral lazurite, although lapis also contains pyrite and calcite, and usually some sodalite and haüyne. The highest-quality material is a deep, dark blue, with minor patches of white calcite and brassy yellow pyrite. A large quantity of modern lapis material originates from mines in Afghanistan, its original source (see below), while lighter blue material is found in Chile, and lesser amounts in Italy, Argentina, Russia, and the US.

Lapis lazuli in history

For many centuries, the only known deposits of lapis lazuli were those at Sar-e-Sang, in a remote mountain valley in Afghanistan, from where it was widely traded across the ancient world. Objects from ancient Egypt containing lapis lazuli date from at least 3100 BCE and include scarabs, pendants, inlays in gold and silver, and beads. Powdered lapis lazuli was used as a cosmetic—the first eye shadow (along with malachite)—as a blue pigment, and as a medicine. Outside ancient Egypt, the tomb of Sumerian Queen Pu-abi (2500 BCE) contained numerous gold and silver jewelry pieces richly adorned with lapis, and the Chinese and the Greeks were carving lapis lazuli as early as the 4th century BCE.

Specification

Chemical name Sodium, calcium, alumino-silicate (Lazurite)
Formula $Na_3Ca(Al_3Si_3O_{12})S$ | **Colors** Blue | **Structure** Cubic | **Hardness** 5–5.5 | **SG** 2.4 | **RI** 1.5 | **Luster** Dull to vitreous | **Streak** Blue

Locations
1 US **2** Chile **3** Argentina **4** Italy **5** Afghanistan **6** Russia

Key pieces

Ancient Egyptian gold pectoral | Featuring a central scarab carved from lapis lazuli supporting a gold disk that represents the sun, this pectoral was created for the pharaoh Amenemopet of the 21st Dynasty, and found in Tanis.

Gold cherub

17th-century ewer | Created in the Miseroni workshop in Florence, Italy, around 1608, this ewer was carved from two separate pieces of lapis lazuli. The foot, collar, and handle are made from gold, the handle in the form of a cherub.

Diamond brilliants

Lapis lazuli cabochons

Cartier Nouvelle Vague ring | Created by Cartier, Paris, this 18-karat yellow gold ring is set with nine lapis lazuli cabochons and nine chrysoprase cabochons, along with 112 brilliant-cut diamonds.

Rough

Calcite veining

Deep blue material

Lapis rough | This piece of lapis lazuli rough consists of areas of rich blue coloring, along with large cross-veining of calcite and some substantial inclusions of pyrite. This material could be cut into interesting cabochons.

Streaked rough | The rich blue of this large piece of lapis rough—slightly larger than a fist—is accented by streaks and scatters of golden pyrite.

Lapis rough | Some lapis rough is almost devoid of pyrite, as in this specimen. It is often preferred for use in inlaid carvings and small cabochons.

The therapeutic stone

Lapis lazuli in ancient treatments

The Greek physician Dioscorides noted in around 55 CE that lapis was an antidote for snake venom; even earlier than this, the Assyrians used it as a cure for melancholy. Another widespread ancient belief was that it protected the wearer from evil because it resembled the night sky, the dwelling place of God. Similarly, a medieval treatise suggests that "meditation upon stone carries the soul to heavenly contemplation," while to the Buddhists of antiquity, lapis lazuli brought peace of mind.

Ancient remedy Ground lapis lazuli, which was sometimes administered as a "medication."

Cut

Polished lapis | Lapis polished in irregular forms, such as this wedge-shaped piece, are desirable decorative objects even without a jewelry setting.

Imitation lapis | Created by Gilson, this cabochon is cut from imitation lapis. It can be identified as synthetic by its uniform color and unnatural scatter of pyrite.

Chilean cabochon | While ancient lapis came from Afghanistan, the New World source is in Chile. Its material tends to be lighter in color, such as in this cabochon.

Settings

Lapis ring | The square cabochon set in this 18-karat gold ring has fine color, and includes a beautiful scatter of bright gold pyrite.

Ancient eagle | Originating from the Sumerian civilization around 2650 BCE, this lion-headed eagle is created from lapis lazuli, gold, copper, and bitumen.

Copper antlers

Bust of Mehurt | From the Old Kingdom of ancient Egypt (c.1539–1075 BCE) comes this head of Mehurt, the celestial mother, made of lapis, copper, and gold.

The **Latin term "sapphirus"** probably referred to **lapis lazuli,** with the **modern term** derived from the Arabic word "lazaward," meaning "heaven" or "sky"

Golden handle

Calcite veining

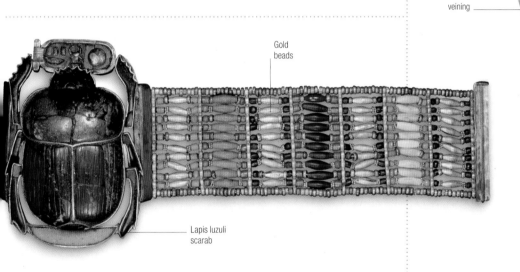

Gold beads

Lapis luzuli scarab

Ancient Egyptian bracelet | Recovered from the tomb of Tutankhamun, this beaded bracelet features a scarab beetle carved from Afghan lapis lazuli with a turquoise inlay.

Vase | This elegant vase of turned lapis is mounted with gold fittings set with garnets in the base and a top knob of garnet.

Polished lapis lazuli

18-karat gold

Cufflinks | This pair of gold and lapis cufflinks was produced by the House of Bulgari. The inset lapis lazuli material is flecked with sparkling pyrite.

Victorian masterwork | Created in the mid-19th century, this gold pendant features a central lapis cluster surrounded by split pearls and lapis beads.

Tiffany bangle | This 18-karat white gold bangle from Tiffany & Co. is set with inlaid lapis lazuli cut in irregular shapes. It originates from the 1980s.

Checkered bangle | Also created for Tiffany & Co. around 1980, this wavy, 18-karat gold bangle is inlaid with mother-of-pearl, black onyx, and lapis lazuli.

Jewels of
ancient Egypt

When archaeologist Howard Carter opened the solid gold coffin of Tutankhamun in 1924, he lifted the lid on the culture of the ancient Egyptians, which dates back as far as 5000 BCE. Egyptian royal burial chambers were steeped in gold, in honor of the gods, and jewelry was placed on the bodies of the dead. Jewels often took the form of a Wedjat—a symbolic Eye of the god Horus—and animals that had religious symbolism.

Funerary mask of Tutankhamun
Inlaid with lapis lazuli and obsidian, this gold mask, c.1336–1327 BCE, was said to protect the pharaoh so that his soul could be reborn.

Vulture collar
This gold vulture, c.1550–1298 BCE, is clutching a *shen* (a ring with a short bar), the symbol for eternity.

Gold swivel ring
This ring features a hinged carving of a sphinx and symbols intended to protect the wearer.

Scarab pectoral
This gold, lapis lazuli, carnelian, and turquoise pectoral from c.1361–1352 BCE takes the form of a scarab beetle.

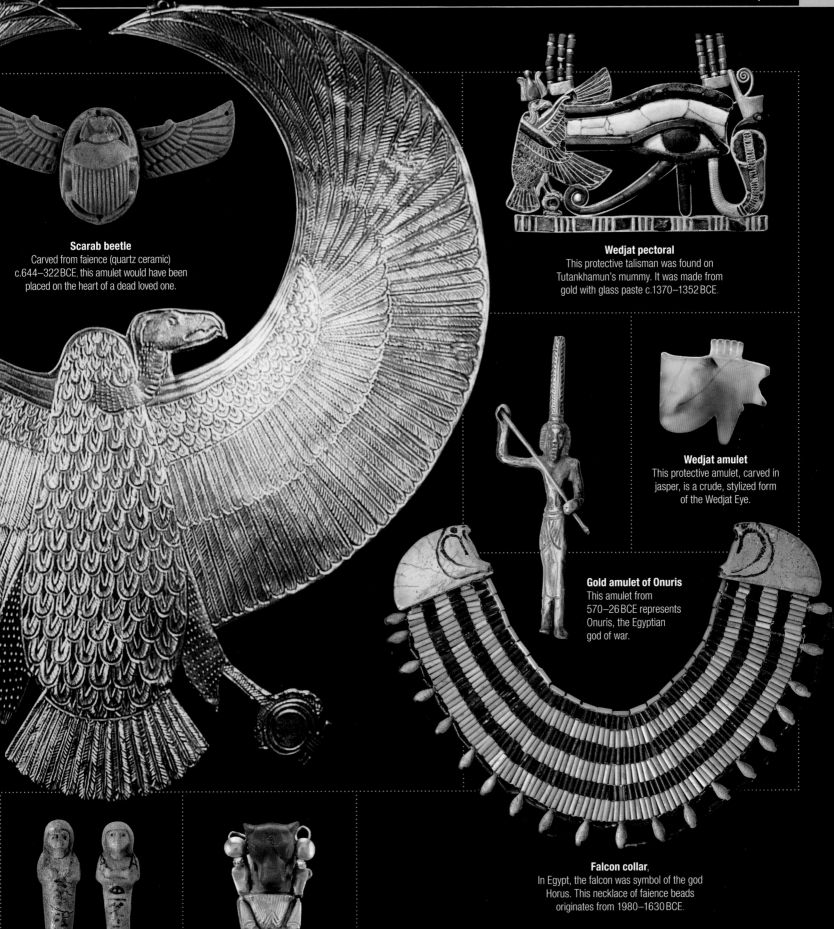

Scarab beetle
Carved from faience (quartz ceramic)
c.644–322 BCE, this amulet would have been
placed on the heart of a dead loved one.

Wedjat pectoral
This protective talisman was found on
Tutankhamun's mummy. It was made from
gold with glass paste c.1370–1352 BCE.

Wedjat amulet
This protective amulet, carved in
jasper, is a crude, stylized form
of the Wedjat Eye.

Gold amulet of Onuris
This amulet from
570–26 BCE represents
Onuris, the Egyptian
god of war.

Falcon collar,
In Egypt, the falcon was symbol of the god
Horus. This necklace of faience beads
originates from 1980–1630 BCE.

Shabti servant figurines
Faience figurines such as these,
c.1292–1190 BCE, were placed
in the tombs of the wealthy.

Bull's head
This lapis lazuli carving in gold
mount, c.1070–656 BCE, represents
the bull god, Apis.

Sodalite

△ **Faceted oval cabochon** of semitranslucent sodalite

Sodalite is the mineral sometimes mistaken for lapis lazuli (see pp.174–77). It can also be one of the constituents of lapis lazuli, but is visually different, since the latter contains small crystals of pyrite. It is one of only a handful of minerals whose only use is as a gemstone—often veined with calcite, it is favored by carvers for its interesting patterns. It is usually cut *en cabochon*, but rare transparent material from Mont Saint-Hilaire, Canada, is faceted for collectors. Single pieces can weigh many pounds.

Specification

Chemical name Sodium aluminosilicate chloride | **Formula** $Na_4Al_3Si_3O_{12}Cl$ | **Colors** Gray, white, blue | **Structure** Cubic | **Hardness** 5.5–6 | **SG** 2.1–2.4 | **RI** 1.48 | **Luster** Vitreous to greasy | **Streak** White to light blue | **Locations** Canada, Russia, Germany, India, Canada, USA

White veining

Uncut sodalite | **Rough** | This fine piece of sodalite rough demonstrates a good blue coloring with a minimum of the mineral's characteristic white veining.

White groundmass

Sodalite in rock | **Rough** | This specimen features a vivid blue sodalite scattered within the rock groundmass of another, white, feldspathoid.

Pink fluorescence

Fluorescent sodalite | **Color variety** | When illuminated under ultraviolet light, many sodalite pieces fluoresce (see pp.186–87), as can be seen in this specimen from India.

Yellow fluorescence

Yellow fluorescing sodalite | **Color variety** | Under ultraviolet light, sodalites from different localities fluoresce in different colors (see pp.186–87). This example is Russian.

Sodalite was named in 1811 for its high sodium content

Cabochon | **Cut** | Sodalite is mostly cut *en cabochon*, with the cutter orienting the stone to get the best color or pattern, as in this example.

Mottled patterning

Sodalite ornament | **Carved** | Sodalite is reasonably brittle, but in the hands of a skilled carver it can be shaped into pieces that are both attractive and amusing, such as this pig carved from unusually patterned material.

Haüyne

△ **Faceted haüyne** with a modified brilliant cut

Haüyne is one of the components of lapis lazuli, along with pyrite, lazurite, calcite, and sodalite. Blue is the most common color and occurs in lapis lazuli, but haüyne also comes in white, gray, yellow, green, or pink. Single crystals are sometimes found, and can be faceted only with difficulty—the mineral has perfect cleavage (planes of breakage), which makes it hard to cut without shattering the material. Facet-grade haüyne crystals tend to be small, with faceted stones usually weighing five carats or less.

Specification

Chemical name Sodium, calcium aluminosilicate with sulfate
Formula $Na_3Ca(Al_3Si_3O_{12})(SO_4)$ | **Colors** Blue, white, gray, yellow, green, pink | **Structure** Cubic | **Hardness** 5.5–6 | **SG** 2.4–2.5
RI 1.49–1.51 | **Luster** Vitreous to greasy | **Streak** Blue to white
Locations Germany, Italy, US, Serbia, Russia, Morocco, China

Gem crystal

Rock matrix

Crystals in matrix | **Rough** | A number of small but good-quality, transparent patches of haüyne crystals are contained within this large rock.

Gem crystal

Rock groundmass

Gemmy crystals | **Rough** | In this specimen, a group of small, intensely colored gemmy crystals of haüyne has developed in a rock groundmass.

Flawed gem | **Cut** | Gem-quality haüyne is so rare in pieces over one carat that even stones cut from slightly flawed rough are acceptable, as here.

René Haüy's name is one of only 72 that are inscribed on the Eiffel Tower

Superb color | **Color variety** | Although far from flawless, the stunning color of this 0.82-carat, pear-shaped German gem makes it a highly desirable stone.

Napoleon's professor

The father of crystallography

Haüyne is named after René Just Haüy (1743–1822), who grew interested in crystallography when he noticed that fragments of a broken calcite crystal cleaved along straight lines that met at constant angles. He was the first to show that a crystal is built up of tiny, identical units. In 1802 (after imprisonment in the French Revolution), Haüy was appointed Professor of Mineralogy at the Museum of Natural History in Paris. At Napoleon's request, he wrote a book on crystallography.

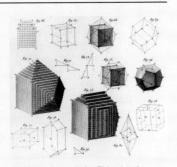

Illustrated crystals This plate is from *Treatise of Crystallography*, by René Just Haüy, 1822.

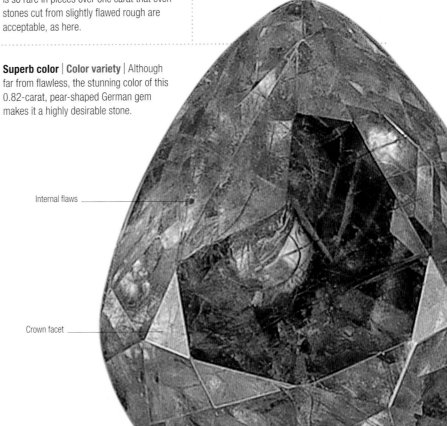

Internal flaws

Crown facet

Kiani Crown | 12½ x 7½ in (32.5 x 19.5 cm) | Pearls, rubies, emeralds, spinels, diamonds | Shown here in a portrait of Fath-Ali Shah (1762–1834), c.1805

Kiani
Crown

△ **Aga Muhammad Khan Qajar**, who created the Kiani Crown in 1796 (shown here around 1820)

Mohammad Ali Shah Qajar
(1872–1925) with the crown

A piece of royal regalia like no other, the Kiani crown was the coronation crown in the Persian crown jewels and was used throughout the Qajar dynasty (1796–1925). It was a powerful symbol of royal authority, and is unique in its lavish decoration and for the sheer number of pearls used in its design.

The crown is 12½ in (32 cm) high without its detachable aigrette (plume), and 7½ in (19.5 cm) wide. The base of the crown is made from red velvet and is stitched with 1,800 small pearls of about ¼ in (7–9 mm) in diameter. Around 300 emeralds decorate the crown, mainly on the aigrette, the largest of which is 80 carats; there are also approximately 1,800 rubies and spinels, the largest weighing 120 carats. Numerous diamonds stud the crown, with a 23-carat diamond as the centerpiece.

The Kiani Crown was created by Agha Mohammad Khan, founder of the Qajar dynasty, in 1796 and modified by Fath-Ali Shah (who reigned from 1797 to 1834), the second of the Qajar kings. Fath-Ali Shah was renowned for his patronage of Persian art, as well as for having 1,000 wives and fathering over 100 children. Five more kings wore the crown at their coronation.

The crown was present, although unworn, at the 1926 coronation of Reza Shah, who seized power in a coup. This brought an end to the Qajar dynasty and the last of the line, Ahmed Shah, fled to Europe. Reza ordered the creation of a new crown, and the Kiani crown became a museum piece.

120-carat spinel

23-carat diamond

The Kiani Crown, Iran's symbol of royal and religious power for about 130 years, displayed at the National Treasury, Tehran

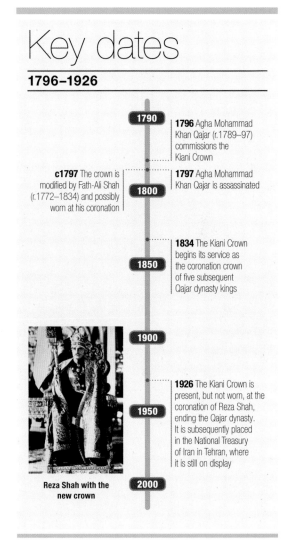

Key dates
1796–1926

1790

1796 Agha Mohammad Khan Qajar (r.1789–97) commissions the Kiani Crown

c1797 The crown is modified by Fath-Ali Shah (r.1772–1834) and possibly worn at his coronation

1797 Agha Mohammad Khan Qajar is assassinated

1800

1834 The Kiani Crown begins its service as the coronation crown of five subsequent Qajar dynasty kings

1850

1900

1926 The Kiani Crown is present, but not worn, at the coronation of Reza Shah, ending the Qajar dynasty. It is subsequently placed in the National Treasury of Iran in Tehran, where it is still on display

1950

Reza Shah with the new crown

2000

It was entirely composed... as to form a mixture of the most beautiful colours

R. Ker **Porter**
Travel writer, on the crown

Scapolite

△ **Fine, oval, brilliant-cut** yellow scapolite gem

Originally believed to be a single mineral, scapolite is the name now given to a group of minerals related by structure, and it is still used in the gemstone trade to refer to any members of the scapolite mineral group cut as gemstones. It is distinctly pleochroic and stones vary in color when viewed from different angles—violet stones appear dark or light blue, and violet and yellow stones appear pale yellow and colorless. Some scapolites cut *en cabochon* show signs of chatoyancy, a streak of light in the form of a cat's eye.

Specifications

Chemical name Sodium, calcium silicate chloride or sulfate
Formula $Na_4(Al_3Si_9O_{24})Cl$—$Ca_4(Al_6Si_6O_{24})(CO_3SO_4)$ | **Colors** Colorless, white, gray, yellow, orange, pink | **Structure** Tetragonal
Hardness 5–6 | **SG** 2.5–2.7 | **RI** 1.54–1.58 | **Luster** Vitreous
Streak White | **Locations** Myanmar, Canada, US, Tanzania

Prismatic crystals

Rock groundmass

Scapolite crystals | **Rough** | The hollow in this rock groundmass is filled with a cluster of numerous prismatic scapolite crystals, each with four sides of similar length.

Blue scapolite

Blue crystals | **Rough** | Scapolite comes in a number of colors, but usually forms in metamorphic rocks. These pale, almost translucent, blue crystals are prismatic.

Mineral inclusions

Scapolite cabochon | **Cut** | This highly polished cabochon displays a hazy purple-violet coloring and features visible mineral inclusions.

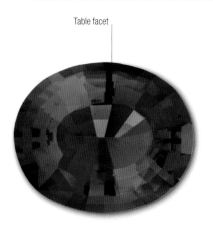

Table facet

Crown facets

Cushion-cut gemstone | **Color variety** | This cushion-cut scapolite stone is notable for its deep brownish purple coloring. It weighs 2.95 carats.

Mixed-cut scapolite | **Cut** | This beautifully clear example of colorless scapolite is given a striking brilliance by the many facets of its unusual mixed cut.

Pear-shaped table facet

Museum quality | **Cut** | This 113-carat, yellow, pear-shaped scapolite (shown actual size) is of exceptional clarity and brilliance, and would grace any museum collection.

Blue scapolite

Scapolite earrings | **Set** | These delicate 18-karat gold earrings are carved from frosted rock crystal, and crowned with pear-shaped, faceted blue scapolites.

Pollucite

△ **Round, mixed-cut** pollucite gem

Discovered in 1846, pollucite is one of two minerals named after Castor and Pollux, the Gemini twins in Greek mythology (see box, below), although the other, castorite, has since been renamed petalite (see p.196). Pollucite is only found in rare element-bearing deposits, where it occurs with other gem minerals such as spodumene, petalite, quartz, and apatite. Facet-grade material tends to be very small, but crystals up to 24 in (60 cm) across have been found at Kamdeysh in Afghanistan. It also occurs in Italy and the US.

Specifications

Chemical name Cesium sodium aluminosilicate
Formula $(Cs,Na)(AlSi_2)O_6H_2O$ | **Colors** Colorless, white, pink, blue, violet | **Structure** Cubic | **Hardness** 6.5–7
SG 2.85–2.94 | **RI** 1.51–1.525 | **Luster** Vitreous to greasy
Streak White | **Locations** Afghanistan, Elba, Italy, US

Gem-quality crystal

Pollucite rough | Rough | Only an expert cutter will be able to see through the rough, waterworn exterior to glimpse the fine, gem-quality material inside.

Golden-yellow crystals

Well-worn, rounded edges

Massive pollucite | Rough | This piece of massive pollucite was found in Buckfield, Maine, in the US. This is a broken fragment; well-formed crystals are rare and are especially prized in the jewelry trade.

Triangular facet

Mixed cut | Cut | This pendeloque pollucite gemstone features triangular facets on the crown and rectangular, step-cut facets on the pavilion.

Step-cut facets

Octagonal step cut | Cut | The rectangular, cushion step cut, faceted to reveal the gem's vitreous luster, emphasizes the pale blue of this pollucite.

Unusual cut | Cut | The cutter of this 2.69-carat, oval pollucite has faceted the gem in an unusual way, accentuating its radiance and exquisite peach hue.

Castor and Pollux

Mythological warriors

In Greek and Roman mythology, Castor and Pollux were twin brothers renowned for their horsemanship. The Romans believed that their victory at the Battle of Lake Regillus was aided by the mythological twins, and built the Temple of Castor and Pollux in Rome's Forum to honor them. Each year on July 15, the 1,800 members of Rome's elite cavalry paraded through the city to commemorate the military victory.

Castor and Pollux The brothers are portrayed in these Roman statuettes from the 3rd century CE.

Benitoite
This specimen from California, US, turns blue under shortwave UV light, as seen here.

Gypsum on rock groundmass
Fluorescence is common in gypsum. This specimen from Paris, France, fluoresces a rich yellow.

Aragonite
This specimen from Sicily, Italy, fluoresces pink. Aragonite material can also fluoresce yellow, blue, or green.

Zinc ore
This specimen contains willemite, franklinite, and calcite, displaying green, black, and pink respectively. It also contains zincite, which does not fluoresce.

Scapolite
This specimen from Canada is fluorescent, but not all scapolite has this quality.

Calcite
The columnar crystals of this specimen fluoresce blue white. Trace elements in calcite can cause other fluorescent colors.

Manganoan calcite
Many types of calcite fluoresce, including this one from Arizona, US.

Fluorescent
minerals

When seen under ultraviolet (UV) light, some crystals glow in eerie, psychedelic colors. First noted in 1824 in fluorite, the phenomenon is called fluorescence. It is unpredictable, since some specimens of a mineral fluoresce, while others, even from the same locality, do not. UV light is produced in long and short waves, and minerals may fluoresce in only one or the other or both.

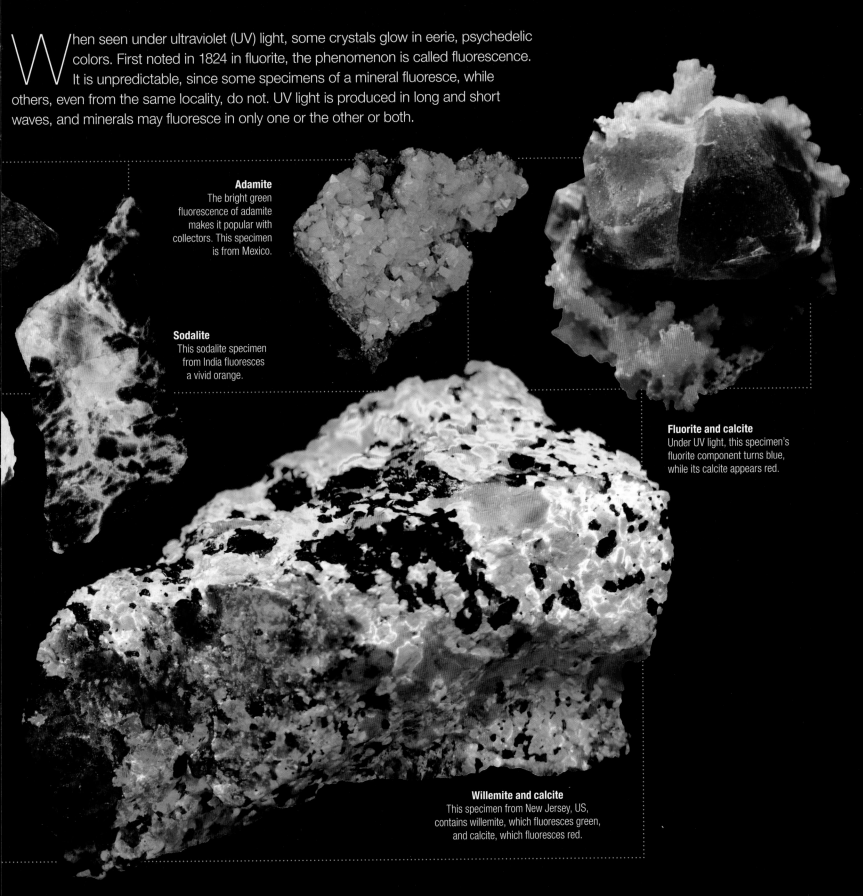

Adamite
The bright green fluorescence of adamite makes it popular with collectors. This specimen is from Mexico.

Sodalite
This sodalite specimen from India fluoresces a vivid orange.

Fluorite and calcite
Under UV light, this specimen's fluorite component turns blue, while its calcite appears red.

Willemite and calcite
This specimen from New Jersey, US, contains willemite, which fluoresces green, and calcite, which fluoresces red.

INDIAN JEWELS

For thousands of years, jewelry has played an integral part in the history of India, not only as an art form but also as a spiritual talisman, a signifier of social position, and a means of diplomatic leverage. It was also a motive for political and military conflict, especially during the age of the Mughal emperors (1526–1707). Early jewels were made from stone beads, but Hindu texts from the 1st century BCE refer to a magical jewel called the Syamantaka that originally belonged to Surya the sun god. Able to produce gold and protect whoever possessed it, the jewel sparked clashes among the nobility in their efforts to claim ownership. Historians speculate that the gem was a diamond, possibly the Koh-i-noor, which is now part of the British Crown Jewels (see pp.58–59).

Diamonds have long been the most coveted gems in Indian cultures. Hindu god Krishna is said to have given a diamond to his lover Radha, so that it would reflect her beauty on moonlit nights. Legend says that diamonds were created when lightning struck rock, and they were believed to have healing powers—for example, the wealthy would sprinkle diamond powder on their teeth in the belief it would ward off lightning strikes and prevent tooth decay.

The entire universe is suspended from me as my necklace of jewels

Sri **Krishna**
Bhagavad Gita, 5th–2nd century BCE

Detail of a painting depicting Krishna's youth, 19th century
This painting shows the young Krishna playing the flute and bedecked in jewels, even though he is only a lowly cowherd. The elaborate jewelry elevates him to divine status, the luster of the jewels outshining the sun.

Serpentine

△ **Serpentine specimen** from Lowell, Vermont, US

Serpentine is not just one mineral, but a group of at least 16 white, yellowish, green, or gray-green magnesium minerals with complex chemistry and similar appearance. Serpentine minerals generally occur as masses of tiny, intergrown crystals, and are named in allusion to their mottled appearance, which resembles a snake's skin. Gem-quality serpentine, often with a jadelike appearance, is cut *en cabochon*. Soft enough to be engraved, the mineral is also used in carvings. It is widespread, and there are huge quarrying operations in various parts of the world.

Specification

Chemical name Magnesium, iron, or nickel silicate | **Formula** $(Mg,Fe,Ni)_3Si_2O_5(OH)_4$ | **Colors** White, gray, yellow, green, green blue | **Structure** Monoclinic | **Hardness** 2.5–5.5 | **SG** 2.5–2.6 | **RI** 1.56–1.57 | **Luster** Subvitreous to greasy, resinous, earthy, dull | **Streak** White | **Locations** Worldwide

High-grade serpentine | **Rough** | The translucency of this example of fine green serpentine shows how it can easily be mistaken for jade.

Internal fractures

Rock groundmass

Serpentine with white chrysotile | **Rough** | Chrysotile is one of the minerals commonly called "asbestos," and is also one of the serpentine minerals.

Carved shell-like ridges

Seashell carving | **Color variety** | Serpentine can come in many colors, such as the light-green serpentine of this finely executed carving of a seashell.

Williamsite cabochon | **Cut** | Williamsite is one of the serpentine varieties used as gems, and produces interesting cabochons, as seen in this example.

Serpentines quarried as ornamental stones are sometimes called serpentine marble

Neolithic carving | **Carved** | Mysterious serpentine carvings such as this are found in archaeological sites across northern Britain, and date from the 2nd millennium BCE.

Fine carved details

Bowenite pendant | **Carved** | Bowenite is a variety of antigorite, a type of serpentine. Like much serpentine, it can be carved in fine detail, as demonstrated here.

Soapstone

△ **Soapstone** "chop" (seal) from Korea

Far into prehistory, soapstone was used for carvings, ornaments, and utensils. Aside from flint, it may be mankind's oldest lapidary material. Today, translucent, light-green talc soapstone carvings are widely sold in China, lacquered to improve their hardness and color, and to make them appear more like jade. The name soapstone is used to describe compact masses of various minerals that have a soapy or greasy feel, the most common of which is talc. A dense, high-purity talc called steatite is sought after for carving.

Specification

Chemical name Magnesium silicate hydrate | **Formula** $Mg_3Si_4O_{10}(OH)_2$ | **Colors** White, colorless, green, yellow to brown | **Structure** Triclinic or monoclinic | **Hardness** 2.2–2.8 | **SG** 2.8 | **RI** 1.54–1.59 | **Luster** Pearly to greasy | **Streak** White | **Locations** US, Canada, Germany, China

Uncut steatite | Rough | Steatite is a compact form of the mineral talc. The finest examples can be colorful and translucent, as in this piece of rough.

Foliation (leaflike layers of rock)

Talc specimen | Rough | This uncut soapstone from Roxbury, Connecticut, US, includes areas of compact mineral suitable for carving.

Relief carving

Steatite cup | Carved | This steatite cup from the ancient city of Ur dates back to the 3rd millennium BCE; it is carved in relief with scorpions.

Lathe-turned form

Tumbler | Carved | This ancient vessel was carved from one of the many steatites used by the craftsmen in Ur in the 3rd millennium BCE.

Incision for mouth

Rhinoceros ornament | Carved | Soapstone is still a popular material for modern tribal art, such as this polished animal sculpture from Kenya, because it can be carved and incised easily with simple tools.

From ancient to modern

Steatite in history

In the ancient Middle East, steatite was made into bowls, pots, seals, reliquaries, and statues. It absorbs and distributes heat evenly, so it was a good material for cooking utensils and smoking pipes; ancient peoples also carved molds out of steatite for metal casting. The mineral is still widely carved into bird and animal figures by the Inuit peoples of Canada and Alaska.

Inuit carving This steatite and ivory owl figurine was carved by Inuit craftsmen from Cape Dorset in northern Canada.

Pezzottaite

△ **Unusual crystal** of pezzottaite showing a trapiche structure

Pezzottaite was only formally recognized as a new mineral in 2003, having previously been considered a variety of red beryl (see pp.236–41). Although similar to the crystals found in Utah, US, its chemical elements differ, and unlike beryl, it crystallizes in the trigonal crystal system. Nonetheless, it has been marketed as raspberyl or raspberry beryl, and it ranges in color from raspberry red through to orange red and pink. Most pezzottaite gems are small, between 1 and 2 carats. About 10 percent will show chatoyancy, or "cat's-eye."

Specification

Chemical name Cesium, lithium, beryllium silicate
Formula $Cs(Be_2Li)Al_2Si_6O_{18}$ | **Colors** Raspberry red, orange red, pink | **Structure** Hexagonal, trigonal
Hardness 8 | **SG** 3.10 | **RI** 1.601–1.620 | **Luster** Vitreous
Streak White | **Locations** Afghanistan, Madagascar

Classic hexagonal form

Transparent surface

Pezzottaite rough | Rough | This fine, gem-quality, 8.40-carat pezzottaite crystal from a pegmatite found at Ambatovy, Madagascar, shows hexagonal form.

Color darkens at base

Pezzottaite crystal | Rough | Displaying a distinctive form, this pezzottaite crystal is known as an "hourglass" crystal. It features an attractive raspberry tint.

Emerald cut | Cut | A fine-quality, pinkish-lavender pezzottaite weighing 0.71 carats, this stone features an emerald cut. Due to pezzottaite's rarity, cut stones are usually small.

New discoveries

One mineral, many names

Pezzottaite is one of many recent gem and mineral discoveries. These are often new types of an already well-known gemstone, such as the transparent blue variety of zoisite (known as tanzanite—see p.253), to the discovery of entirely new minerals such as pezzottaite. Other discoveries have revealed previously unknown variations in existing materials. For example, gem tourmaline is thought to be 11 different minerals, all of which are still referred to as tourmaline in the gem trade.

Rutile needles An inclusion deep within this rose-colored pezzottaite crystal is a spray of rutile needles.

Cat's-eye | Color variety | This richly colored Madagascan cabochon weighing 3.46 carats displays superb chatoyancy—a cat's-eye effect created by internal fibers.

Madagascan pezzottaite | Cut | This 4.15-carat oval-cut gem is from Ambatovy in Fianarantosa Province, Madagascar, one of the few locations where pezzottaite is mined.

Sepiolite

△ **Rough specimen** of sepiolite (meerschaum)

Sepiolite is perhaps best known by its popular name—meerschaum, from the German for "seafoam." It is compact, earthy, claylike, and often porous. Because it is usually found in nodular masses of interlocking fibers, it has a toughness that belies its mineralogical softness. This allows it to be intricately carved, most often in the form of smoking pipes. Meerschaum sepiolite is soft when first extracted and easily carved, but hardens upon drying. The most important commercial sepiolite deposit is near Eskişehir, Turkey, where it is found as irregular nodules.

Specification

Chemical name Magnesium silicate hydrate
Formula $Mg_4Si_6O_{15}(OH)_2.6H_2O$ | **Colors** White, gray, pinkish
Structure Orthorhombic | **Hardness** 2–2.5 | **SG** 2.1–2.3
RI Opaque | **Luster** Dull to earthy | **Streak** White
Locations Turkey, US, Italy, Czech Republic, Spain

Cracked surface

Sepiolite rough | **Rough** | This sepiolite (meerschaum) specimen reveals a compact surface made up of microscopic needles that form a light material able to float on water.

Surface abrasions

Meerschaum rough | **Rough** | The compact, claylike nature of light and porous meerschaum (sepiolite) can be seen in this specimen of gray rough.

Fine detailing on carved hair

Silver rim around mouthpiece

Claw feet

Cigar holder | **Carved** | This elaborately crafted cigar holder with claw feet and silver fittings is a fine example of meerschaum carving. A porous mineral, meerschaum develops a delicate brown patina over time.

Finely carved

Meerschaum beads | **Carved** | Although meerschaum is mainly carved into pipes, it can be made into virtually anything, including jewelry, as these fine beads attest.

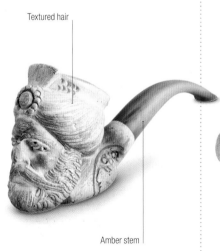

Textured hair

Amber stem

Meerschaum pipe | **Carved** | Intricately carved with the head of a bearded gentleman wearing a turban, this classic meerschaum pipe has a pipestem made from amber.

Gold T-bar, set with turquoise and diamond, for passing through buttonhole

Gold and diamond set crown

Fob seal with a gold, leaping horse

Gold chain of textured knots and turquoise- and diamond-set, barrel-shaped links

Crown with acorn and leaf decoration

Pavé-set diamond horse's head on blue Bavaria background

I want to remain an eternal mystery to myself and others

King **Ludwig II**
of Bavaria

Ludwig II's pocket watch | c.1880 | Gold, diamonds, rubies, turquoise, enamel

Ludwig II's
pocket watch

△ **The front of the watch casing**, showing the king's monogram

This 19th-century pocket watch bearing a crown and monogram on the front, and a horse's head on the reverse, was most likely made for King Ludwig II of Bavaria, an eccentric and controversial figure who died in strange circumstances.

The pocket watch was made in around 1880 and consists of gold, enamel, diamond, and ruby, with a matching chain and fob seal. An applied gold-mounted, diamond-set entwined monogram of Ludwig II, surmounted by his crown, adorns the front of the case. On the reverse of the watch is a silver-mounted, pavé-set diamond horse's head, with ruby-set eye: this elaborate

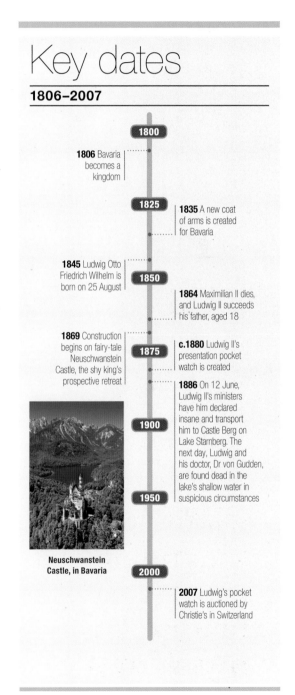

Ludwig II of Bavaria, who ruled the country from 1864 until his deposition in 1886

decoration acknowledges Ludwig's passion for horses. The gold-link chain of textured knots and four turquoise- and diamond-set, barrel-shaped links, representing the Bavarian coat-of-arms, terminates in a fob seal of a gold horse leaping from a turquoise cabochon.

Ludwig II was just 18 years old when he took the throne. Young and good-looking, he was popular in Bavaria, but was little concerned with affairs of state, instead commissioning fanciful constructions such as the fairy-tale Neuschwanstein Castle. His ministers deposed him by having him declared mentally unsound in 1886, and Ludwig was removed to Castle Berg on Lake Starnberg on 12 June. The next day, his body was found in shallow water with that of his minder, Dr. von Gudden. His watch had stopped at 6:54pm. Although suicide by drowning was reported as the cause of death, no water was found in his lungs, and his death remains unexplained. This watch—not the one present at his death—was auctioned in 2007.

Key dates

1806–2007

1800

1806 Bavaria becomes a kingdom

1825

1835 A new coat of arms is created for Bavaria

1845 Ludwig Otto Friedrich Wilhelm is born on 25 August

1850

1864 Maximilian II dies, and Ludwig II succeeds his father, aged 18

1869 Construction begins on fairy-tale Neuschwanstein Castle, the shy king's prospective retreat

1875

c.1880 Ludwig II's presentation pocket watch is created

1886 On 12 June, Ludwig II's ministers have him declared insane and transport him to Castle Berg on Lake Starnberg. The next day, Ludwig and his doctor, Dr von Gudden, are found dead in the lake's shallow water in suspicious circumstances

1900

1950

Neuschwanstein Castle, in Bavaria

2000

2007 Ludwig's pocket watch is auctioned by Christie's in Switzerland

The music room at Neuschwanstein Castle, the lavish 19th-century castle, still open to visitors today, that was one of Ludwig II's pet projects and where he intended to hide from the public eye

Chrysocolla

△ **Opalized** chrysocolla cabochon

The term chrysocolla was first applied by the Greek philosopher Theophrastus in 315 BCE (see box, below) to various materials used in soldering gold, derived from the Greek *chrysos*, meaning "gold," and *kolla*, meaning "glue." Chrysocolla forms as a decomposition product of other copper minerals, mainly in arid regions, and is often intergrown with harder minerals such as quartz, chalcedony, or opal, yielding a more resilient gemstone variety. It is usually cut *en cabochon*, and translucent, richly blue-green chrysocolla is particularly prized.

Specification

Chemical name Copper hydrosilicate | **Formula** $Cu_2H_2(Si_2O_5)(OH)_4.nH_2O$ | **Colors** Blue, blue green | **Structure** Orthorhombic | **Hardness** 2–4 | **SG** 2.0–2.4 | **RI** 1.46–1.57 | **Luster** Vitreous to earthy | **Streak** Pale blue, tan, gray | **Locations** UK, Israel, Mexico, Czech Republic, Australia, DR Congo, US

Blocky surface

Uncut chrysocolla | Rough | This freshly broken chrysocolla rough has a granular outside surface, probably concealing fine gem material within.

Egg form | Carved | This ornament cut from chrysocolla stands 17½ in (44.5 cm) tall, potentially making it the world's largest egg carved from this material.

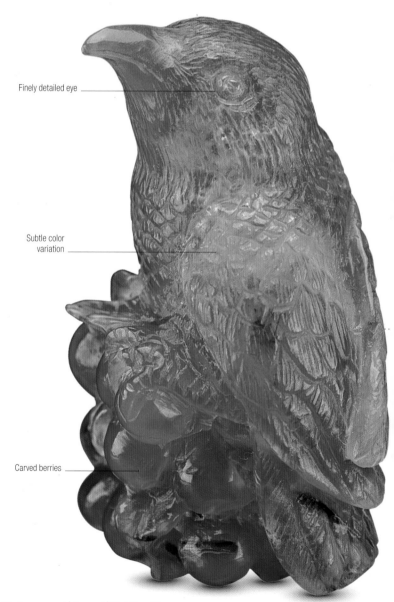

Finely detailed eye

Subtle color variation

Carved berries

Bird ornament | Carved | Weighing 69 carats, this carving by Ronald Stevens of a bird resting on a cluster of berries is one of the all-time finest chrysocolla pieces.

Theophrastus

The man who first described chrysocolla

Theophrastus came to Athens at a young age and studied under Plato. He is often said to be the father of botany due to his work on plants—however, he carried out equally important studies of minerals, and his treatise *On Stones* was used by mineralogists until the Renaissance. He was the first scholar to attempt a systematic classification of gems and minerals and, although his work was superseded, he can be considered the forerunner of modern, scientific mineralogy.

Theoprastus A pupil of Plato, he was the first person in the ancient western world to write about minerals and rocks.

Petalite

△ **Fine, cushion** mixed-cut petalite

Petalite's name comes from the Greek word for "leaf," a reference to its tendency to break into thin, leaflike layers. It is usually found as masses of small crystals and, rarely, as individual crystals. In its massive form it is cut *en cabochon*, and its colorless, transparent crystals are faceted for collectors only. Because it is brittle and easily split, it requires extreme care in the faceting process, and it is too fragile to wear. Facet-grade petalite is found principally in Brazil, which produces collectors' gemstones of up to 50 carats in weight.

Specification

Chemical name Lithium aluminum silicate | **Formula** $LiAlSi_4O_{10}$
Colors Colorless to grayish white, pink, green | **Structure** Monoclinic | **Hardness** 6–6.5 | **SG** 2.4 | **RI** 1.50–1.52
Luster Vitreous | **Streak** White | **Locations** Brazil, Sweden, Italy, Russia, Australia, Zimbabwe, Canada

Irregular surface

Unusual form | **Rough** | This striking grouping consists of several petalite crystals, naturally acid etched, but retaining some gem material.

Cabochon grade | **Rough** | This rough specimen will be sliced and polished into cabochons since the raw material needs to be of higher quality for faceting.

Polished petalite | **Carved** | In an unusual decision, a lapidary has shaped this petalite specimen into a baroque form, and then polished it to a smooth finish.

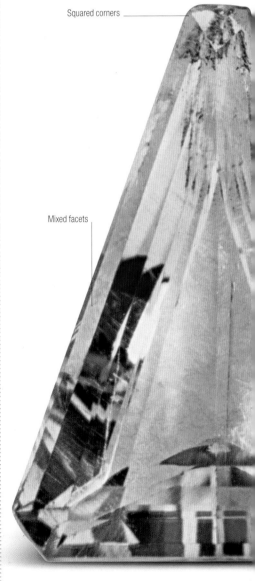

Squared corners

Mixed facets

Curved edges

Color change | **Color variety** | This rare 6.22-carat stone from Myanmar changes from olive green in sunlight to fiery red in incandescent light.

The chemical element lithium was first discovered in petalite, which is still an important ore of the element

Table facet

Burmese petalite | **Cut** | This flawless 25.20-carat smoky brown, mixed-cut cushion petalite, also from Myanmar, is part of the Smithsonian Institution's gem collection.

Mixed cut | **Cut** | The corners of this yellow, triangular, mixed-cut petalite have been squared off to prevent breakage of the brittle material.

Prehnite

△ **Grapelike prehnite crystal clusters** on a rock groundmass

Prehnite is usually found as globular, spherical, or stalactitic aggregates of fine to coarse crystals. Rare individual crystals tend to be short and stumpy, and have square cross sections, while some pale yellowish-brown fibrous material shows a cat's-eye effect when cut *en cabochon*. Prehnite is sometimes faceted, but the stones are almost always translucent rather than transparent; faceted stones tend to be small and are cut only for collectors and museums. Semitransparent prehnite comes from Australia and Scotland, where the occasional almost transparent piece is found.

Specification

Chemical name Calcium aluminum silicate | **Formula** $Ca_2Al_2Si_3O_{10}(OH)_2$ | **Colors** Green, yellow, tan, white **Structure** Orthorhombic | **Hardness** 6–6.5 | **SG** 2.8–2.9 **RI** 1.61–1.67 | **Luster** Vitreous | **Streak** White | **Locations** Canada, Portugal, Germany, Japan, US, Australia, Scotland

Gemmy crystals

Prehnite crystals | Rough | This specimen consists of a number of single crystals of prehnite of a yellowish color, resting on a groundmass of rock.

Low dome

Square cabochon | Color variety | Prehnite is not often found in light blue but, as can be seen in this translucent square cabochon, it can be a subtle pastel color.

Color gradation

Step-cut stone | Cut | Facet-grade prehnite is relatively rare, and is still somewhat cloudy when cut, as can be seen in this step-cut stone.

Rounded edges of table facet

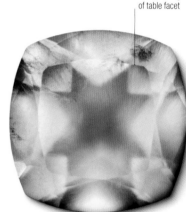

Square cushion | Cut | The cloudiness of this green, square-cut faceted prehnite gemstone lends it a mystical appearance.

Colonel Hendrik von Prehn

The man who discovered prehnite

Prehnite was named after Colonel Hendrik von Prehn (1733–85), and was first described in 1788 at Cradock, Eastern Cape Province, South Africa. Information on Colonel Prehn is scarce, but he is listed as commander of the military forces of the Dutch colony at the Cape of Good Hope from 1768 to 1780, and as the governor of the Cape Colony. He is also the reputed discoverer of the mineral that bears his name.

Battle off the Cape of Good Hope South Africa was the site of Anglo-Dutch military rivalry in the 17th and 18th centuries.

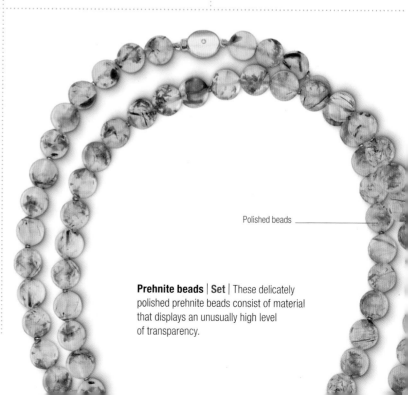

Polished beads

Prehnite beads | Set | These delicately polished prehnite beads consist of material that displays an unusually high level of transparency.

Phosphophyllite

△ **Piece of superb-colored**, facet-grade phosphophyllite rough

Phosphophyllite is a rare mineral and an even rarer gemstone, highly prized by museums and collectors. Crystals with a delicate bluish-green color are the most sought after. Phosphophyllite is brittle and fragile, and can be faceted only with the greatest difficulty. It is rare as a gem partly because crystals that are large enough to be cut are also too valuable to be broken. The finest crystals, and the ones that provided most of the existing faceted stones, came from the deposit at Potosí, Bolivia, which is now exhausted.

Specification

Chemical name Hydrated zinc phosphate | **Formula** $Zn_2(Fe^{2+}Mn^{3+})(PO_4)_2 \cdot 4H_2O$ | **Colors** Colorless to deep bluish green | **Structure** Monoclinic | **Hardness** 3–3.5 | **SG** 3.08–3.10 | **RI** 1.59–1.62 | **Luster** Vitreous | **Streak** White | **Locations** US, Australia, Germany, Bolivia

Pyrite groundmass

Gemmy crystals

Crystals | Rough | This specimen features a cluster of light-blue phosphophyllite crystals resting on a groundmass of pyrite. The gemmy crystals are of fine quality and would make good faceting material.

Plane of fracture

High-quality phosphophyllite | Rough | Facet-grade phosphophyllite of this transparency and turquoise color is rare, and will make a first-rate collector's gem.

Exceptional color | Color variety | These phosphophyllite crystals in parallel growth are a stunningly rare color, and will be sought after by serious gem cutters.

Crown facets

Emerald cut | Cut | This light-blue-green phosphophyllite gem faceted in an emerald cut has unusual clarity, adding to its value.

Emerald cut | Cut | An unusually transparent phosphophyllite is faceted here into an emerald step cut to emphasize its color and high transparency.

Phosphophyllite is named after the Greek words for "phosphorus-bearing" and "cleavable"

THE GEM INDUSTRY

For thousands of years, diamonds were incredibly rare, but the discovery and mining of huge diamond deposits in South Africa in 1870 not only changed the availability of the coveted rocks, but marked the beginnings of the modern gem industry. Thanks to these mines, world diamond production exceeded 1 million carats per year for the first time in 1871; by 1907 it had reached 5 million carats, and it steadily increased throughout the century, hitting 126 million carats in 2000. Supplies were overflowing as new mines in Angola, DR Congo, West Africa, Botswana, Russia, Australia, and Canada contributed to rising output.

Abundant supply should have driven prices down, but the diamond producers acted quickly to maintain the value of their precious commodity. The British conglomerate behind the South African mines formed the De Beers cartel in 1888 to control every aspect of the new diamond trade, from production to marketing. The gem trade as it now exists was born as a result, with market prices dictated by the diamond cartel's strict control of supply and demand, while De Beers' slogan, "A Diamond is Forever" also helped to make diamonds synonymous with love and marriage.

> **There was no brand name to be impressed on the public mind. There was simply an idea**
>
> N. W. **Ayer**
> De Beers Advertising Agency

Workers sort diamonds at the De Beers' mine in South Africa
Before 1870, diamonds came from riverbed deposits, mostly in India, Indonesia, and Brazil, but the De Beers mines were primary deposits, yielding diamonds directly from their source in the Earth.

Enstatite

△ **Piece of massive enstatite** suitable for tumble-polishing

While it is an important rock-forming mineral, enstatite's major commercial use is as a gemstone, in colorless, pale yellow, or pale green hues. The most popular color is the emerald-green variety, known as chrome-enstatite, and its green color is caused by traces of chromium—hence the name. All colors are relatively rare, and gemstones are faceted or cut *en cabochon*. Mysore, India, produces star-enstatite; Canada produces iridescent enstatite; and the gem gravels of Myanmar and Sri Lanka yield good-quality facet-grade material.

Specification

Chemical name Magnesium silicate | **Formula** $Mg_2Si_2O_6$
Colors Colorless, yellow, green, brown, black | **Structure** Orthorhombic | **Hardness** 5–6 | **SG** 3.2–3.3 | **RI** 1.66–1.67
Luster Vitreous | **Streak** Gray to white | **Locations** India, Canada, Myanmar, Sri Lanka

Enstatite crystal | Rough | Enstatite is rarely found in well-formed crystals such as this, a large specimen that shows the mineral's prismatic form.

Plane of fracture

Rough specimen | Rough | This piece of enstatite demonstrates its occurrence as a rock-forming mineral. It could be cut and polished, but would not yield a gem.

Intergrown enstatite crystals

Polished surface

Polished enstatite pebble | Cut | This tumble-polished baroque of enstatite shows the multicolored patterning—in this case in a patchworklike formation—that sometimes occurs when enstatite is found in conjunction with other minerals.

Enstatite in space

Minerals of the early solar system

The presence of enstatite in certain meteorites implies that it was one of the first-formed silicate minerals in the solar nebula that created the Earth and solar system. It occurs in about 10 percent of meteorites known as chondrites. These are agglomerations of mineral grains that never became part of large enough bodies to undergo melting and recrystallization. It is thought that they were first formed near the center of the solar system.

Stony iron meteorite Known as chondrites, this variety of iron meteorite sometimes carries quantities of enstatite.

Cat's-eye cabochon | Cut | Enstatite is one of the minerals that will sometimes produce a star or "cat's-eye" when polished *en cabochon*, as in this example.

Oval brilliant | Color variety | This fine-quality, highly transparent yellow enstatite gemstone has been faceted in a modified brilliant oval cut.

Diopside

△ **Emerald-cut** diopside gem

Diopside is one of the minerals found alongside diamonds in some kimberlites, an igneous rock, generally occurring as dark bottle-green, light green, brown, blue, or colorless stones. Rich green diopside, colored by chromium and known as chrome diopside, is faceted as a prized collector's gem, and would potentially rival emerald in popularity but for its softness. Stones of another rich hue, a violet blue colored by manganese, are found in Italy and the US—these are sometimes called violane and are highly prized collector's gems.

Specification

Chemical name Calcium, magnesium silicate | **Formula** $CaMg(Si_2O_6)$ | **Colors** White, pale to dark green, violet blue
Structure Monoclinic | **Hardness** 5–6 | **SG** 3.2–3.4
RI 1.66–1.72 | **Luster** Vitreous | **Streak** White to pale green
Locations Italy, US, Myanmar, Austria, Canada, Pakistan, Sri Lanka

Diopside crystals

Diopside crystals | **Rough** | This specimen consists of a number of prismatic, green crystals of diopside embedded in a groundmass of quartz.

Violane | **Rough** | Rich purple diopside is sometimes called violane, and is shown here in massive form suitable for cutting into cabochons.

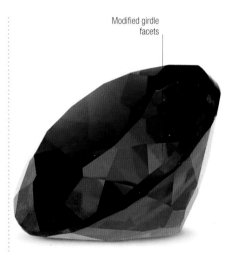

Modified girdle facets

Modified brilliant | **Cut** | Viewed side-on, this richly colored diopside can be seen to have modified girdle facets on the crown, and main facets on the pavilion.

Chrome diopside | **Color variety** | This deep green chrome diopside gemstone features a particularly fine color, emphasized by its clean emerald cut.

Fibrous diopside material can be cut **into** cabochons to show cat's-eye patterning

Interesting flaws | **Cut** | Although this long, rectangular step-cut diopside features a number of flaws, they add to the character and interest of the gemstone.

White gold setting

Ring | **Set** | Dark green diopside is considered the most desirable color, but lighter green, as in these matched stones set in a ring, has its own beauty.

Hypersthene

△ **Banded hypersthene rough** with crystal striations

he name **hypersthene is** still retained as a gemstone variety, even though it has been largely abandoned as a mineral name (the mineral is now considered a middle member of a related series of silicates called pyroxenes). Hypersthene is commonly gray, brown, or green in color. As a gem, it is noted for its copper-red iridescence, caused in part by inclusions of the minerals hematite and goethite. It is most frequently cut *en cabochon* because it can be too dark to facet. When faceted, stones are often somewhat cloudy, although intense in color.

Specification

Chemical name Magnesium, iron silicate | **Formula** (Mg,Fe) (Si_2O_6) | **Colors** Gray, brown, green | **Structure** Orthorhombic **Hardness** 5.5 | **SG** 3.35 | **RI** 1.65–1.67 | **Luster** Vitreous **Streak** White | **Locations** India, Germany, Norway, Greenland

Double-terminated crystal

Rare, high-quality crystal | Rough | Hypersthene occasionally forms well-developed crystals, such as this double-terminated crystal resting on a rock groundmass.

Striated surface

Hypersthene rock | Rough | Hypersthene in its rough form, as here, can resemble the mineral hornblende, and so the two are often confused. Hypersthene is the harder mineral.

Tumble polish | Cut | Like all gemstones, hypersthene comes in a number of grades. The lowest-grade material is often tumble polished, as in this example.

Once a gem, always a gem

Pyroxene or hypersthene?

As scientific instrumentation evolves, previously unknown chemical nuances in minerals are revealed. This can affect the naming conventions that apply to them, particularly in terms of the different requirements of science and commerce. For example, now that hypersthene is known to be a member of the mineral series pyroxene, it no longer technically needs a separate name. To the commercial market, however, "hypersthene" is a well-known gem name, and remains in use.

Scientific analysis Close examination of hypersthene has led to it being mineralogically reclassified.

Faceted hypersthene | Cut | This rare, step-cut hypersthene shows good color and clarity, but characteristically lacks brilliance. The corners are rounded to prevent chipping.

Shaped cabochon | Carved | This leaflike carving of the hypersthene variety bronzite glows from platy inclusions (flat, thin mineral crystals), giving it a subtle metallic sheen.

Bronzite

△ **Tumble-polished bronzite** with typical bronze coloration

Bronzite is a member of the mineralogical chemical series that includes hypersthene. The color of bronzite is green or brown, with schiller (a metallic sheen) that gives it a bronzelike appearance. For gemstone use, it is usually cut *en cabochon*, and it is also carved for small ornamental items. Some bronzite has a fairly distinct fibrous structure, and when this is pronounced, the sheen has a resemblance to cat's-eye (see pp.84–85). The general use of bronzite as a gemstone is less extensive than that of its mineralogical cousin, hypersthene.

Specification

Chemical name Iron, magnesium silicate | **Formula** $(Mg,Fe)(Si_2O_6)$ | **Colors** Green, brown, bronze | **Structure** Orthorhombic | **Hardness** 5.5 | **SG** 3.35 | **RI** 1.65–1.67 | **Luster** Vitreous | **Streak** White to gray | **Locations** India, Germany, Norway, Greenland

Bronze coloration

Bronzite rock | **Rough** | The characteristic bronze sheen that gives the stone its name can be seen on the surface of this specimen of bronzite rough.

Tumbled bronzite | **Cut** | Tumble polishing is a popular means of revealing the bronzelike appearance of the stone, as seen in this piece.

Smooth pebble | **Cut** | The best coloration in bronzite is not always evident from the rough. This specimen has been polished to a smooth finish, revealing its rich color.

Bronzite is named for its color, which resembles polished bronze

Bronzite cabochon | **Cut** | This rectangular bronzite cabochon has been cut with rounded corners to help avoid chipping. The high dome and smooth surface reveal the bronzelike character and translucency of the material.

Corners rounded off

Pattern, texture, and
inclusions

Some gems grow in patterns that have a natural beauty, enhanced by a skillful cutter. The many varieties of quartz, including agates and chalcedonies such as jasper, are particularly renowned for their intrinsic decorative qualities. Other species of gem have unusual surface texture or inclusions within the stone that add beauty.

Apatite crystal
This large apatite crystal displays a number of small incusions in its interior, giving it a fractured appearance.

Stained agate
This cabochon of agate has been stained to highlight the patterns created by its mineral inclusions.

Baryte crystals
This cluster of tabular baryte crystals displays a jagged pattern and a waxy texture on its surfaces.

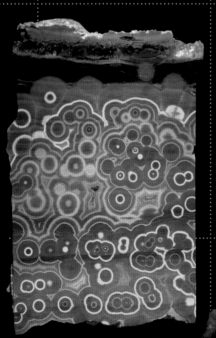

Orbicular agate
This stunning ring-shaped pattern features an "eye" in the center of each small circle, seen here in a colorful cross-sectional polished slice of agate.

Landscape agate
In this distinctive naturally occurring pattern, images resembling a scene from nature are created by iron-oxide dendrites.

Shell
This striking spiral shell shows perfect geometry in its concentric circles.

Rhodochrosite
This pebble displays natural striations with a banded appearance.

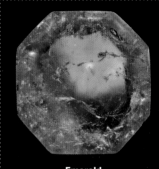

Emerald
This cut stone features the kind of small inclusions and internal cracks often found in emeralds.

Jasper
This specimen's reddish pattern comes from presence of hematite inclusions in varying amounts.

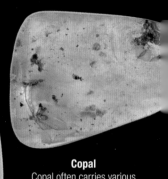

Copal
Copal often carries various small plant or animal inclusions, usually insects or leaves.

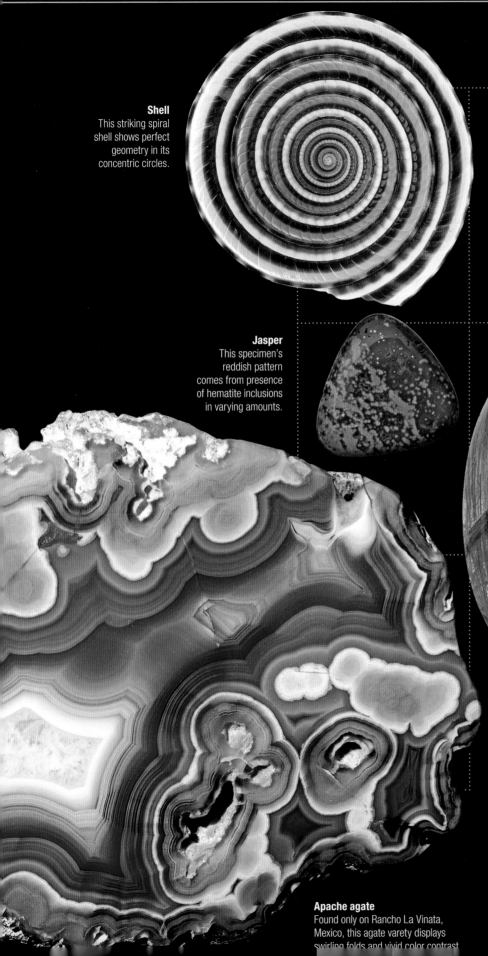

Rutilated quartz
This transparent quartz cabochon has a number of fine rutile needle inclusions, contrasting with its smooth, polished exterior.

Apache agate
Found only on Rancho La Vinata, Mexico, this agate varety displays swirling folds and vivid color contrast

Hiddenite

△ **Emerald-cut gem** of the spodumene variety hiddenite

his green variety of the lithium mineral spodumene was discovered by a geologist who had been commissioned by the inventor Thomas Edison to search for any sources of platinum in North Carolina, US. Mining began in the 1890s at the discovery site, where hiddenite occurred alongside emerald and was, for a time, called "lithia emerald." Hiddenite crystals are small, rarely exceeding 1 in (25 mm) in length, and are strongly pleochroic, showing green, bluish green, and yellowish green when viewed from different directions.

Specification

Chemical name Lithium aluminosilicate | **Formula** LiAl(Si$_2$O$_6$)
Colors Green, blue green, yellow green | **Structure** Monoclinic
Hardness 6.5–7 | **SG** 3.0–3.2 | **RI** 1.66–1.68 | **Luster** Vitreous | **Streak** Colorless | **Locations** US, Brazil, China, Madagascar

Elongated crystal | Rough | While emerald green is the preferred color for hiddenite, lighter green crystals, as here, can also be faceted for jewelry.

Hiddenite in gneiss | Rough | Crystals have grown here in a gneiss matrix, although hiddenite is usually found in pegmatites (coarse-grained quartz and feldspar mixes).

Vitreous luster

Gem-quality crystal | Rough | This hiddenite rough is transparent and of a fair green color. It is likely to be cut into an elongated stone (see right).

Pointed oval shape called navette

Navette cut | Cut | Hiddenite crystals tend to be elongated, lending themselves to fancy cuts like this 4.96-carat navette cut with a mix of triangular and rectangular faces.

The tiny settlement of White Plains, North Carolina, was renamed Hiddenite after the mineral was discovered there

Elongated gem | Cut | The clean lines of the facets emphasize the pale blue-green coloring of this rectangular step-cut hiddenite gem.

Faceted rectangle | Cut | This 31.60-carat round-cornered rectangular cut stone from Afghanistan displays excellent luster and clarity, and is unusual for its large size.

Kunzite

△ **Scissors-cut**, 17-carat gem of the spodumene variety kunzite

Kunzite is the pink variety of the mineral spodumene, and is named after the American gemologist G. F. Kunz, who first described it in 1902. Faceting-grade kunzite shows strong pleochroism—two different shades of the body color when viewed from different directions. Gems must be carefully oriented to show the best color through the top surface; additionally, kunzite tends to be splintery, and slivers are likely to break off during cutting if the stone is not oriented correctly. Kunzite and other spodumene gems are almost always faceted.

Specification

Chemical name Lithium aluminosilicate | **Formula** $LiAl(Si_2O_6)$ | **Color** Pink | **Structure** Monoclinic | **Hardness** 6.5–7 | **SG** 3.0–3.2 | **RI** 1.66–1.68 | **Luster** Vitreous | **Streak** White | **Locations** Afghanistan, Brazil, Madagascar, US

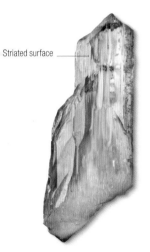

Striated surface

Gem-grade kunzite crystal | Rough | This piece of kunzite rough displays excellent color and transparency, and can be cut into a fine and valuable gem.

Elongated crystal | Rough | This crystal is 4¼ in (11 cm) long, and it can be cut into a desirable stone. The color is light and delicate rather than intense.

Step cut

Fancy heart shape | Cut | The finest kunzite, when cut by a master cutter, produces some of the finest gems, as this high-quality specimen demonstrates.

Diamond surround

Kunzite ring | Set | Kunzite gems are often set among colorless diamonds to emphasize their color—here, the faceted stone is surrounded by white gold and diamonds.

Kunzite earrings | Set | Set with diamonds and smaller pink sapphires in 18-karat white gold, these pendant earrings feature two oval-cut kunzites each.

Picasso necklace | Set | Designed by Paloma Picasso, this necklace of baroque pearls features a dazzling 396.3-carat kunzite from Afghanistan mounted in gold and accentuated by diamonds.

Baroque pearls

18-karat gold

Two thumb-sized sapphires, one 50.8 and the other 42.45 carats, form the clasp

A style icon who... wore so many jewels they weighed her tiny body down

Daily **Mail** on Daisy Fellowes

Thirteen briolette-cut blue sapphires border the necklace

Diamond "stems" sprout "leaves" and "fruit" made of emeralds, sapphires, and rubies

Crescent "bib" design

Tutti Frutti necklace by Cartier | 1936 | Length (open) 17in (43cm) | Briolette-cut sapphires, diamonds, emeralds, rubies, platinum, white gold

Tutti Frutti
necklace

△ **Detail** of a briolette-cut sapphire from the necklace

This necklace is one of the most spectacular pieces of Indian-inspired jewelry that Cartier produced. Colorful and exotic, the necklace, which the jewelry house designed for the French-American socialite and heiress Daisy Fellowes in 1936, heralded an exciting new look that came to epitomize the glamour and opulence of the Jazz Age. At the time, this style of necklace was known as a *collier hindou* ("Hindu necklace"), but in the 1970s, Cartier rebranded the range "Tutti Frutti" ("All Fruits"), to highlight the resemblance of the colourful cut gems to berries, leaves, and blossoms.

Cartier forged its cultural connection with India in 1901, when Queen Alexandra (wife of King Edward VII) commissioned

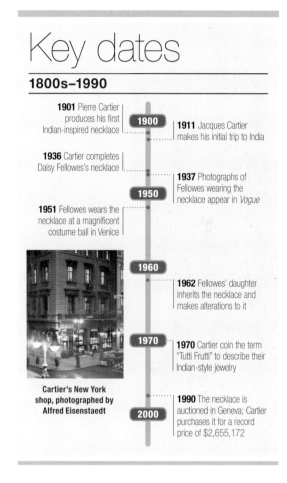

Sarpech clip in the Indian style by Van Cleef & Arpels

Pierre Cartier to design a necklace for her to complement three Indian gowns that she had been given. A decade later, Pierre's brother Jacques travelled to the Indian subcontinent, partly to witness the coronation of King George V, but also to make contact with several maharajas, who wanted their own jewels set with Gallic flair.

These Indian pieces created a sensation in the West, with Cartier's clients queuing up for similar items—Daisy Fellowes's necklace was loosely based on one that Cartier had designed for the Maharaja of Patna. Daisy was famed in society circles at the time, with gossip columnists lingering over details of her scandalous affairs and extravagant lifestyle, but she was also admired as a genuine fashion icon. For her *collier hindou*, Cartier recycled 785 gems from a necklace and two bracelets that she already owned, and added a further 238 diamonds and eight rubies. Completed in 1936, the necklace was a resounding triumph.

Key dates
1800s–1990

1901 Pierre Cartier produces his first Indian-inspired necklace

1900

1911 Jacques Cartier makes his initial trip to India

1936 Cartier completes Daisy Fellowes's necklace

1950

1937 Photographs of Fellowes wearing the necklace appear in *Vogue*

1951 Fellowes wears the necklace at a magnificent costume ball in Venice

1960

1962 Fellowes' daughter inherits the necklace and makes alterations to it

1970

1970 Cartier coin the term "Tutti Frutti" to describe their Indian-style jewelry

Cartier's New York shop, photographed by Alfred Eisenstaedt

2000

1990 The necklace is auctioned in Geneva; Cartier purchases it for a record price of $2,655,172

Daisy Fellowes wearing the Tutti Frutti necklace at a celebrated costume ball in Venice in 1951, where she masqueraded as the "Queen of Africa"

Jewelry that gleams with wicked memories

New York **Times** on the Tutti Frutti necklace

Jade

△ **Polished jadeite stone** in its most typical color

There are two distinct minerals both called jade—nephrite and jadeite. They have very different textures: jadeite is made of interlocking, blocky, granular crystals whereas nephrite is fibrous. Nephrite comes only in cream and shades of green, while jadeite comes in many other colors, and its pure form is white. The most valuable is emerald green, which is colored by chromium and known as Imperial Jade. The name jade comes from the Spanish *piedra de hijada*, "loin stone," named in the belief that it cured kidney ailments.

Specification (jadeite)

Chemical name Sodium, iron, aluminium silicate
Formula $Na(Al,Fe)Si_2O_6$ | **Colors** White, green, lavender, pink, brown, orange, yellow, red, blue, or black | **Structure** Monoclinic
Hardness 6–7 | **SG** 3.2–3.4 | **RI** 1.66–1.68 | **Luster** Vitreous to greasy | **Streak** White | **Locations** Myanmar, Japan

Rough

Classic color mottling

Sawn block | This jadeite block is translucent with good internal color patterning and a typical sugary, granular texture. Typically, a block in this condition would be in a fit state for a lapidary to carve into decorative shapes or cut up into cabochons.

Veins of lighter material

Slice of veined jadeite | The white streaks in this dark jadeite create an interesting contrast. The green coloring comes from the presence of iron.

Cut

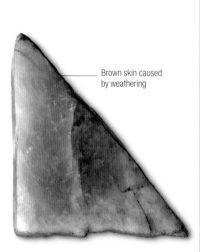

Jadeite oval cabochon | The elongated cut and smooth dome of this oval cabochon emphasizes its translucency and blue-green color cast.

Olmec jade

Jadeite or nephrite

The Olmecs were the first of the Mesoamericans (native people of Mexico and Central and South America) to discover and carve jadeite, but until the late 16th century, virtually all European jade was nephrite. The Spanish discovered that the Aztecs of Mexico prized a green stone thought to be the same. In 1863, a jade carving was analyzed and found to be different. It was named jadeite. It had a similar cultural value for Mesoamericans as nephrite in China, and was more prized than gold.

Votive ax-head This Olmec jadeite ax-head was carved between 1200 and 400 BCE.

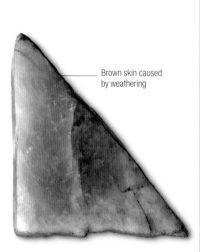

Brown skin caused by weathering

Polished jadeite slice | The lavender of this example (caused by manganese) is among the most prized of jadeite colors. The brown skin can be incorporated into carvings.

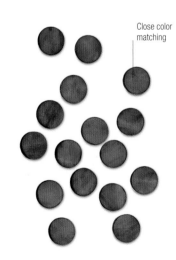

Close color matching

Nephrite cabochons | The round cabochons in this group are well matched for size and color, so that they can be mounted together in a jewelry setting.

Settings

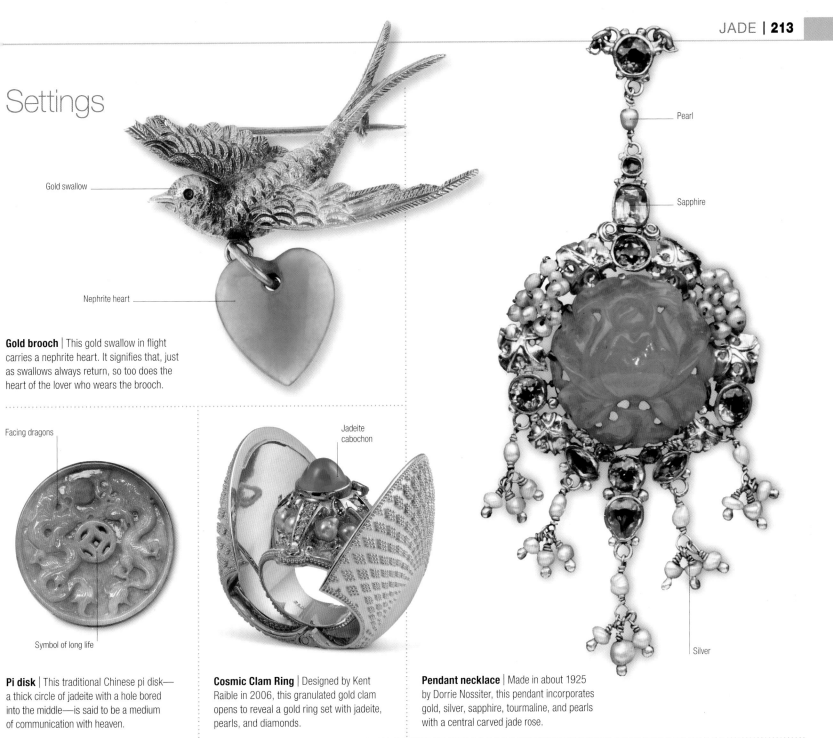

Gold swallow

Nephrite heart

Gold brooch | This gold swallow in flight carries a nephrite heart. It signifies that, just as swallows always return, so too does the heart of the lover who wears the brooch.

Facing dragons

Symbol of long life

Pi disk | This traditional Chinese pi disk— a thick circle of jadeite with a hole bored into the middle—is said to be a medium of communication with heaven.

Jadeite cabochon

Cosmic Clam Ring | Designed by Kent Raible in 2006, this granulated gold clam opens to reveal a gold ring set with jadeite, pearls, and diamonds.

Pearl

Sapphire

Silver

Pendant necklace | Made in about 1925 by Dorrie Nossiter, this pendant incorporates gold, silver, sapphire, tourmaline, and pearls with a central carved jade rose.

Oval cabochon shape

Diamonds

Nephrite ring | This white-gold man's ring is mounted with a cabochon of mid-toned nephrite and is framed by diamond-set shoulders.

Classic color mottling

Translucent surface

Nephrite bowl | The carving of this thin-walled bowl shows the color mottling within the more limited color range—usually shades of green—typical of nephrite.

Better to be shattered jade than unbroken pottery

Chinese proverb

Portrait of a woman with children, pets, and birdcage | Early 18th century, Qing Dynasty (1644–1912)

Chinese **birdcage**

△ **Ch'ien Lung**, 18th-century Emperor of China

This ornate antique birdcage from China, festooned with treasures, is more decorative than practical. Featuring finely carved materials and precious stones, it would have introduced an airy aesthetic to a wealthy home.

The cage is 24¾ in (63 cm) high with a diameter of 13 in (33 cm), and is constructed from carved wood. The base is lacquered ebony inlaid with bone and ivory, and the cage features carved ivory details, porcelain water and food bowls, and amber and jade decorations. The cage was primarily produced in the era of Chinese ruler Ch'ien Lung (1735–96), with later additions around 1880–1910.

Around the 17th century, bird keeping was in vogue, and by the 18th century, birdcages were lavish interior design pieces indicating wealth and status. Bird keeping

Bird owner in Beijing, China, taking his bird out in its cage c.1930s–40s, photographed by German documentary photographer Hedda Morrison

dates back to 300 BCE in China, and was a popular pastime among Chinese nobility. Songbirds were particularly valued, and some cages were intended to accompany their owner around the house providing a pleasant soundtrack, much like a modern-day stereo. Some owners took this a step further, and the ornate cage and bird were taken outside for a "walk," a custom that continues today. The cage is swung lightly to encourage the bird to cling to its perch, exercise that helps to maintain its plumage.

In China, "birdcage holder" was a derogatory term for an idle person

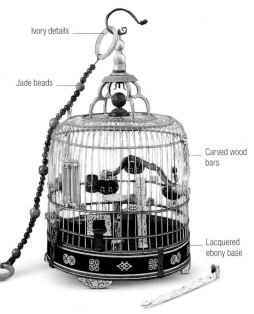

Ivory details

Jade beads

Carved wood bars

Lacquered ebony base

Qing dynasty birdcage, thought to have been made around 1735–96, decorated with amber, jade, ivory, bone, and lacquered ebony

Key dates
1735–1916

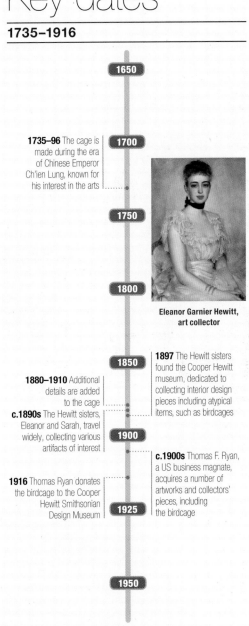

1650

1735–96 The cage is made during the era of Chinese Emperor Ch'ien Lung, known for his interest in the arts

1700

1750

Eleanor Garnier Hewitt, art collector

1800

1850

1897 The Hewitt sisters found the Cooper Hewitt museum, dedicated to collecting interior design pieces including atypical items, such as birdcages

1880–1910 Additional details are added to the cage

c.1890s The Hewitt sisters, Eleanor and Sarah, travel widely, collecting various artifacts of interest

1900

c.1900s Thomas F. Ryan, a US business magnate, acquires a number of artworks and collectors' pieces, including the birdcage

1916 Thomas Ryan donates the birdcage to the Cooper Hewitt Smithsonian Design Museum

1925

1950

Rhodonite

△ **Tumble-polished** rhodonite gemstone

Rhodonite takes its name from the Greek *rhodon*, meaning "rose." It is a source of manganese and is relatively widespread, but it is usually mined as a semiprecious gem and ornamental stone. It is typically a pink color; however, material streaked with black veins is also favored by carvers and cutters. Massive rhodonite is relatively tough, and so is excellent for carving; it is primarily cut *en cabochon* for gems and as beads. Transparent crystals are sometimes found, but they must be faceted with great care, and are strictly for collectors.

Specification

Chemical name Manganese, calcium silicate | **Formula** (Mn,Fe,Mg,Ca)SiO₃ | **Colors** Pink to rose red | **Structure** Triclinic | **Hardness** 5.5–6.5 | **SG** 3.4–3.7 | **RI** 1.72–1.76 | **Luster** Vitreous | **Streak** White | **Locations** Brazil, Canada, Sweden, Russia, England, US

Rhodonite crystals | **Rough** | In this specimen, well-developed rhodonite crystals rest on a rock groundmass; crystals such as these are rare.

Granular surface

Rhodonite rough | **Color variety** | This rough specimen of finest-quality rhodonite displays a typical granular form and a superb color.

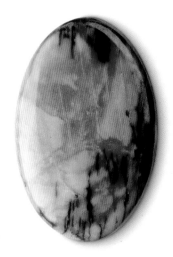

Rhodonite cabochon | **Color variety** | Material with black veins and streaks, as seen in this cabochon, is favored by some cutters over the solid rose color.

Finely carved, smooth surface

Quartz layering

Rhodonite box | **Carved** | This spectacular decorative box has been carved from rhodonite that is black streaked and interlayered with inclusions of quartz.

Rhodonite bear | **Carved** | This black-streaked rhodonite carving of a bear catching a salmon was created in a German gem-cutting center.

Black veining

Unusually large crystals of rhodonite have been discovered in Franklin, New Jersey, US

Walking stick | **Carved** | This head of a walking stick has been carved from fine rhodonite, and is decorated with a platinum royal monogram inset with diamonds.

Pectolite

△ **Tumble-polished** larimar (pectolite) gem

Pectolite is widely found in Canada, England, and the US, but gem-quality crystals are relatively scarce. Rare faceted stones tend to have internal veils and cloudy areas, and a banded Peruvian variety is sometimes cut *en cabochon*. The type most widely used as a gem is known as larimar, a blue- to blue-green variety found only in the Caribbean. Other forms of pectolite are found in many locations, but none has the unique coloration of larimar (see box, below). Most pectolite jewelry uses silver but, occasionally, high-grade larimar is set in gold.

Specification

Chemical name Sodium calcium silicate | **Formula** $NaCa_2(Si_3O_8)$ (OH) | **Colors** Colorless, white, bluish, greenish | **Structure** Triclinic | **Hardness** 4.5–5 | **SG** 2.8–2.9 | **RI** 1.59–1.64 **Luster** Vitreous to silky | **Streak** White | **Locations** Dominican Republic, Canada, England, US, Greenland, Russia

51.31-carat larimar cabochon

Larimar specimen | Rough | The swirling patterns and intense color banding typical of larimar rough material can be clearly seen in this piece.

Polished larimar | Cut | This tumble-polished nugget of larimar features patches of intense color and angular patterning.

Larimar statuette | Carved | The larimar museum in Santo Domingo, Dominican Republic (see box, below), is home to rough and carved pieces such as this.

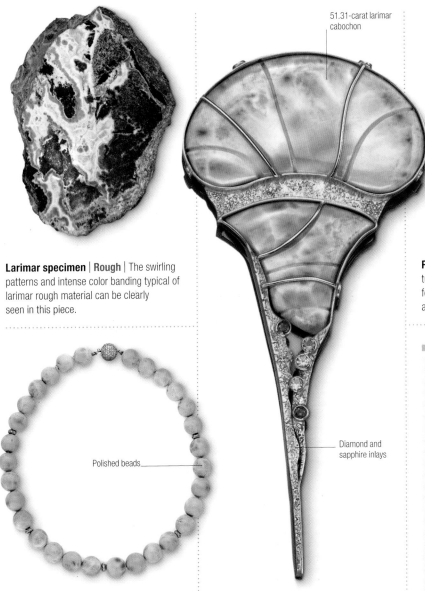

Polished beads

Diamond and sapphire inlays

Necklace | Set | The subtle, pastel blue coloring unique to larimar, seen in this exquisite bead necklace, is one of the reasons for its popularity.

Lotus flower brooch | Set | This gold brooch is set with diamonds, sapphires, and two larimar stones—of 51.31 carats and 19.72 carats.

The origins of larimar

Central American sea stone

The inhabitants of the Dominican Republic called the gem "blue stone," believing it came from the sea. It was given its modern name in 1974 by Miguel Méndez, who combined his daughter's name—Larissa—and the Spanish for "sea" (*mar*), into the word "larimar." At first, Méndez and his companion found a few stones on a beach, washed into the sea by the Bahoruco River. They followed the larimar trail upstream, and discovered outcrops of the mineral; these formed the basis of the first mine.

Central America A 17th-century map showing the Dominican Republic, where larimar was first discovered.

DESIGNERS' HEYDAY

NEW JEWELS

The late 19th century marked the beginning of a creative outpouring in the jewelry world that reached its height in the 1930s. One factor was the abundant supply of large gems on the market from newly opened mines in South Africa. These larger gemstones required lighter settings, challenging jewelers to develop new styles. Britain's Queen Alexandra was another catalyst. Wearing extravagant jewels by day, she personified the *belle époque*, an age of flamboyant, frivolous fashions, and jewels—it was the perfect time to be a jewelry designer.

Among the names to make their mark in the 1930s were the French designers René Lalique, Cartier, Mauboussin, and Boivin, becoming some of the most sought-after design houses in the jewelry world. René Boivin founded his Paris workshop in 1890, but it was not until the 1930s, under the control of Boivin's wife Jeanne, that the house became known for its bold, original pieces. Around this time, Hollywood also bought fine jewelry to the public's attention. Marlene Dietrich commissioned a pair of emerald and diamond bracelets from Mauboussin, one of which she wore in the 1936 movie *Desire*; her patronage helped seal the jeweler's fortunes.

> ## It was a toss-up whether I'd go in for diamonds or sing in the choir. The choir lost

Mae **West**
Hollywood actress

An illustration by Georges Lepape for Vogue, 1933 This illustration showcases pieces by some of the leading jewelry houses of the 1930s— Cartier, Mauboussin, and René Boivin. They were all founded in the 19th century, but rose to fame in the 1930s with the help of Hollywood.

Dioptase

△ **Fine dioptase crystals** on a rock groundmass

Bright green dioptase would make a superb gemstone to rival emerald in color, were it not for its softness and easily set-off cleavage. It is very popular with mineral collectors, but its extreme fragility means that, although it can be cut into a collector's gemstone, these are very susceptible to mechanical shock, and will shatter if exposed to ultrasonic cleaning; even mineral specimens of dioptase must be carefully handled and stored. The name dioptase refers to the gem's highly transparent crystals, from the Greek for "to see through."

Specification

Chemical name Copper silicate | **Formula** $CuSiO_2(OH_2)$
Colors Emerald green to blue green | **Structure** Hexagonal/trigonal | **Hardness** 5 | **SG** 3.3 | **RI** 1.67–1.72 | **Luster** Vitreous to greasy | **Streak** Pale greenish blue | **Locations** Kazakhstan, Iran, Namibia, Congo, Argentina, Chile, US

Dioptase crystals

Dioptase on quartz | Rough | This spectacular specimen of dioptase crystals on quartz shows why it is a favorite with mineral collectors as well as gemologists.

Large crystals on quartz | Rough | In this specimen, dioptase crystals of remarkable form are highlighted by overgrowths of the mineral plancheite.

Diamonds

Gold mount

Brooch/pendant | Set | This piece can be worn as a brooch or a pendant. It features 14-karat gold mounting a natural cluster of dioptase highlighted with diamonds.

Superb crystals | Color variety | This striking group of dioptase crystals displays unusually fine crystallization and a deep blue-green color.

Gemstone with inclusions | Cut | Although this dioptase gem has a large number of internal inclusions, its skilled cut and rich color still make it a fine gem.

Dioptase crystals, mistakenly identified as emeralds, were given to Tsar Paul I in 1797

Bird's nest pendant | Set | This intricately textured gold pendant in the shape of a nest has a centrally mounted group of natural dioptase crystals.

Sugilite

△ **Group of top-grade pieces** of sugilite rough

Sugilite was discovered in 1944, but it was only recognized as a mineral in 1976. It is most commonly found in massive or granular form, and rarely found as crystals; when crystals do occur, they are small—less than ¾ in (2 cm) across. Its color can be pale to deep pink, brownish yellow, or purple, with deep hues of the latter commanding the most value. It is always cut *en cabochon* when used as a gemstone, while pebbles are sometimes polished in rock tumblers. Sugilite is named after its co-discoverer, Japanese petrologist Ken-ichi Sugi.

Specification

Chemical name Potassium, sodium lithosilicate | **Formula** $KNa_2(Fe,Mn,Al)_2Li_3Si_{12}O_{30}.H_2O$ | **Colors** Pink, brown yellow, or purple | **Structure** Hexagonal | **Hardness** 5.5–6.5 **SG** 2.73–2.79 | **RI** 1.60–1.61 | **Luster** Vitreous **Streak** White | **Locations** Canada, Japan, South Africa, Italy

Rock groundmass

Sugilite in rock | **Rough** | Within this specimen, a layer of vividly colored, gem-quality sugilite can be seen sandwiched between two layers of rock.

Manganese minerals

Sugilite slice | **Rough** | This slice of sugilite rough shows good color along with black manganese minerals. The cutter will select the best areas for a cabochon.

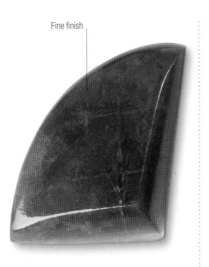

Fine finish

Polished piece | **Cut** | This polished piece of high-grade sugilite can be used as a cabochon, or simply enjoyed as a fine mineral specimen by a collector.

Cabochon | **Cut** | The finest gem-quality sugilite is usually cut into cabochons, such as this elongated oval, high-domed example. It contains a number of small inclusions.

Rare carving | **Carved** | Carved in Idar Oberstein, Germany, from the highest-grade South African sugilite, these herons stand on a base of calcite and have yellow jasper beaks.

Jasper beak

Silver legs

The forgotten gem

Sugilite's long road to recognition

Several decades passed between the time when Professor Sugi first discovered sugilite in 1944 and the time by which gem material was found. The original Japanese finds were tiny yellow crystals with no gem value. In 1955, some dark pink crystals were discovered in India, which were also identified as sugilite, but these were not cuttable. Finally, in 1975, a seam of rich purple sugilite was found at a manganese mine in South Africa, the first commercial source.

Tumbled gem The highest-quality sugilite can be fashioned into exquisite cabochons and carvings.

Iolite

△ **Unusual**, round, step-cut iolite gem

G em-quality blue cordierite is known as iolite, derived from a Greek word meaning "violet," a reference to its color. As a gemstone, it is particularly noted for its pleochroism, appearing intense blue in one direction, yellowish gray or blue in another, and almost colorless as the stone is turned in the third direction. It is almost always faceted, and cutters need to take careful note of the orientation of the stone to obtain the best color. Another informal name applied to iolite is "water sapphire," again because of its color.

Specification

Chemical name Magnesium, iron aluminosilicate | **Formula** (Mg,Fe)$_2$Al$_4$Si$_5$O$_{18}$ | **Colors** Blue | **Structure** Orthorhombic | **Hardness** 7–7.5 | **SG** 2.6 | **RI** 1.53–1.55 | **Luster** Vitreous to greasy | **Streak** Colorless | **Locations** Sri Lanka, India, Canada, Myanmar, Madagascar

Iolite in matrix | **Rough** | This matrix specimen contains numerous gem-quality, dark iolite crystals studded within a quartz groundmass.

Oval brilliant | **Cut** | The fine, sapphire-blue color of iolite is beautifully displayed in this oval brilliant, and illustrates its informal name of "water sapphire."

Finely carved eye

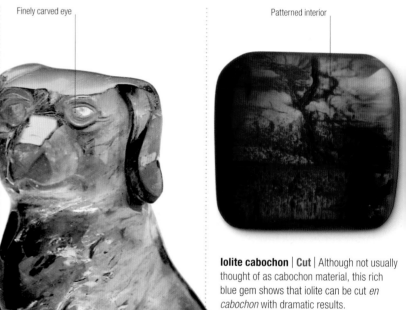

Tiny marks on surface

Patterned interior

Iolite cabochon | **Cut** | Although not usually thought of as cabochon material, this rich blue gem shows that iolite can be cut *en cabochon* with dramatic results.

The many names of iolite

Obsolete and current

Iolite is a good example of how gemstone names evolve. "Water sapphire" may have originated because the stone was usually found in water, for example in the gem gravels of Sri Lanka and Myanmar. Another old name is dichroite, a Greek word meaning "two-colored rock," a reference to its pleochroism. Yet another obsolete name for iolite is steinheilite, after Fabian Steinheil, the Russian military governor of Finland, who first observed that it was a mineral distinct from quartz.

Sri Lankan scene A print showing Kelani Ganga River in Sri Lanka, a source of iolite gem gravels.

Iolite carving | **Carved** | Iolite is rarely used as a carving material. This charming dog weighs 39.16 carats and is part of a set of various gemstone animal carvings.

Tourmaline "petals"

Iolite earrings | **Set** | These flower earrings set in 18-karat yellow gold each feature a central, step-cut iolite, with pink tourmalines forming the "petals."

Benitoite

△ **Deep blue** benitoite gem

Benitoite was discovered in 1906 near the San Benito River in California, from which it takes its name. It is reputed to have been found by a prospector looking for mercury and copper deposits, who came across some brilliant blue crystals he mistook for sapphires. Its best blue color is seen through the side of its crystals, a fact that imposes a size limitation on cut stones, which rarely exceed three carats. Benitoite has exceptionally strong dispersion: its "fire" is similar to that of diamond, though it is often masked by the intensity of the stone's color.

Specification

Chemical name Barium titanium silicate | **Formula** $BaTiSi_3O_9$ | **Colors** Blue, colorless, pink | **Structure** Hexagonal | **Hardness** 6.5 | **SG** 3.7 | **RI** 1.76–1.80 | **Luster** Vitreous | **Streak** White | **Locations** US, Belgium, Japan

Uncut benitoite | **Rough** | Specimens of benitoite are rarely found in rough form weighing over 5 carats—this unusually fine piece weighs nearly 7 carats.

Benitoite crystals | **Rough** | Although the benitoite crystals in this specimen are not gem quality, they are accompanied by calcite, which is typical of the mineral.

Unusual faceting

Fine gem | **Cut** | The deep color and natural fire of this benitoite gemstone have been enhanced by the use of a cut that features a large number of facets.

Gold setting

Benitoite gem

Butterfly brooch | **Set** | Made to celebrate the California state gemstone, this brooch is set with blue and colorless benitoites, with heat-treated orange benitoites for eyes.

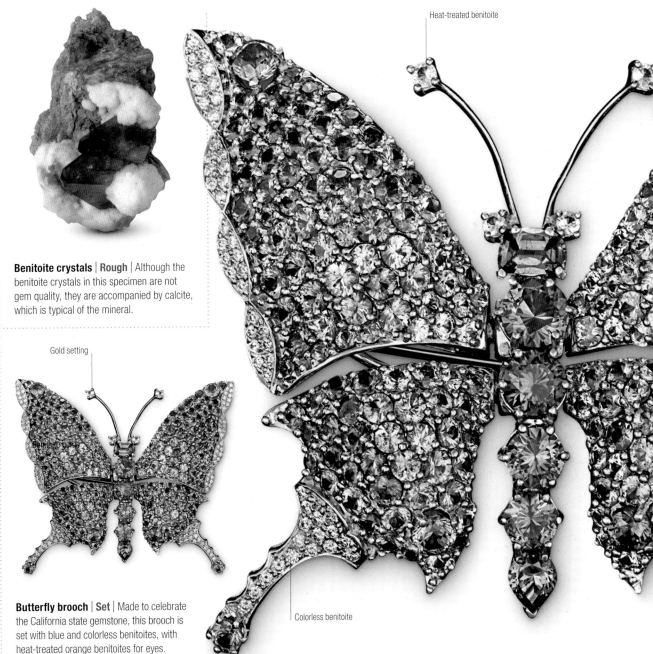

Heat-treated benitoite

Colorless benitoite

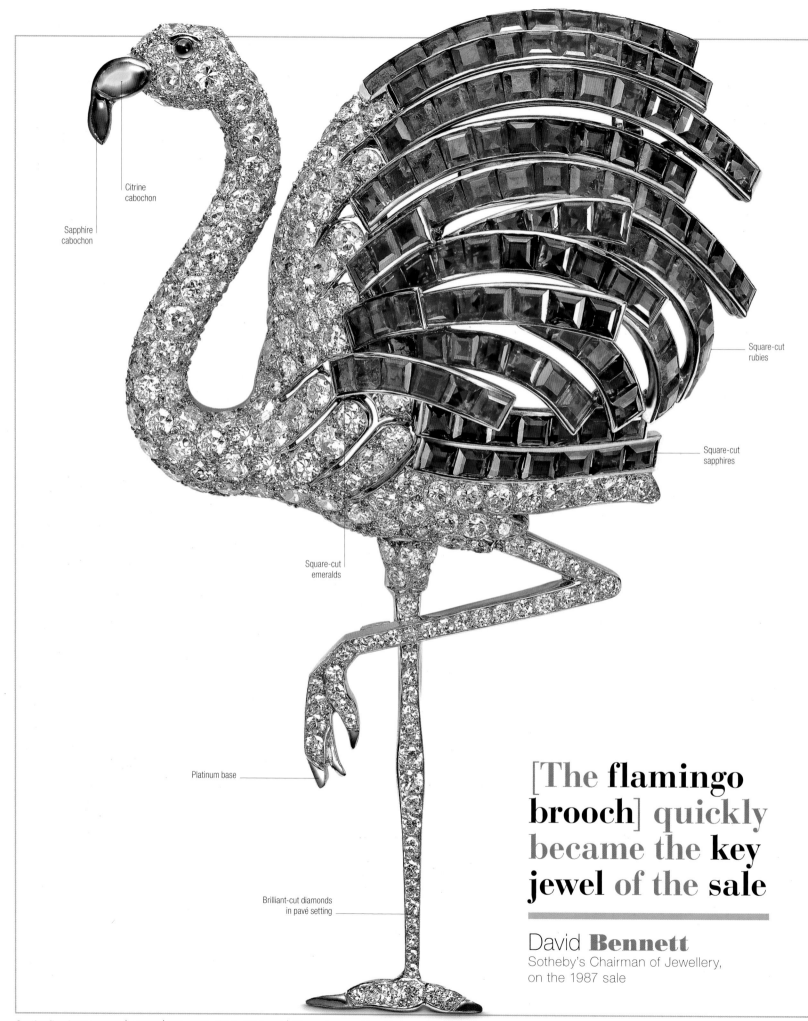

Citrine
cabochon

Sapphire
cabochon

Square-cut
rubies

Square-cut
sapphires

Square-cut
emeralds

Platinum base

Brilliant-cut diamonds
in pavé setting

[The **flamingo
brooch**] **quickly
became the key
jewel of the sale**

David **Bennett**
Sotheby's Chairman of Jewellery,
on the 1987 sale

Cartier flamingo brooch | c.1940 | 3¾ x 3¾in (96.5 x 95.9mm) | Diamonds, emeralds, rubies, sapphires, citrine, gold, platinum

Duchess of Windsor's Cartier flamingo brooch

△ **Wallis Simpson**, later the Duchess of Windsor, 1936

This Cartier flamingo was commissioned by the Duke of Windsor for Wallis Simpson, the woman for whom he gave up the British throne. It is one of the most famous of the numerous jewels that he bestowed on her.

The brilliant-cut, calibrated (uniformly sized) diamonds have a pavé setting in the platinum and yellow-gold base to create the glittering body and legs of the flamingo. The plumage of the wings and tail is composed of step-cut emeralds, rubies, and sapphires. The eye is a single sapphire cabochon (a polished rather than faceted stone) while the beak is formed from citrine and sapphire cabochons.

American socialite Wallis Simpson had already been twice divorced when she captured the attention of the heir to the British throne, Edward VIII, in 1934. The prince was determined to marry her at any cost, causing a crisis when the reigning monarch King George V

Duchess of Windsor's emerald and ruby 20th-anniversary Cartier brooch

died. Edward was now the head of the Church of England, which did not permit divorcées to remarry. Unable to bring himself to relinquish Simpson, Edward abdicated in 1936, and his younger brother George VI became king, granting Edward the title "Duke of Windsor."

Edward commissioned the flamingo brooch for his wife three years after their wedding. As materials, he supplied one of her necklaces and four of her bracelets to Cartier's director Jeanne Touissant in Paris. Touissant, together with her design partner, Peter Lemarchand, used the reclaimed gems to complete the flamingo jewel in 1940.

The Duke presented the brooch to his wife, now the Duchess of Windsor, for her birthday that year and it became one of her most treasured pieces. After her death it was owned by private collectors, and was exhibited by Cartier in 2013.

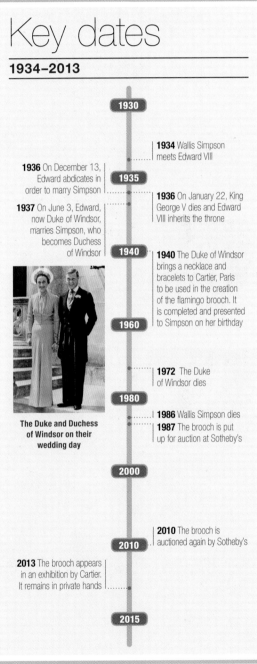

Key dates

1934–2013

1930

1934 Wallis Simpson meets Edward VIII

1936 On December 13, Edward abdicates in order to marry Simpson

1935

1936 On January 22, King George V dies and Edward VIII inherits the throne

1937 On June 3, Edward, now Duke of Windsor, marries Simpson, who becomes Duchess of Windsor

1940

1940 The Duke of Windsor brings a necklace and bracelets to Cartier, Paris to be used in the creation of the flamingo brooch. It is completed and presented to Simpson on her birthday

1960

1972 The Duke of Windsor dies

1980

The Duke and Duchess of Windsor on their wedding day

1986 Wallis Simpson dies

1987 The brooch is put up for auction at Sotheby's

2000

2010 The brooch is auctioned again by Sotheby's

2010

2013 The brooch appears in an exhibition by Cartier. It remains in private hands

2015

Duchess of Windsor's gem and diamond Cartier bracelet

Bulgari Gemma watch | Perhaps the ultimate in jeweled watches, this extravagant 18-karat pink gold watch has a face set with tourmalines, diamonds, and amethysts. The band features emerald beads and brilliant-cut diamonds.

Gold frame

Diamonds

Pink gold

Native Americans have used pink and green tourmalines as funeral gifts for centuries

Pink and green tourmaline beads

Tourmaline

△ **Faceted 7.79-carat** indicolite tourmaline, side view

ourmaline refers to a family of borosilicate minerals of variable composition, but all with the same basic crystal structure. There are more than 30 mineral species in the tourmaline group, including elbaite, dravite, and schorl. However, while mineral names are based on chemistry, gemstone names are based on color and take no notice of tourmaline species. These include indicolite (blue), achroite (colorless), and rubellite (pink or red). The crystals generally form pencil-like prisms, with a rounded-triangular cross section, and, unlike the rocks in which they often form, tourmaline minerals are resistant to weathering. As a result, they tend to accumulate in gravel deposits; the origin of the name is the Singhalese word *turamali*—"gem pebbles."

Variety of colors

There is no simple correlation between chemical composition and color. Most gemstone tourmaline material comes from the species elbaite, which is usually green, although it can occur in many other colors. Emerald green is fairly rare and thus valuable; until the 18th century, it was often confused with emerald. The most dramatic tourmalines are the color-zoned gems called "watermelon" tourmaline: when sliced across the crystal, this variety shows a red or pink center surrounded by a rim of green. The deepest color is always seen when looking down the length of the crystal, so it is important to position rough material correctly when cutting gems.

Specification

Chemical name Complex boron silicate | **Formula** $Na(Li_{1.5}Al_{1.5})Al_6(BO_3)_3[Si_6O_{18}](OH)_3(OH)$ (elbaite) | **Colors** Various **Structure** Trigonal | **Hardness** 7–7.5 | **SG** 2.8–3.3 **RI** 1.635–1.675 | **Luster** Vitreous | **Streak** White

Emerald   Mixed   Marquise

Pendaloque   Step   Cabochon

Locations
1 US **2** Brazil **3** Czech Republic **4** Italy **5** Nigeria
6 Namibia **7** South-eastern Africa **8** Madagascar
9 Afghanistan **10** Pakistan **11** Sri Lanka **12** Australia

Key pieces

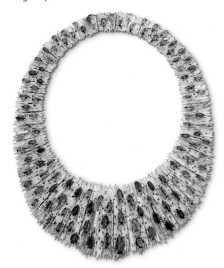

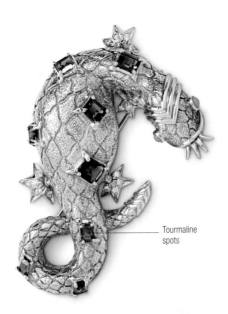

Tourmaline spots

Gold and tourmaline necklace | This 18-karat gold necklace is set with pear-shaped tourmalines weighing a total of approximately 100 carats, accented with brilliant-cut diamonds.

Tiffany brooch | Created for Tiffany & Co. by Jean Schlumberger, this 18-karat textured gold salamander brooch is set with rectangular-cut green tourmalines, diamond feet, and turquoise cabochon eyes.

Cartier earrings | These fanciful orchid earrings, set in 18-karat gold, are studded with faceted pink tourmalines, pink sapphires, rhodolite garnets, and 24 diamonds. A single briolette-cut rose quartz drop completes the earrings.

Rough

Yellow-green rough | Yellow green is the most common tourmaline color, although the above piece of facet rough is slightly more yellow than most.

Indicolite rough | Blue is a less common color for tourmaline and is called indicolite. It can vary from light to deep blue in hue. This piece of rough has particularly good color.

Concentric color zoning

Color zoning | Tourmaline can have concentric color zoning. When the center is pink and the outer is green, as here, it is called watermelon tourmaline.

Varieties

Achroite | Achroite is the color-name given to colorless tourmaline, seen here shaped as a 12.24-carat, rectangular, brilliant-cut cushion.

Dravite tourmaline | Dravite is brown tourmaline, and is not common in facet-grade material. This cushion mixed-cut dravite is particularly fine.

Natural flaws on face of gem

Yellow-green tourmaline | With its color tending more toward green than yellow, this 4.20-carat, brilliant-cut gem has a few natural flaws.

Indicolite tourmaline | Blue tourmaline is called indicolite. This flawless 7.79-carat indicolite gem is faceted as a hexagonal mixed cut.

Rubellite tourmaline | Red or pink-red tourmaline is called rubellite. Shades range from pale pink to shocking red. This emerald step-cut rubellite shows classic color.

Watermelon tourmaline | Occasionally a piece of watermelon tourmaline, such as this emerald-cut stone, is wide enough for a gem to be faceted across the color zoning.

Paraiba tourmaline | Paraiba tourmaline is relatively new to the gem market. It contains copper and has a "neon" look. A green example is shown here.

Yellow-green tourmaline | Strangely, there is no color name for yellow-green tourmaline, although this example, shaped in a trillion cut, is a fine gem.

Schorl tourmaline | Schorl is black tourmaline and always opaque. Of all the tourmalines, it is the most common. It is shown here as tumble polished.

Settings

Bulgari Cerchi earrings | This striking pair of 18-karat gold earrings is set with green tourmaline, peridot, blue topaz, rhodolite garnet, citrine, and diamonds.

White gold

Lizard brooch | One of Cartier's animal-inspired pieces, this brooch is based around a 13.71-carat cat's-eye tourmaline. It is also set with sapphires and diamonds.

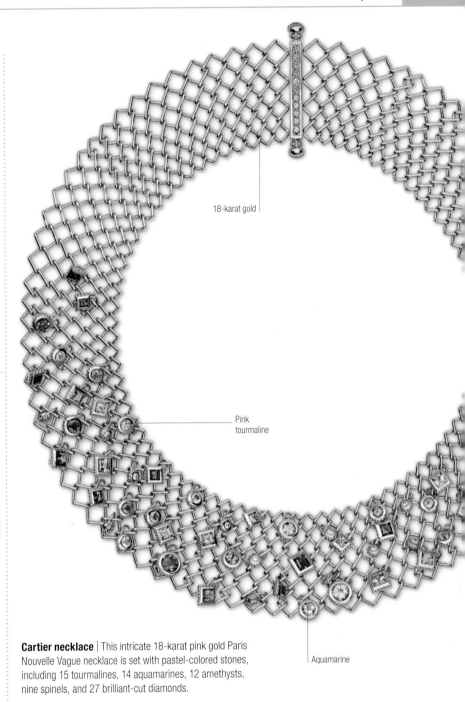

18-karat gold

Pink tourmaline

Aquamarine

Yellow gold earrings | These beautiful earrings are based around a pair of deep red tourmalines with oval cuts. They are set in yellow gold with cut diamonds.

Gold ring | This unusually set 19-karat gold ring features a custom-cut, triangular-section, bezel-mounted pale pink tourmaline cabochon.

Cartier necklace | This intricate 18-karat pink gold Paris Nouvelle Vague necklace is set with pastel-colored stones, including 15 tourmalines, 14 aquamarines, 12 amethysts, nine spinels, and 27 brilliant-cut diamonds.

Bulgari bracelet | Made by the House of Bulgari, this extravagant bracelet is set with diamonds, cabochon emeralds, rubies, amethysts, and pink tourmalines.

Arts and Crafts brooch | Created in 1912 by Georgie and Arthur Gaskin, this brooch is set with blue opal, pink tourmaline, silver, and gold.

> ## The **chemistry of [tourmaline]** is more like a **medieval doctor's prescription** than the making of a **respectable mineral**

John **Ruskin**
Artist and art critic

Cross

Orb

Arches

Emeralds inlaid throughout

Atahualpa Emerald

Gold bodywork

Crown of the Andes | c.1590s | 13½ in (34.5 cm) tall, 20½ in (52 cm) circumference; 4¾ lbs (2.18 kg); Atahualpa Emerald: ½ x ¾ in (15.8 x 16.15 mm) | 18–22-karat gold, over 450 emeralds

Crown of the
Andes

△ **Atahualpa Emerald**, the crown's centerpiece

The Crown of the Andes is a spectacular religious object, featuring the oldest collection of emeralds on a single artifact in the world. It was fashioned by Spanish craftsmen in the 16th century in Popayán (in present-day Colombia). When the conquistadors came to plunder Inca gold, they brought with them European diseases, and in 1590 a virulent strain of smallpox swept through the region. The faithful of Popayán prayed to the Virgin for deliverance and, miraculously, they were spared. In gratitude, they decided to create a fabulous crown for the statue of the Virgin in their cathedral.

Atahualpa, the last Inca emperor and guardian of the Atahualpa Emerald

The oldest parts of the crown are the orb and cross at the top. The rest was added, year by year, with donations from the congregation. The centerpiece is the Atahualpa Emerald, named after the last of the Inca emperors and reputedly seized after his defeat by the Spanish conquistador Francisco Pizarro. The crown was displayed once a year during the majestic processions in Holy Week, but word of its splendor soon spread, and so, to protect it from treasure hunters, the church set up a clandestine group of local nobles called the Confraternity of the Immaculate Conception. At the first sign of trouble, its members were entrusted with dismantling the crown and hiding the sections in the jungle.

The group kept the crown safe until 1936, when the local clergy sold it to pay for a new hospital and orphanage. The buyers were a syndicate of American gem dealers, who wanted to break up the crown and sell its jewels. However, it proved such a popular attraction, including at the 1939 World's Fair, that this decision was reversed. It is now displayed intact at the Metropolitan Museum of Art, New York, US.

Key dates

1532–2015

1532 Francisco Pizarro captures Atahualpa, the ruler of the Incas, and seizes the Atahualpa Emerald

1500

1550

1590 Citizens of Popayán pray to the Virgin for help when an epidemic sweeps through the region

1593 Twenty-four goldsmiths begin work on the crown

1599 The statue of the Virgin in Popayán Cathedral is "crowned"

1600

1650 English privateers briefly seize the crown, but it is recovered after bloody street fighting

1700

Simón Bolívar, 19th-century Venezuelan leader

c.1770 The final sections —the intersecting arches— are added to the crown

1800

1812 Simón Bolívar captures the crown, but returns it to Popayán

1914 Pope Pius X grants permission for the sale of the crown

1900

1939 The crown is exhibited at the New York World's Fair

1936 The crown is sold to a syndicate of American dealers

2000

2015 The crown is sold to the Metropolitan Museum of Art, New York

Virgin Mary, shown here in a Peruvian painting, c.1680, was an important figure in postconquest, early Christian religious life in the Andes

[The crown is] extraordinary for its rarity and its richness

Ronda **Kasl**
Curator, Metropolitan Museum of Art

Emerald

△ **Emerald stone** featuring the signature emerald cut

One of the most desirable gemstones, emeralds are the rich green variety of beryl, the mineral found in igneous, metamorphic, and sedimentary rocks. Most emeralds have numerous inclusions and internal flaws, and these imperfections are unique to each stone. For jewelry, the brittle gem is usually faceted in its signature emerald cut. This is a step, or trap, cut, which combines a rectangular shape with shortened corner facets, maximizing the emerald's distinctive green color, and protecting it from external damage and internal stress.

Specification

Chemical name Aluminum beryllium silicate | **Formula** $Be_3Al_2(SiO_3)_6$ | **Colors** Emerald green to green, yellow green to blue | **Structure** Hexagonal | **Hardness** 7.5–8 | **SG** 2.7–2.8 | **RI** 1.565–1.602 | **Luster** Vitreous | **Streak** White **Locations** Colombia, Zambia, Brazil, Zimbabwe

Rough

Columbia emerald | This finely formed, richly colored, hexagonal emerald crystal is the size of a walnut. It originates from Santa Fe de Bogota, Columbia.

Internal cracks

Emerald rough | The reddish staining on this specimen of emerald rough follows the lines of internal cracks, helping the cutter to assess its suitability.

Cut

Extra face | The cutter of this octagonal, step-cut emerald has added an extra pavilion facet to aid the removal of a particularly bad internal flaw.

Translucent surface

Synthetic emerald | This synthetic pendeloque specimen has the same crystal structure as a natural emerald but can be purchased at a much lower price.

Internal flaws

Emerald cut | Despite being relatively flawed, this octagonal emerald has been specifically shaped and cut in the signature style to minimize loss.

Emerald cut

Settings

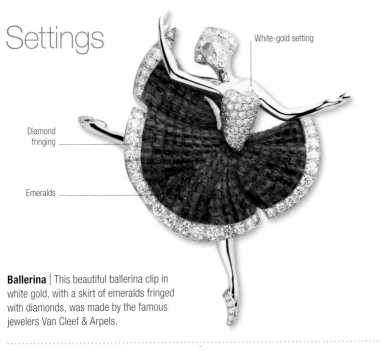

White-gold setting

Diamond
fringing

Emeralds

Ballerina | This beautiful ballerina clip in
white gold, with a skirt of emeralds fringed
with diamonds, was made by the famous
jewelers Van Cleef & Arpels.

Demantoid
garnets

Emerald cross | This stunning white-gold
cross is set with 11 emeralds weighing 24
carats in total, surrounded by diamonds and
demantoid garnets.

Platinum setting

The Hooker Emerald | At 75.47 carats, this
emerald is one of the largest known. Bought
by Tiffany & Co. in 1911, it was initially set in a
tiara, and then in this platinum brooch setting.

Baguette diamonds

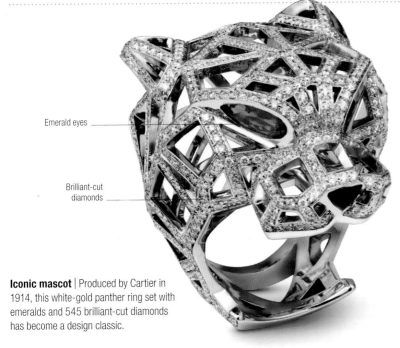

Emerald eyes

Brilliant-cut
diamonds

Iconic mascot | Produced by Cartier in
1914, this white-gold panther ring set with
emeralds and 545 brilliant-cut diamonds
has become a design classic.

When **pure,** beryl is
colorless. Traces of
chromium or vanadium
in the mineral cause it
to develop a green color
and become an **emerald**

Painting of Persian ruler Nadir Shah, c.1740 | Intended recipient of the Topkapi dagger

Topkapi
emerald dagger

△ **Sultan Mahmud I**, who commissioned the dagger

This celebrated emerald dagger is the star attraction of the Topkapi Palace Museum in Istanbul, Turkey. It is one of the finest objects of its kind, but its origins are inextricably linked with bloodshed and treachery.

It was made in Istanbul in the mid-18th century by the royal craftsmen of the Ottoman ruler, Sultan Mahmud I, most likely as a diplomatic gift for the Persian leader, Nadir Shah. Later known as "the Napoleon of Persia," Nadir Shah was the most powerful military figure in the region, and he had recently waged a bitter war against the Ottomans.

The two countries made peace in 1746 and exchanged gifts. Mahmud's contribution included the spectacular dagger. This was a shrewd choice, since Nadir's fondness for jewels was well known—he had seized many during his campaigns in India, including the Koh-i-noor diamond (see pp.58–59).

The Topkapi dagger, comprising gold set with emeralds and diamonds

The dagger is dominated by the huge emeralds in its handle. In the Islamic world, emeralds were highly prized and exotic—these are thought to originate from the Muzo mines in Colombia. The main upper and lower emeralds are pear shaped, while the middle one has a rectangular cushion cut. Another octagonal emerald on top of the handle lifts to reveal a watch. The handle and sheath consist of gold set with diamonds, with enamel and mother-of-pearl decoration.

Nadir Shah never lived to see the dagger: while the gifts were in transit, he was assassinated in his bed. Upon hearing the news, the escort party returned home and the dagger was placed in Topkapi Palace, where it is still on display. Its popularity grew in 1964, when the heist movie *Topkapi* depicted a fictitious plot to steal the dagger, winning actor Peter Ustinov an Oscar.

Key dates
1739–1964

1743–46 Persia is at war with the Ottoman Empire

1747 In May, Sultan Mahmud I reciprocates by sending the dagger and other gifts to Nadir Shah

1747 Nadir Shah is assassinated on June 20th by members of his own guard

1924 The Topkapi Palace is converted into a museum

1700

1739 Nadir Shah carries off the Koh-i-noor diamond to Persia

1746 After the peace treaty is signed in September, Nadir Shah sends lavish gifts to Mahmud I

1750

1800

THE OSCAR WINNERS

Join us—we'll cut you in on the theft of the century!

Topkapi
(where the jewels are!)

1850

1900

Poster for the movie featuring the dagger

1964 The dagger is central to the plot of the Hollywood movie *Topkapi*, which was based on a novel by Eric Ambler

1950

2000

Topkapi Palace, where the dagger is still displayed

The emerald-set cover at the top of the handle opens to reveal a gold watch

... the **famous Istanbul dagger** contains the four world's most priceless emeralds

Topkapi movie
1964

Emerald set in cartouche

Gold link set
with pearl

**Ancient Egyptian
civilizations
regarded emeralds
as a symbol of
life and fertility**

Large cabochon
emeralds

Pendant | This 19th-century Spanish pendant is in the form
of a hippocamp with a female figure. The body is set with
cabochon emeralds and hangs from a chain with four pearls.
The cartouche is set with an emerald and a suspended pearl.

Beryl

△ **Fine, octagonal step-cut** aquamarine with excellent clarity

Beryl provides some of nature's most beautiful gemstones. Although it is colorless in its pure form, it is perhaps best known for its colored varieties, which include aquamarine and emerald—indeed, its name comes from the Greek *beryllos*, meaning "green stone." The colorless form of beryl is known as goshenite, and its clarity is such that it was used to make lenses for some of the earliest eyeglasses during the late Middle Ages.

The colors of beryl

Where colors do occur in beryl, they are caused by minute chemical impurities, and this is sometimes reflected in the varieties' names. The green color of emerald, for example, is caused by traces of chromium. Morganite is colored pink, rose lilac, peach, orange, or pinkish yellow by the presence of manganese, and its crystals sometimes show color banding, with a sequence from blue near the base to nearly colorless in the center, to peach or pink at the tip. It is almost always faceted, and stones with a yellow or orange tinge are sometimes heat treated to emphasize their pink tones. Manganese is also the coloring agent in the rare red beryl, sometimes called red emerald or scarlet emerald. The colors in blue and green aquamarine (meaning "seawater"), and yellow to golden heliodor (from the Greek *helios*, meaning "sun"), result from traces of iron. Much greenish blue aquamarine is heated to produce an intense blue color.

Specification

Chemical name Beryllium aluminium silicate | **Formula** $Be_3Al_2Si_6O_{18}$ | **Colors** Colorless, red, blue, green, yellow
Structure Hexagonal | **Hardness** 7.5–8 | **SG** 2.6–2.8
RI 1.57–1.60 | **Luster** Vitreous | **Streak** White

Round brilliant Oval brilliant Emerald

Step Marquise Pendaloque Scissors

Locations
1 US **2** Colombia **3** Brazil **4** Ireland **5** Norway **6** Sweden
7 Germany **8** Austria **9** South Africa **10** Zambia
11 Mozambique **12** Madagascar **13** Russia

Key pieces

Fabergé egg | This fabulous egg was made by Carl Fabergé (see pp.278–79). It is crafted from gold, platinum, and silver, and is set with aquamarines and diamonds and holds a gold model of a cruiser.

Cartier clip brooch | Designed by Cartier in 1935, the platinum and diamond setting of this clip brooch holds an unusual late 17th-century or early 18th-century Indian carved emerald in the form of a flower.

Gold piping

Aquamarine brooch | Made in 1967 by British jeweler and goldsmith John Donald, this gold brooch features unusually cut aquamarine—part faceted, part cabochon. The stone is ringed with a striking gold setting consisting of sections of tubing.

Rough

Aquamarine
crystals

Prismatic crystals | This outstanding mineral specimen displays numerous prismatic aquamarine crystals on a rock groundmass.

Aquamarine crystal | In general, beryl crystals of all colors tend to have fairly flat terminations, so the end faces on this crystal are unusually large.

Well-formed
faces

Aquamarine
crystal

Host
rock

Hexagonal
outline

Goshenite crystal | This blocky, hexagonal goshenite crystal shows good clarity, a typically colorless interior, and a fine geometric form.

Well-formed
planes

Morganite crystal | Morganite is the pink variety of beryl. Like the other beryls, it forms fine, well-developed crystals such as this specimen.

Classic crystal | Still retaining some of its host rock at the base, this gem-quality aquamarine crystal has a classic prismatic form, with flat end faces.

Red beryl
crystal

Red beryl | Red beryl is very rare in cut stones, since its crystals are both scarce and small. This crystal from Utah, US, rests on a groundmass of rhyolite.

Hexagonal
face

Emerald crystal | This large and fine emerald crystal exhibits classic hexagonal form and emerald-green color. It is approximately the size of a walnut.

Gem-quality
crystal

Heliodor | The yellow variety of beryl is called heliodor. The stunning clarity of this crystal is evidenced by its rock groundmass, which is clearly visible through it.

Cuts and colors

Large table facet

Girdle

Specimens of **Brazilian morganite** crystal have reached up to **55 lb (25 kg)** in weight

Flaws in interior

Faceted aquamarine | Certain nonstandard cuts are used on large stones to enhance brilliance. The cut of this substantial 25.70-carat aquamarine stone is known as a "Portuguese" cut.

Emerald-cut aquamarine | Aquamarine can have either blue or green coloring, as seen in this ethereal, pale green stone with an emerald cut.

Emerald | This octagonal emerald is cut with a combination of cabochon rounding and flat facets. It is designed to emphasize the center of the stone—the area with the fewest flaws.

Extra pavilion facets

Colorless interior

Pavilion visible through table facet

Minor inclusions

Faceted heliodor | The cutter of this square-cushion heliodor has used a mixture of facet styles to emphasize both its color and its brilliance.

Goshenite | This fine goshenite gemstone features a modified emerald cut. It is completely colorless, although such stones can appear blue when photographed.

Morganite | The delicate pink of this morganite gem is one of the lighter shades of the material, which usually occurs in rose-pink to red-pink hues.

Red beryl | This red beryl is approximately 1 carat in weight, but is still relatively large for this very rare gemstone. It is brilliant-cut and has only a few inclusions.

Settings

Aquamarine

Diamond links

Multigem necklace | This lavish white gold necklace is highlighted with faceted stones of beryl, aquamarine, peridot, and diamond. Its diamond-set links have been crafted to resemble crossed strands.

Beryl and amethyst earrings | Faceted stones of amethyst and golden beryl are set in white gold in these cheerful earrings by Colleen B. Rosenblat. The golden beryls have a combined weight of 7.15 carats.

Golden beryl

Paris Nouvelle Vague necklace | Set in 18-karat gold, this necklace features a cartouche of morganite suspending strands of pearls and 66 faceted spinels.

Morganite stone

Ear pendants | Intricately crafted in silver and gold, these triangular ear pendants are set with morganites, rubies, and diamonds.

Blue tassle necklace | In an unusual display, faceted aquamarines cascade from a gold cone set with round aquamarines and champagne diamonds.

Green beryl ring | This ring is set with a light green octagonal-cut beryl weighing 22.35 carats, flanked on each side by five brilliant-cut diamonds.

Aquamarine earrings | This pair of 18-karat yellow gold earrings features four cabochon aquamarines, 28 oval sapphires, and 38 brilliant-cut diamonds.

Fine transparency

Maximilian Emerald | The 21.04-carat emerald displayed in this modern platinum ring setting was once set in a ring belonging to Hapsburg Emperor of Mexico Maximilian.

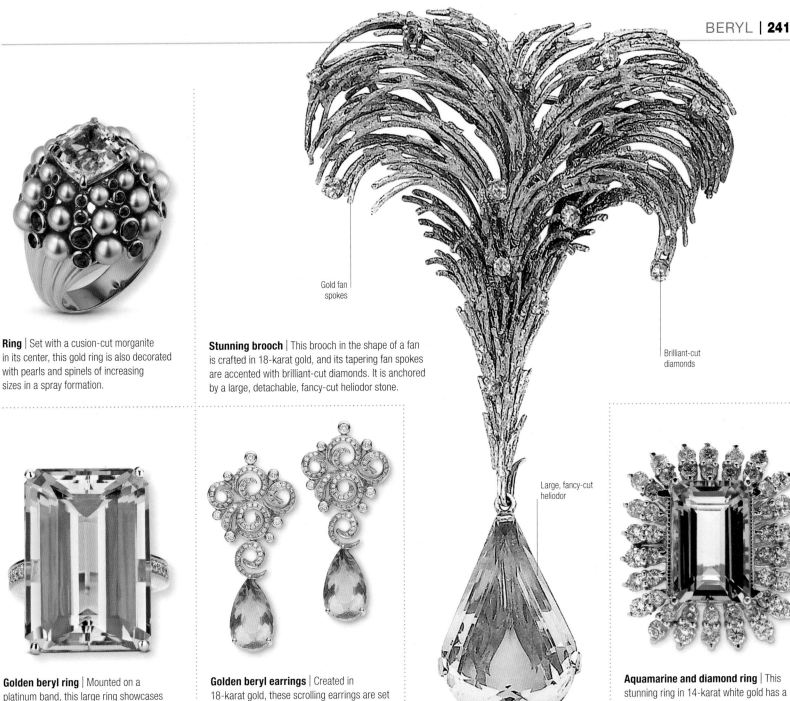

Ring | Set with a cusion-cut morganite in its center, this gold ring is also decorated with pearls and spinels of increasing sizes in a spray formation.

Stunning brooch | This brooch in the shape of a fan is crafted in 18-karat gold, and its tapering fan spokes are accented with brilliant-cut diamonds. It is anchored by a large, detachable, fancy-cut heliodor stone.

Gold fan spokes

Brilliant-cut diamonds

Large, fancy-cut heliodor

Golden beryl ring | Mounted on a platinum band, this large ring showcases a 28.15-carat golden beryl that has been faceted in an emerald cut.

Golden beryl earrings | Created in 18-karat gold, these scrolling earrings are set with diamonds, and suspend drops of golden beryl weighing around 8 carats each.

Aquamarine and diamond ring | This stunning ring in 14-karat white gold has a central aquamarine weighing 7.32 carats, surrounded by 2.20 carats of diamonds.

Parrot brooch | From the Cartier Flora and Fauna collection, this diamond-set parrot brooch surmounts a tourmaline gem. The parrot also features emerald "eyes"

Beryl earrings | These earrings, made around 1940, can also be worn as dress clips. They are set with diamonds, rubies, and a pair of rectangular-cut yellow beryls.

The mineral **beryl** is the **source** of one of the modern world's **most important** metals – beryllium

Dom Pedro Aquamarine | 1980s | 13¾ in (35 cm) tall; 4 in (10 cm) across the base | 10,363 carats, obelisk form

Dom Pedro
Aquamarine

△ **Gem artist** Bernd Munsteiner, who cut the gem

The Dom Pedro is the largest known aquamarine gem in the world. It was fashioned out of an enormous crystal, discovered by three *garimpeiros* (independent prospectors) at Pedra Azul, in the Minas Gerais mining region of Brazil. Before the *garimpeiros* could decide what to do with it, however, they dropped it and the crystal broke into three pieces. The largest of these—which was around 2 ft (60 cm) in length and weighed about 60 lb (27 kg)—was eventually transformed into the Dom Pedro.

From the outset, there was a battle to preserve the crystal. In purely commercial terms, the most profitable outcome would have been to cut it up into small gems to be sold off, and this was the intention of the original Brazilian owner. However, the crystal came to the notice of Jürgen Henn, a German gem dealer. Immediately struck by the exceptional size, clarity, and color of the piece, he

organized a consortium of investors to purchase the crystal and transport it to Idar-Oberstein, a famous gem-cutting center in southern Germany. There, he took it to his friend, the gem artist Bernd Munsteiner, knowing that Munsteiner could turn the crystal into something truly remarkable.

Coming from a long line of gem cutters, Munsteiner is known as the "father of the fantasy cut." Instead of using traditional flat facets, he incorporates grooves and cleverly curved facets into his designs. Munsteiner worked on the aquamarine crystal by hand for more than six months, making a series of tapering, lozenge-shaped cuts. The result is this magnificent, obelisklike gem sculpture that reflects the light in such a way that it almost seems to glow from within. As a tribute to its Brazilian origins, he named it the Dom Pedro, after the country's two emperors who ruled during the 19th century.

Full view of the Dom Pedro Aquamarine

What Mother Nature has made large and beautiful, we should not make small

Jürgen **Henn**
Gem dealer

Key dates

1822–2012

1822 Dom Pedro I (1798–1834) becomes the first emperor of Brazil

1800

1841 Dom Pedro II (1825–91) is crowned as the second and final emperor of Brazil

1960

1960s Bernd Munsteiner pioneers the "fantasy cut," a radically new form of gemstone carving

1980

Dom Pedro I

Late 1980s Three prospectors unearth a huge aquamarine crystal in Pedra Azul, Brazil

1991 A German gem dealer, Jürgen Henn, examines the crystal and photographs it

1990

1992 A consortium organizes the purchase of the crystal and transports it to Germany

1992–93 In "the project of his life," Bernd Munsteiner transforms the crystal into a spectacular obelisk

1995

1993 The Dom Pedro is revealed to the public at a gem fair in Basel, Switzerland

1996 Jane Mitchell brings the obelisk to the US, exhibiting it in Palm Beach

1999 Jane Mitchell and Jeffery Bland purchase the Dom Pedro

2000

2012 The Dom Pedro is placed on permanent display in the National Gem Collection Gallery at the Smithsonian

2010

2011 The obelisk is donated to the Smithsonian Institution, in Washington, DC,

2015

Modern carving and **engraving**

C arving a gem is a step beyond faceting—it involves cutting the stone into a three-dimensional shape. Gems can also be engraved by incising decorative lines. The first task is to select a rough gem that is of high quality and large enough to withstand loss of material during carving. Carved gems can be sculptures in their own right or turned into jewelry.

Rock crystal flaçon
This amethyst and gold stopper was made by Tom Munsteiner, son of Bernd Munsteiner (see p.243).

Sign sculpture
Gem artist Tom Munsteiner also created this bold rock crystal sculpture.

Summer Snow earrings
These carved earrings by Alice Cicolini consist of gold, pavé diamonds, amethyst, rock crystal, and rose quartz.

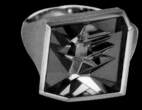

Tourmaline ring
This Munsteiner ring is based around a carved 9.11-carat tourmaline. It is set in yellow gold.

Vortex Mexican opal
This gem sculpture by Michael Dyber is a 299.45-carat opal in a flowing, free-form cut.

Pushkar ring
This Cartier ring showcases carved mandarin and tsavorite garnets, tanzanites, opal cabochons, and brilliant-cut diamonds.

Southwest Sunset
Utilizing two gems naturally occurring together, this 443-carat piece by Sherris Cottier Shank consists of carved ametrine in rose quartz.

18-karat gold pin
Carved from distinctively-colored Bolivian ametrine, this pin is set in gold and studded with diamonds.

Palm sculpture
This piece is cut in Bolivian ametrine, weighing 287.68 carats and featuring a mixture of geometric and organic shapes.

Obelisk
This Dyber ametrine obelisk features optical illusions carved on its front and back, using the

Danburite

△ **Transparent**, single crystal of danburite

Danburite crystals are glassy prisms that resemble topaz, but in roughs they are distinguishable by their poor cleavage. Danburite is named after the US city of Danbury, Connecticut, where it was first discovered as a distinct species in 1839. It is usually colorless, but it can also be amber, yellow, gray, pink, or yellow brown. Danburite is cut as a gemstone, both faceted and *en cabochon*, but is generally considered a collector's stone. Large gem danburites have been found in Dalnegorsk, Russia, in crystals up to 12 in (30 cm) long.

Specification

Chemical name Calcium borosilicate | **Formula** $CaB_2Si_2O_8$
Colors Colorless, yellow, pink, yellow brown | **Structure** Orthorhombic | **Hardness** 7–7.5 | **SG** 3 | **RI** 1.63–1.64 | **Luster** Vitreous to greasy | **Streak** Colorless | **Locations** Switzerland, Russia, Myanmar, Slovakia, US, Mexico, Madagascar, Tanzania

Surface worn by water

Yellow rough | **Rough** | This stream-rounded rough weighs about 23 carats. It comes from Tanzania, where a large amount of yellow material has been found.

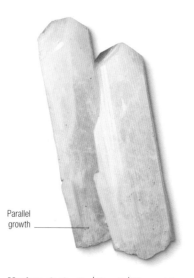

Parallel growth

Mexican danburite | **Rough** | These white crystals from San Luis Potosi, Mexico, are in parallel growth and show perfect prismatic form and classic terminations.

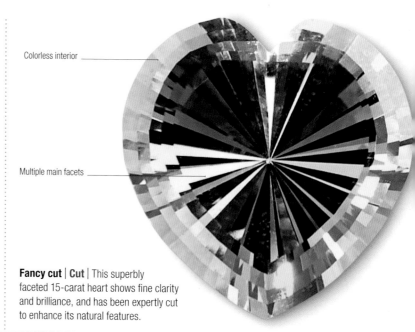

Colorless interior

Multiple main facets

Fancy cut | **Cut** | This superbly faceted 15-carat heart shows fine clarity and brilliance, and has been expertly cut to enhance its natural features.

The Danbury mineral zone

A New England gem belt

The small town of Danbury, in Fairfax County, Connecticut, US, is the center of a local mineral belt running for several miles, which has produced over 50 different mineral types. It is an area of complex geological structure, with folding, faulting, and metamorphism, resulting in its rich mineralogy. In addition to producing danburite, the area has also been an important source of zircon, baryte, celestine, moonstone, sphene, diopside, rutile, garnet, quartz, and pyrite.

Danburite crystal The Connecticut town of Danbury lends its name to the mineral that was discovered there.

Complex faceting

Round brilliant | **Cut** | Danburite often rivals topaz in its brilliance and clarity, as can be seen in this flawless, round brilliant-cut gemstone.

Mixed earrings | **Set** | This earring set's pair of faceted danburites from Mexico contrasts with its pink tourmaline cabochons from Myanmar.

Axinite

△ **Emerald-cut axinite** from Mexico

Axinite refers to a group of four minerals, which in rough form are virtually indistinguishable and structurally identical. The name derives from the Greek *axine*, meaning "ax," which refers to the sharp, hard crystals. The most common color is clove brown; varieties can also be gray to bluish gray, honey brown, gray brown, or golden brown; pink, violet blue, yellow, orange, or red. Its gems are easily chipped, so are usually only faceted for collectors. Axinite is piezoelectric and pyroelectric, meaning it generates electricity when stressed, or rapidly heated or cooled, respectively.

Specification

Chemical name Calcium-, iron-, manganese- aluminium borosilicate | **Formula** $(Ca_2Fe, Mn,Al_2)(BSi_4O_{15})(OH)$ | **Colors** Various | **Structure** Triclinic | **Hardness** 6.5–7 | **SG** 3.2–3.3 | **RI** 1.67–1.70 | **Luster** Vitreous | **Streak** Colorless to light brown | **Locations** US, Russia, Australia, Mexico, France, Sri Lanka

Excellent clarity

Axinite rough | **Rough** | This facet-grade piece of axinite rough has good color and clarity, and retains some of its original crystal form. The tabular structure is typical.

Axinite in matrix | **Rough** | This rock matrix holds a number of red-brown axinite crystals showing classic ax-head crystal forms.

Classic crystal shape

Axinite crystals | **Rough** | These two crystals, each of which exhibits classic and perfect axinite form, have intergrown at one of the facial angles.

Prismatic crystal | **Rough** | Although axinite is typically found as thin, hard ax-head-shaped crystals, it can also occur in blocky form, as in this specimen.

Termination face

Oval cushion | **Cut** | Although there are some natural inclusions in one end of this step-cut oval cushion, it is still a desirable gem.

Blue oval | **Color variety** | This oval brilliant is unusually fine both in clarity and color: most axinite is golden brown or reddish brown, so this hue is relatively uncommon.

Mexican axinite | **Color variety** | This cushion-cut axinite gem displays a fairly typical reddish-orange color. It weighs 4.29 carats and has good transparency.

JEWELRY SHOPPING

The Industrial Revolution changed the way people bought jewelry. Although many pieces were still made by hand, the spread of mass-production brought jewelry within the reach of the new middle class, and many designers opened stores to sell their pieces to the general public for the first time. The Art Nouveau movement, with its focus on using gems for their pleasing aesthetics, not necessarily for their value, also helped to make jewelry more affordable.

This can be seen in the work of Parisian designer Georges Fouquet, whose spectacular store on the Rue Royale offered a luxurious shopping experience for the public. The splendid interior, created by Alphonse Mucha, harmonized perfectly with Fouquet's pieces. Jewelry was placed in bubble-shaped display cases, and Mucha skillfully weaved strong, opaque colors throughout the room to emphasize the gemstones.

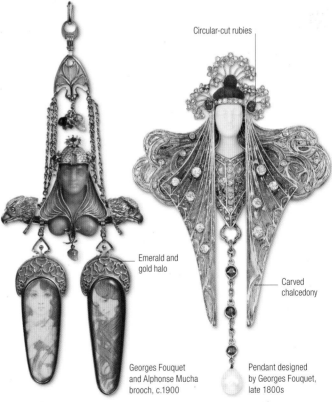

Circular-cut rubies

Emerald and gold halo

Carved chalcedony

Georges Fouquet and Alphonse Mucha brooch, c.1900

Pendant designed by Georges Fouquet, late 1800s

Alphonse Mucha's store interior was a celebration of beauty within nature, with vividly-colored stained-glass panels adorning the walls, and two spectacular peacock sculptures surveying the room. The room's curved lines and jewel-toned colors perfectly complement Fouquet's pieces.

Vesuvianite

△ **Vesuvianite** cut into a brilliant cushion

Vesuvianite is the new name for the mineral formerly called idocrase—indeed, transparent, gemstone vesuvianite is still sometimes referred to as idocrase. Its crystals are usually colored green or chartreuse, but a number of other colors are also found. Vesuvianite can incorporate various elements in its structure; for example, an unusual bismuth-bearing vesuvianite from Långban, Sweden, is bright red; and a greenish blue, copper-bearing vesuvianite is called cyprine.

Specification

Chemical name Calcium, iron, magnesium aluminosilicate | **Formula** $Ca_{10}(Mg,Fe)_2 Al_4(SiO_4)_5(Si_2O_7)_2(OH,F)_4$ | **Colors** Yellow, brown, green, red, black, blue, purple | **Structure** Tetragonal or monoclinic **Hardness** 6.5 | **SG** 3.3–3.4 | **RI** 1.70–1.72 | **Luster** Vitreous to resinous | **Streak** White | **Locations** Italy, Russia, US

Well-formed crystal

Vesuvianite crystals | Rough | This group of substantial transparent, gem-quality, yellow-green vesuvianite crystals would be suitable material for faceting.

Tumble polished | Cut | Vesuvianite, and vesuvianite intermixed with grossular garnet, is popular for tumble-polished stones, as in this specimen.

Vesuvianite

Translucent vesuvianite | Cut | Vesuvianite that is not quite transparent enough for faceted gems is cut into attractive cabochons, as shown here.

Vesuvianite is sold under different trade names— "californite" is massive, jadelike vesuvianite

Dark green interior

Dark green cabochon | Color variety | Vesuvianite, such as this dark green, opaque specimen, has been marketed as California jade in the past.

Moonstone cabochon

Necklace and earrings | Set | This gold necklace and earring group is set with vesuvianite cabochons, and accented by blue cabochon moonstones.

Dark brown vesuvianite | Cut | This dramatic stone features an emerald cut and a number of inclusions, which combine with its color to create a brooding appearance.

Epidote

△ **Highly transparent**, step-cut, oval epidote gem

Although epidote is widespread and abundant in metamorphic and granitic rocks, it is less well known as a gemstone. It frequently forms well-developed, transparent crystals that are strongly pleochroic, usually varying in shades of green when viewed from different angles. This requires the cutter to take the orientation of an epidote crystal into consideration when faceting it. It is a fairly fragile mineral with a distinct cleavage, so faceted stones are unsuitable to be made into jewelry and are cut only for collectors.

Specification

Chemical name Calcium aluminium ferrosilicate | **Formula** $Ca_2(Fe,Al)_3(SiO_4)_3(OH)$ | **Colors** Pistachio, mottled pink and green (unakite) | **Structure** Monoclinic | **Hardness** 6–7 | **SG** 3.3–3.5 | **RI** 1.73–1.77 | **Luster** Vitreous | **Streak** Colorless or grayish | **Locations** Myanmar, France, Norway, Peru, US, Pakistan

Epidote in matrix | Rough | This elongated specimen consists of a series of long, thin epidote crystals that have grown in a quartz matrix.

Multiple growths

Prismatic crystal

Pistachio epidote | Rough | This cluster of long, prismatic, pistachio-green epidote crystals originates from Peru, a source of large amounts of the mineral.

Parallel growths

Gem crystals | Rough | The well-formed epidote crystals in this cluster are transparent and would make excellent gem-quality material for faceting.

Epidote

Mottled unakite | Color variety | Rock made primarily of epidote, such as this mixture of epidote and feldspar, may be polished or tumbled and sold as unakite.

Brown epidote | Color variety | Brown epidote is an uncommon color, and is even more uncommon in faceted gems, as in this rectangular step cut.

Unakite

A colorful variety

Unakite is an altered granite made up of pink orthoclase feldspar, green epidote, and generally colorless quartz. It is also referred to as epidotized, or epidote, granite. Found in various shades of green and pink, it is usually mottled in appearance; it takes a good polish and is thus used as beads or cabochons, and as eggs, spheres, and animal carvings. Some material called unakite lacks the feldspar and is called epidosite; this is also used as beads and cabochons.

Various beads Strands of colorful semiprecious stones hang in a store, including those of unakite, top center.

Kornerupine

△ **Fine, rectangular step-cut** greenish-brown kornerupine

Kornerupine is a rare borosilicate mineral, named in honor of the Danish geologist, Andreas Nicolaus Kornerup. Its crystals can resemble tourmaline prisms, and are found in shades of brown, green, and yellow all the way through to colorless; of all these, emerald green and blue are the most highly valued. It is still relatively rare in faceted stones. When faceting, the lapidary must orient the stone carefully in order to obtain the best color, with the table facet parallel to the prism faces of the crystal.

Specification

Chemical name Magnesium-aluminium borosilicate | **Formula** $Mg_3Al_6(Si,Al,B)_5O_{21}(OH)$ | **Colors** Green, white, blue **Structure** Orthorhombic | **Hardness** 6.5–7 | **SG** 3.3–3.5 **RI** 1.66–1.69 | **Luster** Vitreous | **Streak** White | **Locations** Madagascar, Sri Lanka, Canada, Greenland, Norway, Russia

Striations

Kornerupine specimen | Rough | This small but excellent quality piece of kornerupine faceting rough originates from Mogok, Myanmar.

Prismatic crystal

Kornerupine crystals | Rough | This specimen consists of a number of prismatic kornerupine crystals combined in a groundmass of rock.

Blue cabochon | Cut | Cut from a Tanzanian rough, this 4.38-carat blue cabochon contains a few internal flaws, but is still a desirable stone.

Kenyan kornerupine | Cut | This kornerupine from Kenya in an intense shade of green is far from flawless, but this is compensated by its bold emerald cut.

Kornerupine was first described and named in 1884, but nearly 30 years passed before the first gem-quality material was discovered

Scissors cut | Cut | This flawless kornerupine gemstone features a scissors cut crafted to emphasize both its clarity and its brilliance.

Table facet

Oval brilliant | Cut | Kornerupine is such a rare gem that a few internal flaws such as the healed fractures within this 7.43-carat Sri Lankan stone are acceptable.

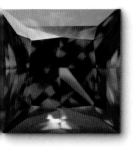

Zoisite

△ **Mixed-cut** tanzanite gemstone

T he mineral name **"zoisite"** may not be particularly well known, but its gemstone variety, tanzanite, is popular among collectors. Found in lilac-blue to sapphire-blue coloring, it was named after its place of discovery, Tanzania. Tanzanite crystals have strong pleochroism, and show gray, purple, or blue depending on the angle from which they are viewed. Another, pink, variety is called thulite, from Thule, the name of an ancient island now thought to be Norway. Ordinary zoisite is usually massive and can be carved as a decorative stone, and as beads or cabochons.

Specification

Chemical name Calcium aluminosilicate | **Formula** $Ca_2Al_3(SiO_4)_3(OH)$ | **Colors** Blue, pink, white, light brown, green, gray | **Structure** Orthorhombic | **Hardness** 6.5–7 | **SG** 3.2–3.4 | **RI** 1.69–1.70 | **Luster** Vitreous | **Streak** White | **Locations** Tanzania, Norway, Italy, Spain, Germany, Scotland, Japan

Irregular broken surface with inclusions

Crystal form

Fractured edges

Girdle facet

Uncut thulite | **Rough** | Displaying a light pink color and dense texture, this piece of thulite rough would be suitable for carving or cutting into cabochons.

Tanzanite rough | **Rough** | This piece of tanzanite rough shows excellent color and transparency, and still retains much of its original crystal form.

Thulite cabochon | **Cut** | The color of thulite tends to be subtle and soft. This thulite cabochon cut with a low dome has a delicate pink color.

Huge tanzanite | **Cut** | Tanzanite gems over five carats are uncommon. This stunning, curved triangular brilliant-cut gem weighs 15.34 carats.

Cushion-cut tanzanite

Brilliant-cut diamonds

Cluster ring | **Set** | This ring features a cushion-cut tanzanite weighing 5.46 carats, encircled by a surround of brilliant-cut diamonds, with baguette diamonds on the openwork sides.

Ruby in zoisite

Naturally occurring patterns

Anyolite is the name of a brilliant green variety of zoisite sprinkled through with rubies. The bright red rubies are often distorted and irregularly spread throughout the sea of massive green zoisite, and can vary in size from a few millimeters to several centimeters. The rubies are not of gem quality, but their color provides a striking contrast to the green zoisite. It is popular as a carving and ornamental stone, with the rubies greatly enhancing the decorative pieces carved from it.

Red on green The contrast of the bright red ruby with the green zoisite is clear in this specimen.

Peridot

△ **Polished pebble** of peridot

The name "peridot" is French, possibly derived from the Arabic word *faridat*, meaning "gem." This variety of gem-quality olivine has been mined for over 3,500 years—the Red Sea island of Zabargad (now St. John's Island) was the main source of peridot for ancient Mediterranean civilizations. The Greeks and Romans called this island Topazios, and so they named this stone "topaz," although it has nothing to do with the gem of the same name. Peridot can range from pale golden green to brownish green in color; rich green is the most valued of all.

Specification

Chemical name Magnesium, iron silicate | **Formula** $(Mg,Fe)_2SiO_4$
Colors Pale green to brownish green | **Structure** Orthorhombic
Hardness 6.5–7 | **SG** 3.32–3.37 | **RI** 1.64–1.69 | **Luster** Vitreous to greasy | **Streak** White to greenish | **Locations** China, Myanmar, Norway, US, Canary Islands, Australia, Sierra Leone

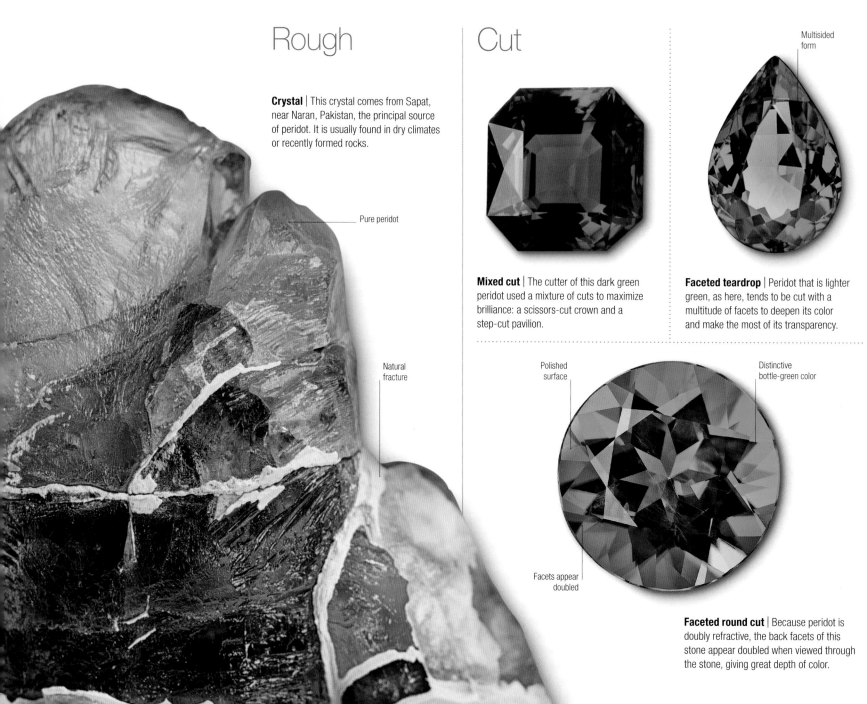

Rough

Crystal | This crystal comes from Sapat, near Naran, Pakistan, the principal source of peridot. It is usually found in dry climates or recently formed rocks.

Pure peridot

Natural fracture

Cut

Multisided form

Mixed cut | The cutter of this dark green peridot used a mixture of cuts to maximize brilliance: a scissors-cut crown and a step-cut pavilion.

Faceted teardrop | Peridot that is lighter green, as here, tends to be cut with a multitude of facets to deepen its color and make the most of its transparency.

Polished surface

Distinctive bottle-green color

Facets appear doubled

Faceted round cut | Because peridot is doubly refractive, the back facets of this stone appear doubled when viewed through the stone, giving great depth of color.

Settings

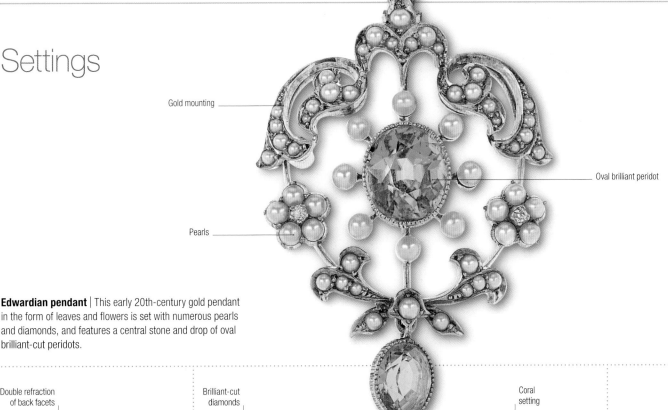

Gold mounting

Pearls

Oval brilliant peridot

Edwardian pendant | This early 20th-century gold pendant in the form of leaves and flowers is set with numerous pearls and diamonds, and features a central stone and drop of oval brilliant-cut peridots.

Double refraction of back facets

Signet ring | The central bezel of this striking gold ring is set with a large, bright, central peridot, surrounded by a circle of smaller diamonds.

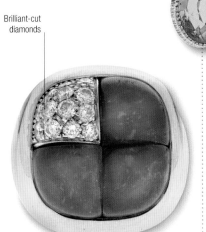

Brilliant-cut diamonds

Ring | This unusual ring makes a feature of asymmetry, with three peridot segments set off by a contrasting wedge of diamonds, encased in a gold band.

Coral setting

Caterpillar brooch | This quirky brooch maximizes the contrast of complementary colors, pairing coral with peridot and adding diamond eyes for good measure.

Flawless stone

Gold pendant | Here a flawless, octagonal step-cut, bottle-green peridot is surrounded by diamonds, with a diamond-encrusted suspension loop.

The August born without this stone, 'tis said, must live unloved alone

Traditional birthstone rhyme for peridot

Twisted strands of peridot beads

Necklace | Beads of peridot were threaded together into chains and then twisted to form thick, textured bands to make up this gold-clasped necklace.

Peridot and emerald

Green gemstones in history

Peridot has long been confused and compared with its famous fellow green gemstone. Cleopatra's collection of emeralds is now thought to have been peridot, and the ancient Romans described the gem as "emerald of the evening" for the way it caught the dimmest of light. For centuries, the 200-carat peridots topping the lavish shrine of the Three Holy Fathers in Cologne, Germany were also thought to be emeralds.

Shrine of the Three Holy Fathers The five huge peridots along the top of this cathedral shrine were once prized as emeralds.

Shwedagon Pagoda, Yangon, Myanmar | c.6th–10th century CE | Plated with gold and set with diamonds, rubies, sapphires, and other precious stones

Shwedagon
Pagoda

△ **Statue of Buddha** at the Shwedagon Pagoda, Myanmar

The Shwedagon Pagoda in Yangon, Myanmar's capital city, is a stupa (Buddhist reliquary) that was built to house eight of the Buddha's hairs, as well as other relics. It is one of the most sacred Buddhist pagodas and, with its gold plating and precious stones, is also one of the most opulent.

The pagoda rises 326 ft (99 m) from a hill above the city. The lower part is covered with 8,688 gold plates and 13,153 cover the upper part. The top of the stupa, too high to see clearly from the ground, is set with 5,448 diamonds, a mixture of 2,317 rubies, sapphires, and other precious stones, and 1,065 golden bells. It is tipped with a huge, 76-carat diamond.

The stupa dazzles in the sunshine and emits a golden glow when illuminated at night. It is traditionally said to be 2,600 years old, making it the world's oldest stupa, although evidence suggests it is more recent, possibly from around the 6th–10th centuries CE. According to legend, two brothers, merchants from Balkhin

Gold plating

Schwedagon Pagoda with the Great Bell of Dhammazedi, once thought to be the largest bell in the world, in the foreground. The bell has since been lost in the Yangon River, Myanmar

(in present-day Afghanistan), met the Lord Gautama Buddha and were presented with eight of his hairs, which they later brought to Burma. With the help of local ruler King Okkalapa, they traveled to Singuttara Hill where three relics of other Buddhas preceding Gautama were also enshrined. The relics were placed in a chamber filled knee-deep with jewels, covered with a stone slab, and entombed when the stupa was built around them. Since then the pagoda has been rebuilt, ransacked, and restored, but throughout it has remained a crucial site of veneration.

Shwe Dagon dominates the city physically, aesthetically and spiritually

Win **Pe**
Author and artist

Top of the stupa set with diamonds, rubies, sapphires, and other precious stones

Key dates

6th century CE–2012

500	**6th–10th century CE** The pagoda is built by the Mon people, an ethnic group from Myanmar
1300s King Binnya U rebuilds the derelict pagoda to 59 ft (18 m) in height	
1485 King Dhammazedi donates a great bell weighing 300 tons **1500**	**1400s** Queen Binnya Thau raises the pagoda to 131 ft (40 m), assigning slaves to maintain it
1600	
1608 Portuguese adventurer Filipe de Brito e Nicote steals the bell, but it sinks into the Bago River	
1700	
	1768 An earthquake brings down the top of the stupa. King Hsinbyushin later raises it to 326 ft (99 m)
1800	**1824** Pillage and vandalism occur during the First Anglo-Burmese War when the British occupy the Pagoda
	1852 British reoccupy the pagoda during the Second Anglo-Burmese War
1900	
General Aung San, politician and revolutionary	
1946 General Aung San addresses a crowd at the stupa, demanding independence from the British	
2000	**1988** General Aung's daughter, Aung San Suu Kyi, addresses 500,000 people at the stupa, demanding democracy
	2012 Devotees celebrate the annual Shwedagon Pagoda Festival for the first time since 1988
2010	

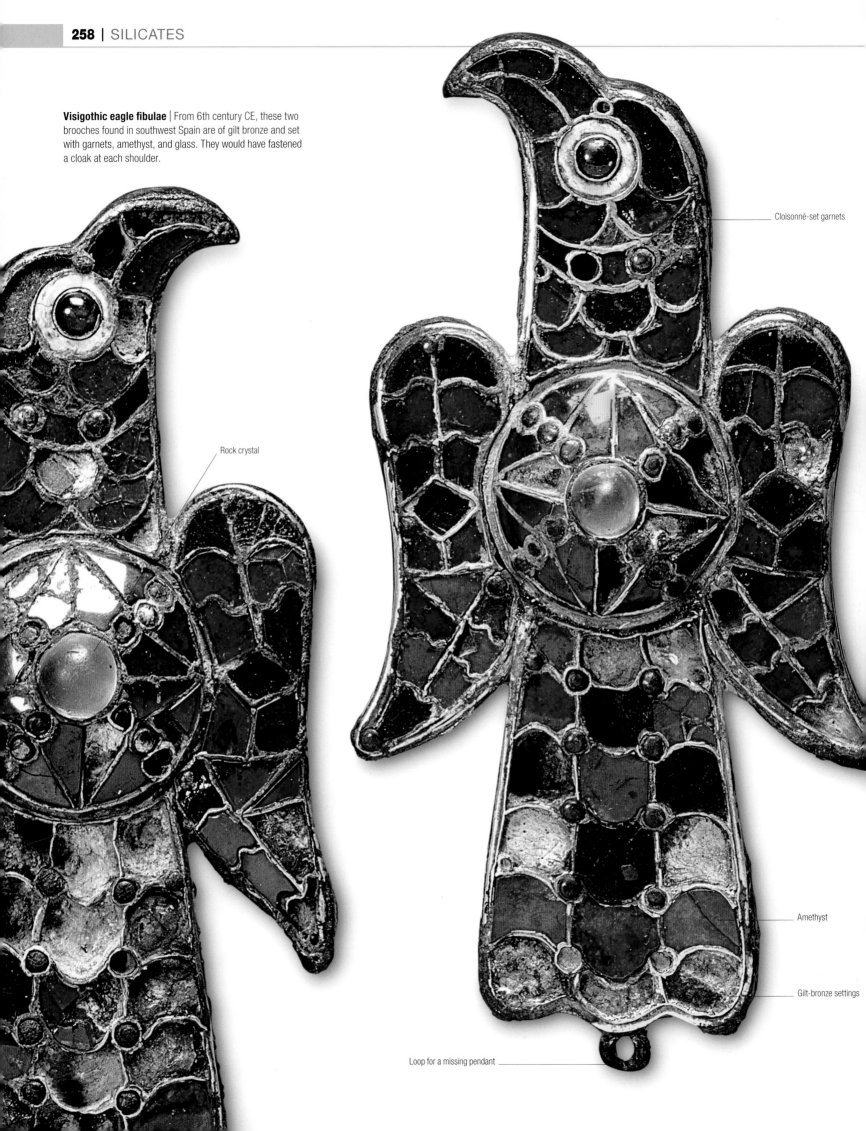

Visigothic eagle fibulae | From 6th century CE, these two brooches found in southwest Spain are of gilt bronze and set with garnets, amethyst, and glass. They would have fastened a cloak at each shoulder.

Rock crystal

Cloisonné-set garnets

Amethyst

Gilt-bronze settings

Loop for a missing pendant

Garnet

△ **Round, brilliant-cut** almandine garnet

Garnets are generally thought of as red, but they can also be orange, pink, green, black, and honey brown. All species of garnets have similar physical properties and crystal forms, but differ in chemical composition. There are over 15 garnet species, of which six varieties are most commonly used as gems: pyrope, almandite, spessartite, grossularite (includes hessonite and tsavorite), andradite (includes demantoid), and uvarovite. Although they are found in many different colors and compositions, garnets are easily recognized because they are generally found as well-developed crystals with a basic—although sometimes modified—dodecahedral form. The name "garnet" is derived from the Latin *granatus*, from granum ("grain, seed"), possibly a reference to the vivid red seed covers of the pomegranate, which are similar in shape, size, and color to some garnet crystals.

A tradition of inlaying

The use of garnets as gemstones dates back to at least the Bronze Age. They were especially used inlaid in gold cells in the cloisonné technique, a style often just called garnet cloisonné. Many consider this the highest point of garnet work, and it can be seen in Anglo-Saxon England at Sutton Hoo and in the Staffordshire Hoard (see pp.264–65). A lesser known use of garnet is as an abrasive, used instead of silica sand in sandblasting. Garnet is used to cut steel and other materials in high-pressure water jets. Garnet sandpaper is a favorite of cabinet makers for finishing bare wood.

Specification

Chemical name (A)Ca/Fe/Mg/Mn (B)Al/Cr/Si/Ti/Zr/Vn Silicate
Formula $A_3B_2(SiO_4)_3$ | **Colors** Black, brown, yellow, green, red, violet, orange, pink | **Structure** Cubic | **Hardness** 7–7.5
SG 3.6–4.3 | **RI** 1.72–1.94 | **Luster** Vitreous | **Streak** White

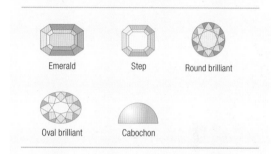

Emerald   Step   Round brilliant

Oval brilliant   Cabochon

Locations
1 Canada **2** US **3** Mexico **4** Germany **5** Czech Republic
6 Italy **7** Namibia **8** South Africa **9** Kenya **10** Tanzania
11 Madagascar **12** Sri Lanka

Key pieces

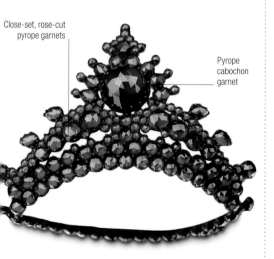

Close-set, rose-cut pyrope garnets

Pyrope cabochon garnet

Antique hairpin | Set with pyrope garnets from Bohemia (now part of the Czech Republic), this gold hairpin was probably crafted in Victorian times. Pyrope, from the Greek *pyropos*, means "firelike."

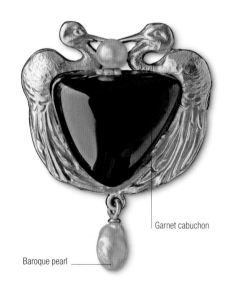

Garnet cabuchon

Baroque pearl

Stork pendant | Set with a large and luxurious garnet cabochon and highlighted with baroque pearls, this gold pendant with facing storks dates from about 1900. Storks often symbolize purity and renewal.

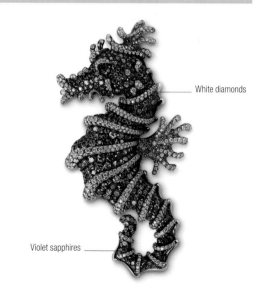

White diamonds

Violet sapphires

Fabergé seahorse brooch | This seahorse wrapped in seaweed is set with green demantoid garnets, tsavorite garnets, alexandrites, tourmalines, sapphires, and diamonds.

Rough

Alamandine garnet crystals — Mica schist

Almandine in matrix | This specimen has a number of classic almandine garnet dodecahedrons embedded in a groundmass of mica schist.

Melanite andradite garnet | This fine crystal of andradite garnet displays the form of a dodecahedron modified by an octahedron.

Grossular garnet

Grossular garnet | Grossular garnets tend to be pink or green, but may be found in other colors. This crystal in a groundmass is deep pink red.

Uvarovite crystals

Uvarovite | Green uvarovite garnet is one of the rarest varieties of the mineral. Here, uvarovite crystals form a crust on a rock groundmass.

Hessonite garnets | Hessonite, informally referred to as "cinnamon stone," is an orange-brown garnet and one of the varieties of grossular garnet. This specimen consists of a cluster of vividly-colored crystals.

Dodacahedron hessonite crystals

Varieties

Grossular brilliant | This flawless, light green grossular faceted into a classic round brilliant is cut from rough material mined in Mali.

Demantoid garnet | Demantoid means "diamondlike", referring to brilliance. This stone is a green variety of andradite and is faceted here with an oval mixed cut.

Demantoid garnet | Faceted in a round standard brilliant, this demantoid variety of andradite has a yellowish tinge to its otherwise green coloring.

Grossular garnet | Grossular garnets are found in a variety of colors. This example, cut in a mixed-cut cushion, is nearly colorless with a greenish tinge.

Green demantoid | The green of demantoid andradite garnets varies from yellow green to the deep, rich green of this triangular fancy-cut stone.

Color-change garnet | Discovered in Madagascar only in 1990, this garnet is a mixture of pyrope and spessartine, and shifts from blue green to purple.

Some Asiatic tribes used garnets as musket balls in the belief that their blood-red color would make them more lethal

Spessartine garnet | Sometimes mistaken for hessonite in cut stones, spessartine is less rare now than in the past, due to abundant new finds. It is shown here in a brilliant-cut oval gemstone.

Inclusions add interest

Cinnamon heart | This hessonite "cinnamon stone" heart is faceted in a mixed cut, and has numerous inclusions of gas-filled bubbles, all of which have been attractively magnified by the faceting.

Hessonite garnet | Faceted in a round, mixed cut, this hessonite variety of grossular is characterized by an unusually deep and rich coloring.

Pyrope garnet | With its deep red coloring, pyrope is sometimes mistaken for ruby. This example has been faceted in a pendaloque cut.

Malaya garnet | This garnet is a mixture of pyrope and spessartine, and is often richly colored, as can be seen in this brilliant-cut cushion gemstone.

Almandine garnet | Rich, purple-red garnets are usually regarded as the best gem-grade almandine garnets, as in this rectangular, cushion mixed cut.

Settings

Almandine ring | The almandine garnet set in this white gold ring is faceted in an unusual checkerboard pattern, and surrounded by citrines.

Tsavorite garnet

White sapphire

Tsavorite and sapphire ring | The central stone in this 14-karat gold ring is an oval-cut tsavorite garnet, flanked by two white sapphires.

Pear-shaped tsavorite garnet

Suspended pendoloque-cut tsavorite garnet

Tsavorite necklace | This spectacular necklace is composed of 14 large, pear-shaped tsavorite garnets of varying sizes with a total weight of 30.79 carats. Tsavorite is one of the most rare and most prized varieities of garnet.

Antique earrings | Made c.1890, these feature pendaloque-cut hessonite garnets suspended below cushion-cut hessonites, all circled by diamonds.

Diamond cluster

Cartier ring | This 18-karat white gold ring from the Paris Nouvelle Vague collection is set with chalcedonies, garnets, tourmaline, diamonds, and aquamarines.

Trio brooch | Designed in three clusters, each centers on an oval-cut garnet, surrounded by rubies, and with diamonds in a trefoil pattern.

Small rose-cut garnets

Cocktail ring | Seen closeup, this spectacular white gold cocktail ring features dozens of rose-cut garnets closely set in a floral pattern.

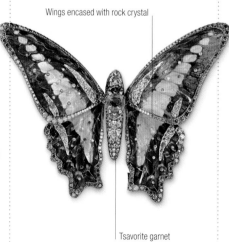

Wings encased with rock crystal

Tsavorite garnet

Butterfly-clip brooch | Unusually crafted in titanium, this intricate butterfly brooch has a body set with a tsavorite garnet and yellow diamonds.

Garnet locket | Made in England in 1852, this antique locket features an intricately patterned gold heart surrounding a polished garnet cabochon.

Rose-cut garnet

Antique cross | This Victorian silver cross is set with 10 larger rose-cut garnets, surrounded by a number of smaller rose-cut garnets.

Rhodolite and diamond brooch | Set with blood-red, rectangular and square-cut rhodolite garnets, this striking Art-Deco brooch was made around 1930.

Diamond border

Belle-Epoque pendant brooch | This brooch, dating from around 1910, features a yellow sapphire surrounded by green demantoid garnets.

Demantoid crab brooch | This vivid, 18-karat gold brooch in the shape of a crab is pavé-set with demantoid garnets and features old-cut diamonds.

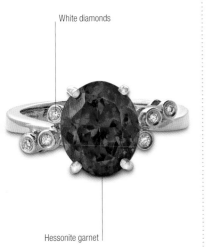

White diamonds

Hessonite garnet

Hessonite and platinum ring | This platinum ring mounts an oval-cut hessonite garnet flanked by diamonds set in the form of a knot.

Brilliant-cut diamonds

Cartier ring | This 18-karat gold ring from the Paris Nouvelle Vague collection, shown here from above, is set with 120 yellow garnets and yellow sapphires.

Chrysoprase

Hessonite garnet

Spessatite garnet

Spessatite pendant | Set in platinum, this pendant features an oval-cut spessatite surrounded by 14 round-cut diamonds, with a suspended diamond.

Bulgari watch | This 18-karat gold watch has three circles of diamonds set with amethyst, aquamarine, chrysoprase, tourmaline, and two hessonite garnets.

Staffordshire hoard | c.6th century CE | More than 11 lb (5 kg) of gold, 3 lb (1.4 kg) of silver, and 3,500 garnets | Discovered in a field in the UK near an ancient Roman road

Staffordshire
hoard

△ **Gold and garnet** decoration for a larger item

One summer's day in July 2009, metal detectorist Terry Herbert set out across the Staffordshire countryside near Hammerwich, UK, with permission from local farmer Fred Johnson to search his fields. By the end of the day, Herbert had uncovered thousands of richly decorated gold and silver fragments.

Buried less than a finger's depth below the soil, the finely wrought metal pieces were later identified by archaeologists at Birmingham University as the world's largest hoard of Anglo-Saxon gold. Herbert and Johnson sold the find to museums in Birmingham and Stoke-on-Trent for £3.3 million (around $5 million), splitting the proceeds between them. Excavation of the site in 2012 revealed another batch of fragments, bringing the total to more than 4,000 remnants of the armor, weapons, and battle dress of Anglo-Saxon men. Conservationists have assembled

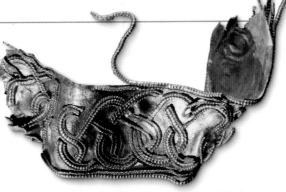

Gold hilt collar from a sword, featuring detailed decoration consisting of fine gold strands wound into coils and set in knot-work patterns

more than 80 sword pommels (the counterweight at the end of a sword's handle). One of the most significant items in the hoard is a silver Anglo-Saxon warrior's helmet, one of only five in Britain. Its recreation involved assembling 1,500 scraps of silver-gilt foil, many measuring less than ¼ in (1 cm) across. The Staffordshire hoard displays an extraordinary level of craftsmanship, featuring fine threads of gold wound into tight coils and used to make swirling filigree patterns. Other pieces are inlaid with red garnet, and blue Roman and Saxon glass. Museum conservationists have dubbed the hoard "warrior bling."

Rise up, O Lord and may thy enemies be scattered and those who hate thee be driven from thy face

Biblical inscription (translated from Latin) on a silver-gilt strip in the Staffordshire hoard

Gold relic from the hoard with cloisonné and garnet decoration

Key dates

5th century CE–2013

5th century CE
Anglo-Saxons from northwestern Europe invade Great Britain and settle

400

7th–8th centuries CE
Anglo-Saxon metal workshops produce beautifully decorated military wear

700

800

Anglo-Saxon helmet, c.7th century CE

2000

Birmingham Museum and Art Gallery

September 2009 Part of the hoard goes on display at Birmingham Museum and Art Gallery

2009

July–August 2009
Gold and silver fragments are found by Terry Herbert near Hammerwich, England. The site is excavated by the University of Birmingham's Archaeology Department

November 2009
The hoard is sent to the British Museum

January 2010
A nationwide appeal is launched to purchase the hoard for the nation

2010

March 2010 Fundraising is completed, and the hoard is bought from Herbert and Johnson

November 2012
A second batch of fragments is found close to the original site

2012

2013 A three-year touring exhibition is launched across the West Midlands

2015

Diamond RI 2.42
Cut diamonds have a very high RI and great dispersion, giving the gems their distinctive brilliance and shine.

Sphalerite RI 2.36–2.37
Sphalerite is extremely difficult to facet, but cut stones exhibit a high RI and good fire.

Cassiterite RI 2.00–2.10
Cassiterites are dichroic, displaying different colors when moved.

Fire and
brilliance

The term "fire" is used to describe the flashes of light that make a gemstone sparkle when it is moved. As with a prism, when white light enters a gem its component colors are dispersed: the greater the dispersion of white light, the greater the fire. The refractive index (RI, see p.23) is a measure of dispersion. Diamonds have high dispersion and are valued for their brilliance; however, gems with low RI can be valued for other reasons.

Scheelite RI 1.92–1.93
Faceted scheelite is rare. However, it exhibits good dispersion of light when cut.

Demantoid garnet RI 1.85–1.89
This variety is the most highly sought-after type of andradite garnet; it has greater color dispersion than diamond.

Zircon RI 1.81–2.02
Zircon's high RI and excellent dispersion come close to diamond in terms of fire and brilliance.

Sphene RI 1.84–2.11
Transparent sphene crystals, where they occur, have good fire and brilliance.

Jadeite RI 1.65–1.68
Jadete varies in color, and is white in its pure form. It has a moderate amout of dispersion.

Chrysocolla RI 1.46–1.57
Chrysocolla is generally blue green and massive. Its fire and brilliance are moderately low.

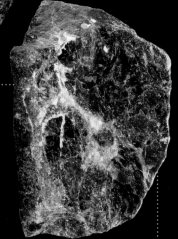

Ruby RI 1.76–1.78
The best ruby material has excellent brilliance, as well as its striking red coloring.

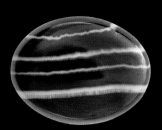

Onyx RI 1.54–1.55
Onyx can be black or brown with white color banding, giving a fairly restrained dispersion.

Sodalite RI 1.48
Transparent sodalite is rare, and its RI is low; however, good quality gemstones can be cut.

Spessartine garnet
RI 1.79–1.81
Gem-quality spessartine crystals are rare, but have good dispersion.

Obsidian RI 1.45–1.55
Despite its deep black coloring, obsidian is more refractive than some gems.

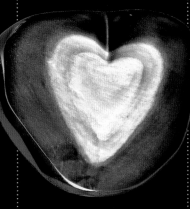

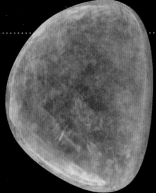

Opal RI 1.37–1.52
Opal has a distinctive color play, which comes from diffraction caused by tiny silica spheres.

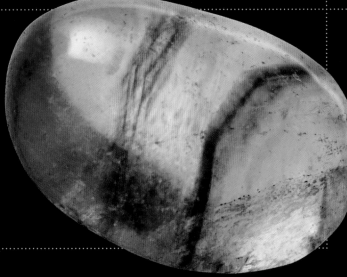

Fluorite RI 1.43
With one of the widest color ranges of any mineral, fluorite has moderate brilliance.

Zircon

△ **Cushion cut**, 10-carat zircon from Myanmar

S ome zircon material is 4.4 billion years old, making it the oldest-known mineral on Earth. It is a colorful gem with high refraction and fire. Colorless zircon is known for its luminescence and reflective flashes of multicolored light, and is often used in jewelry as a substitute for diamonds. Vibrant blue zircons are produced by heat-treating the more common brown stones. The mineral sometimes contains traces of uranium and thorium, and this natural radioactivity can disrupt the crystal structure, causing changes to color, density, RI, and double refraction.

Specification

Chemical name Zirconium silicate | **Formula** $ZrSiO_4$
Colors Reddish brown, yellow, green, blue, gray, colorless
Structure Tetragonal | **Hardness** 6.5–7.5 | **SG** 3.9–4.7
RI 1.81–2.02 | **Luster** Vitreous to brilliant sheen | **Streak**
White | **Locations** Australia, Myanmar, Cambodia, Tanzania

Rough

Zircon crystals | Zircon crystals can form in many different types of rock. This specimen is set on a groundmass of pegmatite, a common zircon occurrence.

Classic crystal | This reddish-brown specimen is a typical example of a tetragonal prism capped on both ends by fine pyramid-shaped terminations.

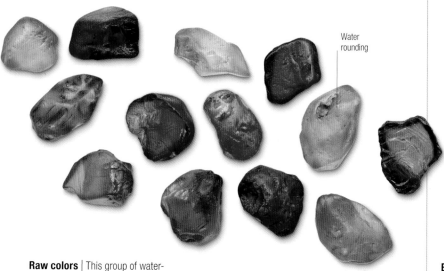

Water rounding

Raw colors | This group of water-rounded, yellow to red-brown zircons demonstrates the variance between specimens when found in their natural state.

Colors and cuts

Crown star facet

Pavilion facets show double refraction

Blue pendeloque | At 15 carats, this magnificent zircon has been heat-treated to intensify its color. It is pendeloque-cut and displays superb dispersion.

Side view | Viewed side-on, this 7.28-carat, white specimen exhibits a classic brilliant cut, exposing the many different facets and revealing the mineral's refractive quality.

Table facet

Crown main facet

Pavilion main facet

Settings

White gold setting

Blue zircon

Blue zircon ring | Crafted by jewelry designer Karina Brez, this ring is based around a 10.60-carat blue zircon, set in white gold with white diamonds.

Champagne color | The soft color of this brilliant-cut gem is highly unusual, and is most likely an unexpected response to heat treatment.

Changing color | Not all heat-treated brown zircons turn a brilliant dark blue—some, like this faceted example, turn a very pleasing, almost transparent, mid-blue.

Openwork brooch | A cluster of circular-cut, pale blue zircons sit at the center of this scrolling gold brooch, with ruby and diamond highlights.

Stunning earrings | Set in white gold, the vibrant blue zircons in these attractive earrings are perfectly complemented by the yellow sapphires at the top of each.

Blue gems | This group of fine blue, step-cut zircons with deep pavilions displays the blue color characteristic of heat treatment.

Natural state | Some zircons do not go through heat treatment. This beautiful emerald-cut stone has been kept in its natural reddish-brown state.

Clear zircon resembles diamond, but the two can be distinguished: zircon exhibits double refraction, while diamond does not

Portrait of Princess Leonilla Bariatinskaia, 1843 | Owner of the Black Orlov diamond in the early 20th century, and part of the inspiration behind the curse

Black Orlov diamond

△ **Nadezhda Petrovna Orlov**, one of the diamond's former owners

The Black Orlov diamond, also known as the Eye of Brahma, is famous for its distinctive color and notorious for its curse, which is said to bring doom to its owners.

The Black Orlov's color is not true black but a gunmetal dark gray. The original rough 195-carat stone was later cut to a 67.50-carat cushion cut. It is currently set in a pendant, surrounded by a leaf motif of 800 smaller white diamonds, and hangs from a platinum necklace set with 124 small white diamonds.

The diamond's history is uncertain. It is said to have originally been an eye in a statue of the god Brahma in India, but became cursed after it was stolen by a traveling monk. Supposedly, American diamond buyer J. W. Paris purchased the stone in 1932, sold it, and jumped to his death from a New York skyscraper shortly afterward. Later, according to the story of the curse, Russian princess Leonilla Bariatinskaia fell to her death in Rome in

The Black Orlov diamond set with hundreds of white diamonds

1947 after acquiring the stone—and one month after that, the stone's new owner, Princess Nadia Vyegin-Orlov, also fell to her death from a building in Rome.

As with many "cursed" jewels, the truth is debatable. The story of its origin is suspiciously similar to that of the Orlov diamond, a white diamond stolen from an Indian idol and owned by the Orlovs. Mr. Paris's plunge went unrecorded, and "Princess Nadia Vyegin-Orlov" is not known as a historical figure. Of the real princesses who owned the stone, Princess Leonilla died in 1918 at 101, and Nadezhda Petrovna Orlov, who gave the diamond its name, died in 1988 at around 90. The curse may be baseless, but it has done much for the Black Orlov's mystique.

I'm pretty confident that the curse is broken...

J. Dennis **Petimezas**
Owner, 2004–06

Key dates

Pre-1800s–2006

Unknown The diamond is supposedly stolen from the eye of Brahma in a shrine in Pondicherry, India, by a monk and is cursed — **1800**

1900

Early 1900s The Black Orlov's necklace reaches Russia and passes through the hands of "Princess Nadia Vyegin-Orlov" and Princess Leonilla Bariatinskaia — **1910**

Hindu pagoda in Pondicherry, c.1867

1918 Princess Leonilla Bariatinskaia actually dies in France at 101

1932 J. W. Paris buys the diamond, sells it, and, shortly after, jumps from skyscraper in New York — **1920**

1947 "Princess Vyegin-Orlov" and Princess Leonilla Bariatinskaia allegedly leap to their deaths in Rome, just one month apart

1950

c.1950s Charles F. Winson, a New York City gem dealer, buys the stone from persons unrecorded. It is later set in a pendant

1980

1988 The real Princess Nadezhda Petrovna Orlov, on whom Princess Nadia may be based, dies at around 90 in Switzerland

1990

1995 The diamond is auctioned to an anonymous collector for $1.5 million

2006 The Natural History Museum's "Diamonds Exhibition" in London featuring the diamond is closed early due to threat of robbery

2004 J. Dennis Petimezas, a jeweler and diamond dealer, acquires the diamond from anonymous private collector — **2000**

The Hindu god Brahma, who is said to have placed a terrible curse on the Eye of Brahma jewel after a traveling monk stole it from a statue of the deity

Topaz

△ **Well-formed** topaz crystal

Once, it was believed that all yellow gems were topaz, and that all topaz was yellow—however, neither statement is true. Some topaz is yellow, but it can also be colorless, blue, green, sherry, and its most valuable color variation, pink. It is fairly refractive, splitting light into its constituent colors; as a result, colorless topaz resembles diamond, and has often been mistaken for it. Additionally, some blue topaz is almost indistinguishable from aquamarine. A certain number of gems on the market have been treated by heat and irradiation to change their color.

Specification

Chemical name Aluminium fluorosilicate | **Formula** $Al_2SiO_4(F,OH)_2$
Colors Yellow, golden, orange, pink, green, blue, colorless
Structure Orthorhombic | **Hardness** 8 | **SG** 3.5–3.6 | **RI** 1.62–1.63 | **Luster** Vitreous | **Streak** White | **Locations** Brazil, Russia, Germany, Nigeria, Afghanistan, US, Pakistan, Japan

Rough

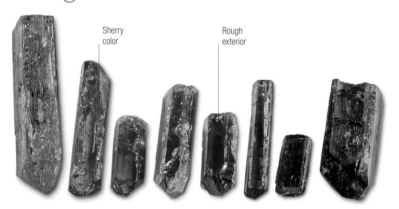

Sherry color

Rough exterior

Pegmatite

Topaz crystal

High level of transparency

Rough crystals | This group of gem-quality topaz crystals shows a clear color gradation from yellow and sherry to a dark, almost red hue. The outer texture of these crystals obscures their transparent interiors.

Topaz in pegmatite matrix | This finely crystallized, light blue topaz is resting on a groundmass of pegmatite, a mineral host in which it frequently occurs.

Topaz as a megagem

Gemstone heavyweights

Topaz is found in well-formed, prismatic crystals with a lozenge-shaped cross section. Although most gemstone material is found in alluvial deposits as water-worn pebbles, a number of very large crystals have also been found in situ. The world's largest preserved topaz crystal weighs 596 lb (271 kg), and in the 1980s a faceted gem weighing 22,892.5 carats—10.1 lb (4.6 kg)—was cut from a Brazilian rock fragment.

Sherry topaz crystal This prismatic crystal shows good form and color. It comes from the Ouro Preto deposit in Minas Gerais, Brazil, the largest commercial source of sherry topaz.

Termination faces at end of crystal

Brown topaz | A fine, gem-quality topaz, this rough has an unusual brown cast to its color. Its lightness and transparent surface reveal internal blemishes.

Brazilian topaz | The rich, red-brown color and excellent transparency of this superb topaz crystal from Brazil make it a gem-cutter's dream.

Cut

Settings

Fancy cut | The brilliant-cut heart shape—one of the most difficult cuts—of this 12.77-carat blue topaz shows the highest degree of craftsmanship.

Pavillion facets reflect through table

Neckace

Oval-cut stones on bracelet

Intricate facets

Enhanced color

Imperial topaz | This oval mixed-cut imperial topaz uses a combination of triangular facets on the crown and rectangular facets on the pavilion to spectacular effect.

Emerald cut | The emerald cut of this large, 55.68-carat gem showcases its deep color, most likely obtained by heat treatment and irradiation of natural topaz.

Earrings

Pendant, which can be worn as a brooch

Antique necklace set | This set, dating from around 1830, features a necklace with accompanying pendant, bracelet, and earrings. The stones comprise matching oval-cut topaz and citrines, and the pendant can also be detached and worn as a brooch.

Shallow crown with deep pavillion

Side of stone intricately faceted

Brilliant cut | The masterful cutting of this 81.30-carat brilliant-cut cushion is revealed in a side view, in which the layering of facets on the sides of the stone can be seen.

Good transparency

"Basket" mounting

Blue oval ring | The blue topaz gemstone in this white gold ring setting is cut as an oval brilliant, and set in a particularly deep "basket" mounting.

In ancient times, the name "topaz" was mistakenly applied to peridot crystals

Andalusite

△ **Blocky** andalusite crystals grouped in a matrix

Crystals of andalusite are pleochroic, meaning they appear to be different colors when viewed from different angles. Andalusite is named after Andalusia, the Spanish region where it was first discovered. It is an aluminium silicate, closely related to both silimanite and kyanite, with which it shares the same chemical composition, but it has a different crystal structure. A strikingly beautiful but relatively lesser-known gem type, andalusite is most often opaque or translucent, with transparent specimens being extremely rare.

Specification

Chemical name Aluminium silicate | **Formula** $Al_2(SiO_4)O$
Colors Pink, brown, white, gray, violet, yellow, green, blue
Structure Orthorhombic | **Hardness** 7½ | **SG** 3–3.2
RI 1.63–1.64 | **Luster** Vitreous | **Streak** White
Locations Belgium, Australia, Russia, Germany, US

Four chiastolite crystals

Cross section | **Rough** | These andalusite crystals are of the variety chiastolite, with a cross-shaped cluster of elongated and tapered crystals.

Polished cross sections

Smooth cabochons | **Cut** | Chiastolite is often tumble polished into rounded gems. These fine examples highlight the typical cross-shaped twinning.

Yellow-brown colors

Octagonal step cut | **Cut** | This fine octagonal, step-cut gem presents a pleasing blend of yellow-tinted brown, highlighted by its bold cut.

Facets show the different colors

Faceted oval | **Cut** | This yellow-tinted andalusite gemstone has an oval step cut, emphasizing the exceptional clarity and briliance of the stone.

Andalusite slab | **Rough** | This specimen is another example of chiastolite—andalusite that forms cross-shaped patterns.

Single crystal

Dark graphite impurities

Andalusite is known as the "seeing stone"

Faceted oval stone

Ring with andalusite | **Set** | This disctinctively asymmetrical ring features an oval-cut andalusite gemstone surrounded by a swirl of diamonds.

Titanite

△ **Classic** wedge-shaped titanite crystals on a rock matrix

Formerly known as sphene, (the Greek for "wedge"), titanite is a common mineral in many igneous rocks and in metamorphic rocks such as gneiss and schist. It occurs as translucent or transparent crystals. It is found in numerous locations, and can occur as reddish brown, gray, red, yellow, or green monoclinic crystals. Its "fiery" color results from its high level of dispersion and high refraction index. In addition to its use as gems, it is also a source of titanium dioxide, which is used in pigments.

Specification

Chemical name Calcium titanium silicate | **Formula** $CaTiSiO_5$
Colors Yellow, green, brown, black, pink, blue | **Structure** Monoclinic | **Hardness** 5–5½ | **SG** 3.5–3.6 | **RI** 1.84–2.11
Luster Vitrous to greasy | **Streak** White | **Locations** Europe, Madagascar, Canada, US, Brazil, Russia, Pakistan

Crystals emerge from the rock matrix

Highly refractive facets

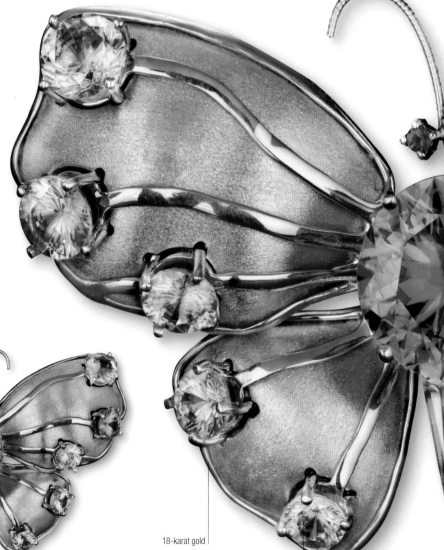

Crystal on rock groundmass | Rough | Lozenge-shaped titanite crystals cover the top of the rock matrix in this superb collector's specimen.

Faceted oval | Cut | This oval brilliant-cut titanite stone has been expertly faceted. Its naturally dark yellow coloring gives its cut a dense appearance.

Fine antennae

18-karat gold

Natural titanite sphere gems

Rectangular titanite | Color variety | This rectangular, step-cut gemstone has a lower iron content, resulting in the clear, yellow-green color seen here.

Butterfly with titanite | Set | Madagascar is the source of the 11 fine-quality, brilliant titanite gems that provide the sparkle for this 18-karat gold butterfly brooch with sapphire eyes.

Sillimanite

△ **Superbly transparent**, mixed-cut cushion sillimanite

Although mainly an industrial material, transparent sillimanite is the basis of attractive faceted gemstones. Cabochons are cut from a form of sillimanite called fibrolite, so named because the mineral resembles bunches of fibers twisted together. Crystals are long, slender, and glassy, or in blocky prisms. Blue and violet are the most prized colors for gemstones. Sillimanite is distinctly pleochroic: yellowish green, dark green, and blue can be seen within the same stone from different angles. It is a common mineral in some metamorphic rocks.

Specification

Chemical name Aluminium silicate | **Formula** Al$_2$OSiO$_5$
Colors Colorless, blue, yellow, green, violet | **Structure** Orthorhombic | **Hardness** 7 | **SG** 3.2–3.3 | **RI** 1.66–1.68
Luster Vitreous or silky | **Streak** White | **Locations** Myanmar, India, Czech Republic, Sri Lanka, Italy, Germany, Brazil, US

Fibrous rock | Rough | This specimen of fibrous sillimanite is typical of its natural occurrence—it is rare to find gem-quality examples of the mineral.

Acicular (needlelike) crystals

Sillimanite crystals in rock | Rough | These elongated, prismatic sillimanite crystals are contained within a matrix of muscovite mica.

Faceted sillimanite | Color variety | This Burmese gem, with a brilliant-cut crown, is a perfect example of the gem's pleochroism, showing bluish-violet and pale yellow colors.

Large oval | Cut | Found in Brazil, this stone is exceptionally large, weighing just over 21 carats, and has been faceted to enhance its yellowish-green coloring.

Mixed-cut cushion | Cut | The skillful faceting of this fibrolite sillimanite gemstone brings out its particularly fine clarity, transparency, and hue.

Pavilion facets visible through table facet

Cabochon, emerald, and scissors are the cuts most commonly used for sillimanite

Visibly fibrous cat's-eye effect

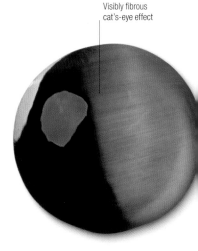

Cat's-eye cabochon | Cut | Sillimanite's own fibrous nature sometimes yields a cat's-eye effect when cut *en cabochon*, as illustrated in this fibrolite.

Dumortierite

△ **Richly colored**, tumble-polished dumortierite

The most prized colors of dumortierite are an intense deep blue to violet. Although sometimes found in small crystals, it is best known in its massive form, when it is used for gemstones cut *en cabochon*, and in carvings. Crystals show pleochroism from red to blue to violet, and have, on rare occasions, been faceted for collectors. Dumortierite occurs in pegmatites, in aluminium-rich metamorphic rocks, and in rocks metamorphosed by boron-bearing vapor from intruding bodies of granite.

Specification

Chemical name Aluminium-iron borosilicate | **Formula** $Al_7(BO_3)(SiO_4)_3O_3$ | **Colors** Blue, violet, brown, green
Structure Orthorhombic | **Hardness** 7–8.5 | **SG** 3.2–3.4
RI 1.68–1.69 | **Luster** Vitreous | **Streak** White | **Locations** US, Madagascar, Japan, Canada, Sri Lanka, South Africa, Italy

Blue crystals

Dumortierite in matrix | **Rough** | The intense blue acicular dumortierite crystals contrast with the whitish-brown rock in this matrix specimen.

Rich color

Dumortierite rock | **Rough** | This dumortierite rough shows a vivid blue coloring. This specimen could be cut into highly desirable cabochons.

Tumbled gem | **Color variety** | Even when tumble polished, this dumortierite stone is valuable for its strong color, heightened by the smooth and shiny finish.

Oval cabochon | **Cut** | The streaks of the white mineral within this high-domed dumortierite cabochon lend texture and interest to the stone.

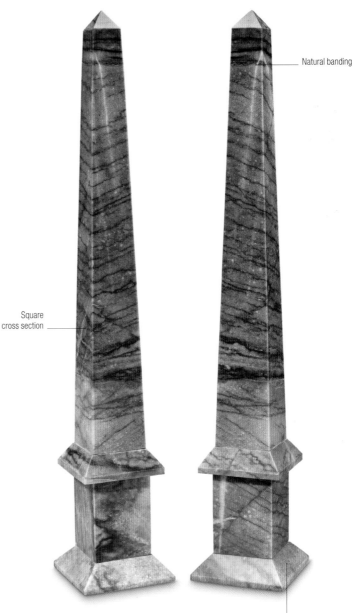

Natural banding

Square cross section

Chamfered base

Rare obelisks | **Carved** | Cut from massive Brazilian dumortierite, this stunning pair of 28 in (71 cm) high obelisks are designed as a household ornament. The cut makes the most of the banded color variation of the original specimen.

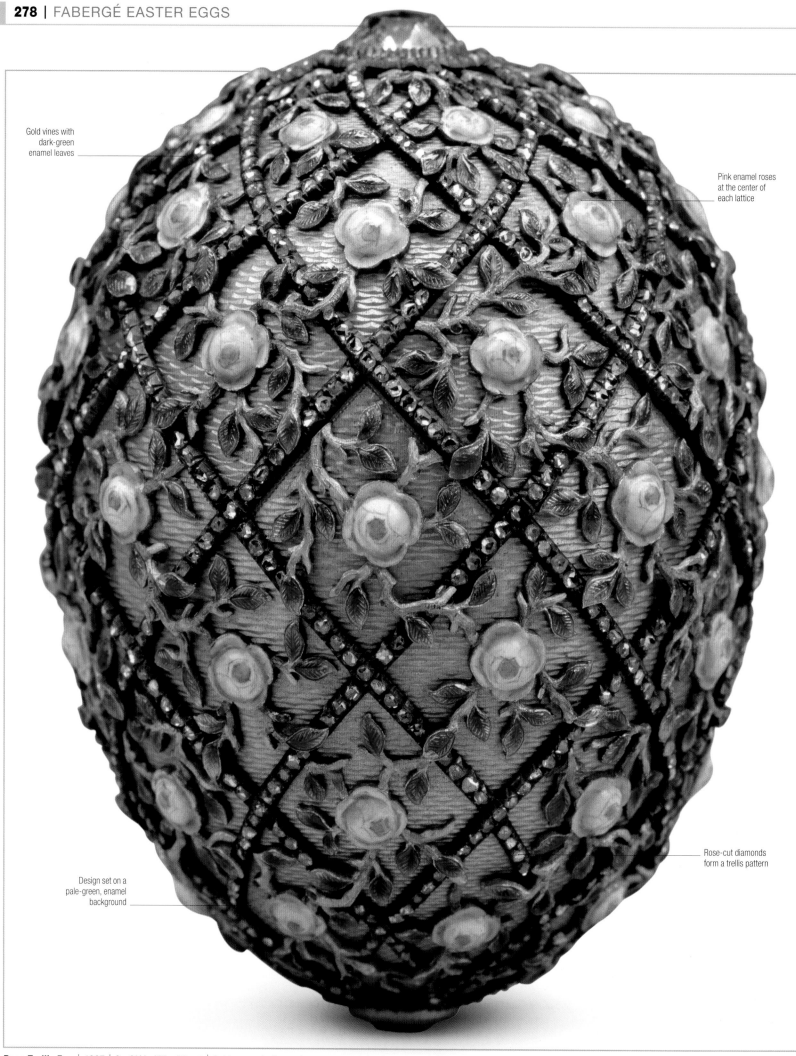

Gold vines with
dark-green
enamel leaves

Pink enamel roses
at the center of
each lattice

Rose-cut diamonds
form a trellis pattern

Design set on a
pale-green, enamel
background

Rose Trellis Egg | 1907 | 3 x 2¼ in (77 x 59 cm) | Gold, enamel, diamonds

Fabergé
Easter eggs

△ **Coronation Egg**, which contained a replica of Tsarina Alexandra's coronation coach

n the Russian Orthodox Church, Easter was the most important date in the calendar. After fasting for weeks during Lent, the faithful could look forward to the climax of the celebrations on Easter Sunday, when eggs—one of the prohibited foods—would be exchanged. These ranged from real, hand-painted eggs to artificial ones, produced as presents for ladies. The most sumptuous of all were the jeweled eggs created by Carl Fabergé for the tsarinas (empresses) of Russia.

Fabergé designed his first imperial egg in 1885, when Alexander III commissioned one as a gift for his wife. From the outset, Carl was determined to do more than simply create an aesthetic arrangement of valuable gems. Instead, he hoped to delight his royal client by placing a surprise within a surprise. Inside his plain, enameled Hen Egg, there was a golden yolk, containing a tiny golden hen. This in

Lily of the Valley Egg, a gift from Nicholas to Alexandra featuring Art Nouveau styling

turn could be opened up to reveal two further surprises—a miniature, diamond crown and a ruby pendant.

The Hen Egg proved a huge success and Fabergé was engaged to create a similar gift each year. This became something of a royal tradition, lasting for over 30 years, until the outbreak of the Revolution. The most exquisite example, perhaps, is the Coronation Egg, which was ordered as a gift for the newly-crowned empress, Alexandra. The "surprise" was a perfect, miniature replica of the coach that was used in the ceremony, while the color scheme of the eggshell echoed the design of her dress. A decade later, in April 1907, the Tsarina received the Rose Trellis Egg to commemorate the birth of her first and only son, Alexei. Decorated with pink enamel roses and a lattice of rose-cut diamonds, the egg contained a diamond necklace and a portrait of the young Tsarevich Alexei.

Monsieur Fabergé's work reaches the limits of perfection

Carl Fabergé's workshop in St. Petersburg, Russia, pictured around 1910, after Carl and his brother Agathon expanded their business to increase capacity

Review of Fabergé's eggs at the Paris Exposition Universelle, 1900

Key dates

1793–2013

1800

1793 Catherine the Great commissions the coach that appears in the Coronation Egg

1885 Carl Fabergé produces his first Easter Egg—the Hen Egg—for a Russian Emperor

1896 The coronation of Nicholas II and Alexandra takes place in Moscow

1900

1897 At Easter, Nicholas presents the Coronation Egg to the Empress

1907 Nicholas presents the Rose Trellis Egg to Alexandra to mark the birth of Tsarevich Alexei

1910

Tsarina Alexandra, Empress of Russia

1918 Nicholas and Alexandra are murdered by Bolsheviks in the aftermath of the Russian Revolution

1925

1927 Joseph Stalin sells several of the eggs to enable him to acquire foreign currency. Many are taken to the West

1950

Bolshevik poster from the Russian Revolution

2000

2013 Viktor Vekselberg, owner of the largest collection of Fabergé eggs in the world, opens the Fabergé Museum in St. Petersburg

2007 The Fabergé family is reunified with the Fabergé brand, having lost their rights to the name in 1920

2015

Kyanite

△ **Gem kyanite rough** showing unusual thickness

Usually blue or blue gray, and generally mixed or zoned within a single crystal, kyanite can also be green, orange, or colorless. It mainly occurs as elongated blades that are often bent, and less commonly as radiating, columnar aggregates. Kyanite is formed during the metamorphism of clay-rich sediments. It occurs in mica schists, gneisses, and associated hydrothermal quartz veins. Until recently it was not considered a gem mineral, but in the last decades transparent material has been found. Stones cut from it rival blue sapphire in intensity of color.

Specification

Chemical name Aluminuim silicate | **Formula** Al_2SiO_5 | **Colors** Blue, green, orange, colorless
Structure Triclinic | **Hardness** 4½–7 | **SG** 3.5–3.7
RI 1.71–1.73 | **Luster** Vitreous | **Streak** Colorless
Locations Brazil, Switzerland, US

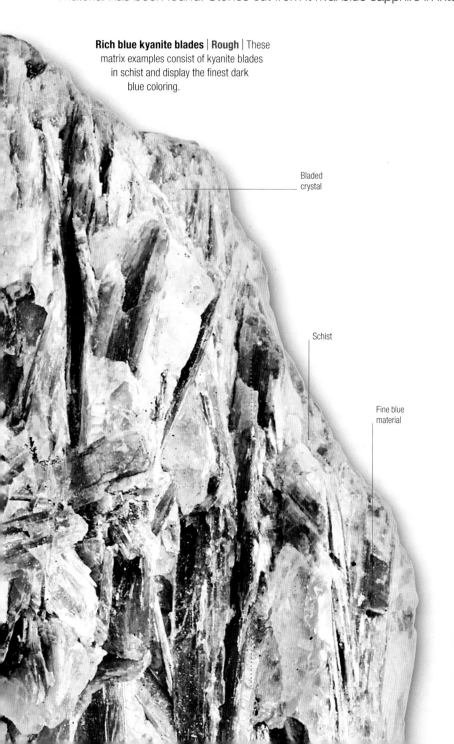

Rich blue kyanite blades | **Rough** | These matrix examples consist of kyanite blades in schist and display the finest dark blue coloring.

Bladed crystal

Schist

Fine blue material

Bladed crystal

Bladed kyanite crystals | **Rough** | Kyanite classically forms relatively thin, bladed crystals, such as these gem-quality specimens in a matrix.

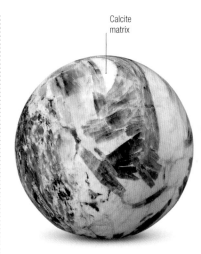

Calcite matrix

Kyanite sphere | **Carved** | The skill of the lapidary is evident in this carved sphere, since the blue kyanite is set in calcite, which is much softer than the kyanite.

Fine oval | **Color variety** | While it is not the deepest kyanite blue, the color of this oval brilliant gem is close to that of fine Burmese sapphire.

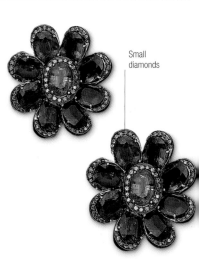

Small diamonds

Ear clips | **Set** | This pair of flower head ear clips is set with rich blue kyanite oval-cut stones, each of which is surrounded by a fringe of tiny diamonds.

Staurolite

△ **Staurolite in schist** from Russia showing cross-shaped twinning

Staurolite is hydrous iron magnesium aluminium silicate. It occurs with garnet, tourmaline, and kyanite or sillimanite in mica schists and gneisses and other metamorphosed aluminium-rich rocks. Staurolite is reddish brown or yellowish brown, or nearly black, and normally occurs as prisms which are hexagonal or diamond shaped in section. Staurolite is named from the Greek *stauros*, "cross," and *lithos*, "stone," for its crosslike twinned form. These cross-shaped crystals are frequently set in silver for use in religious jewelry.

Specification

Chemical name Aluminuim silicate | **Chemical formula** $(Fe,Mg)_4Al_{17}(Si,Al)_8O_{45}(OH)_3$ | **Color** Brown | **Structure** Monoclinic | **Hardness** 7–7½ | **SG** 3.7 | **RI** 1.74–1.75 | **Luster** Vitreous to resinous | **Streak** Colorless to gray | **Locations** US, France, Brazil

Gemmy crystals

Small crystals

Staurolite and kyanite specimen | **Rough** | Staurolite and kyanite often occur together, as in this specimen of matrix of muscovite mica schist.

Staurolite schist | **Rough** | Staurolite commonly occurs in a groundmass of mica schist. In this example, the crystals are rich brown and gemmy.

Twin staurolite crystal | **Rough** | This specimen of cross-shaped crystal of staurolite displays typically geometric lines caused by twinning.

Single crystal | **Rough** | This twin of staurolite has been separated from its matrix. These crystals are frequently mounted as pendants.

Carved sphere | **Carved** | Small crystals are set in a feldspar and mica matrix in this rare sphere. The material originates from the Kola Peninsula of Russia.

Phenakite

△ **Large, finely-formed** gem quality phenakite crystals on matrix

Phenakite's name originates from the Greek for "deceiver"—see box, below. It can be colorless, but more often is translucent gray or yellow and, occasionally, pale, rose red. Phenakite is found in high-temperature pegmatites and in mica schists, often accompanied by quartz, chrysoberyl, apatite, and topaz. Its crystals are mainly rhombohedrons, and somtimes short prisms. Transparent crystals are faceted for collectors. Its indices of refraction are higher than topaz, and its brilliance approaches that of diamond.

Specification

Chemical name Beryllium silicate | **Formula** Be$_2$SiO$_4$ | **Colors** Colorless, white | **Structure** Trigonal | **Hardness** 7.5–8 | **SG** 2.9–3 | **RI** 1.65–1.67 | **Luster** Vitreous | **Streak** White | **Locations** Russia, Norway, France, US

Large crystal | Rough | This large single crystal of phenakite with adhering matrix at the base shows perfect phenakite crystal form.

Brazilian phenakite | Cut | Cut in a fancy cushion with multiple, layered facets, this stunning Brazilian stone weighs 29.80 carats.

Flawless interior

Table facet

Phenakite and quartz

The great deceiver

Phenakite has a well-deserved reputation for deception, also the basis for its name—it is difficult to to distinguish it from colorless quartz, both in its appearance and its technical specifications. Mineralogists can use various solutions to tell the two substances apart, including testing their specific gravity (quartz's is slightly lower, at 2.65, while phenakite's is 3) or their hardness (phenakite is slightly harder). The latter can be tested by conducting a scratch test on a piece of quartz.

Clear quartz crystals This specimen of rough rock crystal could potentially be mistaken for phenakite.

Burmese phenakite | Cut | This extremely fine, totally colorless 25.57-carat brilliant oval-cut phenakite from Myanmar is nearly an inch in length.

Euclase

△ **Single euclase crystal** from Chivor, Columbia, weighing 46.2 carats

uclase is beryllium aluminium silicate. It is generally white or colorless but can also be pale green or pale to deep blue—a color for which it is particularly noted. It forms grooved prisms, often with complex end faces, and can also be found in masses and fibers. In faceted gems, pale to rich aquamarine is favorite, but other colors are also cut. It is relatively uncommon in gem material, and is principally cut for collectors. Euclase takes its name from the Greek *eu*, "good," and *klasis*, "fracture," in reference to the way in which it breaks in perfect planes.

Specification

Chemical name Beryllium aluminium silicate | **Formula** $BeAlSiO_4(OH)$ | **Colors** Colorless, white, blue, green
Structure Monoclinic | **Hardness** 7.5 | **SG** 3
RI 1.65–1.68 | **Luster** Vitreous | **Streak** White
Locations Brazil, US

Pyramidal termination

Prismatic colorless euclase crystal | **Rough** | This single, perfectly formed euclase crystal displays a fine prismatic form. It is technically colorless, although it appears to have a yellowish tinge to its interior.

Blue crystal

Matrix specimen | **Rough** | Aside from being gem quality, this blue euclase crystal, resting in a groundmass of quartz sprinkled with pyrite, is a fine specimen.

Dark inclusions

Octagonal gemstone | **Cut** | Full of dark inclusions yet a desirable stone, this colorless euclase is faceted in deep step cut.

Brazilian stone | **Cut** | Featuring an emerald cut and a medium blue green color, this Brazilian gemstone originates from the Minas Gerais region.

Cushion-cut euclase | **Cut** | This cushion-cut gem also come from Brazil, and weighs 7.17 carats. It displays a grayish blue coloring.

Napoleon Diamond necklace | Commissioned in 1811 | About 7¾ in (20 cm) wide | 234 diamonds weighing about 263 carats, seen here in a portrait of Marie Louise of Austria

Napoleon diamond
necklace

△ **Emperor Napoleon I** in a portrait by François Gérard (detail), c.1805–15

Commissioned by Napoleon I of France in 1811 for his wife Archduchess Marie Louise of Austria to celebrate the birth of their son, this necklace is composed of 234 diamonds. The single thread is set with 28 mine-cut (the earliest form of brilliant-cut) diamonds. A second tier has nine pendeloques and 10 briolettes (teardrop-shaped cut).

Napoleon had divorced the Empress Josephine, who had failed to give him an heir, and married Archduchess Marie Louise of Austria in 1810. Napoleon's son was born within a year, and he duly commissioned the Parisian jewelers Nitot et Fils to produce the 376,274-franc necklace, a sum equal to the Empress's entire annual household budget. Marie Louise wore the necklace in several contemporary portraits and kept it until her death.

Maria Theresa of Portugal, who inherited the necklace in 1914

Maria Theresa, a princess of Portugal, eventually inherited it and, in 1929, decided to sell. She engaged two agents, "Colonel Townsend" and "Princess Baronti," to sell the necklace for $450,000. Since the stock market had just crashed, this figure proved unrealistic and the agents began offers at $100,000, enlisting Archduke Leopold of Hapsburg, Maria Theresa's penniless grandnephew, to give assurances of authenticity to buyers. The necklace eventually sold for $60,000, but the agents and Archduke Leopold claimed a collective fee of $53,730 as expenses. Maria Theresa took the matter to court, recovered the necklace and Leopold was jailed. The "Townsends" evaded capture, however, and their true identities remain a mystery.

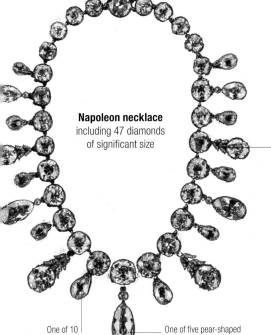

Napoleon necklace
including 47 diamonds
of significant size

One of 10
briolette diamonds

One of five pear-shaped
pendeloques

Motif set with 23 small
diamonds attached to each
of four ovaline pendeloques

Key dates

1811–1962

June 1811 Napoleon commissions the necklace to celebrate his son's birth

1800

March 1811 Napoleon's son, Napoleon François-Joseph Charles, is born to Marie Louise

1872 On Sophie's death, the necklace is inherited by her sons, Archdukes Karl Ludwig, Ludwig Viktor, and Franz Joseph of Austria

1850

1847 Marie Louise dies. The necklace passes to Sophie of Austria. Two diamonds are removed for earrings, since lost

1900

1914 The necklace passes to Karl Ludwig's third wife, Maria Theresa of Portugal, on his death

1920

1929 Maria Theresa tries to sell the necklace, but recovers it after an attempt to swindle her out of the proceeds

Paul-Louis Weiller and his wife at the Academy of Fine Arts, Paris, 1965

1940

1944 Maria Theresa dies

1948 The Hapsburg family sells the necklace to French industrialist Paul-Louis Weiller

1960

1960 Harry Winston buys the necklace from Weiller and later sells it to Marjorie Merriweather Post

1962 Post donates the necklace to the Smithsonian Institution. It remains on display at the National Museum of Natural History, Washington DC

1980

Thirteen... diamonds are type IIa [almost completely pure]... consistent with the jewel's imperial pedigree

Drs. **E. Gaillou** and **J. Post**
National Museum of Natural History, Smithsonian Institution

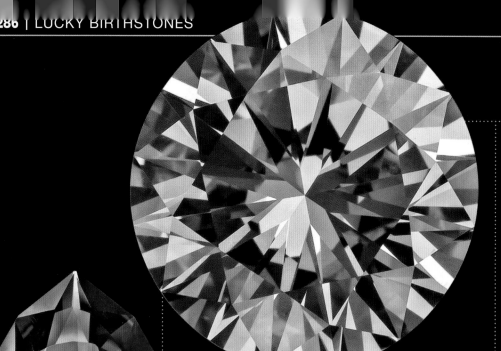

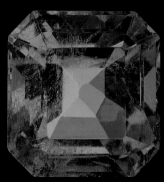

May—Emerald
Western birthstone tradition has roots in the Bible, which links gemstones, including emerald, to zodiac signs.

April—Diamond
Assigned to April in the modern tradition, diamond is also said to improve relationships for those with April birthdays.

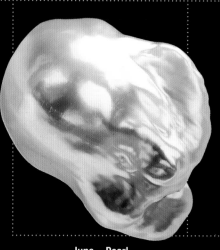

February—Amethyst
Associated with royalty and wine, amethyst is the birthstone for February in modern and ancient lists.

March—Aquamarine
In 1952, aquamarine was designated as March's birthstone. It is said to bring calm.

June—Pearl
Pearl, representing purity, is the most traditional gem for June, but moonstone and alexandrite are also popular.

January—Garnet
Garnet was associated with January in the ancient Ayurvedic tradition, as well as in modern Western lists.

Lucky
birthstones

Each of the 12 signs of the zodiac has long been associated with a gem, which is thought to strike a chord with the character of someone born under that sign, and so bring the person good luck. Later, the gemstones were linked to months rather than astrological signs. The major stones occur in most cultures, though how they are assigned can vary. In this modern set of European associations, March can also be linked to bloodstone; moonstone is an alternative to pearl for June; sardonyx is lucky for August and topaz for November, while turquoise can be a December birthstone.

Rubies ensure health, wealth, and a cheerful nature to the owner

Old Hindu belief

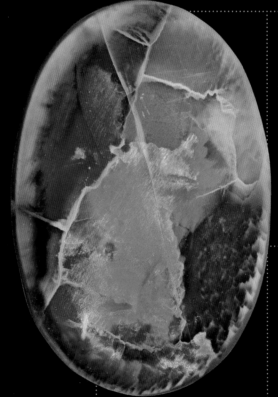

December—Zircon
Zircon was recognized as one of the birthstones for December in 1952, when it replaced lapis lazuli.

October—Opal
Opal was designated as October's birthstone by the American National Association of Jewelers in 1912.

July—Ruby
Ruby is both the modern and traditional birthstone for July; it is associated with passion.

November—Citrine
This is one of the more recent additions to the list, put forward by American jewelers in 1952.

August—Peridot
Prior to 1900, August's birthstone was variously sardonyx, carnelian, moonstone, or topaz.

September—Sapphire
Sapphire is thought to protect loved ones from envy and from harm. It is also the zodiac stone for Taurus.

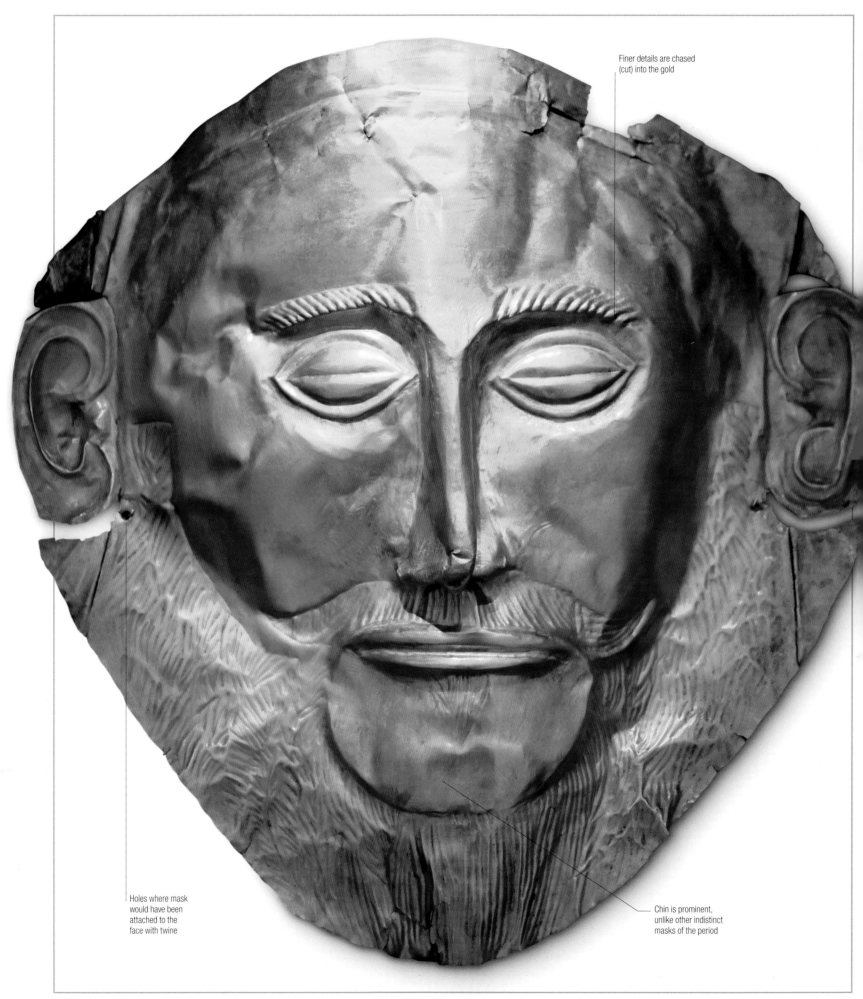

Finer details are chased
(cut) into the gold

Holes where mask
would have been
attached to the
face with twine

Chin is prominent,
unlike other indistinct
masks of the period

Mask of Agamemnon | c.1500 BCE | Gold mask, repoussé method | Discovered in a burial shaft designated Grave V, at the site Grave Circle A, Mycenae

Mask of
Agamemnon

△ **Agamemnon** depicted in a painting c.1633

A gold death mask, the Mask of Agamemnon is one of the world's most famous—and controversial—archaeological artifacts. It was found covering the face of a body in shaft graves in Mycenae, Greece, in 1876.

Made from a thick sheet of gold that has been hammered over a wooden mold, it depicts the face of a bearded man. Fine textures, such as the eyebrows and beard, were added using a sharp tool. Heinrich Schliemann, the archaeologist who found it, claimed he had "gazed on the face of Agamemnon," legendary king and leader of the famous attack on ancient Troy. This claim was later refuted when it emerged that the graves, from around 1500 BCE, predated the Trojan War by around three centuries, although the site was still identified as Troy. The name has stuck, however, and the mask is accepted as a genuine artifact by the National Archaeological Museum of Athens, where it remains a highlight today.

Typical ancient Greek death mask

However, critics point to the atypical features—facial hair, separate ear flaps, and distinct eyebrows—none of which are present on other masks of the same period at the site. Schliemann had been suspected of "salting" his previous finds with treasures from other sites, and detractors suggested he may have planted a forgery, or reworked an ancient mask.

Calls have been made to test the age of the mask—ancient gold is impure and contains alloyed minerals that corrode over time, which can help establish age. However, the Archaeological Museum of Athens regards doubts about the mask as unfounded, and it remains one of the world's most intriguing pieces of precious metalwork, as well as a stunning artifact in its own right.

I have gazed on the face of Agamemnon

Heinrich **Schliemann**
in a telegram to a Greek newspaper upon finding the mask

Scene from the Trojan War in a 16th-century Italian fresco, featuring the Trojan horse, with which Agamemnon tricked the Trojans and took the city of Troy

Key dates
c.1500 BCE–1983

c.1550–1500 BCE
The mask is created and buried in a grave shaft

1500

c.1260–1180 BCE
The modern date range of the Trojan War, and time of mythic figure Agamemnon, are too late to coincide with the mask

1200

0

Archaeologist Heinrich Schliemann

1800

1871 Heinrich Schliemann begins work at Troy (modern-day Hisarlik) at the behest of British archaeologist Frank Calvert

1850

1876 Schliemann discovers the mask and cables the King of Greece

1900

1983 The Central Archaeological Council rejects a request for the mask to be tested on advice from the National Archaeological Museum, Athens, where the artifact still resides

1950

1972 Archaeologist David Calder criticizes Schliemann's work and other critics follow, casting doubt on his finds

2000

Organic gems

"Sword" set with diamonds

Pearl forms "body"

Baroque pearl

The Canning Jewel | This Italian Renaissance pendant is in the form of a merman holding a gorgon's head. His "body" is a large blister pearl set in enameled gold, with rubies, table-cut diamonds, and baroque pearls.

In **Japan, pearl hunters have been diving without breathing equipment for around 2,000 years**

Pearl

△ **Spherical pearl** showing iridescence

Pearls are natural gems produced by the pearl oyster and the freshwater pearl mussel. Although other kinds of mollusk can produce a "pearl," these have little value since they are not composed of nacre (the same substance as mother-of-pearl). Nacre is secreted in response to a microscopic irritant in the mollusk's soft tissue. The concentric rings of nacre around the particle create the particular iridescence of pearls due to the way the overlapping layers diffract light waves. A pearl's color is described in terms of body color and overtone—the most common body color is white, though it can range widely, and the overtone is the color that seems to appear only on the surface of the pearl.

Natural and man-made

Natural pearls that form in the wild are rare and valuable, and pearl divers have to open hundreds of pearl oysters before chancing on a specimen. Although diving for natural pearls still occurs in Bahrain and Australia, today's pearls are largely cultured, making them much more affordable: an artificial nucleus, such as a round shell bead, is placed in the oyster or mussel for the nacre to form around. Freshwater pearls are cheaper because the freshwater mussel can produce up to 20 pearls at a time, whereas the smaller saltwater oyster can create just one. Saltwater pearls are differentiated by region: South Sea pearls are most valued due to their size, Tahitian pearls are next for their colors (black among them), with Akoya the least prized, being the most common.

Specification

Chemical name Calcium carbonate | **Formula** $CaCO_3$
Colors white, pink, silver, cream, brown, green, blue, black, yellow | **Structure** Amorphous | **Hardness** 2.5–4.5
SG 2.60–2.85 | **RI** 1.52–1.69 | **Luster** Pearly

Locations
1 Coastal waters of Japan **2** Coastal waters of China
3 Coastal waters of Australia

Key pieces

Gold support

Roman earring | Crafted in a style common in the 3rd century CE, this extravangant Roman earring is set with garnets cut *en cabochon* and five natural pearls suspended on gold mounts.

Enamel decoration

The Hope Pearl | One of the former owners of the Hope Diamond (see pp.62–63), 19th-century collector Henry Hope also acquired this stunning 450-carat bronze-to-white baroque pearl capped with gold and enamel.

The Baroda Necklace | Originally a seven-strand necklace belonging to Maharajah Khande Rao Gaekwad of Baroda, India, this piece was reduced in size in the mid-20th century. It is still the world's most valuable pearl necklace.

Rough

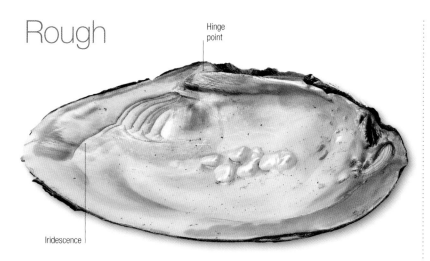

Hinge point

Iridescence

Mother-of-pearl | The lining of pearl-producing mollusks, mother-of-pearl, is the same material as pearl—a mixture of aragonite in conchiolin called nacre. It can be seen lining the interior of this shell with an iridescent sheen.

Queen conch pearl | Of the mollusks that produce pearls, the queen conch is among the rarest. Its pearls have a distinctive look, as seen in this deep pink pearl.

Settings

Gold and pearl pin | Created in the form of a eight-pointed star, this gold pin is set with a large central pearl, pearls on each point, and pearls on "rays."

Irregular lines

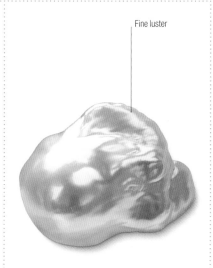

Fine luster

Baroque pearl | While perfectly round pearls are considered the most desirable by some, baroque pearls such as this black pearl offer more creativity for the jeweler.

Freshwater baroque white pearl | This baroque, white pearl is the kind that could provide the centerpiece for a whimsical sculptured gold jewelry piece.

Double pearl

Freshwater pearls | Freshwater pearls such as these have exactly the same makeup and luster as saltwater pearls, and were more accessible to ancient man.

Diamonds

Baroque necklace | This Van Cleef & Arpels necklace uses baroque pearls—11 drop-shaped pearls suspended from a necklace of diamond-set gold beads.

The Palawan Princess

The world's second largest pearl

Found in the coastal waters off the Island of Palawan in the Philippines, the Palawan Princess is the world's second largest known pearl, weighing 5 lb (2.27 kg)—equivalent to 11,340 carats. The product of the giant clam shell species *Tridacna giga*, the pearl is not considered a gem pearl because it is non-nacreous and lacks the luster of gem pearls. Nevertheless, it was valued at around $300,000–$400,00 in 2009.

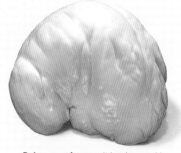

Palawan princess It has been said that the pearl's shape bears an uncanny resemblance to a human brain.

"Pink" pearl

Cultivated pearls | This group of four pearls are all cultivated, and display the color variations possible depending on the growth environment.

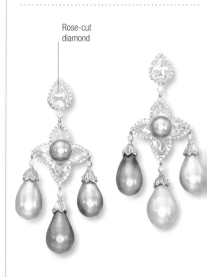

Rose-cut diamond

Multicolored pearl earrings | These pendant earrings set in 18-karat white gold feature pearl drops of three different colors suspended from diamond-set mountings.

Ancient Egyptian queen Cleopatra was said to dissolve pearls in vinegar and drink the mixture

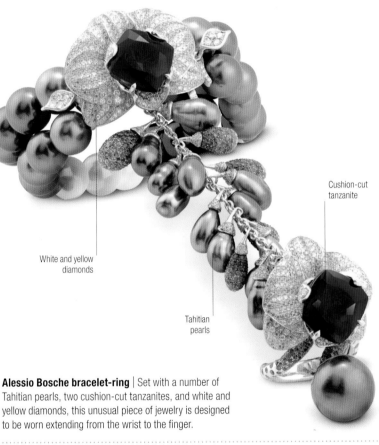

White and yellow diamonds

Cushion-cut tanzanite

Tahitian pearls

Alessio Bosche bracelet-ring | Set with a number of Tahitian pearls, two cushion-cut tanzanites, and white and yellow diamonds, this unusual piece of jewelry is designed to be worn extending from the wrist to the finger.

Multicolored pearl cluster

Cartier trinity ring | Featuring an eye-catching cluster of white, gold, and pink freshwater pearls, this trinity ring made by Cartier is based around interlocking bands of white and yellow gold set with pavé diamonds.

"Leaves" set with diamonds

Pearl brooch | Set in white gold, this brooch takes the form of diamond-set leaves supporting a group of three drop-shaped pearl dangles.

Pearl centerpiece

Fortune ring | Set with a white South Sea cultured pearl, this 18-karat white gold ring by Mikimoto also features diamond-studded "leaves."

Black pearl necklace | Produced by YOKO, London, this black pearl necklace transitions from dark Tahitian pearls, through gray, to silvery Australian South Sea pearls.

Cartier pearl necklace or bracelet set | In this Cartier piece, accented with tanzanite and diamonds, the second strand of pearls is detachable for use as a bracelet.

ANNO DNT   1 5 4 4.

LADI MARI   DOVGHTER TO
THE MOST   VERTVOVS PRINCE
KING HENRI   THE EIGHT

THE AGE OF   XXVIII YERES

La Peregrina | 1 x ¾ in (25.5 x 17.9 mm) | 50.56 carats (55.95 carats in original form) | Hangs from a necklace worn by Queen Mary I in a portrait from 1544

La Peregrina

△ **La Peregrina**, a drop-shaped, 50-carat natural pearl

Although not the largest natural pearl in the world (that honor goes to the Pearl of Allah), La Peregrina's almost perfect pear shape and its bright white luster have made it one of the most celebrated gems of the past 500 years. The other reason for the pearl's reputation is its provenance—it was worn by Queen Mary I of England in the 16th century, stolen by Napoleon's brother Joseph in the early 19th century, and owned by actress Elizabeth Taylor.

According to 16th-century Peruvian writer Inca Garcilaso de la Vega, the son of a Spanish aristocrat and Incan noblewoman, the pearl was found in the early 1550s by an African slave working in a Panama fishery—he was given his freedom in return. The largest known pearl at the time, it was brought to Spain as a gift for King Philip II and become part of the Spanish crown jewels. "Its circumference, at the largest part, was the same as a large pigeon's egg," wrote de la Vega. It was named La Peregrina ("The Incomparable") because it was so rare. King Philip gave the pearl, hanging from a necklace, to Mary Tudor (later Queen Mary I) as an engagement present. After Mary's death, La Peregrina was returned to Spain, where it was worn by generations of royalty before Joseph Bonaparte snatched it in 1813. It later resurfaced in England where, in 1969, actor Richard Burton bought it for his wife, Elizabeth Taylor. On one occasion, her dog is said to have picked up the pearl in its mouth—fortunately, La Peregrina escaped unharmed.

La Peregrina in Elizabeth Taylor's chosen design

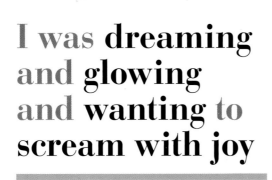

Elizabeth Taylor wearing La Peregrina in her role on the set of the movie *Anne of the Thousand Days*

I was dreaming and glowing and wanting to scream with joy

Elizabeth **Taylor**
on receiving La Peregrina

Key dates

1513–2011

1500

1513 A slave discovers the pearl in a Panama fishery and presents it to Philip II of Spain

1558 After Mary's death, the pearl is returned to Spain

1550

1554 Mary Tudor of England's future husband, Philip II, gives her the pearl

1598–1621 It is owned by Philip III and his wife Margaret of Austria, whose portraits by Diego Velázquez show them wearing the pearl

1600

1700

Queen Margaret of Austria

1813 Joseph Bonaparte, appointed ruler of Spain in 1808 by his brother Napoleon, takes La Peregrina when he flees the country

1800

1844 On Joseph's death, the pearl is bequeathed to his son Louis

1848 Louis sells the pearl to the Duke and Duchess of Abercorn

1850

1900

1972 The pearl is set as a pendant hanging from a diamond and ruby necklace designed by Taylor and Al Durrante of Cartier

1969 Richard Burton buys the pearl at auction as a gift for Elizabeth Taylor

2000

2011 La Peregrina is sold at auction to an anonymous buyer

Shell

△ **Common** spider conch shell

eashells are the exoskeletons of mollusks, composed primarily of a
mineral secretion, as opposed to living cells. Seashells have a long history
of use in body ornamentation, as well as for currency, the cowrie in particular.
Tortoiseshell is of a different substance—the scutes (plates) of the shell used
decoratively are of keratin, the protein that also forms fingernails and hair.
Tortoiseshell is actually from the Hawksbill turtle, which is now protected. It
is a natural plastic, meaning it can be heated and molded into new shapes.

Specification (seashell)

Chemical name Calcium carbonate, aragonite
Formula $CaCO_3$ (calcium carbonate) | **Colors** white, pink, silver,
cream, brown, green, blue, black, yellow | **Structure** Amorphous
Hardness 3–4 | **SG** 2.60–2.78 | **RI** 1.52–1.66
Luster Pearly | **Locations** Worldwide

Pink conch | **Rough** | Adult conch shells
can grow up to around 12 in (30 cm) in size.
Parts of shells were used by tribes in North
America and the Caribbean to make tools.

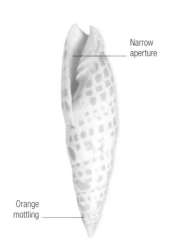

Narrow
aperture

Orange
mottling

Episcopal miter shell | **Rough** | This is
the shell of a large sea snail of the species
Mitra mitra. It is traditionally said to resemble
a bishop's headdress.

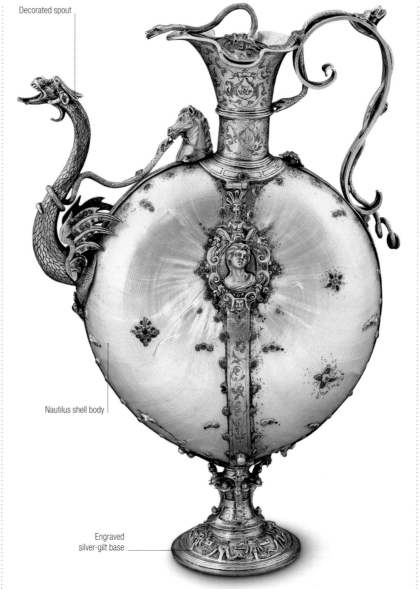

Decorated spout

Nautilus shell body

Engraved
silver-gilt base

Shell pitcher | **Set** | This magnificent pitcher from the
Argenti Museum in Florence, Italy, is made from nautilus
shells. It is also set with pearls, rubies, and turquoise,
and is mounted with gilt silver.

Tortoiseshell handle

Tortoiseshell comb | **Set** | Although it is
now prohibited, tortoiseshell was once used as
an organic gem. This comb features an ornate
tortoiseshell handle with imitation pearls.

**Seashells
have been
used as
a form of
currency
by various
cultures
throughout
history**

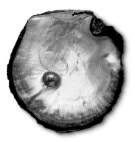

Mother-of-pearl

△ **Black pearl** in its shell of mother-of-pearl

Mother-of-pearl is the name given to nacre, the substance that lines the interior of some mollusks, but chiefly pearl oysters and freshwater pearl mussel shells. It is also the material from which pearls are formed. Prized for its iridescence, nacre is used decoratively in jewelry, clothing, architecture, and art. It is also of great interest to scientists due to its overlapping bricklike structure at a microscopic level, which allows impacts to be absorbed by dispersing them over a wide area—a property that could be applied to the creation of resilient materials.

Specification

Chemical name Calcium carbonate, calcium phosphate, amorphous silica | **Formula** $CaCO_3$ (calcium carbonate) | **Colors** All **Structure** Amorphous (prismatic, nacreous, crossed lamellar, foliated, homogeneous) | **Hardness** 3.5 | **SG** 2.70–2.89 | **RI** 1.530–1.685 | **Luster** Greasy to pearly | **Locations** Worldwide

Patterned shell

Turban shell | **Rough** | South African turban shells such as this have been used in tribal art for millennia, and also occasionally as currency.

Mother-of-pearl coating

Nautilus shell | **Rough** | This stunning example of a large nautilus shell is coated with a fine layer of iridescent mother-of-pearl.

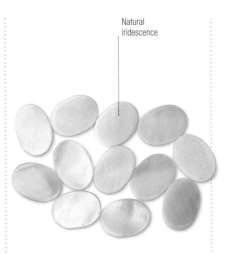

Natural iridescence

Mother-of-pearl beads | **Cut** | These flat oval beads show the natural shimmer and delicate finish that make mother-of-pearl such a popular decorative material.

Mother-of-pearl inlay

Mother-of-pearl pendant | **Set** | This fanciful pendant in the shape of a window consists of mother-of-pearl, synthetic sapphire, and diamond.

Nacreous shell | **Rough** | This shell displays excellent iridescence, with different colors appearing according to the angle of viewing.

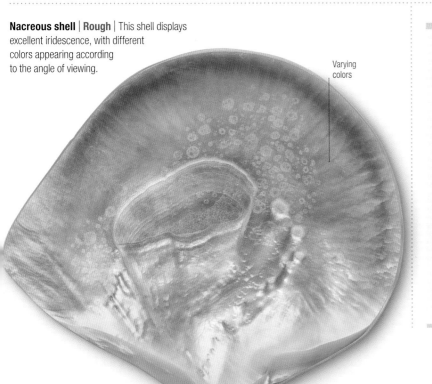

Varying colors

Art in Asia

Mother-of-pearl and lacquer

In various Asian cultures between the 8th and 19th centuries, craftsmen produced exquisite lacquered mother-of-pearl decorative pieces. These ranged from small boxes to large screens, and featured designs illustrating various religious or cultural themes. Nacreous shells were boiled and cut into pieces to form the designs. The craftsmen then coated the pieces with many layers of lacquer, a form of resinous tree sap that hardens to a protective, plasticlike finish.

Lacquer panel This piece features mother-of-pearl figures of a man and boy in rural Japan.

Sea monster's face

Infant Hercules

Mouth forms
aperture of cup

Engraved silver gilt

Engravings of Chinese
dragons and birds

Strapwork decdoration

The collector...
might be likened
to a child who
cries for the moon

Baron Ferdinand **Rothschild**

Fruit decorations

Eagle's-claw
stand

Nautilus shell cup | Assembled c.1550 | 10¼ x 6¾ x 4 in (26.1 x 17 x 10.3 cm); 1¾ lb (845 g) in weight | Engraved nautilus shell, silver-gilt mounts

Nautilus
shell cup

△ **Cutaway** of a nautilus shell showing internal chambers

This exquisite object is one of the greatest treasures in the Waddesdon Bequest, a collection of artifacts owned by Baron Ferdinand Rothschild. It consists of a beautiful chambered nautilus shell from Asia, transformed into a goblet in the form of a grotesque sea monster by Western craftsmen. The shell probably came from Guangzhou, China, where its surface had already been decorated with engravings of dragons. In Europe, this type of shell was an exotic novelty only available from the early 16th century, when the Portuguese began trading with Guangzhou. The identity of the Western artist or artists who worked on it is unknown, but experts believe that the work orginated in Padua, Italy. The decoration of the cup reflects its nautical origins as well as its cultural influences: in Chinese mythology,

Example of a German nautilus shell cup set in silver gilt, c.1700

dragons lived in undersea caverns and were revered as rain bringers. The beast is also similar to those portrayed on contemporary European maps, while the boy is Hercules—identifiable by the serpent that he killed in his crib—who later rescued a maiden from a sea monster.

Items of this kind were much sought after by Renaissance connoisseurs, who displayed them in their "Cabinets of Curiosities." These were treasure stores of lavish or unusual objects, which were designed to highlight the wealth, learning, and worldliness of the collector. The cup's owner, Baron Rothschild, aimed to revive this idea with the New Smoking Room, which he built at his home, Waddesdon Manor. Here, he liked to astound his business associates by showing them his amazing collection of antiques after dinner. Rothschild later donated the contents of the room to the British Museum, on the condition that the collection was kept intact.

These... are an extraordinary way of mapping the world

Edmund **de Vaal**
Artist and writer, describing the objects in the Waddesdon Bequest

"Vanitas" still life from 1689 showing a nautilus cup among other luxury possessions. The painting symbolized the transience of wealth

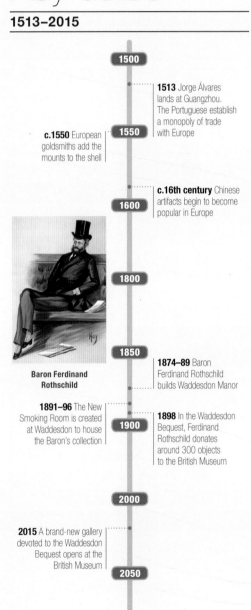

Key dates
1513–2015

1500

1513 Jorge Álvares lands at Guangzhou. The Portuguese establish a monopoly of trade with Europe

c.1550 European goldsmiths add the mounts to the shell

1550

c.16th century Chinese artifacts begin to become popular in Europe

1600

1800

1850

Baron Ferdinand Rothschild

1874–89 Baron Ferdinand Rothschild builds Waddesdon Manor

1891–96 The New Smoking Room is created at Waddesdon to house the Baron's collection

1900

1898 In the Waddesdon Bequest, Ferdinand Rothschild donates around 300 objects to the British Museum

2000

2015 A brand-new gallery devoted to the Waddesdon Bequest opens at the British Museum

2050

HIGH SOCIETY

European aristocrats had always been the primary patrons of luxury jewelry houses, but in the 20th century, with Europe ruined by war and political upheaval, jewelers turned to the new big spenders: the stars, socialites, and heiresses of America. These new clients not only had big budgets, but creative vision, too, since many of the superrich buyers of fine jewelry were also trendsetters. Jeanne Toussaint was head designer of luxury jewels at Cartier from 1933 onward, attracting commissions from some of the leading society women of the day: Wallis Simpson was a devoted client (see pp.224–25), as was Barbara Hutton, heiress to the Woolworths stores.

Rivaling Cartier and other traditional jewelry houses, geologist Harry Winston founded his own jewelry workshop in New York in 1932, and soon made a name for himself with his jaw-dropping gems. In 1944, he became the first jeweler to lend diamonds to an actress for the Academy Awards (Oscar-winner Jennifer Jones), thus securing his name among the stars of Hollywood and high society. Other famous clients included Richard Burton and Elizabeth Taylor (see p.297), and Jackie Kennedy.

People will stare. Make it worth their while

Harry **Winston**
Jeweler

King of Diamonds Harry Winston appealed to the women of American high society by sourcing spectacular gems and setting them in designs that would maximize their brilliance. His philosophy—to let the gemstones dictate the design—set the standard for high-end jewelry in the 1930s.

Carved
mother-of-pearl
inlays

Painted warrior's
face emerges

Plumbate-ware
pottery base

Lid of an effigy jar, showing a Toltec warrior of the Coyote order | 10th to 12th century CE | 5¼ in (13.5 cm) tall | Plumbate-ware pottery, mother-of-pearl, bone

Mother-of-pearl
coyote

△ **Quetzalcoatl**, the Mesoamerican god of wind and learning

This fascinating and striking object is a lid for an effigy jar (a vessel styled as human or animal) from the Toltec civilization of Mesoamerica (c.900–c.1150 CE). According to one theory, the lid portrays the god Quetzalcoatl with human features, but it is more widely thought to represent a Toltec warrior's helmet, which imitated a coyote's head and had an opening for the warrior's face between its jaws.

The lid displays intricate craftsmanship: it is modeled in clay and inlaid with carved mother-of-pearl and bone. The helmet it depicts is that of the Coyote order, one of the military classes of the Toltec, which included the Eagle and Jaguar, among others. Such a helmet indicated a degree of military rank, while also signifying a state of existence between the material world and the animal spirit world. Warriors also dressed in imitation of the animal's body.

The artifact was found in Tula (in modern-day Mexico), once the capital city of the Toltec Empire. The Toltec, who predated the Aztec, were a warlike people

Portrait of a Native American man wearing a coyote headdress—the animal remained a powerful symbol for many cultures

who dominated the region through military force. Religion played an important role in their lives, and human sacrifice to appease the gods was a key component of worship. Evidence of this at Tula includes a *tzompantli*, a skull rack for displaying the heads of sacrificial victims. In addition, three *chacmool* statues were found—reclining warrior figures clutching bowls intended to receive human hearts and other sacrificial offerings for the gods.

Chacmool statue from the Toltec era. The Toltec religion included sacrifice, and these statues were made to hold human organs for the gods

Key dates
250 CE–1970

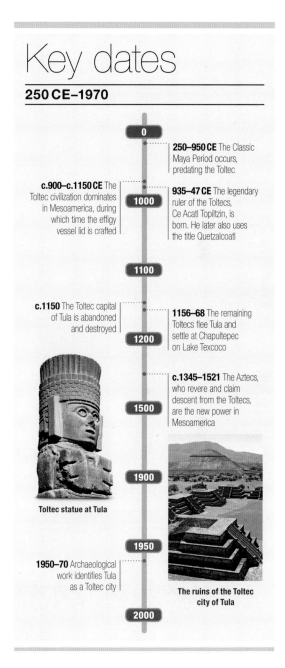

0

250–950 CE The Classic Maya Period occurs, predating the Toltec

c.900–c.1150 CE The Toltec civilization dominates in Mesoamerica, during which time the effigy vessel lid is crafted

1000

935–47 CE The legendary ruler of the Toltecs, Ce Acatl Topiltzin, is born. He later also uses the title Quetzalcoatl

1100

c.1150 The Toltec capital of Tula is abandoned and destroyed

1200

1156–68 The remaining Toltecs flee Tula and settle at Chapultepec on Lake Texcoco

c.1345–1521 The Aztecs, who revere and claim descent from the Toltecs, are the new power in Mesoamerica

1500

Toltec statue at Tula

1900

1950

1950–70 Archaeological work identifies Tula as a Toltec city

The ruins of the Toltec city of Tula

2000

The Toltecs of tradition were chiefly remarkable for their intense love of art

Lewis **Spence**, author

Jet

△ **Sliced lump of jet** showing wood grain

Jet is a type of lignite made from fossilized, compressed driftwood from the Araucariaceae family of trees—Chile pine, also known as monkey puzzle tree. Jet is composed of organic matter and, like coal, is readily flammable. Its color—jet black—never fades, and its polished surface can be used as a mirror, as it was in medieval times. Like amber, when jet is rubbed it produces an electric charge, a property that made it popular as a talisman and earned it the name "black amber." The best quality jet is found in Whitby, UK.

Specification

Chemical name Carbon | **Formula** C | **Colors** Dark brown, black, occasional brassy inclusions of pyrite | **Structure** Amorphous | **Hardness** 2.5–4 | **SG** 1.30–1.34 | **RI** 1.66 | **Luster** Waxy | **Streak** Black to dark brown | **Locations** UK, Switzerland, France, US, Canada, Germany

Rough

Semimetallic luster

Naturally textured surface

Jet block | This piece of high-quality jet shows the characteristic semimetallic luster found only in the finest and densest forms of the gem. The luster is visible in this example both on its natural wood-textured surface and on its flat-sawn ends.

Original wood grain

Raw jet | This piece of beach-recovered jet—a common way of finding it—has a slightly brownish cast and shows some of the original wood-grain structure.

Cut

Oval cabochon | This oval jet cabochon has been polished with a number of flattened surfaces, giving it an appearance that almost looks faceted.

Fashion for mourning

Jet and mourning jewelry

Jet has been in use since the Bronze Age, but saw a huge resurgence during the Victorian era, largely due to Britain's Queen Victoria, who, grieving for her husband Albert, popularized the wearing of jet mourning jewelry. Whitby jet was also the only jewelry permitted at court, and the fashion soon spread to other parts of society, causing a surge in popularity.

Queen Victoria Britain's popular monarch played a part in the vogue for jet jewelry during her reign.

Healed natural crack

Whitby jet | Showing healed natural fractures, this piece of raw jet comes from Whitby on the northern coastline of the UK, which is famous as a source of jet.

Stress crack

Jet bead | This antique jet bead, which was originally hand faceted and drilled, has cracked over time due to the release of internal stresses created by cutting it.

Settings

Intricate carving

High polish

Victorian earrings | The deep and detailed cuts on this pair of Victorian jet earrings fashioned as flowers illustrates the gem's suitability for carving.

Necklace | This modern necklace has highly polished, hand-faceted beads and a free-form drop, all showing the semimetallic luster of fine jet.

Approx 1 in (25 mm) bead

Earrings | The flowing shape of the 15-karat-gold-mounted drops on these fabulous Victorian earrings illustrates the fine carving qualities of jet.

Deeply carved details

Silver gilt setting

Smooth finish

Biconical beads

Rose carving | The intricacy of the carving on this jet brooch shows not only the skill of the craftsman, but the beauty of the mineral as a medium.

Trefoil brooch | A Roger Jean Pierre design, this piece features faceted jet rectangles set off by Swarovski hot pink and opaque pink stones in silver gilt.

Pendant | This three-dimensional jet pendant features a dove carrying a heart in its beak. The fine finish demonstrates the smooth texture that can be imparted by polishing.

Bead necklace | The beads on this Turkish jet necklace display a rounded, biconical form, which highlights their brilliant polish. Jet is often used in necklaces.

Native American eagle | Native American jewelry makes wide use of jet, although not all pieces are as spectacular as this silver-mounted jet eagle with turquoise insets on the wings.

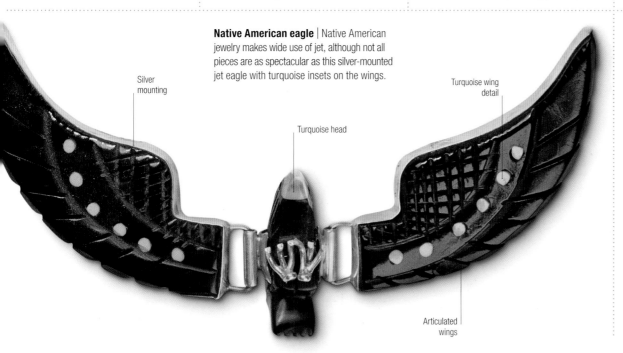

Silver mounting

Turquoise head

Turquoise wing detail

Articulated wings

Jet has been used in the making of decorative objects since the Neolithic period 10,000 years ago

Copal

△ **Translucent** golden copal nugget from New Zealand

Copal is semifossilized tree resin from the copal tree, Protium copal. It differs from amber, which also comes from tree resin, in that it is far younger—copal is less than 100,000 years old, while amber may be millions of years in the making. For this reason copal is more common and thus cheaper, although it is often used to imitate amber. Copal has historically been burned as incense, particularly in offerings to the Mayan gods in Mesoamerica. Europeans later valued it as an ingredient in wood varnish, particularly during the 19th and 20th centuries.

Specification

Chemical name Copal gum | **Formula** $C_{10}H_{16}O$
Colors Light lemon yellow to orange | **Structure** Amorphous | **Hardness** 2–3 | **SG** 1.05–1.10 | **RI** 1.54
Luster Resinous | **Streak** n/a | **Locations** Malaysia, Phillipines, Africa, Colombia, New Zealand

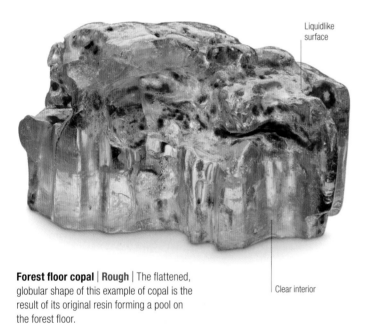

Liquidlike surface

Clear interior

Forest floor copal | **Rough** | The flattened, globular shape of this example of copal is the result of its original resin forming a pool on the forest floor.

Gemmy copal | **Color variety** | This group of gemmy, light to dark honey-colored copal pieces shows a variety of different shades and colors.

Copal is still burned as a form of incense in sweat lodge ceremonies in Mexico and Central America

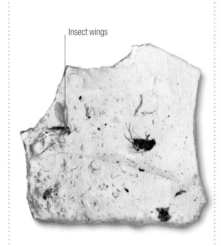

Insect wings

Copal slice | **Rough** | This thin slice of copal contains a few well-preserved insects, and is notable for its country of origin, Madagascar.

Trapped insects | **Cut** | This polished copal example is dotted with trapped insects, pollen, and seeds, in a similar manner to pieces of amber.

Dominican copal | **Cut** | This striking piece of polished copal from the Dominican Republic is also populated with trapped flies, spiders, and gnats.

Anthracite

△ **Specimen** of anthracite showing its semimetallic luster

Anthracite is the purest, most carbonized type of coal, consisting almost entirely of carbon. Like bituminous coal, it is composed of organic matter, but is older, much more highly compressed, and does not leave behind any residue on the hand when touched. It is used in beads and carvings, although its main use is as a fuel—though difficult to ignite, once lit it produces a lot of heat and burns slowly. Anthracite fires combust with a small blue flame that is smokeless, making it a good fuel for indoor use, but its expense means it is less widely used on an industrial level.

Specification

Chemical name Anthracite | **Formula** $C_{240}H_{90}O_4NS$
Colors Metallic black | **Structure** Amorphous | **Hardness**
2.75–3 | **SG** 1.4 | **RI** 1.64–1.68 | **Luster** Submetallic
Streak n/a | **Locations** Russia, Ukraine, North Korea, South Africa, Vietnam, UK, Australia, US

Semimetallic luster

Compact form | Rough | The typical density of anthracite can be seen in this specimen, as can its characteristic, almost metallic luster.

Rock groundmass

Bright luster | Rough | The irregular surface of this specimen of anthracite shows an unusually bright luster and inclusions of rock groundmass.

Blocky breakage | Rough | Because anthracite is hard and brittle, its surface tends to break in sharply angled blocks, as in this specimen.

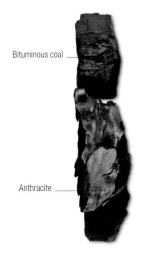

Bituminous coal

Anthracite

Contrasting specimens | Rough | The upper specimen is bituminous or ordinary household coal, while the lower specimen consists of anthracite.

Weathered anthracite | Rough | When exposed to weathering, the outer layers of anthracite blocks oxidize and deteriorate, as in this specimen.

High sheen

Polished anthracite | Cut | This irregularly shaped piece of anthracite is polished to a sheen, showing how it can sometimes be used as a jet substitute.

Slow burn

The Centralia mine fire

An underground fire has been burning for decades in an anthracite mine in Centralia, Pennsylvania, US. The fire started in 1962, and came to a head in 1981 when a 12-year-old fell into a 50 yd (46 m) sinkhole caused by the fire, which opened up beneath him (he survived, hauled out by his cousin with a rope). The fire is still burning and Centralia is now a ghost town.

The Centralia fire Anthracite burning in the old mine can be seen here breaking through the surface of the ground.

Amber

△ **Polished amber** containing a preserved spider

The fossilized tree resin from a prehistoric pine tree that was common in the Baltics, amber is also found in a few other locations. True amber is around 25–60 million years old, and specimens can function as tiny time capsules, preserving long-extinct plants and insects—these are highly valued. The Greeks noted how amber becomes charged when rubbed with fur or wool; the word for amber in Greek, *elektron*, is the root of the word "electricity." Amber has a low density and can float on salt water, so it is often found along saltwater coastlines.

Specification

Chemical name Oxygenated hydrocarbon | **Formula** Organic
Colors White, yellow, orange, red, brown, blue, black, green
Structure Amorphous | **Hardness** 2–2.5 | **SG** 1.05–1.09
RI 1.54 | **Luster** Resinous | **Streak** White | **Locations**
Eastern Europe, Dominican Republic, US

Rough

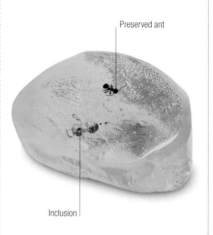

Opaque surface

Amber rough | Like many amber roughs, the luminously transparent interior of this piece of amber is visible behind its opaque surface.

Broken rough | This amber nodule has been broken, revealing the fine-quality amber within the dull, textured exterior that is typical of amber in its natural state.

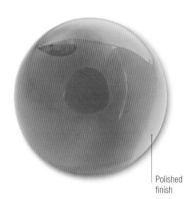

Preserved ant

Inclusion

Clear amber | This piece of clear amber contains a preserved insect and other inclusions. It has been smoothed to a natural finish by the elements.

Cut

Polished finish

Amber sphere | This finely polished spherical amber bead shows a deep orange color and opaque texture. It originates from the southeastern coast of the Baltic.

Polished specimen | Preserved insects are clearly visible in this piece of amber from Playa del Carmen, Mexico. Its smooth surface reveals its inner transparency.

Insect preserved in the amber

Transparent surface

Unusual coloration

Faceted amber | Amber is rarely faceted because of its extreme fragility. The cutter of this 2.36-carat emerald-cut green amber stone was unusually skillful.

Rounded edges

Varied tones and colors

Visible inclusions

Settings

Amber necklace | This substantial necklace is composed of elongated, polished beads in a fan pattern, and features many small inclusions. Amber's light weight makes it suitable for use in large jewelry pieces such as this.

Amber cabochon

Aquatic pendant | Featuring an amber cabochon set in silver, this German pendant from around 1930 was made by Louis Vausch, who was known for his use of fish motifs.

Silver frame

Earrings | These two teardrop-shaped amber cabochons with inclusions have been framed in silver and suspended in a pair of dangle earrings.

Inclusions

Amber ring | The centerpiece of this silver ring is a fine piece of gem amber, showing numerous trapped air bubbles and inclusions of organic matter.

Grieving gods

Amber's mythological past

The ancient Greeks made reference to amber in one of their myths. Demigod Phaethon lost control while driving his sun-god father's fiery chariot, scorching the Earth's surface. To stop him, Zeus struck him dead with a thunderbolt, causing his body to fall into a river. The river's nymphs buried Phaethon's body on the shore, and his sisters, the three Heliades, wept over it night and day. Eventually, their grieving bodies took root as trees, and their tears hardened into droplets of amber.

Phaethon in Apollo's chariot
This Greek vase depicts the story of Phaethon and the Heliades.

Amber Room (replica) | 1701–16 (original) | More than 6 tons (5.4 tonnes) (original panels) | Wall panels of carved amber, backed with gold leaf, and mosaics made with quartz, jasmine, jade, and onyx

Russian
Amber Room

△ **Prussian coat of arms** on the south wall

The fate of the Amber Room is one of the great mysteries of modern times. This spectacular chamber was originally commissioned by King Friedrich I of Prussia in 1701. It consisted of a series of richly carved amber panels adorned with semiprecious stones and Florentine mosaics, depicting the five senses. A German sculptor, Andreas Schlüter, and a Danish amber specialist, Gottfried Wolfram, collaborated on the work. In 1716, the panels were presented to the Russian Tsar, Peter the Great, after he admired them during a state visit. He installed them in St. Petersburg where they remained until 1755, when the Tsarina Elizabeth had the room enlarged and redesigned so that it would fit into her palace.

The Amber Room remained a prized possession of the Russian state until World War II, when looting German forces

Crown-shaped detail carved in amber

dismantled it. The stolen panels were packed into 27 crates and carried off to Königsberg Castle on the Baltic coast. The trail ends there. The panels may have been destroyed by Allied bombing, or by a fire in the castle—in either case, they had disappeared by the end of the war. A replica of the room was completed in 2003 in St. Petersburg, Russia.

Treasure hunters have never given up the search for the original panels, and fanciful theories abound. Some believe the Nazis buried the loot in an underground bunker; others claim the remains of Hitler were interred with the panels. Periodically, there have been claims of its rediscovery. In 2015, Polish treasure hunters located an armored train believed to be packed with Nazi booty, buried in tunnels near Ksiaz Castle in Poland. The same year, a German search team explored old copper mines at Deutschneudorf, near the Czech border. Neither find came to anything, and the hunt goes on.

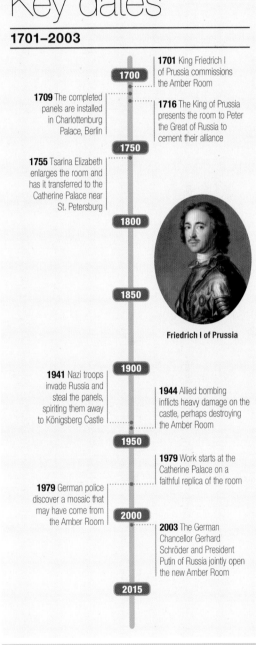

Key dates

1701–2003

1700

1701 King Friedrich I of Prussia commissions the Amber Room

1709 The completed panels are installed in Charlottenburg Palace, Berlin

1716 The King of Prussia presents the room to Peter the Great of Russia to cement their alliance

1750

1755 Tsarina Elizabeth enlarges the room and has it transferred to the Catherine Palace near St. Petersburg

1800

1850

Friedrich I of Prussia

1900

1941 Nazi troops invade Russia and steal the panels, spiriting them away to Königsberg Castle

1944 Allied bombing inflicts heavy damage on the castle, perhaps destroying the Amber Room

1950

1979 Work starts at the Catherine Palace on a faithful replica of the room

1979 German police discover a mosaic that may have come from the Amber Room

2000

2003 The German Chancellor Gerhard Schröder and President Putin of Russia jointly open the new Amber Room

2015

Rococo clock with a highly decorative base mounted on an amber-inlaid table, in the rebuilt Amber Room

The Amber Room has enormous emotional significance for both Germany and Russia

Friedrich **Spath**, chairman of Ruhrgas, corporate donors to the reconstruction project

Coral

△ **Red coral** from the Mediterranean sea

Precious (red) coral includes species such as *Corallium rubrum* and *Corallium japonicum* and is found in tropical and semitropical waters. Precious coral is valued over other types of coral for its toughness and attractive pink to red hues. Coral is the exoskeleton of the marine polyp—a small creature that secretes calcium carbonate to form branchlike structures. Because the branches tend to be fine and narrow, material is usually sourced from the thicker forks. Coral has been used decoratively since prehistoric times.

Specification

Chemical name Calcium carbonate | **Formula** $CaCO_3$
Colors Pale pink (angel skin), orange, red | **Structure** Crystalline
Hardness 3.5 | **SG** 2.6–2.7 | **RI** 1.48-1.66 | **Luster** Vitreous, waxy | **Streak** White | **Locations** Warm seas around Japan and Malaysia, Mediterranean, African coastal waters

Rough

Natural coral | The sections of coral most useful to jewelers are the thickest parts where two branches meet, or the widest part of a limb. The scale is often small—this piece of raw coral is about 2¼ in (6 cm) in width.

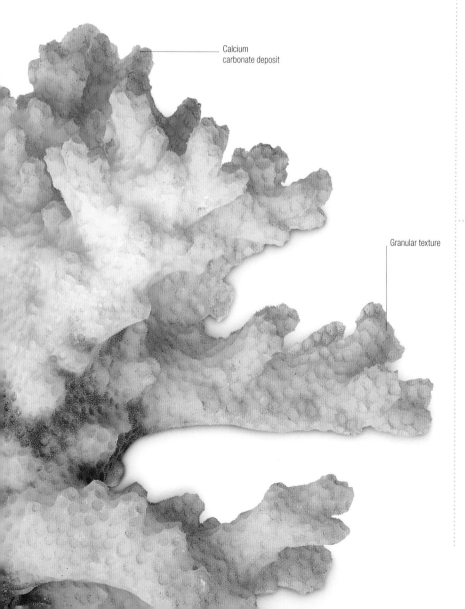

Calcium carbonate deposit

Granular texture

Cut

Polished surface

Red coral slice | This coral cross section reveals the intricate, banded structure of the material. The luster is dull when harvested, but polishing makes it shine.

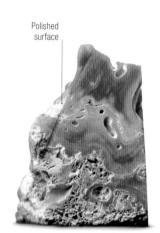

High polish gives vitreous luster

Coral oblong cabochon | Because it is soft and opaque, coral is often cut *en cabochon*. Then it can be polished and the color shown off to its advantage.

Wood-grain pattern

Branch of red coral | The longitudinal striations (grooves) give this natural piece of uncut coral a typical pattern that resembles natural wood grain.

Smooth gloss shows off color

Oval cabochon | The simple shape and cut emphasize the color of this cabochon, often referred to as "angel skin," a term that can apply to hues from pale pink to salmon.

Settings

Coral ring | For the centerpiece of this gold ring, angel skin coral was carved into rose blossom petals. The flushes of color mimic the tints of a real rose.

Delicately radiating petals

Coral set earrings | A specimen of deep red coral has been delicately carved to create matching rose shapes for these small stud earrings.

Gold head

Emerald eye

Fanned-out coral wings

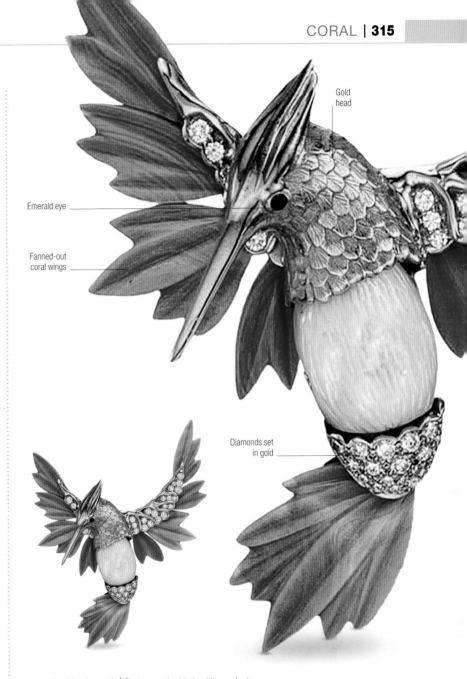

Teardrop cabochon

Maple-leaf pin | This gold-plated pin is set with oval and teardrop-shaped coral cabochons of differing sizes slotted within the serrated edges of the "leaf."

Intricately carved headpiece

Coral carving | Coral is associated with the safeguarding of children, and this miniature carving may have been a gift intended to bring protection to the wearer.

Diamonds set in gold

Hummingbird brooch | This hummingbird, with coral wings and tail outstretched in flight, was produced c.1975 by the jewelers Kutchinsky. It also incorporates diamonds, gold, mother-of-pearl, and emerald.

In the classical world, coral was worn as an amulet to protect against the evil eye

Gold branch

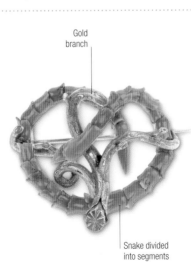

Snake divided into segments

Snake brooch | Designed as a stepped coral snake, the carving of this late 19th-century comprises 30 sections, intertwined with a scrolling branch.

Blood of the Gorgon

Gory beginnings

In Greek mythology, the hero Perseus beheaded Medusa, the Gorgon who turned people to stone by looking at them. Perseus then used Medusa's severed head to petrify a sea monster, Cetus—Medusa's gaze in death was still lethal. Afterward, he set the head down on a riverbank where the blood ran into the water, transforming seaweed into red coral. *Gorgeia*—after Gorgon—is the Greek word for coral.

Looks can kill This coral carving, from the 2nd or 1st century BCE, Bactria (in present-day Afghanistan and Tajikistan), is of Medusa.

Red coral antlers

Removable head of
stag doubles as
drinking vessel

Goddess of the hunt,
Diana, holds customary
bow and arrow

Silver statuette is
only partially gilded

Repoussé technique (metal
beaten from interior)

Hunting dogs
accompany Diana

Base, now empty,
contained wheels
and clockwork
mechanism

Diana with stag | 17th century | 14 in (32.5 cm) high | Silver parcel-gilt, coral, repoussé

Diana with stag
centerpiece

△ **Altarpiece** by Matthäus Walbaum of Augsburg, Bavaria

This statuette depicting Diana, goddess of the hunt, is not merely a decorative centerpiece, but also a drinking vessel and automaton used in early 17th-century party games.

The statuette, 14 in (32.5 cm) in height, is made from silver parcel-gilt (partially gilded) with repoussé work, and red coral forming the stag's antlers. It depicts the Greek goddess Diana riding a stag, accompanied by two hunting dogs. This example was likely to have been made by goldsmith Matthäus Walbaum of Augsburg, Bavaria, or one of his circle. The base of the statuette once contained

Another automaton depicting Diana, Eltz Castle, Germany

wheels powered by a clockwork mechanism, now missing, which could be wound with a key in a slot on the side. The automaton was placed on the table at dinner parties, wound up, and then released, at which point it wheeled off, making many arbitrary turns before stopping in front of one of the guests. The stag, which is hollow, would have been filled with wine by the host and the guest was required to drain it, removing the head and using it as a cup.

These windup sculptures proved popular in the 1600s and 1700s, and "Diana on a stag" was a favorite theme of the renowned Augsburg goldsmiths; around 30 such statues survive, all produced by three of the city's goldsmiths. Walbaum probably created the earliest example. Other variations of Diana from this period, such as the automaton at Eltz Castle, Germany, reserved the stag for the men, with a smaller, hollow hound for the ladies. Chains attached to the cups meant participants had to drink in close proximity to one another.

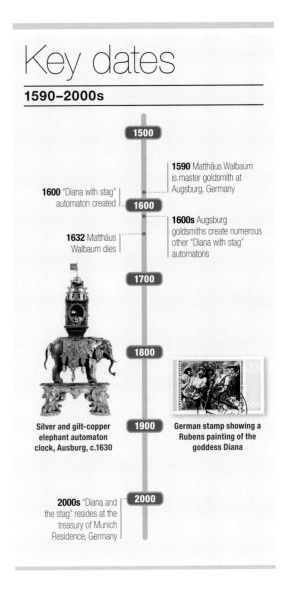

Key dates
1590–2000s

1500

1590 Matthäus Walbaum is master goldsmith at Augsburg, Germany

1600 "Diana with stag" automaton created

1600

1600s Augsburg goldsmiths create numerous other "Diana with stag" automatons

1632 Matthäus Walbaum dies

1700

1800

Silver and gilt-copper elephant automaton clock, Ausburg, c.1630

1900

German stamp showing a Rubens painting of the goddess Diana

2000s "Diana and the stag" resides at the treasury of Munich Residence, Germany

2000

Statue of Diana, which stood in Madison Square Garden, New York, until 1925. The day it was removed, it attracted crowds of onlookers

One of the **most productive goldsmith's workshops in Augsburg**

Grove Encyclopedia of Decorative Arts on Matthäus Walbaum's workshop

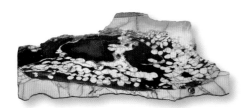

Peanut wood

△ **Slice of peanut wood**, 12 in (30 cm) long

Also called Teredo wood, peanut wood is a petrified wood with white ovoid markings on its surface, giving it the appearance of peanut brittle. It comes from a conifer that grew around 120 million years ago, mainly in Australia. Its curious appearance is the work of ancient marine animals. A wood-eating clam, *Teredo*, tunneled into the conifer driftwood, which later sank to the seafloor. White sediment from the shells of tiny plankton then settled over the wood, filling the boreholes with concentrated levels of silicate and creating white tubes when the wood petrified.

Specification

Chemical name Silicon dioxide (white areas), iron oxide (colored areas) | **Formula** SiO_2 | **Colors** White markings on brown, gray, green | **Structure** Amorphous | **Hardness** 6.5–7 | **SG** 2.58–2.91 | **RI** 1.54 | **Luster** Vitreous **Location** Australia

Good balance of color

Uncut peanut wood | **Rough** | This chunky piece of peanut wood rough comes from the Gascoyne region of Western Australia. It has an even contrast of light and dark areas.

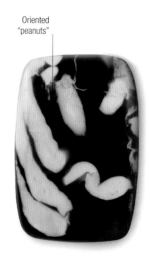

Oriented "peanuts"

Round-cornered cabochon | **Cut** | This rectangular cabochon of peanut wood has been cut along the "peanut" markings to highlight them.

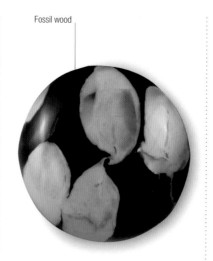

Fossil wood

Round cabochon | **Cut** | This circular gem has been cut to emphasize its unusual 3-D-effect markings. The "peanut" shapes are especially evident.

Silver mounting

Smoky quartz

Bracelet | **Set** | This silver bracelet is bezel set with six Australian peanut wood cabochons and accented by a faceted smoky quartz that suits the color scheme.

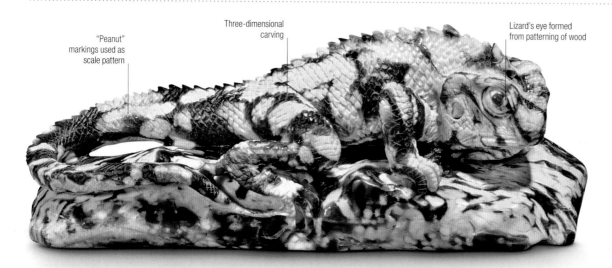

"Peanut" markings used as scale pattern

Three-dimensional carving

Lizard's eye formed from patterning of wood

Lizard ornament | **Carved** | When its pattern is used artistically, as here, peanut wood is a carving medium to rival any. Because it is a form of chalcedony, it is hard, durable, and takes a good polish.

Ammolite

△ **Ammonite fossils** in an unusual grouping

Ammolite is the lining of the shell from the ammonite, a mollusk that became extinct around 66 million years ago, around the same time as the dinosaurs. Its iridescent colors cross the spectrum, but green and red are most common, with gold or purple being more rare. It is found in many parts of the world, but the best examples come from Alberta, Canada, where it is mined. The First Nation people of Alberta, the Blackfoot people, know it as Iniskin, or Buffalo Stone, and believed that it could attract buffalo close enough for them to be hunted.

Specification

Chemical name Aragonite polymorph | **Formula** $CaCO_3$
Colors All spectral colors—red, orange, yellow, green, blue, indigo, violet | **Structure** Orthorhombic | **Hardness** 3.5–4 | **SG** 2.75–2.85 | **RI** 1.52–1.68 | **Luster** Vitreous
Locations Canada, US

Heavy ribbing

Coiled ammonite | Rough | This fossil of the ammonite *Dactyliosarus* shows classic coiling and ribbing. It is from the Jurassic Period, up to 200 million years ago.

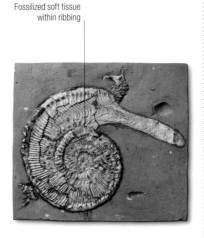

Fossilized soft tissue within ribbing

Unusually preserved specimen | Rough | This fossil ammonite *Kosmoceras duncani* was found in Jurassic Oxford Clay in England; it shows some fossilized soft tissue.

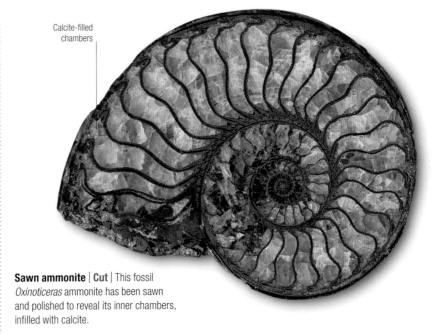

Calcite-filled chambers

Sawn ammonite | Cut | This fossil *Oxinoticeras* ammonite has been sawn and polished to reveal its inner chambers, infilled with calcite.

Rainbow color play

Ammolite | Cut | A section of the fossilized outer shell of an ammonite, this ammolite gem weighs 23.7 carats with a maximum length of 1¼ in (33 mm).

Carved snake's head

Snakestone | Carved | Following the legend of St. Hilda turning snakes to stone, modern lapidaries have started carving snakes' heads onto ammonites, as here.

Snake or ram

Animal links

Ammonite is named for its similarity in appearance to a ram's horn. The 1st-century CE Roman writer and natural philosopher Pliny the Elder called the shells *Ammonis cornua* ("horns of Ammon"), after the Egyptian god Ammon (Amun), who was usually represented with ram's horns. In medieval Europe, the fossils were called snakestones or serpentstones and were believed to be petrified snakes, the work of saints such as St. Patrick or St. Hilda of Whitby.

Egyptian god Ammon as a ram's head This stela (stone slab) would have served as a gravestone.

Rock gems and rocks

Moldavite

△ **Brilliant** oval-cut, faceted moldavite

Moldavites were formed 15 million years ago, when a meteorite struck near modern Ries, Bavaria, melting the local sandstone. The substance is a tektite, the name for minerals formed when large meteorites hit Earth: the terrestrial rock melts on impact, is splashed into the air, and quickly cools to form the glasslike substance. Tektites are found on most continents, but moldavite is local to the Ries impact point. It is typically olive green to dull greenish yellow, and is found in sizes ranging from less than a millimeter to several centimeters across.

Specification

Chemical name Silicon dioxide | **Formula** $SiO_2(+Al_2O_3)$
Color Mossy green, greenish yellow | **Structure** Amorphous
Hardness 5.5 | **SG** 2.40 | **RI** 1.48–1.54 | **Luster** Vitreous | **Streak** n/a | **Locations** Germany, Czech Republic

Irregular surface

Colored layers

Uncut moldavite | Rough | The substantial, chunky form of this moldavite rough suggests that it cooled extremely rapidly after being thrown from the meteorite crater on impact.

Ridges resulting from rapid cooling

Dramatic form | Rough | Some moldavites were still fluid when they were flung from the crater, and as a result tend to assume frothy, "splatter"-like forms such as this.

Pavilion faces upward

Faceted moldavite | Cut | This particularly dark-green moldavite gemstone has been faceted in a free-form shape, viewed here from the pavilion.

Crown facets

Moldavite takes its name from the town of Moldauthein in the Czech Republic

Spindle shape | Rough | During the flight of molten moldavite through the air, it assumes a number of shapes, such as this elongated form, called a spindle.

Excellent clarity | Cut | Certain moldavite gemstones are exceptionally transparent, such as this highly refractive, brilliant-cut example.

Horse head | Carved | The carver of this moldavite shaped in the form of a horse's head has left the natural surface of the mineral as the horse's mane.

Obsidian

△ **Tumble-polished** obsidian piece

bsidian is a natural volcanic glass that forms when lava solidifies so quickly that mineral crystals do not have time to grow. Technically, obsidian can have any chemical composition, although it is usually the product of silica-rich magmas. It is typically jet black; hematite (iron oxide) contained within it can result in red and brown varieties, and the inclusion of tiny gas bubbles can create a golden sheen. In snowflake obsidian, clusters of light-colored, needlelike crystals of cristobalite on broken surfaces resemble snowflakes.

Specification

Chemical name Silicon dioxide | **Formula** Composed of SiO_2, MgO, and Fe_3O_4 | **Colors** Black, red, brown | **Structure** Amorphous | **Hardness** 5–6 | **SG** 2.35–2.60 | **RI** 1.45–1.55 **Luster** Vitreous | **Streak** White | **Locations** Europe, North America, South America, Australasia, Japan

Conchoidal fracture

Mexican obsidian | Rough | This gemmy, highly reflective piece of black obsidian originates from the central highlands of Mexico.

Snowflake obsidian | Rough | As this varety of obsidian cools, snowflakelike white crystals of cristobalite are formed, creating its distinctive appearance.

Iridescent colors

Rainbow obsidian | Cut | As with sheen obsidian (see right), rainbow obsidian contains small, oriented platelets, giving it an iridescence when polished.

Surface sheen

Sheen obsidian | Cut | During its formation, platelets of other minerals form within the obsidian, giving it a sheen when polished, as seen here.

"Snowflakes" of albite

Polished surface

Cat carving | Carved | Although obsidian is brittle and glasslike, with careful carving, attractive ornaments can be produced from the material, such as this sculpture of a wistful cat in patterned snowflake obsidian. Its surface has been polished to a high shine.

Obsidian blades

Ancient cutting tools

When obsidian breaks, it can form an edge sharper than that of a steel scalpel. It was often used throughout antiquity to fabricate cutting tools and weapons, and was a prized material, widely traded across vast distances. It was used from the Stone Age onward in civilizations including pre-Colombian Mesoamericans, ancient Egyptians, Native Americans, and others.

Obsidian spearhead and knife These blades originate from the Admiralty Islands off New Guinea, from around 1900. They retain parts of their painted handles.

Limestone

△ **Pterodactyl fossil**, beautifully preserved in limestone

imestone is largely made up of calcium carbonate and, depending on its formation, can be clastic, crystalline, granular, or massive. A sedimentary rock, most limestone forms in calm marine waters, occurring when marine organisms die and fragments of shell, skeletal debris, and coral break down into sediment. Minerals then cement the sediment together, turning it into limestone. The material was used in many ancient carvings, and is regularly used as a construction material, a base for roads, a white pigment or a filler in paints, plastics, and toothpaste.

Specification

Rock type Marine, chemical, sedimentary | **Fossils** Marine and freshwater invertebrates | **Major minerals** Calcite
Minor minerals Aragonite, dolomite, siderite, quartz, pyrite
Colors White, gray, pink | **Texture** Fine to medium, angular to rounded | **Locations** Worldwide

Ornately carved headdress

Veil with scallop detailing

Limestone statue | Carved | Finely detailed mortuary statues, such as this one from 2nd-century CE Palmyra, were carved throughout the Roman world. This high-relief bust of a woman is richly adorned with necklaces, bracelets, brooches, and rings.

Surface contains fossils

Fossiliferous rock | Rough | Limestone bearing fossils such as this specimen containing bryozoans is often sawn and polished, and used as cladding for buildings.

Freshwater limestone | Rough | Limestone is generally formed in saltwater, but it can also occur in fresh water, as with this fossiliferous example.

Sphinx and base carved from one complete piece

Sphinx | Carved| Just like the Great Sphinx in Giza, this smaller version is also carved from limestone—a favorite carving medium in antiquity, when this sphinx was made.

Homeosaurus fossil | Cut | Fine-grained limestone provides the perfect environment for fossil preservation, as shown by this Jurassic Homeosaurus.

Sandstone

△ **Head of a reclining** Buddha statue carved in sandstone

Sandstone is found throughout the world and is one of the most common varieties of sedimentary rock. The tiny, sand-sized grains of mineral, rock, or organic material that make up its composition are reduced in size by weathering, and then compacted together over long periods of time. Mineral grains in sandstones are typically quartz or feldspar: these grains crystallize around the sand grains, cementing them together. Sandstone can be any color, but is usually brown, yellow, red, gray, pink, or white; it has been used in carvings and architecture for centuries.

Specification

Rock type Continental, detrital, sedimentary | **Fossils** Vertebrates, invertebrates, plants | **Major minerals** Quartz, feldspar | **Minor minerals** Silica, calcium carbonate | **Colors** Cream to red | **Texture** Fine- to medium-grained, angular to rounded | **Locations** Worldwide

Grainy surface | **Rough** | The rough, textured surface of this piece of rust-red sandstone is coated with tiny individual sand grains.

Sandstone boulder | **Rough** | This sandstone boulder could be split along its horizontal bedding planes to produce fine carving material.

Multicolored layering

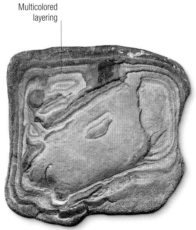

Layered sandstone | **Rough** | Sandstone with different colored layers can be sawn across the bedding to reveal sand "pictures," as seen in this specimen.

Indian statue | **Carved** | This intricately detailed, red sandstone carving from 1st–2nd-century CE India has been coated with gesso before being painted.

Lotus carving | **Carved** | Sandstone can vary in hardness and some material can be difficult to carve, but when fine-grained, as here, it can be sculpted with great detail.

Detailed flower buds

Petra

The Rose City

Petra is a famous archaeological site carved into pink sandstone cliffs in southwestern Jordan. Nicknamed the "Rose City" after the rich rose color of the sandstone, the intricately carved, tombs and temples date back to 300 BCE, when Petra was the capital of the Nabatean kingdom. Petra's most famous temple is Al Khazneh or "the Treasury," a building with an ornate façade, which is accessed via a narrow gorge more than ½ mile (1 km) long and flanked on either side by 250 ft- (80 m-) high cliffs.

Rose-colored columns The façade of this tomb, with its ornate columns, is carved into sandstone.

Interior of the Palace of the Lions | 14th century | Sandstone, stucco, wood | Found in the Palatial City, the Alhambra, Granada, Spain

Spanish
Alhambra

△ **Courtyard** of the Lions

The Alhambra in Granada, Spain, is recognizable from afar for its distinctive from afar for its distinctive brickwork made from red clay and gravel: its name means "red castle." The interior is a marvel of stonework and decoration in sandstone, stucco, and wood. Sandstone has long been an important building material. Its durability as both a building and a sculptural material is unsurpassed among sedimentary rocks. Stucco is a fine plaster used to coat walls and molded into decoration. Used together in the Alhambra, they suffuse the interior with a warm glow and a wealth of textural detail.

In its heyday, the Alhambra was a citadel and palace, but it has since been used as a barracks, prison, and Roma settlement, with farm animals roaming the ruined halls. The Romantics rediscovered it in the 19th century, inspired by its former glories. One famous visitor was Washington Irving, author of *Rip Van Winkle*. Irving invented a history for the citadel's Hall of the Two Sisters, involving a pair of Muslim princesses who fell in love with their Christian captives. One eloped with her suitor, while the other remained behind, a forlorn spinster. The reality is more prosaic—the "sisters" are two large marble flagstones on the floor. This hall was part of the residential quarters, where the *Sultana* (ruler) lived with her children. Its outstanding feature is the muqarnas dome in the center. Muqarnas is an intricate form of tiered vaulting, made of painted stucco. Based on a geometric design, it has thousands of overlapping, stalactite-like "cells"—possibly a visual reference to the cave in which Muhammed received the revelation of the Koran.

Engraving showing the Alhambra, 1890

It absolutely appears to me like a dream

Washington **Irving**
Author

Ceiling of the Hall of the Two Sisters: this worm's-eye view of the interior of its muqarnas dome shows its cell-like patterning in stucco

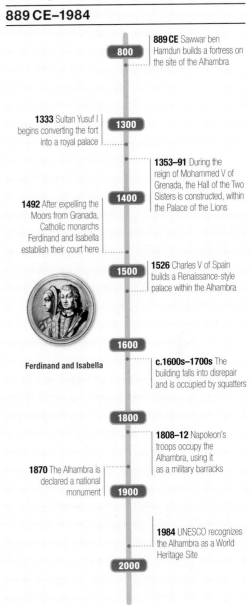

Key dates

889 CE–1984

889 CE Sawwar ben Hamdun builds a fortress on the site of the Alhambra

1333 Sultan Yusuf I begins converting the fort into a royal palace

1353–91 During the reign of Mohammed V of Grenada, the Hall of the Two Sisters is constructed, within the Palace of the Lions

1492 After expelling the Moors from Granada, Catholic monarchs Ferdinand and Isabella establish their court here

1526 Charles V of Spain builds a Renaissance-style palace within the Alhambra

Ferdinand and Isabella

c.1600s–1700s The building falls into disrepair and is occupied by squatters

1808–12 Napoleon's troops occupy the Alhambra, using it as a military barracks

1870 The Alhambra is declared a national monument

1984 UNESCO recognizes the Alhambra as a World Heritage Site

Marble

△ **Breccia marble**, shattered and re-cemented

A granular rock derived from limestone or dolomite, marble consists of a mass of interlocking calcite or dolomite grains. Pure marble is white. Other types take their common names from their color or mineral impurities. These impurities tend to occur as layers of other minerals thinly interbedded in the original limestone, so may be present as bands or swirls. Other veined and patterned marbles are created when an existing example is cracked or shattered, and the spaces between the fragments fill in with calcite or other minerals.

Specification

Rock type Regional or contact metamorphic | **Temperature** High | **Pressure** Low to high | **Structure** Crystalline | **Major minerals** Calcite | **Minor minerals** Diopside, tremolite, actinolite | **Color** White, pink | **Texture** Fine to coarse | **Protolith** Limestone, dolomite

Hyalophane crystal | Pyrite

Marble specimen | **Rough** | This specimen consists of minerals—hyalophane crystal and pyrite—enclosed within dolomitic marble.

Patterned inclusions

Marble sphere | **Carved** | This patterned sphere has been polished to a fine finish, highlighting its "tiger stripe" appearance caused by inclusions.

Polished finish

Marble sculpture | **Carved** | The flowing lines of this sculpture cut from marble are emphasized by the simplicity of the carving. It has fine white colouring.

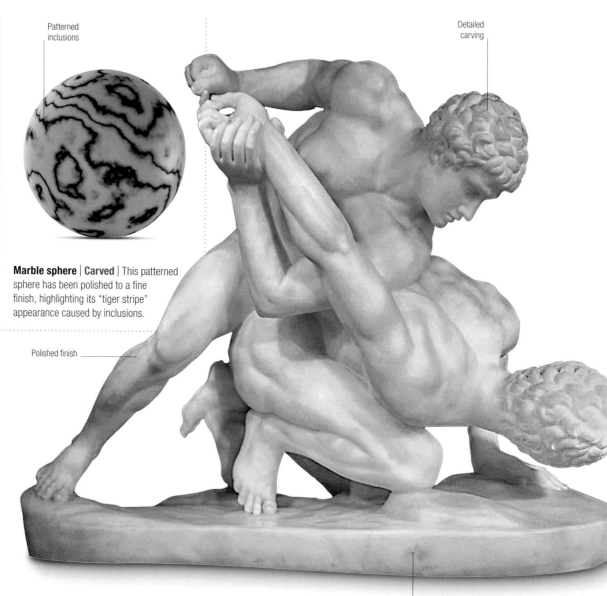

Detailed carving

Marble statue | **Carved** | This early 19th-century Italian, large-scale, translucent marble statue depicts wrestlers raised on a rock-shaped base. Fine craftsmanship and a smoothly rendered finish emphasize the translucence of the material. The piece was inspired by an ancient Roman bronze.

Raised base anchors figures

Granite

△ **Classic pink granite** containing quartz, feldspar, and mica

amiliar as a mottled pink, white, gray, and black ornamental stone, granite is the most common intrusive rock in the Earth's continental crust. Granite's three main minerals are feldspar, quartz, and muscovite, or biotite mica. Of the three principal minerals, feldspar predominates, and quartz (see pp.132–39) usually accounts for more than 10 percent. Granite has been a favorite stone for carving and building for at least four millennia. Wherever it has been available, its strength and durability have made it a first choice for everything from temples to millstones.

Specification

Rock type Felsic, plutonic, igneous | **Major minerals** Potassium feldspar, quartz, mica, sodium | **Minor minerals** Sodium plagioclase, hornblende | **Color** White, light gray, gray, pink, red | **Texture** Medium to coarse

Granite host | Rough | Granites are often host to a variety of gemstones—this granite groundmass is covered in tourmaline crystals.

White plagioclase

Granodiorite | Rough | Commercial "granites" come in a range of colors and textures, depending on their exact makeup. This specimen tends toward granodiorite.

Boulder with microcline | Rough | This granite boulder takes its pink color from the large amount of microcline feldspar that comprises its makeup.

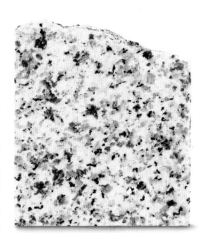

Commercial granite | Cut | This sample of kitchen countertop material is commercially sold as "white granite," but is closer to granodiorite in makeup.

Black hornblende

Granite bead | Carved | The patterns of granite fascinated ancient civilizations, which created carvings such as this bead. The black portion is hornblende.

Coarse, crystalline texture

"Black granite" | Cut | This sample of kitchen countertop material is commercially sold as "black granite," although in reality its composition is closer to diorite.

Ancient elephant | Carved | Despite its hardness, granite has been a popular material for carvings thoughout history. This granite elephant originates from India.

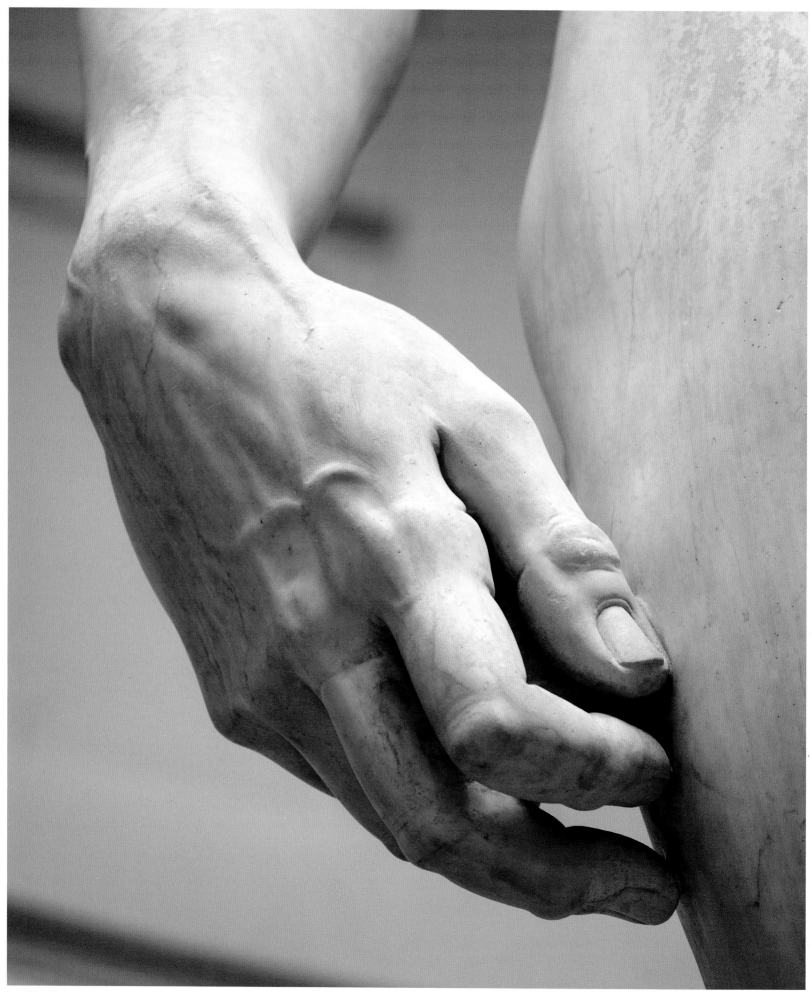

David by Michelangelo (detail of hand showing exaggerated proportion) | 1501–04 | 16 ft 11 in (5.16 m) high, 12,478 lb (5,660 kg) in weight | Solid Carraran marble

Michelangelo's
David

△ **Michelangelo** (1475–1564), after a self-portrait

One of the masterpieces of the Renaissance or, arguably, of any era, Michelangelo's *David* is exceptional for its lifelike rendering of the male anatomy, vast scale, and unusual treatment of its subject matter. The sculpture represents the biblical David, Israelite slayer of the Philistine giant Goliath. David is carved from solid marble and stands at over 16 ft (5 m), weighing more than 5.5 tons (5 tonnes). He holds a sling in one hand and a stone in the other. Michelangelo's sculpture is unprecedented in that Goliath is absent and, rather than representing David's victory, shows him poised in the moments before battle. Michelangelo's great achievement is capturing David's prebattle tension in the protruding veins of

David with Goliath's Head, circle of Caravaggio, c.1600, typically presenting the moment of victory

his hands, the tautness of his neck, and the focus in his gaze—all at monumental scale. Contemporaries were amazed, even though one, Piero Soderini, declared the nose too wide—prompting Michelangelo to make a pretence of altering it, complete with marble dust.

The statue was intended for the battlements of Florence Cathedral, which may account for the unusually large head and hands, to allow for perspective from below. However, a committee of Florentines, including Leonardo da Vinci and Botticelli, considered the work too exquisite (and heavy) to be displayed there, so it was placed outside the Palazzo della Signoria, the town hall. The position had political significance as the figure of David gazed toward Rome— he was intended to represent Florence, which had recently thrown off the Medici family's rule.

Anyone who has seen Michelangelo's David has no need to see anything else by another sculptor, living or dead

Giorgio **Vasari**
Painter and artists' biographer, 1511–74

Sculpture of David as it is today, in Florence

Key dates

1400–2014

1464 Agostino di Duccio is commissioned to create a sculpture of David, and a huge block of marble is provided

1400

1400 Authorities plan 12 large Old Testament sculptures for the cathedral buttresses

1466 Agostino ceases work for unknown reasons, having started to shape legs, feet, and torso

1476 Antonio Rossellino resumes work on the block, but is released from his contract soon after

1500 Authorities determine to find a sculptor to finish the statue

1500

August 1501 Michelangelo, just 26, wins the contract

January 1504 Committee of Florentines decide not to place completed *David* on the cathedral buttresses

September 1501 Michelangelo begins work on the sculpture

June 1504 *David* is moved to the public square of Palazzo della Signoria (Palazzo Vecchio)

1800

1873 *David* is moved to Florence's Galleria dell' Accademia to protect it

Statue protected by bricks during World War II

1900

1910 A replica is placed on the old site

1939–45 The statue is enclosed in bricks to protect it from bombings

1991 A man damages the toes of the statue's left foot in an attack with a hammer

2000

2010 The Italian Culture Ministry claims ownership of *David*, which the city of Florence disputes

2010

2014 Concerns arise over microfractures in the stump supporting the statue, as well as in the legs

Record
breakers

Big, beautiful, and almost perfect—these are the qualities that characterize most of the world's record-breaking gems. Some are kept in the vaults of royal collections or distinguished museums, while others are recent finds fresh from the Earth. Here is a selection of the gems that have set new records for size or quality, both new discoveries and established treasures.

De Beers diamond
This yellow diamond formed the centerpiece of the Patiala Necklace (see pp.90–91.)

Lucapa diamond
The rare Type IIa status was awarded to this almost flawless rough diamond.

Sweet Josephine
When it sold for $28.5 million in 2015, the 16.08-carat, cushion-cut Sweet Josephine gemstone set a record-breaking price for pink diamonds.

Gachala emerald
One of the largest uncut emeralds in the world, the Gachala weighs 858 carats.

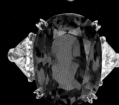

Carmen Lucia ruby
This 23.1-carat Burmese ruby has uniquely fine color and clarity.

Star of Adam
This gem, the world's largest blue star sapphire, was found in 2015 in Ratnapura, Sri Lanka.

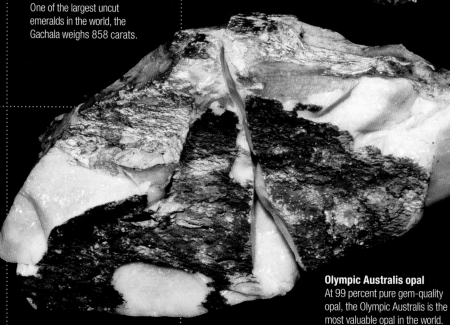

Olympic Australis opal
At 99 percent pure gem-quality opal, the Olympic Australis is the most valuable opal in the world.

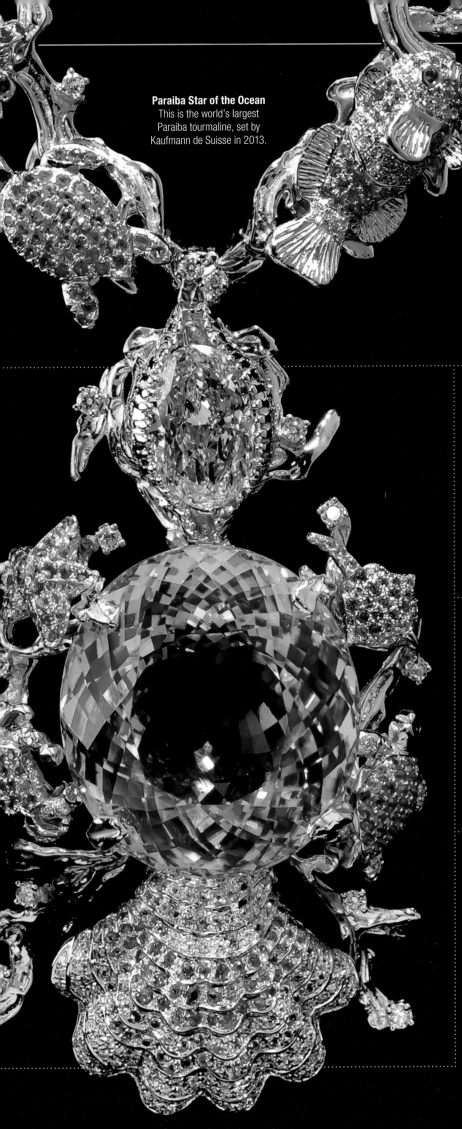

Paraíba Star of the Ocean
This is the world's largest
Paraíba tourmaline, set by
Kaufmann de Suisse in 2013.

Blue giant of the Orient
Found in 1907, this gem still
holds the record as the largest
sapphire in the world.

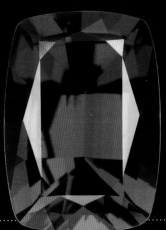

Alexandrite
Notable for its quality, this alexandrite
stone from the Smithsonian's
collection weighs 17.08 carats.

Strawn-Wagner diamond
The only perfect diamond ever
found, the Strawm-Wagner has
an AGS grade of 0/0/0.

MODERN BRANDS

The influence of marketing and advertising from the mid-19th century onward transformed the way consumers perceived jewelry. It was no longer simply an asset with a value based on gem quality, metal content, and rarity: the value now came from its brand association, too. Jewelry was presented to the consumer as a symbol of a particular lifestyle that extended beyond rings and watches to encompass fragrance and homewares, culture and the arts, exclusive sports events, and celebrities on the red carpet.

Tiffany & Co. took the lead in the US in 1845, publishing its Blue Book jewelry catalogue, which was one of the first of its kind. Refining the color theme, in 1878 it introduced its signature blue color to packaging and advertising, which became integral to the brand's image. Cartier, meanwhile, used its links with European royalty to create an identity of prestige and refined taste. It launched a line of lifestyle products, Les Must de Cartier, and sponsored events such as polo matches to cement its image of exclusivity. Other brands, from upscale Van Cleef & Arpels to mass-market Pandora, have since projected their identities to the public in a similar way.

> I've **never** thought of my jewelry as **trophies…** we are only **temporary custodians** of beauty

Elizabeth **Taylor**
Actress

Panthère de Cartier watch Playful and powerful, and representing elegance, power, and luxury, the big cat has been incorporated in many of Cartier's luxury watches, such as this lavish example, and other decorative jewelry pieces in its Panther collection.

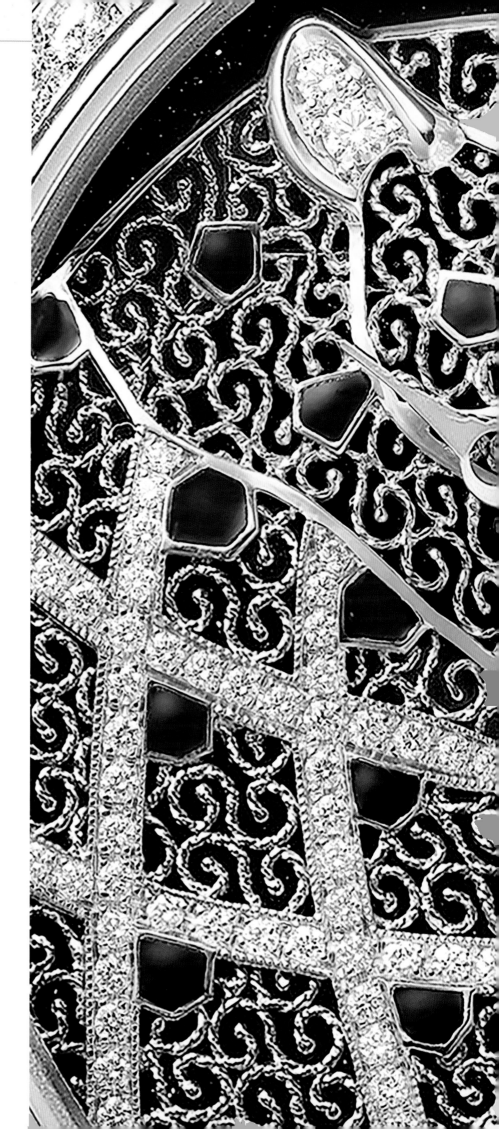

Color guide

Color guide

This directory loosely groups minerals according to their main color. Where multiple colors are available, these are listed in the text.

Diamond | See pp.52–57 | Diamond can be colorless, white to black, yellow, pink, red, blue, or brown. It is transparent to opaque with an adamantine luster.

Quartz (Namibian) | See pp.132–39 | Colorless, yellow, pink, or green, this variety of quartz is transparent to opaque, with a vitreous luster.

Quartz (rutilated) | See pp.132–39 | Rutilated quartz is colorless with gold, red, or green needles of rutile. It has a vitreous luster and is transparent.

(Quartz) rock crystal | See pp.132–39 | Naturally occurring in a colorless form only, rock crystal has a vitreous luster. It is transparent.

Selenite | See p.123 | Selenite can be colorless, white, yellow, or light brown. It is transparent to translucent, with a vitreous or pearly luster.

Pollucite | See p.185 | Pollucite may be colorless or gray, blue, or violet. It ranges from transparent to opaque, and has a vitreous luster.

Danburite | See p.246 | Ranging from colorless to yellow, brown, or pink, danburite is transparent with a vitreous luster.

Celestine | See p.121 | Celestine occurs as colorless, white, red, green, blue, or brown. It is transparent to translucent with a vitreous luster.

Amblygonite | See p.117 | Amblygonite is colorless, white, yellow, pink, brown, green, or blue. It is transparent with a vitreous or pearly luster.

Phenakite | See p.282 | Phenakite is colorless, yellowish, pink, or greenish blue, and is either transparent or translucent with a vitreous luster.

Tourmaline (achroite) | See pp.226–29 | Achroite, a variety of tourmaline, is colorless with a vitreous luster. It is transparent to translucent or opaque.

Albite | See p.172 | Albite may be colorless, greenish, bluish, or black. With a vitreous or pearly luster, it ranges from transparent to opaque.

Euclase | See p.283 | With colorless, white, blue, or green varieties, euclase is transparent to translucent and has a vitreous luster.

Platinum | See pp.44–45 | This precious metal is silver white in color. Platinum has a metallic luster and is opaque.

Silver | **See pp.42–43** | This popular precious metal is silver white, tarnishing to black. It has a metallic luster and is opaque.

Pyrite | **See p.66** | This gem naturally occurs with a silver or pale brass-yellow coloring. Pyrite has a metallic luster and is opaque.

Pearl | **See pp.292–95** | White, cream, black, blue, yellow, green, or pink varieties of pearl may be found. It is opaque with a pearly luster.

Marble | **See p.328** | This opaque stone is found in a wide range of colors, with violet, red, blue, or white veins. It can be dull, pearly, or subvitreous.

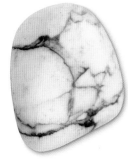

Howlite | **See p.127** | Occurring as off-white, often with a grey or black "spiderweb" matrix, howlite is opaque with a vitreous or dull luster.

Petalite | **See p.197** | Ranging from colorless to pink or yellowish in color, petalite has a pearly luster and is transparent.

Quartz (aventurine) | **See p.134** | This quartz may be gray, green, red brown, or gold brown. It is translucent or opaque with a vitreous or waxy luster.

Quartz (chatoyant) | **See p.137** | Greyish in color, with a weak cat's-eye effect, the chatoyant quartz is translucent with a greasy luster.

Bytownite | **See p.173** | Yellowish or reddish brown, bytownite is either transparent or translucent and may have a vitreous or dull luster.

Quartz (smoky) | **See p.137** | Smoky quartz is clear brown, varying from a light brown to dark brown. It has a vitreous luster and is transparent to opaque.

Chrysoberyl (Alexandrite) | **See pp.84–85** | Alexandrite can be transparent or translucent with a vitreous luster. In gem form, it is pleochraic.

Jet | **See pp.306–307** | Jet ranges from dark brown to deep black. It has an opaque appearance with a luster that may be waxy or dull.

Garnet (melanite) | **See pp.258–63** | A deep black in color, this variety of garnet has a vitreous or subadamantine luster. It is translucent or opaque.

Onyx | **See pp.154–55** | Black onyx is black with white layers, which may appear as straight color bands. It has a waxy luster and is opaque.

Tourmaline (schorl) | **See pp.226–29** | Schorl is black, blue black, or brown black. It may be translucent or opaque, with a vitreous or resinous luster.

Obsidian | **See p.323** | Translucent with a vitreous luster, obsidian ranges from black, bluish, and mahogany, to golden or peacock, among other hues.

Anthracite | **See p.309** | Anthracite is opaque, ranging from black to steel gray in color, and shines with a submetallic luster.

Peanut wood | **See p.318** | This opaque fossilized wood is dark brown to black, with white to cream peanut-sized, ovoid shapes. It is vitreous or greasy.

Enstatite | **See p.202** | Found in brown, gray, white, green, or yellow, enstatite may appear translucent or opaque. It has a vitreous luster and a grey streak.

Epidote | **See p.251** | Epidote is found as brown, pistachio green, yellow, or greenish black. Vitreous to resinous in luster, it can be transparent to nearly opaque.

Axinite | **See p.247** | Found as brown, yellowish green, green, bluish green, or blue material, axinite is transparent with a vitreous luster.

Bronzite | **See p.205** | With a brown or greenish hue, bronzite may be transparent, translucent, or opaque, and has a submetallic luster.

Hypersthene | **See p.204** | With dark hues of black to black brown or black green, this gem is vitreous or silky in luster and transparent to opaque.

Copper | **See pp.48–49** | This metal is brown to copper red, tarnishing to black or green. It has an opaque appearance and a metallic luster.

Agate | **See pp.152–53** | Red, yellow, green, reddish brown, white or bluish white, with varied banding, agate is waxy in luster and translucent to opaque.

Rutile | **See p.94** | With an adamantine to submetallic luster, rutile is brown, red, pale yellow, pale blue, violet, or black, ranging from transparent to opaque.

Chalcedony (sard) | **See pp.146–47** | This gem is a brownish red (sard) and is translucent through to opaque, with a waxy luster.

Chalcedony (carnelian) | **See pp.146–47** | This variety of chalcedony is brownish red to orange, and translucent or opaque, with a waxy or resinous luster.

Fire opal | **See pp.158–59** | This red, orange, or yellow variety of opal has a vitreous luster and runs from transparent to translucent to opaque.

Chalcedony (jasper) | **See pp.146–49** | Jasper occurs in all colors, most commonly in reddish hues, with most examples striped or spotted. With a vitreous luster, it is opaque.

Calcite | **See p.98** | From transparent to opaque, calcite can be orange, white, yellowish, pink, bluish, or colorless. It has a vitreous or resinous luster.

Aragonite | **See p.99** | Commonly banded, this gem is reddish, yellowish, white, greenish, bluish, or violet. Transparent to opaque, it has a vitreous luster.

Onyx (sardonyx) | See pp.154–55 | This stone is brownish red with white or black parallel stripes. It is translucent with a vitreous, silky luster.

Amber | See pp.310–11 | Occurring in yellow, white, red, green, blue, brown, or black form, amber is transparent to opaque with a luster that is resinous.

Tourmaline (dravite) | See pp.226–29 | Dravite is dark yellow, yellow brown, or brownish black. It has vitreous luster and is transparent to opaque.

Cassiterite | See p.88 | Commonly found in brown or black tones with color bands, cassiterite has an adamantine luster and is transparent to opaque.

Quartz (citrine) | See p.137 | This quartz variety is light yellow to dark yellow or gold brown. It is transparent to translucent with a vitreous luster.

Baryte | See p.120 | Yellow, colorless, white, brown, gray, black, or with red, blue or green tints, it is transparent to opaque, with a vitreous luster.

Copal | See p.308 | In shades of yellow, white, red, green, blue, brown, or black, copal is transparent to opaque with a resinous luster.

Scheelite | See p.126 | Scheelite can be yellow, yellowish white, colorless gray, orange, or brown. It is transparent, with an adamantine luster

Scapolite | See p.184 | Scapolite occurs as yellow, rose pink, violet, or colorless material. It is transparent and has a vitreous luster.

Gold | See pp.36–39 | Displaying a distinctive rich yellow color, paling to a whitish yellow, gold is opaque and has a metallic luster.

Garnet (topazolite) | See pp.258–59 | Yellow to yellow brown, topazolite ranges from transparent to translucent with a subadamantine or vitreous luster.

Andalusite | See p.274 | Andalusite may be yellowish green to green, brown, pink, or colorless, with a vitreous luster. It runs from transparent to opaque.

Titanite | See p.275 | Titanite may be yellow, green, or brown, or mixtures of these. From transparent to opaque this gem has an adamantine luster.

Beryl (Heliodor/golden beryl) | See pp.240–41 | Heliodor is lemon to golden yellow, with a greenish tinge. It is transparent to opaque, with a vitreous luster.

Brazilianite | See p.116 | With green, yellowish green or golden hues and a vitreous luster, brazilianite has a transparent appearance.

Apatite | See p.118 | Apatite is transparent and is found in a range of colors, from yellow, green, and colorless, to blue and violet. It has a vitreous luster.

Tourmaline (elbaite) | See pp.228–29 | This is green, yellow, red, orange, colorless or blue, transparent or translucent, and has a vitreous or resinous luster.

Chrysoberyl | See pp.84–85 | Of vitreous luster, this gem can be shades of green, gold, yellow, red, or brown. It is transparent to opaque.

Garnet (andradite) | See pp.258–63 | From transparent to translucent, andradite is green, yellow, black, or colorless with a subadamantine or vitreous luster.

Prehnite | See p.198 | Prehnite occurs in a greenish or oily yellowish color, and has a translucent appearance with a luster that is vitreous or pearly.

Serpentine | See p.190 | Green, yellowish green, white, yellow brown, red brown or black, serpentine's luster is greasy. It is translucent to opaque.

Moonstone | See pp.164–65 | Green, colorless, white, adularescent (with a milky or blue glow), brown, or red, it is transparent or translucent and vitreous.

Jade | See pp.212–13 | Green is most prized, but jade also occurs in other colors. It is vitreous in luster and translucent through to opaque.

Chalcedony (chrysoprase) | See pp.146–149 | Chrysoprase is green or yellowish green, with translucent to opaque examples and a resinous luster.

Fluorite | See pp.96–97 | Transparent to opaque, fluorite is green, colorless, yellow, pink, red, brown, blue, or violet with a vitreous luster.

Variscite | See p.104 | Variscite ranges from green and yellow green to green blue. It is either translucent or opaque and has a waxy luster.

Diopside | See p.203 | With shades of green, yellow, colorless, brown, or black, diopside is transparent to opaque with a vitreous luster.

Peridot | See pp.254–55 | Green, yellow green, or brown green in color, peridot has a vitreous, greasy luster and a transparent appearance.

Hiddenite | See p.208 | Of vitreous luster, this gem occurs in shades of emerald green, yellow green, and green yellow. It is transparent.

Common opal | See pp.158–61 | The common opal offers a wide variety of colors, but green is most common. It has a waxy to resinous luster and is translucent to opaque.

Serpentine | See p.190 | Green, yellowish green, white, yellow brown, red brown, or black, serpentine's luster is greasy. It is translucent to opaque.

Malachite | See p.107 | A deep, saturated green with banding, malachite's luster may range from vitreous to silky to dull. It is opaque in appearance.

Garnet (demantoid) | See pp.258–63 | This green to yellowish green variety of garnet is transparent. It also has an adamantine luster.

Tourmaline (watermelon) | See pp.226–29 | Watermelon tourmaline's name derives from its green rims on red or pink cores. It is transparent with a vitreous luster.

Vesuvianite | See p.247 | A green, yellowish green, yellowish brown, or violet gem, vesuvianite is transparent to translucent with a greasy luster.

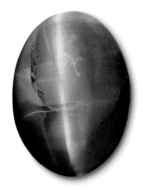

Chrysoberyl (cat's eye) | See p.84–85 | With a chatoyant effect, this greenish yellow to yellow brown gem is opaque with a vitreous to resinous luster.

Moldavite | See p.322 | This gem has a bottle-green to brown-green color. It is translucent to opaque and has a vitreous luster.

Kornerupine | See p.252 | This transparent gem with a vitreous luster can range from green to blue green and mixtures of brown and green.

Ammolite | See p.319 | With a play of mostly green or red color in a mosaiclike pattern, ammolite has greasy luster and appears opaque.

Precious opal | See pp.158–61 | Precious opal exhibits a play of colors that includes all shades. With a vitreous luster, it is transparent to opaque.

Chalcedony (bloodstone/heliotrope) | See pp.146–51 | With bloodlike spots on dark green, bloodstone chalcedony is translucent to opaque with a waxy, resinous luster.

Tourmaline (indicolite) | See pp.226–29 | Indicolite is a dark blue to blue variety of tourmaline. It is transparent to opaque and has a vitreous luster.

Chrysocolla | See p.196 | Green to blue, and exhibiting veins and patches, chrysocolla has a vitreous, waxy luster, and is opaque.

Dioptase | See p.220 | With a vitreous luster, dioptase is translucent and colored a vivid, but dark, emerald green or bluish green.

Emerald | See pp.232–33 | Emerald ranges from emerald green to a slightly yellowish green. Vitreous in luster, it is transparent to opaque.

Microcline | See p.171 | Microline can be blue to green, but is usually white to pale yellow or salmon. It has a vitreous luster and is translucent to opaque.

Smithsonite | See p.105 | Of vitreous or pearly luster, smithsonite may be blue, white, yellow, orange, brown, green, gray, or pink. It is translucent to opaque.

Turquoise | See p.110–11 | Light blue to greenish blue, turquoise may have "spiderweb" inclusions. With a waxy or dull luster, it is translucent to opaque.

Tourmaline (paraiba) | See p. 226–29 |
Paraiba is found in mint green to sky blue,
sapphire blue, and violet to purple. It is
transparent and vitreous.

Beryl (goshenite) | See p.236–41 |
Goshenite is a colorless variety of beryl.
Of vitreous luster, it ranges from transparent
to translucent.

Chalcedony | See pp.146–49 | Found
in all colors, chalcedony, a variety of
quartz, is waxy in luster, ranging from
translucent and opaque.

Quartz (milky) | See p.136 | This cloudy
white quartz variety has semi-transparent,
translucent and opaque examples and shows
a greasy luster.

Phosphophyllite | See p.199 | Occurring
in blue green to colorless forms,
phosphophyllite is translucent with
a vitreous luster.

Pectolite | See p.217 | This gem of
silky luster may be found as light blue,
light green, colorless, or gray. It can be
transparent or translucent.

Lazulite | See p.119 | Lazulite is blue white
to dark blue or green blue with a vitreous
luster. It can be transparent, translucent,
or opaque.

Topaz | See pp.272–73 | Topaz may
be blue, colorless, yellow, brownish,
green, pink, red, or violet. Its luster
is vitreous and it is transparent.

Beryl (aquamarine) | See pp.236–41 |
This blue or greenish blue variety of beryl
has a vitreous to resinous luster, and
is transparent to translucent.

Kyanite | See p.280 | Kyanite has a
vitreous, pearly luster and is blue, green,
brown, yellow, red, or colorless. It ranges
from transparent to translucent.

Zircon | See p. 268–69 | Blue, green,
yellow, brown, red or colorless, zircon
may be transparent to translucent. The
mineral also has a vitreous luster.

Tourmaline (indicolite) | See pp.228–29 |
Indicolite is a blue to dark blue variety of
tourmaline. It is transparent to opaque and
has a vitreous luster.

Azurite | See p.106 | Azurite is
azure blue or dark blue. It can be
transparent, translucent, or opaque,
and has a vitreous luster.

Benitoite | See p.223 | Benitoite is blue,
purple, pink, or colorless. Transparent
or translucent, its luster is vitreous,
subadamantine, or adamantine.

Iolite (or courdierite) | See p.222 | Mostly
occurring in a violet-blue color, Iolite has
a vitreous, greasy luster and may appear
as either transparent or translucent.

Hauyne | See p.181 | Hauyne is azure
blue, green blue, or a blue white. Its luster
is vitreous and it occurs in transparent,
translucent, or opaque form.

Sapphire | **See pp.70–73** | Sapphire occurs in various blues, as well as most other colors. Luster is subadamantine, vitreous, or pearly. It is transparent to opaque.

Kyanite | **See p.280** | Kyanite has a vitreous, pearly luster and is blue, green, brown, yellow, red, or colorless. It ranges from transparent to translucent.

Tanzanite | **See p.253** | Tanzanite occurs in shades of sapphire blue, amethyst, or violet. It is transparent and has a vitreous luster.

Lapis lazuli | **See pp.174–177** | This gem is intense deep blue, violet, or greenish blue and may contain gold-colored pyrite flecks. It has a vitreous, greasy luster and is opaque.

Labradorite | **See p.169** | Dark or black gray, and labradorescent (golden yellow, blue green, purple, bronze), this gem is vitreous and transparent to opaque.

Hematite | **See p.86** | Hematite may be found as black, steel gray, or partially reddish in color. It has a metallic luster and is opaque.

Dumortierite | **See p.277** | Dumortierite is dark blue, violet blue, red brown, or colorless, ranging from translucent to opaque. It is vitreous in luster.

Sodalite | **See p.180** | Sodalite appears as blue or blue violet with a luster that is vitreous or greasy. It is transparent, translucent, or opaque.

Quartz (amethyst) | **See p.136** | Amethyst is a purple, violet, or pale red-violet quartz. It has a vitreous luster and is transparent to opaque.

Quartz (ametrine) | **See p.138** | This variety of amethyst is purple or violet. Like amethyst, it has a vitreous luster and is transparent to opaque.

Sugilite | **See p.221** | Either violet or purple red in color, sugilite has a resinous luster and can be transparent, translucent, or opaque.

Thulite | **See p.253** | This pink to red variety of zoisite is often mottled with white and gray. It has a vitreous luster and an opaque appearance.

Fluorite (blue John) | **See pp.96–97** | This gem is a banded purple and white with a vitreous luster. It ranges from transparent to translucent.

Beryl (red) | **See pp.236–41** | A red to violet-red variety of beryl, this mineral may be transparent to translucent. It has a vitreous luster.

Tourmaline (rubellite) | **See pp.226–29** | Rubellite is a strong dark red to a pinkish red. With a vitreous luster, it is transparent to opaque.

Garnet (almandine) | **See pp.258-63** | Almandine is a red to violet-red variety of garnet. It is transparent and has a vitreous luster.

Ruby | **See pp.76–77** | Occuring in red, deep crimson, and pink shades, ruby has a pearly, subadamantine, or vitreous luster and is transparent to opaque.

Spinel | **See pp.80–81** | Found in red, pink, orange, blue, violet, or blue green, spinel is transparent and has a luster that is vitreous.

Cuprite | **See p.89** | Cuprite is carmine red or dark gray in color, has a transparency that is translucent, and shines with a metallic luster.

Sphalerite | **See p.67** | Red yellow, yellow, green, brown, or black, sphalerite is adamantine and greasy in luster, and transparent to opaque.

Garnet (almandine) | **See pp.258–63** | Almandine is a red to violet-red variety of garnet. It is transparent and has a vitreous luster.

Rhodochrosite | **See p.100** | Rhodochrosite is pinkish to red in color and can have a vitreous or resinous luster. It has a transparent appearance.

Coral | **See pp.314–15** | Coral occurs as red, pink, white, orange, blue, or brown. It it ranges from translucent to opaque and has a vitreous luster.

Rhodonite | **See p.216** | Red, grey red, or orange red, rhodonite's luster is vitreous to dull and it can be transparent, translucent, or opaque.

Sunstone | **See p.168** | Sunstone is found in hues of red and brown, or golden brown. It is aventurescent (with metallic glitter), translucent to opaque, and vitreous.

Orthoclase | **See p.170** | Occurring as yellow or colorless, orthoclase has a vitreous or pearly luster, and a transparent appearance.

Diaspore | **See p.93** | Pink, greenish brown, colorless, white, yellow, or bluish, diaspore has a vitreous or pearly luster and is transparent to translucent.

Zoisite | **See p.253** | Zoisite is red violet, green, brown, or bluish green with a vitreous luster. It can be transparent, translucent, or opaque.

Pezzottaite | **See p.192** | Pezzottaite is rose red to pink with a vitreous luster. It ranges from transparent to translucent.

Quartz (rose) | **See p.136** | Appearing in a range of strong to pale pink colors, rose quartz has a vitreous luster and is translucent.

Taaffeite | **See p.87** | Taaffeite may be pink, violet, colorless, pale green, bluish, or red. It has a vitreous luster and is transparent.

Beryl (morganite) | **See pp.236–41** | Morganite is soft pink to violet or salmon, and transparent with a vitreous luster.

Kunzite | **See p.209** | Kunzite may be found in pink to violet-pink hues, with a vitreous luster. The mineral has a transparent appearance.

Sillimanite | **See p.276** | Occurring in shades of blue and green, gray green, brownish or colorless, it has a vitreous luster and is transparent to opaque.

Cerussite | **See p.101** | With an adamantine luster, cerussite is yellow, brownish, colorless, white, blue green, gray, or black. It is transparent to opaque.

Granite | **See p.329** | Granite ranges from pink to a white or gray color and has a dull luster. It has an opaque appearance.

Seashell | **See p.298** | White, gray, silver, yellow, blue green, pink, red, brown, bronze, or black, seashell has a pearly luster and is translucent to opaque.

Mother of pearl | **See p.299** | Found in most colors, as well as iridescent purples, blues, and greens, mother of pearl has a pearly luster and is translucent to opaque.

Limestone | **See p.324** | Limestone is commonly white, but also brown, yellow, red, blue, black, or gray. It has a dull luster and is opaque.

Alabaster | **See p.122** | Alabaster has a white appearance with a dull luster. It may range in transparency from translucent to opaque.

Scapolite | **See p.193** | This transparent gem may be colorless, or range from yellow to rose pink or violet. It has a vitreous luster.

Staurolite | **See p.281** | Staurolite may have a reddish brown or black appearance with a vitreous luster. It is translucent and, more rarely, transparent.

Sandstone | **See p.325** | Sandstone can be tan, brown, yellow, red, gray, pink, white, or black. It has a vitreous luster and is opaque.

Soapstone | **See p.191** | Soapstone is opaque with a greasy luster, and ranges from greenish to yellowish, white, greenish brown, or reddish.

Color in minerals is caused by the absorption or refraction of light of particular wavelengths

Mineral and Rock Directory

6

Minerals

Entries in this directory follow the standard geological order of groups, and sub-groups within by far the largest group, silicates.

Native elements

DIAMOND

Crystal system Cubic | **Composition** C
Color White to black, colorless, yellow, pink, red, blue, brown | **Form/habit** Octahedral, cubic
Hardness 10 | **Cleavage** Perfect octahedral
Fracture Conchoidal | **Luster** Adamantine
Streak Will scratch streak plate | **SG** 3.4–3.5
Transparency Transparent to opaque | **RI** 2.42

Diamond is pure carbon, and the hardest mineral on Earth. Its hardness comes from the pattern in which its atoms are bonded and the highly uniform arrangement of their component carbon atoms. For this reason diamond crystals are usually well formed, and occur as octahedrons, and cubes with rounded edges and slightly convex faces. Its crystals may be transparent, translucent, or opaque, and range from colorless to black, with brown and yellow being the most common colors. The less common transparent, colorless, or pale blue crystals are commonly cut as gemstones. Red and green have long been considered the rarest colors for use as gemstones, but pure orange and violet are rarer still, and are thus even more valuable. Blue is also considered among the rarest. The color of diamonds can be changed artificially through irradiation or by heat treatment.

PLATINUM

Crystal system Cubic | **Composition** Pt
Color Whitish steel gray | **Form/habit** Cubic
Hardness 4–4½ | **Cleavage** None | **Fracture** Jagged | **Luster** Metallic | **Streak** Whitish steel gray
SG 14–21 | **Transparency** Opaque

The first documented discovery of platinum was by the Spanish conquistadors in the 1500s in the alluvial gold mines of the Río Pinto, Colombia. In the belief that the metal they had found was an impure silver, they named it *platina del Pinto*, platina meaning "little silver." Nonetheless, platinum has been used, unrecognized, for thousands of years. Crystals are rare, and it is found as alluvial flakes or grains in rock, and only rarely in nuggets. Native platinum usually contains some iron and other metals such as iridium, rhodium, and palladium. Today platinum is more important in industry than in ornamentation, although its use in jewelry dates back to the early 20th century. It is used in molecular converters for petroleum refining and in catalytic converters for cars. More importantly, perhaps, recently platinum compounds have found uses in chemotherapy.

Platinum nugget

SILVER

Crystal system Cubic | **Composition** Ag
Color Silver white | **Form/habit** Cubic, octahedral, dodecahedral, wiry, aborescent
Hardness 2½–3 | **Cleavage** None | **Fracture** Jagged | **Luster** Metallic | **Streak** Silver white
SG 9.6–12 | **Transparency** Opaque

Although native silver is widespread in nature, it is still relatively rare in comparison with other metals. Silver crystals are uncommon; but when they occur they are indistinct cubes or wiry masses—scaly, massive, and treelike forms are more common. Next to gold, it is the most malleable and ductile metal. The chemical symbol for silver, Ag, comes from the Latin word for silver, *argentum*, which is itself derived from a Sanskrit word meaning "white" and "shining." The majority of the world's silver comes as a by-product of the refining of lead, copper, and zinc. Even so, deposits of native silver are also commercially important. In the modern world, silver is arguably more important in industry than in ornamentation.

Silver crystals

COPPER

Crystal system Cubic | **Composition** Cu
Color Copper red to brown | **Form/habit** Massive
Hardness 2½–3 | **Cleavage** None | **Fracture** Jagged, ductile | **Luster** Metallic | **Streak** Rose
SG 8.9 | **Transparency** Opaque

When well crystallized, copper is found in cubic or dodecahedral crystals, often arranged in treelike shapes. Native copper seems to be a result of the reaction between copper-bearing solutions and iron-bearing minerals. Copper derives its name from the Latin *aes Cyprium*, meaning "metal of Cyprus," which was eventually shortened to *cyprium* and finally corrupted to *cuprum*. Copper in its free metallic state was probably the first metal to be used by humans: by around 8000 BCE, Neolithic people were using copper as a substitute for stone. Copper in industrial quantities is usually found in large masses weighing up to several tons. Its most common uses are in electrical equipment due to its excellent electrical conductivity, and as a roofing material due to its slow rate of corrosion.

GOLD

Crystal system Cubic | **Composition** Au
Color Golden yellow | **Form/habit** Octahedral, dodecahedral, dendritic | **Hardness** 2½–3
Cleavage None | **Fracture** Jagged **Luster** Metallic | **Streak** Golden yellow | **SG** 15.5–19.3
Transparency Opaque

Gold is only rarely found in well-formed octahedral and dodecahedral crystals, but is more commonly found as fernlike growths, and as grains and scaly masses. Nearly all igneous rocks contain gold in low concentrations as invisible, disseminated grains. Crystals about 1 in (2.5 cm) across were found in California, US, and several masses of over 200 lb (90 kg) have been recovered in Australia. Gold is remarkably inert chemically, and neither bonds nor reacts with most chemicals. For this reason it resists tarnishing—gold artifacts thousands of years old are as bright as the day they were made. Gold's color and brightness are highly attractive, properties that led the ancients to believe this most precious of metals mimicked the sun. Gold is extremely malleable, and is usually found in nature in a relatively pure form, all of which are qualities that have made it exceptionally valuable. Gold was in use in ancient Egypt and Mesopotamia at least 6,000 years ago, when it was nearly always gathered from placer deposits—particles left in river and streambeds due to sedimentary action.

Gold nugget

Sulfides

ACANTHITE

Crystal system Monoclinic | **Composition** Ag$_2$S
Color Black | **Form/habit** Pseudocubic
Hardness 2–2½ | **Cleavage** Indistinct | **Fracture**
Subconchoidal, sectile | **Luster** Metallic | **Streak**
Black | **SG** 7.2–7.4 | **Transparency** Opaque

Acanthite is a form of silver sulfide, and
is the most important ore of silver. It is a
hydrothermal mineral, and forms in veins
with native silver, pyrargyrite, proustite, and
other sulfides such as galena. Acanthite
usually crystallizes from argentite, the
high-temperature form of silver sulfide,
which has cubic or octahedral crystals.
Its name comes from the Greek *akantha*,
meaning "thorn," an allusion to the spiky
appearance of some of its crystals. It is

found in most silver deposits. The
Comstock Lode was an immensely rich
silver deposit, including large amounts
of acanthite, which was discovered in
Nevada, US, in 1859. It was so rich that a
branch of the US mint was established at
nearby Carson City just to coin its output.

Crystal
spikes

Acanthite specimen

CHALCOCITE

Crystal system Monoclinic | **Composition** Cu$_2$S
Color Blackish lead gray | **Form/habit** Short
prismatic, thick tabular | **Hardness** 2½–3
Cleavage Indistinct | **Fracture** Conchoidal
Luster Metallic | **Streak** Blackish lead gray
SG 5.5–5.8 | **Transparency** Opaque

The copper sulfide chalcocite is one
of the most important sources of copper.
Chalcocite is opaque and dark gray to
black with a metallic luster and is usually
found in massive form. On rare occasions,
it occurs as short prisms or blocky
crystals. It can also form twinned crystals
that appear hexagonal. Chalcocite alters
to native copper and other copper ores.
It belongs to a group of sulfide minerals
that form at relatively low temperatures
as alteration products of other copper
minerals such as bornite. These alteration
products often contain more copper than
their original minerals. The name chalcocite
is derived from the Greek word for copper,
but it also has three obsolete names:
chalcosine, copper glace, and redruthite

(an allusion to its occurrence at Redruth,
Cornwall, England). Superb crystallized
specimens come from Russia, and
Cornwall, England. Valuable deposits
occur in the US, Australia, Chile, the
Czech Republic, Peru, Russia, and Spain.

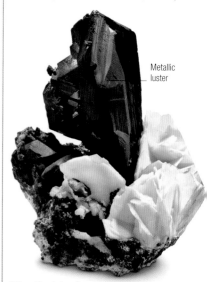

Metallic
luster

Prismatic chalcocite crystals

Bornite rough, also known as "peacock ore" because of its bright colors

◁ BORNITE

Crystal system Tetragonal | **Composition** Cu$_5$FeS$_4$
Color Coppery red, brown | **Form/habit** Usually
massive | **Hardness** 3 | **Cleavage** Poor
Fracture Uneven to conchoidal, brittle | **Luster**
Metallic | **Streak** Pale grayish black | **SG** 4.9–5.3
Transparency Opaque

One of nature's most colorful minerals,
bornite is a copper iron sulfide, resulting in
its common name, "peacock ore." A major
source of copper, it can show iridescent
purple, blue, and red splashes of color on
broken, tarnished faces. Bornite crystals
occur uncommonly, but when found they
are pseudocubic, dodecahedral, or
octahedral, and frequently have curved or
rough faces. Bornite is more commonly
found as compact or granular masses. It
forms principally in hydrothermic veins, and
contact metamorphic zones. Its formal
mineral name comes from the Austrian
mineralogist Ignaz von Born (1742–91),
although other informal names persist,
including "purple copper ore" and
"variegated copper ore." Its natural color is
varying shades of coppery red, coppery
brown, or bronze. It occurs with minerals
such as quartz, chalcopyrite, marcasite,
and pyrite. Major deposits of bornite occur
in Tasmania, Chile, Peru, Kazakhstan,
Canada, and in Arizona and in Butte,
Montana, US. Good-quality crystals are
found in deposits in England.

SPHALERITE

Crystal system Cubic | **Composition** ZnS
Color Brown, black, yellow | **Form/habit**
Tetrahedral, dodecahedral | **Hardness** 3½–4
Cleavage Perfect in six directions | **Fracture**
Conchoidal | **Luster** Resinous to adamantine,
metallic | **Streak** Brownish to light yellow
SG 3.9–4.2 | **Transparency** Opaque to
transparent | **RI** 2.36–2.37

A zinc sulfide, sphalerite takes its name
from the Greek *sphaleros*, meaning
"treacherous," describing its tendency to
occur in a number of forms that can
be mistaken for other minerals. Pure
sphalerite is colorless and rare;
usually iron is present and the color
varies from pale greenish yellow to
brown and black as the iron content
increases. Sphalerite is the most
common zinc mineral and the principal
ore of zinc. It is also found in minor amounts
in meteorites and lunar rock.

Well-formed
crystal

Sphalerite crystals

GALENA

Crystal system Cubic | **Composition** PbS
Color Lead gray | **Form/habit** Cubes, cubo-
octahedrons | **Hardness** 2½–3 | **Cleavage**
Perfect | **Fracture** Subconchoidal | **Luster**
Metallic | **Streak** Lead gray | **SG** 7.2–7.6
Transparency Opaque

There are more than 60 known lead-
bearing minerals, but by far the most
important source is the lead sulfide galena.
In addition to lead, galena often contains
small amounts of silver, zinc, copper,
cadmium, arsenic, antimony, and bismuth.
Galena's crystals are mostly cubes and
cubo-octahedrons that often exceed 1 in
(2.5 cm). Smelting requires nothing more
complex than heating the galena in the
embers of a campfire, and retrieving the
lead from beneath it when it cools. Lead
blobs found in Turkey have been dated to
around 6500 BCE, thus it may well have
been the first mineral to be smelted to

release its metal. Galena can contain so
much silver that it is mined principally for
its silver content. Some Roman lead ingots
bear the inscription *ex arg*, indicating
that the silver has been removed. Its
occurrences are widespread, but major
deposits occur in Serbia, Italy, Russia,
Canada, Mexico, Germany, England,
Australia, and Peru.

Geometric
faces

Octahedral galena crystals

PENTLANDITE

Crystal system Cubic | **Composition** (Ni, Fe)$_9$S$_8$
Color Bronze yellow | **Form/habit** Massive
Hardness 3½–4 | **Cleavage** None | **Fracture**
Conchoidal | **Luster** Metallic | **Streak** Bronze
brown | **SG** 4.6–5.0 | **Transparency** Opaque

Pentlandite is a nickel and iron sulfide,
named in 1856 for the Irish scientist
Joseph Pentland, its discoverer. It is
opaque, metallic yellow in color, and
tarnishes bronze. The chief source of
nickel, it is nearly always found mixed with
pyrrhotite, an iron sulfide, and commonly
with other sulfides, such as chalcopyrite
and pyrite. Its crystals are microscopic; it
is found principally in massive or granular
form. It is usually found in silica-poor rocks,
and has also been found in meteorites.
The Sudbury region of Ontario, Canada, is
thought to have been enriched with nickel
from a meteorite. It is relatively widespread,
but commercial deposits are scarce. The
richest sources are in Russia, South Africa,
Canada, Norway, and the US.

Bronze-colored pentlandite

COVELLITE

Crystal system Hexagonal | **Composition** CuS
Color Indigo blue to black | **Form/habit** Foliated
Hardness 1½–2 | **Cleavage** Perfect basal
Fracture Uneven | **Luster** Submetallic to resinous
Streak Lead gray to black, shiny | **SG** 4.6–4.7
Transparency Opaque

Covellite, also known as covelline, is a rare
copper sulfide mineral. It is generally found
in masses, but when crystals form they
develop as thin, blocky, hexagonal plates,
which, if thin enough, are flexible. It is
indigo blue in color, and is often tinged
with purple iridescence. It typically occurs
as an alteration product of other copper
sulfide minerals, such as chalcopyrite,
chalcocite, and bornite, where it occurs
often as a superficial coating. It was
named in 1832 for the Italian Nicolas
Covelli, who first described it. Rarely, it
occurs around the mouths of volcanoes,
as on Mount Vesuvius, where Covelli
collected it. It is found in Germany,
Australia, Serbia, and the US.

Tabular covellite crystals

GREENOCKITE

Crystal system Hexagonal | **Composition** CdS
Color Yellow to orange | **Form/habit** Pyramidal
Hardness 3–3½ | **Cleavage** Distinct, imperfect
Fracture Conchoidal | **Luster** Adamantine to
resinous | **Streak** Yellow, orange, brick red
SG 4.8–4.9 | **Transparency** Nearly opaque
to translucent

Greenockite is a rare cadmium sulfide,
and is an important source of cadmium,
a vital metal in the control rods for nuclear
reactors due to its absorbtion of neutrons.
Greenockite's crystals are single-ended
pyramids, and can be prisms and blocky
crystals. It is also found as earthy coatings.
Its crystals vary in color from honey yellow
through shades of red to brown. It was
once known as "cadmium ocher," and was
used as a yellow pigment prior to cadmium
being recognized as toxic. Greenockite is
often found with calcite, pyrite, quartz,
prehnite, chalcopyrite, and wavellite.
Localities include Scotland, the Czech

Republic, and the US. It was named in
1840 for Lord Greenock, the British army
officer who discovered it.

Earthy greenockite coating

PYRRHOTITE

Crystal system Monoclinic | **Composition** Fe$_{1-x}$S
Color Bronze yellow | **Form/habit** Massive
Hardness 3½–4½ | **Cleavage** None | **Fracture**
Subconchoidal to uneven | **Luster** Metallic | **Streak**
Dark gray black | **SG** 4.6–4.7 | **Transparency**
Opaque

Pyrrhotite is an iron sulfide in which the
ratio of iron to sulfur atoms is variable, but
is usually slightly less than one to one.
After magnetite it is the most common
magnetic mineral, which distinguishes it
from a number of similarly brassy-looking
sulfide minerals. It is generally found as
masses or grains, but it also forms fine,
hexagonal-appearing crystals at a number
of localities. It is often found where
magmatic separation—the settling of
heavy minerals to the bottom of a
crystallizing magma—has taken place.
It is found in Russia, the US, Romania,
Germany, Australia, Canada, and Japan.

Pseudohexagonal pyrrhotite crystals

CINNABAR

Crystal system Trigonal | **Composition** HgS
Color Cochineal red | **Form/habit** Trigonal
Hardness 2–2½ | **Cleavage** Perfect | **Fracture**
Subconchoidal to uneven | **Luster** Adamantine to
dull | **Streak** Scarlet | **SG** 8–8.2 | **Transparency**
Transparent to opaque

The mercury sulfide cinnabar is the major
source of mercury. Cinnabar is bright
scarlet to deep grayish red, and is usually
found in masses, granular aggregates,
or powdery coatings. Its crystals are rare.
Cinnabar commonly is found associated
with pyrite, marcasite, and stibnite in veins
near recent volcanic rocks, or in deposits
around hot springs. Cinnabar has been
mined for at least 2,000 years at Almadén
in Spain, making it one of the oldest, if not
the oldest, continuously mined deposits in
the world. This deposit is still mined and
still produces excellent crystals. The
mineral's name is derived from the
Arabic *zinjafr* and the Persian *zinjirfrah*,
both of which mean "dragon's
blood." Cinnabar deposits are found
in Peru, Italy, Slovenia, Uzbekistan,
and California, US. Mercury derived
from cinnabar was used to extract
gold in the 19th and early 20th century,
and is responsible for mercury pollution
in old mining areas.

Scarlet
coloration

Massive cinnabar

REALGAR

Crystal system Monoclinic | **Composition** AsS
Color Scarlet to orange yellow | **Form/habit**
Granular | **Hardness** 1½–2 | **Cleavage** Good
Fracture Conchoidal | **Luster** Resinous to greasy
Streak Scarlet to orange yellow | **SG** 3.5–3.6
Transparency Subtransparent to opaque

Realgar is an arsenic sulfide. It is orange
red in color and has also been known as
"ruby sulfur" or "ruby of arsenic." It is not
commonly found in crystalline form, but
when found crystals are short and deeply
lined prisms. It occurs more frequently in
coarse to fine granular masses. Realgar is
also found as thin encrustations, leading to

its name, taken from the Arabic *rahj al
ghar*, meaning "powder of the mine." Its
earliest record in English is in the 1390s.
Specimens of realgar disintegrate on
prolonged exposure to light, forming an
opaque yellow powder, which is principally
orpiment. Similar to orpiment, realgar forms
around volcanic vents, around hot spring
and geyser deposits, and as a weathering
product of other arsenic-bearing minerals.

Rare prismatic
realgar crystal

Scarlet realgar crystals

NICKELINE

Crystal system Hexagonal | **Composition** NiAs
Color Copper red | **Form/habit** Massive
Hardness 5–5½ | **Cleavage** None | **Fracture**
Conchoidal to uneven, brittle | **Luster** Metallic
Streak Black | **SG** 7.5–7.8 | **Transparency**
Opaque

Nickeline or niccolite is nickel arsenide,
containing about 44 percent nickel and 56
percent arsenic. Its color is pale copper
red with a blackish tarnish, but on polished
sections it is white with a strong yellowish
pink hue. It rarely forms crystals and is
usually found in masses or in granular form.
Nickeline was named in 1832 by French
mineralogist François Sulpice Beudant in
reference to its nickel content. In the Middle
Ages it had been known as *Kupfernickel*,
"devil's copper," from the German, since
it was believed that it contained copper
that was impossible to extract, and also
caused smelters to become ill in the
process. It is found in ore deposits
with other nickel minerals, and in vein
deposits containing copper and silver. It
is rarely used as a source of nickel due to
the large amount of arsenic that it contains.
Notable deposits are the Harz Mountains in

Germany, Ontario and the Northwest
Territories in Canada, and Japan, Mexico,
Iran, the US, Russia, and Australia.

Metallic luster

Massive nickeline

CHALCOPYRITE

Crystal system Tetragonal | **Composition** CuFeS$_2$
Color Brassy yellow | **Form/habit** Tetrahedral
Hardness 3½–4 | **Cleavage** Distinct | **Fracture**
Uneven, brittle | **Luster** Metallic | **Streak** Green
black | **SG** 4.1–4.3 | **Transparency** Opaque

Chalcopyrite is a copper and iron sulfide.
Freshly broken chalcopyrite is opaque
and brassy yellow in color, but on
weathering it develops an iridescent
tarnish. Its crystals—tetrahedral in
appearance—reach up to 5 in (10 cm) on
a face; massive aggregates are common,
while botryoidal masses are less common.
Although not rich in copper, it is the most
important source of copper due to its
widespread and extensive occurrences.
Its name comes from the Greek *khalkos*
for "copper," and "pyrite." Chalcopyrite
is commonly found in hydrothermal ore
veins deposited at medium and high
temperatures. It is among the minerals
that have been worked at Rio Tinto, Spain,
since Roman times. Important deposits
are found in the US, England, Tasmania,
Germany, Canada, Spain, and Japan.

Chalcopyrite crystals

MILLERITE

Crystal system Trigonal | **Composition** NiS
Color Brass yellow | **Form/habit** Acicular
Hardness 3–3½ | **Cleavage** Perfect | **Fracture**
Uneven, brittle | **Luster** Metallic | **Streak** Greenish
black | **SG** 5.3–5.6 | **Transparency** Opaque

Millerite is a nickel sulfide and, when found
in high enough concentration, it is a very
important source of nickel. When found
as crystals, it occurs as brassy, needlelike
crystals in cavities in sulfide-rich limestone
and dolomite, or as an alteration product
of other nickel minerals. Crystals can
be free standing as single crystals, tufts,
matted groups, or radiating sprays. It
is found in masses also, often with an
iridescent tarnish. Millerite is found in
iron-nickel meteorites and condensed in
crusts on Mount Vesuvius, Italy. Millerite
was discovered in 1845 and was named
for British mineralogist William Hallowes
Miller. Important deposits are found in
Australia, the US, Canada, Italy, Germany,
Belgium, and New Caledonia.

Millerite needles

STANNITE

Crystal system Tetragonal | **Composition**
Cu$_2$FeSnS$_4$ | **Color** Black | **Form/habit** Massive
Hardness 3–4 | **Cleavage** Indistinct | **Fracture**
Uneven | **Luster** Metallic | **Streak** Black
SG 4.3–4.5 | **Transparency** Opaque

Stannite is a sulfide of copper, iron, and
tin. Zinc and trace germanium may also
be present. It is steel gray to iron black in
color. Crystals are rare but, when found,
they are usually octahedral in appearance.
It is often associated with cassiterite,
pyrite, tetrahedrite, and chalcopyrite, and
in tin veins. Bearing approximately one
quarter tin, it is a source of that metal.
The name comes from the Latin for
tin: *stannum*. Important localities
are England, Canada, Tasmania,
Australia, Bolivia, the Czech
Republic, Russia, and the US.

Orpiment and stibnite deposits, Champagne Pool, Waiotapu, New Zealand's North Island

STIBNITE

Crystal system Orthorhombic | **Composition** Sb$_2$S$_3$
Color Lead gray to steel gray, black | **Form/habit**
Prismatic | **Hardness** 2 | **Cleavage** Perfect
Fracture Subconchoidal | **Luster** Metallic
Streak Lead gray to steel gray | **SG** 4.6–4.7
Transparency Opaque

Stibnite, sometimes called antimonite,
is antimony sulfide. It is lead gray
to steel gray in color, and often
has a black, iridescent tarnish.
Its crystals tend to be elongated
prisms, often deeply grooved
parallel to the prism faces. It is
also found in massive aggregates. Stibnite
is the principal ore of antimony, and its

name comes from its Latin name, *stibium*.
It is often found along with realgar, galena,
pyrite, and others.

Prismatic
crystals

Striated stibnite crystals

△ ORPIMENT

Crystal system Monoclinic | **Composition** As$_2$S$_3$
Color Yellow | **Form/habit** Massive, foliated
Hardness 1½–2 | **Cleavage** Perfect | **Fracture**
Uneven, sectile | **Luster** Resinous | **Streak**
Pale yellow | **SG** 3.5 | **Transparency** Transparent
to translucent

Orpiment is a deep orange yellow colored
arsenic sulfide, and almost always occurs
alongside realgar. It is usually found as
thin, leafy plates, as a powder, or in
columnar masses. Well-formed crystals
are uncommon, but when found they
are short prisms with an orthorhombic
appearance. It is found in low-temperature
veins, hot-spring deposits, and around

volcanic vents. It can also originate from
the alteration of arsenic-bearing minerals,
including realgar. It takes its name from
the Latin *auri* for "golden," and *pigmentum*,
"paint." For centuries, orpiment was one
of the few clear, bright-yellow pigments
available to artists. Unfortunately, it is
extremely toxic, but nonetheless it was
once used as a medicine in China.
Orpiment is unstable when exposed to
light, deteriorating into a powder. It occurs
widely, and notable sources include Sakha
(Yakutia) in Russia; Nye County, Nevada,
US; Copalnic, Romania; Hakkâri in Turkey;
Hokkaido in Japan; St.andreasberg in
Germany; Quiruvilca Mine, Peru; and
Guizhou Province, China.

Pyrite occurs as octehedral crystals on solid pyrite at Tristate, US

HAUERITE

Crystal system Cubic | **Composition** MnS$_2$
Color Red brown to brown black | **Form/habit**
Octahedral | **Hardness** 4 | **Cleavage** Perfect
cube | **Fracture** Subconchoidal to uneven
Luster Adamantine to submetallic | **Streak**
Red brown | **SG** 3.5 | **Transparency** Opaque

Hauerite is a manganese sulfide mineral.
Found in both octahedral and cubo-
octahedral crystals, it can also occur in
globular agglomerations or in masses. Its
color is reddish brown to dark brown, or
black. It was discovered in what is now
Slovakia in 1846, and was named after the
Austrian geologists Joseph Ritter von Hauer
and Franz Ritter von Hauer. It occurs in
sulfur-rich environments in association with
native sulfur, realgar, gypsum, and calcite.
Localities include Texas in the US, the Ural
Mountains in Russia, and Sicily, Italy.

Intergrown hauerite crystals

△ PYRITE

Crystal system Cubic | **Composition** FeS$_2$
Color Pale brass yellow | **Form/habit** Cubic,
octahedral, pyritohedral | **Hardness** 6–6½
Cleavage None | **Fracture** Conchoidal | **Luster**
Metallic | **Streak** Greenish black to brownish black
SG 5.0–5.2 | **Transparency** Opaque

The iron sulfide pyrite is perhaps better
known as "fool's gold." It is brassy yellow
in color and has a relatively high density,
although nowhere near that of gold. Pyrite
commonly forms in cubes; octahedrons
and pentagonal dodecahedrons are also
common. Its crystal faces are often deeply
lined. It is also found as masses, in grains,
and in nodules. Pyrite's name comes from
pyr, the Greek for fire, because it emits
sparks when struck by iron. Since it is
often found in large deposits, pyrite could
potentially be exploited as a source of iron;
however other minerals are better suited to
its extraction. It has been used as a source
of sulfur for manufacturing sulfuric acid.

BISMUTHINITE

Crystal system Orthorhombic | **Composition**
Bi$_2$S$_3$ | **Color** Lead gray to tin white | **Form/
habit** Short prismatic to acicular | **Hardness** 2
Cleavage Perfect | **Fracture** Uneven | **Luster**
Metallic | **Streak** Lead gray | **SG** 6.8–7.2
Transparency Opaque

Bismuthinite is a mineral consisting of
bismuth sulfide, and is comparatively rare.
It is structurally related to the sulfides of
antimony (stibnite) and arsenic (orpiment)
and, like stibnite, it forms prismatic to
acicular crystals, which are often elongated
and striated lengthwise. More commonly,
it is found in leaflike or fibrous masses.
Bismuthinite is formed in high-temperature
hydrothermal veins and in granite
pegmatites, igneous rocks that are formed
during the final stages of the crystallization
of magma. It is sometimes found with
native bismuth, as well as other sulfide
minerals. It is an important source of
bismuth, a major commercial metal that
expands slightly upon solidifying, making

its alloys particularly suitable for the
manufacture of detailed metal castings.
Bismuth solders have low melting points
and are used in fire-safety devices, such
as automatic sprinkler heads and safety
plugs for compressed-gas cylinders.
Bismuth compounds are used to treat
skin infections and digestive disorders.

Bismuthinite crystals

COBALTITE

Crystal system Orthorhombic | **Composition**
CoAsS | **Color** Silver white, pink | **Form/habit**
Pseudocubic or pyritohedral | **Hardness** 5½
Cleavage Perfect basal | **Fracture** Conchoidal or
uneven, brittle | **Luster** Metallic | **Streak** Grayish
black | **SG** 6–6.4 | **Transparency** Opaque

Cobaltite is a cobalt arsenic sulfide with up
to 10 percent iron and variable amounts of
nickel substituting the cobalt in its structure.
Its color ranges from steel gray to silver
white with a reddish tinge. Its crystals are
pink, although it is more commonly found
occurring in grains or in masses. It is
sometimes also known as cobalt glace,

a name it received in 1832 because it contains the element cobalt, of which it is the principal ore. Cobaltite is formed in high-temperature hydrothermal deposits, and as veins in contact-metamorphic zones. It is soluble in nitric acid. Cobalt alloys are used wherever a combination of high strength and heat resistance is needed, such as for the interior components of gas turbines. Notable localities are Cobalt in Ontario, Canada; Tunaberg, Sweden; Siegerland, Germany; Skutterud, Norway; New South Wales, Australia; Colorado, US; Sonora, Mexico; and Bou Azzer, Morocco.

Pseudocubic cobaltite crystal

Cobaltite crystals

MOLYBDENITE

Crystal system Hexagonal | **Composition** MoS_2
Color Lead gray | **Form/habit** Tabular, prismatic
Hardness 1–1½ | **Cleavage** Perfect basal
Fracture Uneven | **Luster** Metallic | **Streak**
Greenish or bluish gray | **SG** 4.7–4.8
Transparency Opaque

A sulfide of molybdenum, molybdenite was originally thought to be lead. In that mistaken belief, its name was derived from the Greek *molybdos*, meaning "lead." It was recognized in 1778 by Swedish chemist Carl Scheele (for whom scheelite is named) as a distinct mineral that contained a new element, but the metal itself was not extracted until 1782. It is a very soft, bluish gray, opaque, metallic mineral that forms platy hexagonal crystals, flaky masses, scales, and disseminated grains. Its greasy-feeling crystals can be confused with graphite, although it has a much higher specific gravity and a more metallic luster. Molybdenum—of which molybdenite is the principal source— imparts high strength, toughness, hardness, and corrosion resistance to alloys, and is commonly used for heating elements in high-temperature electric furnaces, and in some electronic components, such as the filament supports in light bulbs. Molybdenite is found in granites and pegmatites, in high-temperature hydrothermal veins, and

in contact-metamorphic deposits. Commercially viable quantities are known in Japan, England, Tasmania, Canada, Norway, and the US. It closely resembles graphite, and the two can be easily confused unless compared side by side.

Hexagonal foliated molybdenite

Layered molybdenite masses

ARSENOPYRITE

Crystal system Monoclinic | **Composition** FeAsS
Color Silver white to steel gray | **Form/habit**
Prismatic | **Hardness** 5–6 | **Cleavage**
Distinct, indistinct | **Fracture** Uneven | **Luster**
Metallic | **Streak** Dark grayish black | **SG**
5.9–6.2 | **Transparency** Opaque

The most common arsenic mineral, arsenopyrite is an iron arsenic sulfide. Its crystals are frequently twinned, giving the appearance of crystals of another crystal system. It is also found in granular or compact form. It produces a garlic odor when heated, like other arsenic-bearing minerals. These fumes are toxic. It is often found associated with minerals such as gold and cassiterite. Its name is the contraction of its older name, arsenical pyrites, and was once known by its German name, *mispickel*. Its color is silver white to steel gray on freshly broken surfaces, but weathering gives it a brownish or pink color. It is a widespread mineral, with noted localities in Germany, Mexico, Australia, Portugal, the US, Canada, and Cornwall, England.

Arsenopyrite crystals

MARCASITE

Crystal system Orthorhombic | **Composition**
FeS_2 | **Color** Pale bronze yellow | **Form/habit**
Tabular, prismatic | **Hardness** 6–6½
Cleavage Distinct | **Fracture** Conchoidal
Luster Metallic | **Streak** Gray to black
SG 4.8–4.9 | **Transparency** Opaque

Marcasite is iron sulfide, and is chemically identical to pyrite but crystallizing in a different crystal system. It is opaque and pale silvery yellow when fresh, darkening and tarnishing after exposure. It comes in a number of crystal forms, and especially in twinned, curved, sheaflike shapes that resemble a cock's comb. It is also commonly found in nodules with radially arranged fibers. Marcasite is found near the surface where it forms from acidic solutions penetrating downward through shale, clay, limestone, and chalk. Jewelry that was said to be made of marcasite was actually made of other minerals, such as hematite or pyrite.

Marcasite crystals

SYLVANITE

Crystal system Monoclinic | **Composition**
$(Au,Ag)_2Te_4$ | **Color** Silver white to pale yellow
Form/habit Prismatic, tabular, bladed | **Hardness**
1–2 | **Cleavage** Perfect | **Fracture** Uneven,
brittle | **Luster** Metallic | **Streak** Gray | **SG**
8–8.3 | **Transparency** Opaque

Sylvanite is a gold and silver telluride. The tellurides are in the same mineral group as the sulfides, but there are relatively few of them, of which sylvanite is the most important. Its color ranges from a steely gray to nearly white. It forms complex prisms: boxy, or bladed crystals, which are frequently twinned. Some twin forms are treelike and resemble written characters, and are known as graphic tellurium. It is often found with other gold tellurides. In a few instances it occurs in sufficient

quantities to constitute an ore of gold. It is photosensitive and can develop a dark tarnish if exposed to bright light. Important deposits are in Canada, Western Australia, Sweden, New Zealand, and the US.

"Graphic tellurium" sylvanite

GLAUCODOT

Crystal system Orthorhombic | **Composition**
(Co,Fe)AsS | **Color** Gray to white | **Form/habit**
Prismatic | **Hardness** 5 | **Cleavage** Perfect,
distinct | **Fracture** Uneven | **Luster** Metallic
Streak Black | **SG** 6.0 | **Transparency** Opaque

Glaucodot is a cobalt iron arsenic sulfide mineral that takes its name from the Greek *glaukos* for "blue," an allusion to its use in coloring blue glass. The cobalt to iron ratio is typically 3:1, with minor nickel substituting in the structure. It is opaque gray to tin white, and is typically found as massive forms without external crystal form. When it does form crystals they occur as prisms, and are frequently found in cruciform twins. Glaucodot forms in high-temperature hydrothermal veins, where it is often accompanied by pyrite and chalcopyrite. Localities include Huasco, Valparaíso Province, Chile, where it was first discovered in 1849; Tunaberg in Sweden; Cobalt in Ontario, Canada; and Franconia, New Hampshire and Sumpter, Oregon in the US.

Prismatic glaucodot crystals

Sulfosalts

POLYBASITE

Crystal system Monoclinic | **Composition** $(Ag,Cu)_{16}Sb_2S_{11}$ | **Color** Iron black | **Form/habit** Pseudohexagonal | **Hardness** 1½–2 | **Cleavage** Imperfect | **Fracture** Uneven | **Luster** Metallic | **Streak** Black | **SG** 6–6.2 | **Transparency** Opaque

Polybasite is a sulfosalt mineral of silver, copper, antimony and arsenic. It forms iron black, blocky, hexagonal-appearing crystals, often with triangular grooves, and it is also found in masses. It melts at a low temperature. It is found with native silver, other silver minerals, lead minerals, and other sulfides and sulfosalts, such as argentite and galena. Its name comes from the Greek for "many" and "base," in regard to the number of base metals in the mineral. Important occurrences include Germany, the Czech Republic, Honduras, Mexico, and the US.

Iron-black polybasite

ENARGITE

Crystal system Orthorhombic | **Composition** Cu_3AsS_4 | **Color** Grayish black to iron black | **Form/habit** Tabular, columnar | **Hardness** 3 | **Cleavage** Perfect | **Fracture** Uneven, brittle | **Luster** Metallic | **Streak** Black | **SG** 4.4–4.5 | **Transparency** Opaque

Enargite is a copper arsenic sulfosalt. It is colored steel gray and blackish gray to violet black. Crystals are usually blocky,

tabular, or slender prisms, and are sometimes pseudohexagonal or hemimorphic (with the terminations different at each end). They occasionally form star-shaped multiple twins. It is also found in masses and grains. When exposed to light, its luster often dulls from metallic silver gray to black. Enargite is formed in veins and replacement deposits associated with chalcopyrite, covellite, bornite, galena, pyrite, and sphalerite and can be an important ore of copper. It takes its name from the Greek word *enarge*, "distinct," in reference to its perfect cleavage. Notable localities are in the US, Taiwan, Peru, Chile, Sardinia, Italy, Serbia, Tunisia, and Namibia.

Striated enargite crystals

▷ TETRAHEDRITE

Crystal system Cubic | **Composition** $(Cu,Fe)_{12}Sb_4S_{13}$ | **Color** Flint gray to iron black | **Form/habit** Massive | **Hardness** 3–4 | **Cleavage** None | **Fracture** Subconchoidal to uneven | **Luster** Metallic | **Streak** Brown to black to cherry red | **SG** 4.6–5.1 | **Transparency** Opaque

Tetrahedrite is a copper iron antimony sulfosalt that often contains minor amounts of zinc, silver, lead, and mercury. Tetrahedrite is a common mineral, and is probably the most common of the sulfosalts. It is chemically and structurally related to tennantite, in which arsenic replaces antimony in the crystal structure. Tetrahedrite gets its name from the distinctive tetrahedron shape of its crystals, which are steel gray to black metallic, or brassy in color. More commonly it is found in masses and grains. It is found in metal-bearing veins, often associated with galena, pyrite, chalcopyrite, baryte, bornite, and quartz. Major deposits are in the US, Germany, Peru, Australia, and Romania.

TENNANTITE

Crystal system Cubic | **Composition** $(Cu,Fe)_{12}As_4S_{13}$ | **Color** Steel gray to black | **Form/habit** Pseudocubic | **Hardness** 3–4½ | **Cleavage** None | **Fracture** Subconchoidal to uneven, brittle | **Luster** Metallic | **Streak** Black | **SG** 4.6–5.2 | **Transparency** Opaque

Tennantite is a copper iron arsenic sulfide, and was named in 1819 for the English chemist Smithson Tennant. It is chemically and structurally related to tetrahedrite, and forms crystals that are sometimes similar to tetrahedrite's. It is also found in masses and grains. Tennantite is found associated with fluorite, baryte, galena, quartz, chalcopyrite, and sphalerite. Important localities are in Germany, Switzerland, and the US. In antiquity the arsenic component of tennantite caused the metal smelted from it to be harder than that of pure copper, a discovery that is believed to have been a key step toward the Bronze Age.

Rough tennantite

STEPHANITE

Crystal system Orthorhombic | **Composition** Ag_5SbS_4 | **Color** Iron black | **Form/habit** Short prismatic to tabular | **Hardness** 2–2½ | **Cleavage** Imperfect | **Fracture** Subconchoidal to uneven, brittle | **Luster** Metallic | **Streak** Iron black | **SG** 6.2–6.5 | **Transparency** Opaque

Stephanite is silver antimony sulfide, and is sometimes called brittle silver ore or black silver ore. Its crystals are short prisms or blocky, but it can also be found in masses and in grains. It is iron black in color. Crystals are lustrous and brilliant when found, but on exposure to light they soon become dull. It is generally found in small quantities, but when of sufficient size, deposits are an important source of silver. Significant amounts are found in the US, Mexico, Canada, England, Norway, Russia, and Bolivia. It was named stephanite in 1845 in honor of the Archduke of Austria, Stephan Franz Victor of Habsburg-Lorena.

Stephanite crystals

Tetrahedrite crystals

PYRARGYRITE

Crystal system Hexagonal/trigonal | **Composition** Ag_3SbS_3 | **Color** Deep red | **Form/habit** Prismatic, scalenohedral | **Hardness** 2½–3 | **Cleavage** Distinct | **Fracture** Conchoidal to uneven | **Luster** Metallic | **Streak** Purplish red | **SG** 5.8 | **Transparency** Translucent

Pyrargyrite is silver sulfantimonide, and a major source of silver. It is also called ruby silver and dark red silver ore. Pyrargyrite is usually grayish black; thin splinters and small crystals are deep ruby red by transmitted light, hence the informal names. Pyrargyrite itself takes its name from the Greek words *pyros*, meaning "fire," and *argent*, meaning "silver." Its dark-red color becomes darker when exposed to light. Pyrargyrite forms along with other silver minerals, such as proustite (also called, confusingly, ruby silver), galena, sphalerite, and tetrahedrite. Some of the best specimens are found in the Harz Mountains in Germany.

Adamantine luster

Pyrargyrite crystals

BOURNONITE

Crystal system Orthorhombic | **Composition** $PbCuSbS_3$ | **Color** Steel gray | **Form/habit** Short prismatic to tabular | **Hardness** 2½–3 | **Cleavage** Indistinct | **Fracture** Subconchoidal to uneven | **Luster** Metallic | **Streak** Steel gray | **SG** 5.7–5.9 | **Transparency** Opaque

Bournonite is a lead copper antimony sulfide. It is found as heavy, dark crystal aggregates and masses. It is also found in interpenetrating cruciform twins that give it the appearance of a cogwheel, and thus its informal name of cogwheel ore. Bournonite is widespread, and is found associated with galena, sphalerite, chalcopyrite, and pyrite. Particularly prized specimens come from the Harz Mountains of Germany, where a few crystals exceed 1 in (2.2 cm) in diameter. Good specimens also come from Cornwall, England; and

from localities in Italy, France, Bolivia, Peru, Canada, Australia, Romania, Greece, Japan, and the US. It is named for Count J. L. de Bournon, a French mineralogist.

Cruciform twins

Prismatic bournonite crystals

BOULANGERITE

Crystal system Monoclinic | **Composition** $Pb_5Sb_4S_{11}$ | **Color** Bluish lead gray | **Form/habit** Long prismatic to acicular | **Hardness** 2½–3 | **Cleavage** Good | **Fracture** Uneven, brittle | **Luster** Metallic | **Streak** Brownish gray to gray | **SG** 5.8–6.2 | **Transparency** Opaque

Boulangerite is lead antimony sulfide. It is lead gray, brittle, soft, and dense. It occurs in hydrothermal veins and other environments. Boulangerite forms flexible, needlelike crystals or prisms, and is also fibrous or in masses. Sometimes the crystals form a fine feathery mass, which has been called plumosite. It was named in 1837 for French mining engineer Charles Boulanger. It is found widely in small amounts in Germany, Canada, Mexico, Sweden, France, the Czech Republic, and the US. When deposits are of sufficient size, it is used as a lead ore.

Boulangerite specimen

PROUSTITE

Crystal system Trigonal | **Composition** Ag_3AsS_3 | **Color** Scarlet, gray | **Form/habit** Prismatic to short prismatic | **Hardness** 2½ | **Cleavage** Distinct rhombohedral | **Fracture** Conchoidal to uneven, brittle | **Luster** Adamantine to Submetallic | **Streak** Vermillion | **SG** 5.5–5.7 | **Transparency** Translucent

Proustite is a sulfosalt, silver sulfarsenide. Its color is scarlet vermilion, giving it its informal names of light-red silver or ruby silver ore. Its crystals occur as prisms, rhombohedrons, or scalenohedrons, and are light sensitive, turning transparent scarlet to opaque gray when exposed to strong light. Proustite is also found in masses. An important source of silver, it is found with other silver minerals, galena, and calcite. Notable localities are Chañarcillo, Chile; Saxony, Germany; Idaho, US; and Chihuahua, Mexico. It is related to the sulfantimonide pyrargyrite, from which it was distinguished by the chemical analyses of Joseph L. Proust, for whom it is named.

Prismatic crystals

Crystalline proustite

JAMESONITE

Crystal system Monoclinic | **Composition** $Pb_4FeSb_6S_{14}$ | **Color** Steel gray to dark lead gray | **Form/habit** Acicular, fibrous | **Hardness** 2–3 | **Cleavage** Good | **Fracture** Uneven to conchoidal | **Luster** Metallic | **Streak** Grayish black | **SG** 5–6 | **Transparency** Opaque

Jamesonite is a lead iron antimony sulfide, which may also contain manganese. It is a dark gray metallic mineral which forms needlelike, fibrous crystals, and can occur in masses. It is one of the few sulfide minerals to form these kinds of crystals. It is found in veins with other sulfosalt minerals; and in quartz veins with

carbonate minerals, such as rhodochrosite, dolomite, and calcite. It was named for Scottish mineralogist Robert Jameson, and was first identified in 1825. It is widespread in small amounts, with the best specimens from Mexico, Serbia, Romania, England, and Bolivia.

Metallic luster

Fibrous jamesonite

ZINKENITE

Crystal system Hexagonal | **Composition** $Pb_9Sb_{22}S_{42}$ | **Color** Steel gray | **Form/habit** Fibrous, massive | **Hardness** 3–3½ | **Cleavage** Indistinct | **Fracture** Uneven, brittle | **Luster** Metallic | **Streak** Steel gray | **SG** 5.3 | **Transparency** Opaque

Zinkenite is a lead antimony sulfide. It forms needlelike crystals in columns and radiating aggregates, and is often found in masses or in thick mats of hairlike fibers. It usually occurs in quartz associated with other sulfosalts. It is named for German mineralogist and mining geologist, Johann Karl Ludwig Zinken. The best specimens are found in France, Germany, Bolivia, Australia, Canada, Romania, and the US.

Acicular crystals

Zinkenite crystals

Oxides

ILMENITE

Crystal system Trigonal | **Composition** FeTiO$_3$
Color Iron black | **Form/habit** Thick tabular
Hardness 5–6 | **Cleavage** None | **Fracture**
Conchoidal | **Luster** Metallic to submetallic | **Streak**
Black | **SG** 4.5–5 | **Transparency** Opaque

A major source of titanium, ilmenite is an
iron titanium oxide named after the locality
of its discovery in the Il'menski Mountains,
near Miass, Russia. Its crystals are usually
thick and blocky, and it occurs in masses
or scattered grains. Ilmenite is opaque and
metallic gray black, and is often intergrown
with magnetite or hematite. It is frequently
found in kimberlites, with diamond, and in
minor amounts in other igneous rocks.
It is also found in veins, pegmatites, and
concentrated in sands and gravels with
magnetite, rutile, and other heavy minerals.

Ilmenite crystals

PEROVSKITE

Crystal system Orthorhombic | **Composition**
CaTiO$_3$ | **Color** Black, brown, yellow | **Form/habit**
Pseudocubic | **Hardness** 5½ | **Cleavage**
Imperfect | **Fracture** Subconchoidal to uneven
Luster Adamantine/metallic | **Streak** Gray to
colorless | **SG** 4–4.8 | **Transparency** Opaque

Perovskite is calcium titanium oxide;
niobium or cerium can replace a large
amount of the titanium. Although it is
orthorhombic, its crystals may appear to
be cubes or octahedrons. It can be black,
brown, or yellow, its luster depending on
its color. Perovskite is thought to be a
major constituent of the upper mantle
of the Earth, and it can also be found in
carbonaceous chondrite meteorites. At
the Earth's surface, it occurs in magnesium
and iron-rich igneous rocks, in some
schists, and in some contact-metamorphic
rocks. Significant deposits are found in
Greenland, where crystals of up to 3 in
(8 cm) have been found. Italy, Germany,
Brazil, Canada, and the US are other
notable localities for Perovskite. It was
named in 1839 for Russian mineralogist
L. A. Perovski.

Perovskite crystals

◁ HEMATITE

Crystal system Trigonal | **Composition** Fe$_2$O$_3$
Color Steel gray | **Form/habit** Tabular, sometimes
platy, botryoidal | **Hardness** 5½–6½ | **Cleavage**
None | **Fracture** Subconchoidal to uneven | **Luster**
Metallic to dull | **Streak** Cherry red or red brown
SG 5.2–5.3 | **Transparency** Opaque

Hematite is an iron oxide and comes in a
number of different forms, from soft and
fine grained to hard and dense crystalline.
Soft, powdery hematite is called red ocher,
and has long been used as a pigment.
Ground crystalline hematite, called rouge,
is used to polish plate glass and jewelry.
Its steel-gray crystals and coarse-grained
varieties with a brilliant metallic luster are
known as specular hematite. Crystals
in flowerlike forms are called iron roses.
The most important hematite deposits are
sedimentary in origin, either as sedimentary
beds or metamorphosed sediments.

CORUNDUM

Crystal system Hexagonal/trigonal | **Composition**
Al$_2$O$_3$ **Color** Occurs in most colors | **Form/habit**
Pyramidal, prismatic barrel shape | **Hardness** 9
Cleavage None | **Fracture** Conchoidal to uneven
Luster Adamantine to vitreous | **Streak** Colorless
SG 4.0–4.1 | **Transparency** Transparent to
translucent | **RI** 1.76–1.77

Corundum is aluminum oxide. Next to
diamond, corundum is the hardest mineral
on Earth. Its crystals are rough and are

Hematite or Ironstone concretions on bed of Navajo sandstone

generally hexagonal in form, either blocky, tapering barrel-shaped, or in double pyramids. Corundum forms in syenites, certain pegmatites, and in high-grade metamorphic rocks. It is heavy and highly resistant to weathering, thus is often found concentrated in stream gravels. Opaque corundum is used as an abrasive, and can be colorless, gray, or brown. Its transparent gemstone varieties can be red (ruby), pink orange (padparadscha), and colorless, blue, green, yellow, orange, violet, and pink (sapphire). Ruby forms a continuous color succession with pink sapphire; only stones of the darker hues are considered to be ruby. Cat's-eye and star stones of sapphire and ruby contain microscopic rutile arranged in a netlike pattern that reflects light when the stone is polished. The mineral name corundum is likely to be derived from the Sanskrit *kuruvinda*, meaning "ruby," the name given to red corundum.

Red corundum crystals on rock

ICE

Crystal system Hexagonal | **Composition** H_2O
Color Colorless | **Form/habit** Platy, prismatic, dendritic, massive | **Hardness** Varies | **Cleavage** Perfect, difficult | **Fracture** Conchoidal, brittle
Luster Vitreous | **Streak** White | **SG** 1.0
Transparency Transparent to translucent

Ice is an oxide of hydrogen, and is the most abundant mineral exposed at the Earth's surface. It forms crystals that rarely exceed 1/4 in (7 mm) in length in its variety as snow, although in its massive aggregates within glaciers, individual crystals may reach 18 in (45 cm). Other forms include branching, treelike frost, skeletal, hopper-shaped prisms, and rounded bodies with concentric structures (hailstones and icicles) made up of randomly oriented crystals. Its crystals are generally colorless; its common white color is due to gaseous inclusions of air. There are at least nine different crystalline forms of ice, each forming under slightly different

pressure and temperature conditions. The hardness of ice varies with its crystal structure, purity, and temperature. At the low temperatures found in the Arctic it has the same hardness as feldspar, and is hard enough to erode stone when windblown. Liquid water is not classified as a mineral because it has no crystalline form.

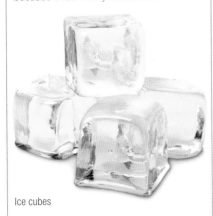

Ice cubes

ZINCITE

Crystal system Hexagonal | **Composition** ZnO
Color Orange yellow to deep red | **Form/habit** Massive | **Hardness** 4–5 | **Cleavage** Perfect
Fracture Conchoidal | **Luster** Resinous
Streak Orange yellow | **SG** 5.4–5.7
Transparency Almost opaque

Zincite is an oxide of zinc, and is sometimes called red oxide of zinc. Its color is orange yellow to deep red. Natural crystals are rare but when crystals do occur, they are hollow and pyramidal in form. It is usually found in masses and grains. Zincite is a rare mineral, but where it does occur, it does so abundantly. It is found mainly as a minor mineral in zinc ore deposits. Crystals are found only in veins or fractures where they have developed from the weathering of other zinc minerals. The most important localities of zincite are Franklin, New Jersey, US, and Tsumeb in Namibia. In both of these occurrences, it is a source of zinc.

Deep red zincite

Zincite specimen

CUPRITE

Crystal system Cubic | **Composition** Cu_2O
Color Shades of red to nearly black | **Form/habit** Pseudocubic | **Hardness** 3½–4 | **Cleavage** Distinct | **Fracture** Uneven, brittle | **Luster** Adamantine, submetallic | **Streak** Brownish-red, shining | **SG** 5.8–6.2 | **Transparency** Transparent to almost opaque

A major ore of copper, cuprite is copper oxide. Its dark crystals with red internal reflections appear as cubes, octahedrons, or dodecahedrons—or as combinations of these. These can turn superficially dark gray on exposure to light. Cuprite was first described in 1845 and the name comes from the Latin *cuprum*, "copper," for its copper content.

MAGNETITE

Crystal system Cubic | **Composition** Fe_3O_4
Color Black to brownish black | **Form/habit** Octahedral | **Hardness** 5½–6½ | **Cleavage** None | **Fracture** Conchoidal to uneven | **Luster** Metallic to semimetallic | **Streak** Black | **SG** 5.2
Transparency Opaque

Magnetite is an iron oxide in the spinel group, and is one of the most widespread of the oxide minerals. Like other spinel group minerals it usually forms octahedral crystals, although it is sometimes found in highly modified dodecahedrons. It has a similar appearance to hematite, but hematite is nonmagnetic and has a red streak. In addition to crystals, it comes in masses and grains and as concentrations in black sand. It occurs in igneous and metamorphic rocks and in sulfide veins. Magnetite is a strong natural magnet and was probably used to magnetize the first compass needles. It is a major ore of iron.

SPINEL

Crystal system Cubic | **Composition** $MgAl_2O_4$
Color Red, yellow, orange red, blue, green, black
Form/habit Octahedral | **Hardness** 7½–8
Cleavage None | **Fracture** Conchoidal to uneven
Luster Vitreous | **Streak** White | **SG** 3.6
Transparency Transparent to translucent
RI 1.71–1.73

A magnesium aluminum oxide, spinel is both a mineral name and the name of a group of minerals, all of which are metal oxides, and all of which have the same crystal structure. Other members of the spinel group include gahnite, franklinite, and chromite. Spinel and other minerals

of its group are usually found as glassy, hard octahedrons, grains, or masses. Spinel is found in many colors, but the most familiar are blue, purple, red, or pink. It can be found in iron and magnesium-rich igneous rocks, aluminum-rich metamorphic rocks, and contact-metamorphosed limestones. Spinel's main use is as a gemstone. It is resistant to weathering, and most gem sources are where it has concentrated in river or stream gravels. It is found in Myanmar, Sri Lanka, Madagascar, Afghanistan, Pakistan, and Australia.

Vitreous luster

Spinel crystals

FRANKLINITE

Crystal system Cubic | **Composition** (Zn,Mn,Fe)(Fe, Mn)$_2O_4$ | **Color** Iron black | **Form/habit** Octahedral | **Hardness** 6–6½ | **Cleavage** None
Fracture Conchoidal to uneven | **Luster** Metallic
Streak Reddish black to iron black | **SG** 5–5.2
Transparency Opaque

Franklinite is a zinc iron oxide, and one of the spinel group of minerals. It usually crystallizes in octahedrons, which often have rounded edges, and also as massive and granular aggregates. In most rocks, it forms as disseminated small black crystals with octahedral faces occasionally visible. It is rarely found as large single crystals. It occurs in zinc deposits in metamorphosed limestones and dolomites. Franklinite is usually accompanied by minerals such as willemite, garnet, and rhodonite. It is named after its discovery at the Franklin Mine in New Jersey, US, and also occurs in Germany, Sweden, and Romania.

Octahedral franklin crystals

CHROMITE

Crystal system Cubic | **Composition** FeCr$_2$O$_4$
Color Dark brown, black | **Form/habit** Granular,
massive | **Hardness** 5½ | **Cleavage** None
Fracture Uneven | **Luster** Metallic | **Streak**
Brown | **SG** 4.5–4.8 | **Transparency** Opaque

Chromite is an iron chromium oxide and
a member of the spinel mineral group.
Crystals of chromite are uncommon, but
when found they occur as octahedrons.
It is usually found as either masses or
as lenses and block-shaped bodies, or
disseminated as granules and streaks. Its
color is dark brown to black, and it can
contain some magnesium and aluminum.
Chromite is most commonly found as
disseminated grains in highly magnesium-
and iron-rich igneous rocks, or
concentrated in the sediments derived
from them. Occasionally, it is found in
thick layers of almost pure chromite.
These rocks are known as chromitites,
and are the most important ores of
chromium. Chromitites are found in the
Bushveld Complex in South Africa and
the Stillwater Complex in Montana, US.
Chromite is sometimes found as a
crystalline inclusion in diamond.

Chromite nodule

CHRYSOBERYL

Crystal system Orthorhombic | **Composition**
BeAl$_2$O$_4$ | **Color** Green, yellow | **Form/habit**
Tabular, stout prismatic | **Hardness** 8½
Cleavage Distinct | **Fracture** Uneven to conchoidal
Luster Vitreous | **Streak** Colorless | **SG** 3.7
Transparency Transparent to translucent
RI 1.74–1.76

Chrysoberyl is beryllium aluminum oxide.
It is hard and durable, inferior in hardness
only to corundum and diamond. Crystals
of ordinary chrysoberyl are not uncommon

and are typically yellow, green, or brown.
Chrysoberyl crystals may form triple twins
called trillings, creating a hexagonal outline.
Because of chrysoberyl's durability, its
crystals weather out of the parent rock
intact and are thus found in streams and
gravel beds, where it tends to concentrate
because of its high specific gravity.
Chrysoberyl usually forms in granites,
granitic pegmatites, or mica schists. It has
several gemstone varieties. Among these
are the color-changing alexandrite, and
cat's-eye chrysoberyl, which has multiple
parallel, needlelike inclusions and shows
a thin eye when cut *en cabochon*.

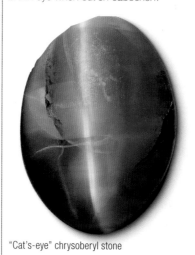

"Cat's-eye" chrysoberyl stone

CASSITERITE

Crystal system Tetragonal | **Composition** SnO$_2$
Color Medium to dark brown | **Form/habit**
Pseudocubic | **Hardness** 7 | **Cleavage** Indistinct
Fracture Subconchoidal to uneven | **Luster**
Adamantine to metallic | **Streak** White, grayish,
brownish | **SG** 6.8–7.1 | **Transparency**
Transparent to opaque | **RI** 2.0–2.1

A tin oxide, cassiterite is named from the
Greek *kassiteros*, for "tin." Its crystals are
heavily grooved prisms and pyramids,
and twinned crystals are quite common.
It is also found in masses, as a botryoidal
fibrous variety (wood tin) or as water-worn
pebbles (stream tin). Colorless when pure,
it is usually brown or black when iron
impurities are present. It forms in
hydrothermal veins associated with granite
rocks, with tungsten minerals including
wolframite, and also with molybdenite,
tourmaline, and topaz. Cassiterite is
durable and relatively dense, thus it
becomes concentrated in stream gravels
and beach sands when eroded from its
primary rocks. Fine crystals come from
Portugal, Italy, France, the Czech Republic,
Brazil, and Myanmar. Cassiterite is virtually
the only source of tin.

GAHNITE

Crystal system Cubic | **Composition** ZnAl$_2$O$_4$
Color Dark green or blue | **Form/habit** Octahedral
Hardness 7½–8 | **Cleavage** Indistinct | **Fracture**
Conchoidal | **Luster** Vitreous | **Streak** Grayish
SG 4.6 | **Transparency** Translucent to nearly
opaque

Gahnite is a zinc aluminum oxide and
a member of the spinel mineral group. Like
other spinel minerals, it forms octahedral
crystals and granular or massive
aggregates. It is dark blue, blue green,
gray, yellow, black, or brown in color.
Gahnite is a minor constituent in granites
and pegmatites, schists, gneisses,
and in contact-metamorphosed
limestones. It is resistant to
weathering and concentrates in
stream gravels. It was named in 1807
for the Swedish chemist J. G. Gahn. The
US, Sweden, Finland, Australia, Brazil, and
Mexico produce good crystal specimens.

Octahedral crystal

Blue gahnite specimen

PYROLUSITE

Crystal system Tetragonal | **Composition** MnO$_2$
Color Steel gray to black | **Form/habit** Massive
Hardness 6–6½ | **Cleavage** Perfect | **Fracture**
Uneven, brittle, splintery | **Luster** Metallic to earthy
Streak Black or bluish black | **SG** 4.5–5.0
Transparency Opaque

Pyrolusite is a common manganese oxide
that rarely forms crystals, and is usually
found as light gray to black massive
aggregates, metallic coatings, crusts,
fibers, and nodules. When found, its
crystals are opaque prisms. Pyrolusite
always forms under highly oxygen-rich
conditions as an alteration product of
other manganese minerals, such as
rhodochrosite. It is found in bogs, lakes,
and marine environments. Pyrolusite
nodules are widespread on the ocean
floor as a result of bacterial action, and
as deposits left by circulating waters. It
is a major source of manganese, used

in the manufacture of glass and steel, and
of saltwater resistant manganese-bronze
for ship's propellers.

Divergent fans
of pyrolusite

Pyrolusite crystals

ANATASE

Crystal system Tetragonal | **Composition** TiO$_2$
Color Various shades of brown, or black, indigo blue
Form/habit Pyramidal | **Hardness** 5½–6
Cleavage Perfect | **Fracture** Subconchoidal
Luster Adamantine to metallic | **Streak** White to
pale yellow | **SG** 3.9 | **Transparency** Transparent
to nearly opaque

Anatase is one of the three mineral forms
of titanium oxide, the others being rutile
and brookite. It was formerly known as
octahedrite. It is always found as small,
isolated, and sharp crystals. Its classic
elongated octahedral crystals are hard
and brilliant, and can be brown, yellow,
indigo blue, green, gray, lilac, and black.
Other, less common, crystal forms are
blocks and prisms. It was named in 1801,
from the Greek *anatasis*, "extension," a
reference to its elongated crystals. It is
often found in association with fluorite,
ilmenite, aegirine, and brookite.

Anatase crystal

BROOKITE

Crystal system Orthorhombic | **Composition** TiO_2
Color Various shades of brown | **Form/habit**
Tabular, elongated | **Hardness** 5½–6 | **Cleavage**
Indistinct | **Fracture** Subconchoidal to uneven
Luster Metallic to adamantine | **Streak** White,
grayish, yellowish | **SG** 4.1–4.2 | **Transparency**
Transparent to opaque

Brookite is one of three titanium dioxide
minerals, along with anatase and rutile. Iron
is almost always present in its structure to
some degree, and brookite with niobium
has also been found. Its crystals are
blocky, or less commonly, pyramids. It is
usually brown in color, but it can also be
red, yellow brown, or black. Well-formed
crystals of up to 2 in (50 mm) are known.
Brookite occurs in hydrothermal veins, in
some contact-metamorphic rocks, and in
sedimentary deposits, where it becomes
concentrated. It was named in 1825 for
English crystallographer H. J. Brooks.

Striated
brookite crystal

Brookite on albite crystals

▷ RUTILE

Crystal system Tetragonal | **Composition** TiO_2
Color Reddish brown to red | **Form/habit** Slender
prismatic | **Hardness** 6–6½ | **Cleavage** Good
Fracture Conchoidal to uneven | **Luster**
Adamantine to submetallic | **Streak** Pale brown
to yellowish | **SG** 4.2–4.3 | **Transparency**
Transparent to opaque | **RI** 2.62–2.90

Rutile, along with anatase and brookite, is
one of the three mineral forms of titanium
oxide. Single rutile crystals are usually
slender needles or prisms, yellowish or
reddish brown, dark brown or black in color,
and often form latticelike structures. When
enclosed in quartz they are pale golden,
needlelike crystals. Rutile is frequently
found as a minor constituent of granites,
pegmatites, gneisses, and schists, and
in veins, and is concentrated in placer
deposits due to its density. Rutile commonly
forms microscopic, oriented inclusions in
other minerals. It is responsible for the
asterism shown by some rose quartz,
rubies, and sapphires. Rutile takes its name
from the Latin *rutilis*, "red" or "glowing."

Rutile needles occurring in rock crystals

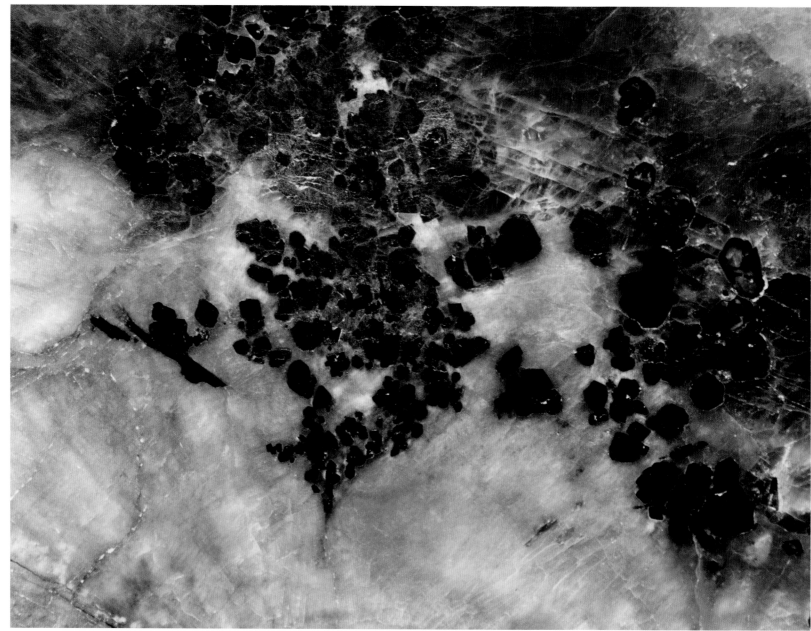

Uraninite in pitchblende form, with zircon

△ URANINITE

Crystal system Cubic | **Composition** UO_2
Color Black to brownish black, dark gray, greenish
Form/habit Octahedral | **Hardness** 4–6
Cleavage None | **Fracture** Uneven to subconchoidal | **Luster** Submetallic, pitchy, dull
Streak Brownish black | **SG** 7.5–10.6
Transparency Opaque

Uraninite is uranium oxide, and a major ore of uranium. It is black to brownish black, dark gray, or greenish in color. When uraninite occurs in masses it is known as pitchblende. It can also be found in botryoidal or granular form, and less commonly in crystals as octahedrons or cubes. Uraninite occurs in pegmatites as crystals along with other uranium-bearing minerals and rare-earth minerals, although these rarely constitute important uranium ores. It also forms in high-temperature hydrothermal veins with cassiterite and arsenopyrite, as well as in medium-temperature hydrothermal veins, where it occurs as pitchblende. It weathers into other uranium minerals, such as carnotite, where it is found in sandstones and conglomerates. These may also be important uranium ores. Uraninite was named in 1792 for its composition. Fine crystals are found in Cordoba, Spain; Saxony, Germany; Chihuahua, Mexico; and Topsham, Maine, in the US. It is a highly radioactive mineral and must be handled and stored carefully. The pioneering work on radioactivity by Pierre and Marie Curie was based on radium extracted from uraninite ores.

FERGUSONITE

Crystal system Tetragonal | **Composition** $YNbO_4$
Color Black to brownish black | **Form/habit** Prismatic to pyramidal | **Hardness** 5½–6½
Cleavage Poor | **Fracture** Subconchoidal, brittle
Luster Vitreous to submetallic | **Streak** Brown, yellow brown, greenish gray | **SG** 4.7–6.3
Transparency Opaque

Named for the 19th-century Scottish mineralogist and politician Robert Ferguson of Raith, fergusonite is a complex oxide consisting of various rare-earth elements where a number of them can substitute in the structure. There are several varieties of fergusonite, each taking its name from its predominant rare earth. In fergusonite-(Y), the most common, yttrium predominates; in fergusonite-(Ce), cesium predominates; and in fergusonite-(Nd) there is a preponderance of neodymium. In other fergusonites, tantalum, cobalt, and numerous others are found. Similar to most other elements in the rare-earth oxides group, fergusonite rarely occurs in sufficient concentrations to permit extraction in large quantities. Nevertheless, rare-earth metals have a range of commercial uses, from nuclear power industry to high-intensity lamps used in film production. Fergusonite crystals are needlelike, prismatic, or pyramidal, and it is black to brownish black in coloration, opaque, and transparent on thin edges. Fergusonite is usually found in granitic pegmatites, and may also occur in placers. It is found in association with monazite, gadolinite, thalenite, euxenite, allanite, zircon, biotite, and magnetite. Distribution is widespread but deposits only occur in small amounts, but notable sites include Hakatamura, Japan; and Ytterby, Sweden.

Fergusonite

Fergusonite crystals in feldspar

ROMANÈCHITE

Crystal system Orthorhombic | **Composition** $(Ba,H_2O)Mn_5O_{10}$ | **Color** Iron black to steel gray **Form/habit** Massive, botryoidal | **Hardness** 5–6 **Cleavage** None | **Fracture** Uneven | **Luster** Submetallic to dull | **Streak** Brownish black, shining **SG** 4.7 | **Transparency** Opaque

Romanèchite is barium manganese oxide, and a major constituent of psilomelane. They are considered synonymous in some sources, but psilomelane is not pure romanechite. A valuable source of manganese, romanèchite is hard, black, and often has a botryoidal surface. It is structureless in that it is a mixture of minerals. It is found associated with hematite, barite, and pyrolusite, and occurs in Europe, the US, and Brazil.

Dull luster

Submetallic luster

Rough and botryoidal romanechite

SAMARSKITE

Crystal system Orthorhombic | **Composition** $(Y,Fe,U)(Nb,Ta)O_4$ | **Color** Black | **Form/habit** Prismatic | **Hardness** 5–6 | **Cleavage** Indistinct **Fracture** Conchoidal, brittle | **Luster** Vitreous to resinous | **Streak** Dark reddish brown to black **SG** 5.7 | **Transparency** Translucent to opaque

Samarskite is a complex oxide of yttrium, ytterbium, iron, niobium, tantalum, and uranium. Two distinct samarskites are recognized: samarskite-(Yb)—samarskite ytterbium and samarskite-(Y)—yttrium samarskite. It is black and opaque, but is more often brown or yellowish brown due to surface alteration. Its crystals are stubby prisms with a rectangular cross section, and opaque. It is translucent in thin fragments. Samarskite is usually found in rare earth bearing granitic pegmatites, and samples are usually radioactive. Yttrium is used to enhance the red phosphors in color television and is used in optical glass and special ceramics. Synthetic Yttrium-aluminum garnet (YAG) is a gem diamond substitute.

Samarskite with iridescent sheen

COLUMBITE

Crystal system Orthorhombic | **Composition** $(Fe,Mn)(Nb,Ta)_2O_6$ | **Color** Iron black to brown black **Form/habit** Short prismatic | **Hardness** 6–6½ **Cleavage** Distinct | **Fracture** Subconchoidal to uneven | **Luster** Metallic | **Streak** Dark red to black | **SG** 5.1–8.2 | **Transparency** Translucent to opaque

Columbite is an iron, manganese niobium tantalum oxide. When tantalum becomes significantly dominant over niobium in the structure, it becomes the mineral tantalite. The two are structurally identical, and a nearly complete chemical series exists between the two. Columbite and tantalite can be found together in granite rocks, granitic pegmatites, and in placer deposits. Columbite is found in three recognized varieties: iron-rich ferrocolumbite, manganese-rich manganocolumbite, and the more rare magnesium-rich magnocolumbite. Columbites are brown or black and are generally found in masses, but may be granular. Crystals

are blocky tabular or short prisms, are often iridescent, and may have complex faceted or rounded terminations. Ferrocolumbite is an important source of both tantalum and niobium, an industrially important mineral that offers increased strength when alloyed with other metals. Localities include Greenland; Germany, Italy, France, Sweden, Greece; and pegmatite regions in Zimbabwe, South Africa, and Uganda.

Rough columbite

PYROCHLORE

Crystal system Cubic | **Composition** $(Na,Ca)_2Nb_2(O,OH,F)_7$ | **Color** Brown to black **Form/habit** Octahedral | **Hardness** 5–5½ **Cleavage** Distinct | **Fracture** Subconchoidal to uneven | **Luster** Vitreous to resinous **Streak** Light brown, yellowish brown | **SG** 3.8–4.6 | **Transparency** Transparent to opaque

Pyrochlore is a complex niobium sodium calcium oxide, and was first described in 1826 from an occurrence in Stavern, Norway. Its name derives from the Greek for "fire" and "green," since some specimens turn green after heating. It is orange, brownish red, brown, or black, and its crystals are typically well-formed octahedra with modified faces. It is also found in masses and grains. Pyrochlore may be radioactive because it often contains traces of uranium and thorium. It is formed in certain types of igneous rocks, in pegmatites, and in small amounts in silica-poor rocks. It often occurs with zircon, apatite, and magnetite. It is a major source of niobium, which is used in low power-consumption electronics, and alone or alloyed with zirconium in claddings for nuclear reactor cores. Notable locations include Veshnovorgorsk, Russia; Mbeya,

Tanzania; St. Peter's Dome, Colorado, US; Brevik, Norway; Alno, Sweden; Oka, Quebec; and Ontario, Canada.

Pyrochlore specimen

TAAFFEITE

Crystal system Hexagonal | **Composition** $Mg_3Al_8BeO_{16}$ | **Color** Violet, colorless, pale green, bluish, pink, red | **Form/habit** Prismatic **Hardness** 8–8½ | **Cleavage** Indistinct **Fracture** Conchoidal | **Luster** Vitreous **Streak** Black | **SG** 3.6 | **Transparency** Transparent | **RI** 1.71–1.73

Taaffeite is beryllium, magnesium, and aluminum oxide. It is a relatively new mineral, only being discovered in 1945. Its colors are pale mauve, green, and sapphire blue. It bears some structural similarities to spinel, such as its appearance, hardness, and density. Taaffeite's geologic origin has yet to be discovered but is believed to be from high magnesium and aluminum-bearing schists. It has been found in the gem gravels of Sri Lanka, China, and South Australia.

Rough surface

Taaffeite

Hydroxides

▽ BAUXITE

Crystal system | Amorphous mixture
Composition Mixture of hydrous aluminum oxides
Color White, yellowish, red, reddish brown
Form/habit Amorphous | **Hardness** 1–3
Cleavage None | **Fracture** Uneven | **Luster**
Earthy | **Streak** Usually white | **SG** 2.3–2.7
Transparency Opaque

Bauxite is a rock rather than a mineral, and is composed of several hydrated aluminum oxides, principally gibbsite $Al(OH)_3$, boehmite $AlO(OH)$, and diaspore, also $AlO(OH)$. It is one of the world's most important ores, since it is the sole source of aluminum. The minerals that make up bauxite are pale colored, but bauxite also contains quartz, clays, hematite, and other iron oxides, so it is variably creamy yellow, orange, pink, and red in color. Some deposits are soft, easily crushed, and structureless; some are hard, dense, and pealike; others are porous but structurally strong, or are layered. Bauxite forms as shallow but quite extensive deposits where aluminum-rich rocks have been heavily weathered in a humid tropical environment. Sophisticated mineralogical and chemical techniques may be required to determine the mineral components of specimens. Most of the world's bauxite is mined in Australia, Jamaica, Brazil, and Guinea.

DIASPORE

Crystal system Orthorhombic | **Composition**
$AlO(OH)$ | **Color** White, gray, yellow, lilac, or pink
Form/habit Thin, platy | **Hardness** 6½–7
Cleavage Perfect, imperfect | **Fracture**
Conchoidal, brittle | **Luster** Vitreous | **Streak**
White | **SG** 3.4 | **Transparency** Transparent
to translucent | **RI** 1.68–1.75

Diaspore is a hydrous aluminum oxide. Its color is white, grayish white, colorless, greenish gray, light brown, yellowish, lilac, or pink. Diaspore's crystals are thin and platy, elongated, blocks, prisms, or needlelike. It can be found in masses and as disseminated grains, and may be strongly pleochroic, exhibiting different colors—from violet blue to asparagus green and reddish plum—when viewed from different angles. It is a relatively widespread mineral, forming in metamorphic rocks, such as schists and marbles, where it is regularly found with corundum, manganite, and spinel. It is also found in bauxite, and aluminous clays, often as the major constituent of bauxite. Its name comes from the Greek *diaspora* for "scattering," in reference to the way diaspore crackles and breaks down when strongly heated. Significant deposits are found in Honshu, Japan; Jordanow, Poland; the Urals, Russia; and North Carolina, Maine, and California, US.

Platy diaspore crystals

Dark red diaspore

Bauxite residue in a pool, Bouches du Rhône, Gardanne, France

BRUCITE

Crystal system Trigonal | **Composition** $Mg(OH)_2$ |
Color White, pale green, gray, or blue | **Form/habit**
Broad tabular | **Hardness** 2½ | **Cleavage**
Perfect | **Fracture** Uneven, sectile | **Luster** Waxy
to vitreous/pearly | **Streak** White | **SG** 2.4
Transparency Transparent

Named for American mineralogist
Archibald Bruce in 1824, brucite is a
magnesium hydroxide. It is usually
white, but it can be pale green, gray,
or blue. Manganese may substitute to
some degree for magnesium, resulting
in yellow to red coloration. Its crystals
are frequently soft, and waxy to glassy,
and occur in the form of blocks or
aggregates of plates, in masses, as
fibers, or granular. A variety of brucite
is nemalite, a fibrous or lathlike form
that occurs in fine, large crystals.
Exceptional, often transparent crystals
come from the Ural Mountains of Russia.
Brucite is found in metamorphic rocks
such as schist, and in low-temperature
hydrothermal veins in marble and chlorite
schists. It is often found with calcite,
aragonite, magnesite, and talc. It is
used as a source of medical magnesia,
and industrial uses include as a lining
material for kilns due to its high melting
point. It is also pyroelectric—when heated,
it generates electricity—and is easily
soluble in acids. Brucite localities include
Ontario, Canada; California, New York,
New Jersey, and Pennsylvania, US;
Austria, England, Russia,
Sweden, and Turkey.

Large crystals

Brucite crystals

GOETHITE

Crystal system Orthorhombic | **Composition**
$FeO(OH)$ | **Color** Orangeish to blackish brown |
Form/habit Prismatic, elongated | **Hardness**
5–5½ | **Cleavage** Perfect | **Fracture** Uneven |
Luster Adamantine to metallic | **Streak** Brownish
yellow to ocher red | **SG** 3.8–4.3 | **Transparency**
Translucent to opaque

The iron oxide hydroxide goethite is a
very common mineral. It is usually black
but can be brownish yellow, reddish
brown, or dark brown depending on
impurities—manganese can substitute
for up to 5 percent of the iron. Its crystals
vary in form: opaque black prisms
vertically grooved; velvety, radiating fibrous
aggregates; flattened tablets or scales;
in stalactites; and also as bubbly-looking
masses, in tufts, or as coatings of
tiny crystals. Goethite is formed as a
weathering product of iron minerals such
as magnetite, pyrite, and siderite in the
weathered capping of iron-ore deposits,
called the gossan, or "iron hat." Goethite
was named in 1806 for the German poet
and author Johann Wolfgang von Goethe,
who was an enthusiastic mineralogist. Fine
specimens are found in Cornwall, England;
Chaillac, France; and the Urals, Russia.

Metallic luster

Goethite crystals

GIBBSITE

Crystal system Monoclinic | **Composition** $Al(OH)_3$ |
Color White | **Form/habit** Tabular | **Hardness**
2½–3½ | **Cleavage** Perfect | **Fracture** Uneven |
Luster Pearly to vitreous | **Streak** White | **SG**
2.3–2.4 | **Transparency** Transparent

Gibbsite is aluminum hydroxide. It is
usually white, but it can also be grayish,
greenish, or yellowish, depending on
impurities. Its crystals are blocky, and can
sometimes appear hexagonal. Gibbsite
is principally an alteration product of
aluminum-rich minerals whose silica has
been leached out, and forms primarily in
tropical and subtropical environments,
often as one of the major components of
bauxite. As such, it is an important source
of aluminum. It was named in 1822 for

George Gibbs, whose minerals formed
the core of the Yale University collection.
Gibbsite is also interesting because it is
often found as a part of the structure of
some clay minerals.

Massive gibbsite

MANGANITE

Crystal system Monoclinic | **Composition**
$MnO(OH)$ | **Color** Steel gray to iron black | **Form/
habit** Prismatic, striated | **Hardness** 4 | **Cleavage**
Perfect, good | **Fracture** Uneven | **Luster**
Submetallic | **Streak** Reddish brown to black |
SG 4.3–4.4 | **Transparency** Opaque

Manganite is hydrated manganese oxide.
It is opaque and metallic dark gray or black
in color. Its crystals, mostly orthorhombic-
appearing prisms—typically with flat or
blunt ends—are often grouped in bundles
and grooved lengthwise. It is hard to
distinguish from other manganese oxides,
such as pyrolusite, when found in masses
or granular forms. It occurs in low-
temperature hydrothermal deposits
associated with baryte, calcite, and
siderite. It is also found in hot-spring
manganese deposits, and in shallow
marine deposits, lakes, and bogs.
Manganite was finally named after its
content in 1827 although it had been
described under a number of different
names since 1772. As a manganese ore
it ranks after pyrolusite and romanèchite.

Striated manganite crystals

LIMONITE

Crystal system Mixture | **Composition** $2Fe_2O_3 \cdot H_2O$, approx |
Color Various shades of brown, yellow |
Form/habit Massive, oolitic, stalactic | **Hardness**
5–5½ | **Cleavage** None | **Fracture** Uneven |
Luster Earthy, sometimes submetallic or dull |
Streak Yellowish brown | **SG** 2.7–4.30 |
Transparency Opaque

Once used as a mineral name, limonite
is now used as a general name for
mixtures of the hydrated iron hydroxide
minerals goethite, lepidocrocite, and
akaganéite, or applied to unidentified
iron oxides and hydroxides. Limonite is
fine grained, and yellow, brownish yellow
or orange brown in color. It does not
form crystals, but is found as concretions,
as stalactites, or as earthy masses.
In addition to hydrated iron minerals,
hematite, clay, and other impurities
may also be present. Limonite is a
secondary product formed from the
oxidation of other iron minerals, and
also occurs by precipitation in the sea
and freshwater, and in bogs. Its name
derives from the Greek *leimon*, meaning
"meadow," in reference to the marshy
localities in which the variety known
as bog iron is found. The use of limonite
as a pigment goes back to ancient
Egypt, and its variously colored forms
yield ochers, siennas, and umbers.
It was also used as an ore of iron in
ancient times.

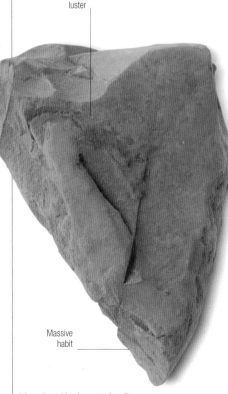

Earthy
luster

Massive
habit

Limonite with sienna coloration

Halides

HALITE

Crystal system Cubic | **Composition** NaCl
Color Colorless to white | **Form/habit** Cubes
Hardness 2½ | **Cleavage** Perfect cubic
Fracture Conchoidal | **Luster** Vitreous
Streak White | **SG** 2.1–2.6 | **Transparency**
Transparent to translucent

Halite is calcium chloride—ordinary table salt. Its crystals are usually cubes, but sometimes it forms "hopper" crystals, where the outer edges of the cube faces have grown more rapidly than their centers, leaving cavernous faces. Most halite is colorless, white, or gray, but it can be orange, brown, bright blue or purple: the orange color is derived from inclusions of hematite; the blue and purple colors derive from defects in the crystal structure. It generally occurs in coarse crystalline masses and is commonly found in massive and bedded aggregates as rock salt. Halite is widespread in large deposits, where it has formed by the evaporation of saltwater or salt pans. It is found worldwide, with large commercial deposits in Germany, Austria, France, Bolivia, and the US. Aside from culinary use, it a source of soda ash, in the manufacture of soap and glass, and as a glaze for porcelain enamels.

Halite crystals on rock groundmass

SYLVITE

Crystal system Cubic | **Composition** KCl
Color Colorless to white | **Form/habit**
Cubes | **Hardness** 2½ | **Cleavage** Perfect cube
Fracture Uneven | **Luster** Vitreous | **Streak**
White | **SG** 2.0 | **Transparency** Transparent
to translucent

Discovered in 1823 on Mount Vesuvius, Italy, sylvite is potassium chloride, and one of the world's major economic minerals.

Millions of tons are mined each year for the manufacture of potassium compounds, such as potash fertilizer. Crystals occur in cubes, octahedrons, or combinations of these, but crusts, grains, and masses are more common. When pure it is colorless to white or grayish, but it can be tinged blue, yellow, purple, or red by impurities. It is found in thick beds along with halite, and gypsum. Its name derives from its old medicinal name, *sal digestivus Sylvii*; meaning digestive salt. Massive bedded deposits occur in the Southwestern US and Canada.

Vitreous luster

Granular sylvite

CHLORARGYRITE

Crystal system Cubic | **Composition** AgCl
Color Pearl gray, greenish, white, colorless
Form/habit Usually massive | **Hardness** 1–2
Cleavage None | **Fracture** Subconchoidal
Luster Dull to adamantine | **Streak** White
SG 5.5–5.6 | **Transparency** Transparent to
nearly opaque

Chlorargyrite is the mineral form of silver chloride. It is usually found in masses and in columnar form. Its crystals, when found, are colorless to variably yellow cubes. Sometimes it forms in hornlike masses, which gives it another name, horn silver. Like many silver minerals it is light sensitive: the color changes to brown or purple when exposed to light. Chlorargyrite is formed as an alteration product of native silver, silver sulfides, and sulfosalts. Its localities include Germany, the Czech Republic, Bolivia, and

the US. The name comes from the Greek *chloros*, for "pale green" and Latin *argentum*, for silver.

Crust of chlorargyrite

Chlorargyrite crust

CALOMEL

Crystal system Tetragonal | **Composition** HgCl
Color White, colorless, yellow gray, or grayish | **Form/habit** Tabular | **Hardness** 1–2 | **Cleavage** Distinct
Fracture Conchoidal | **Luster** Adamantine
Streak Pale yellow white | **SG** 6.4–7.1
Transparency Transparent to nearly opaque

Calomel is mercury chloride and takes its name from the Greek for "beautiful" and "honey," alluding to its sweet taste (although it is in fact toxic). It is white or yellowish white in color, soft, and heavy due to its mercury content, causing it to be referred to as horn quicksilver or horn mercury. Calomel crystals are blocks, prisms, or pyramids, but it is usually found in crusts and masses. From the 16th century until the early 20th century—when its mercury was found to be toxic—it was used as a horticultural fungicide, and historically as a laxative and disinfectant, and in the treatment of syphilis. It is still a source of mercury for industrial processes.

Calomel encrustations

CRYOLITE

Crystal system Monoclinic | **Composition** Na$_3$AlF$_6$
Color Colorless to snow white | **Form/habit**
Pseudocubic | **Hardness** 2½–3 | **Cleavage** None
Fracture Uneven | **Luster** Vitreous to greasy
Streak White | **SG** 3.0 | **Transparency**
Transparent to translucent

Cryolite, sodium hexafluoroaluminate, is arguably one of the most important minerals of our age. Cryolite was the essential ingredient in the first aluminum production. Usually colorless or white, or more rarely brown, yellow, reddish brown, or black, it rarely occurs as crystals and most commonly forms as coarse granular or massive aggregates. Aluminum is the most abundant metal in the Earth's crust, yet prior to 1886 it scarcely existed as a separate metal. It was discovered that if an aluminum oxide (such as bauxite) was dissolved in molten cryolite and an electric current was passed through it, aluminum metal was released. Natural cryolite is too rare for the mass production of aluminum, so synthetic sodium aluminum fluoride is produced from the common halide fluorite.

Massive cryolite

▷ FLUORITE

Crystal system Cubic | **Composition** CaF$_2$
Color Occurs in most colors | **Form/habit**
Cubic, octahedral | **Hardness** 4 | **Cleavage**
Perfect octahedral | **Fracture** Flat conchoidal
Luster Vitreous | **Streak** White | **SG** 3.0–3.3
Transparency Transparent to translucent | **RI** 1.43

Fluorite is calcium fluoride, and is an important industrial mineral. It commonly occurs in well-formed cubes and octahedrons, and in a wide range of colors: violet, green, and yellow are the most common, although it is colorless and transparent when free of color-inducing trace elements. As much as 20 percent cerium or yttrium can replace its calcium. Colors are frequently zoned within the same crystal. It is also found as masses or grains. Fluorite has many industrial uses, such as the production of hydrofluoric acid, and as a catalyst in high-octane fuels. An old name for fluorite is fluorspar; this is now used only in industry to refer to the bulk stone.

CARNALLITE

Crystal system Orthorhombic | **Composition** $KMgCl_3.6H_2O)$ | **Color** Milky white, often reddish **Form/habit** Massive to granular | **Hardness** 2½ **Cleavage** None | **Fracture** Conchoidal | **Luster** Greasy | **Streak** White | **SG** 1.6 | **Transparency** Tranlucent to opaque

Carnallite is potassium magnesium chloride and is one of several minerals from which potash fertilizer is made. Carnallite is white

or colorless, but it may be reddish or yellowish depending on the presence of hematite or goethite impurities. Crystals are rare because they absorb water from the air and dissolve, so it is generally massive to granular. It forms in the upper layers of salt deposits, where it occurs with other potassium and magnesium minerals. It is mined for both potassium and magnesium.

Granular carnalite

ATACAMITE

Crystal system Orthorhombic | **Composition** $Cu_2Cl(OH)_3$ | **Color** Bright green to black green **Form/habit** Slender, prismatic | **Hardness** 3–3½ **Cleavage** Perfect, fair | **Fracture** Conchoidal, brittle | **Luster** Adamantine to vitreous | **Streak** Apple green | **SG** 3.8 | **Transparency** Transparent to translucent

Atacamite is a copper chloride hydroxide, and is named after the Atacama Desert of Chile, where it was first described. Its crystals are bright or dark emerald green, and are typically slender prisms or blocky, with grooves along the length and wedge-shaped ends. It is also found in masses, fibers, and grains. Atacamite is a secondary mineral formed principally under salt-rich, arid conditions from the

weathering of other copper minerals. It is also found in ocean bottom "black smoker" deposits. Atacamite is a major corrosion product on ancient and modern bronzes and copper alloys; for example, the green of the Statue of Liberty in New York is due to atacamite corrosion of its copper alloy.

Atacamite crystals

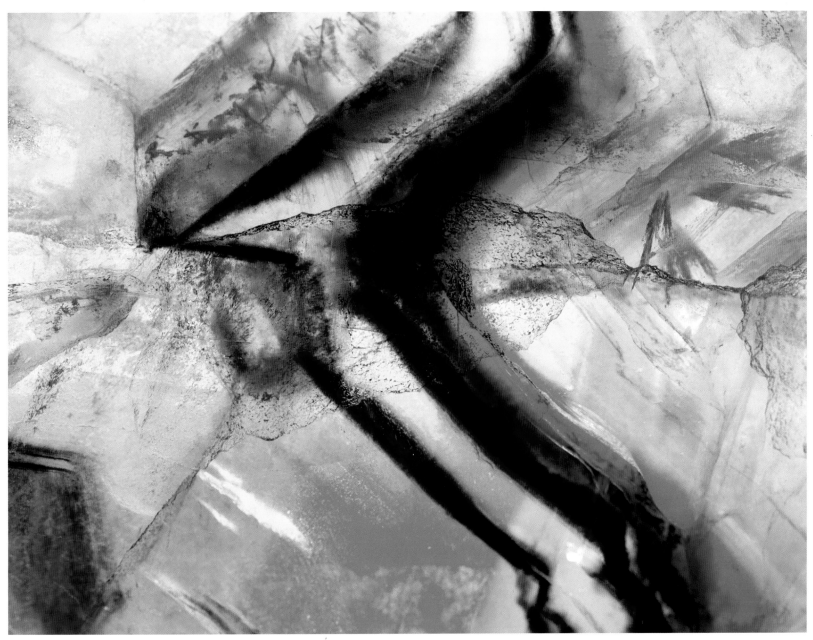

Fluorite with blue-green coloring from trace elements

Carbonates

TRONA

Crystal system Monoclinic | **Composition** Na$_3$(HCO$_3$)(CO$_3$).2H$_2$O | **Color** Colorless to gray, yellow white | **Form/habit** Massive | **Hardness** 2½–3 | **Cleavage** Perfect | **Fracture** Uneven to subconchoidal | **Luster** Vitreous, glistening | **Streak** White | **SG** 2.1–2.2 | **Transparency** Transparent to translucent

Crystals of trona—sodium bicarbonate hydrate—are rare, but when they occur they are elongated prisms, blocks, or fibers. Trona is usually found in masses as bedded deposits in saline lakes, often associated with halite, gypsum, borax, dolomite, glauberite, and sylvite. It also occurs in powdery layers on the walls of mines or in soils in desert regions. What is probably the world's largest deposit is at Sweetwater, Wyoming, US. Other significant deposits are in Chad, Mexico, Mongolia, Canada, Libya, and Tibet. Its name comes from *tron*, an abbreviation of the Arabic *natrun*, meaning "salt."

Trona rough

ARAGONITE

Crystal system Orthorhombic | **Composition** CaCO$_3$ | **Color** Colorless, white, gray, yellowish, reddish, green | **Form/habit** Prismatic, acicular | **Hardness** 3½–4 | **Cleavage** Distinct | **Fracture** Subconchoidal, brittle | **Luster** Vitreous inclining to resinous | **Streak** White | **SG** 2.9 | **Transparency** Transparent to translucent

Aragonite—calcium carbonate—has the same chemical composition as calcite, but crystallizes in a different crystal system. It can be white, colorless, gray, yellowish, green, blue, reddish, violet, or brown in color. Its crystals are blocky, prisms, or needlelike, often with steep pyramid- or chisel-shaped ends. Twinned crystals

are common, often appearing hexagonal in shape. It can also form columnar or radiating aggregates. Aragonite is commonly found in the oxidized zone of ore deposits, in hot-spring deposits, in caves as stalactites, and where it may also form coral-like aggregates, known as flos ferri. Aragonite constitutes the shells of many marine mollusks, and pearls.

SMITHSONITE

Crystal system Trigonal | **Composition** ZnCO$_3$ | **Color** White, blue, green, yellow, brown, pink, colorless | **Form/habit** Botryoidal, rhombohedral, scalenohedral | **Hardness** 5 | **Cleavage** Perfect rhombohedral | **Fracture** Uneven to conchoidal | **Luster** Vitreous to pearly | **Streak** White | **SG** 4.3–4.5 | **Transparency** Translucent to opaque | **RI** 1.62–1.85

With a wide range of colors, smithsonite—zinc carbonate—is most commonly found as grapelike or stalactitic masses, or as honeycombed aggregates called "dry-bone" ore. It rarely forms crystals. When found, they are most commonly rhombohedrons, and generally have curved faces. Smithsonite is a fairly common mineral found in the oxidation zones of most zinc ore deposits. Frequently mined as an ore of zinc, it may have provided the zinc component of brass in ancient metallurgy.

SIDERITE

Crystal system Trigonal | **Composition** FeCO$_3$ | **Color** Yellowish to dark brown | **Form/habit** Rhombohedral | **Hardness** 4–4½ | **Cleavage** Perfect rhombohedral | **Fracture** Uneven or subconchoidal | **Luster** Vitreous to pearly | **Streak** White | **SG** 3.7–3.9 | **Transparency** Translucent

An iron carbonate, siderite takes its name from the Greek *sideros*, meaning "iron." If manganese is present, it can darken to black. It has the same structure as calcite and can form the same rhombohedron crystals, often with curved surfaces. It can also form scalenohedrons and blocky or prismatic crystals. It is usually found in masses or grainy agglomerations. Siderite commonly forms at shallow depths.

Its composition is often related to the deposits of the enclosing sediments, where it occurs in concretions and in thin beds with shale, clay, and coal seams.

Rhombohedral crystals

Siderite in groundmass

MAGNESITE

Crystal system Trigonal | **Composition** MgCO$_3$ | **Color** White, light gray, yellowish, brownish | **Form/habit** Massive | **Hardness** 4–4½ | **Cleavage** Perfect rhombohedral | **Fracture** Conchoidal, brittle | **Luster** Vitreous | **Streak** White | **SG** 2.9–3.1 | **Transparency** Transparent to translucent

A carbonate of magnesium, magnesite takes its name from its composition. It is usually white or gray, but it can be yellow or brown when iron substitutes for some of the magnesium. Magnesite occurs in masses, or as fibers or grains. Crystals are rare, but when found are rhombohedrons or prisms. Magnesite forms principally as an alteration of magnesium-rich rocks, such as peridotites, and through the action of magnesium-containing solutions upon calcite. It also occurs in limestone and talc or chlorite schists. Australia and Brazil are major sources. Magnesite was detected in meteorite ALH84001 and on Mars.

Magnesite crystals in groundmass

RHODOCHROSITE

Crystal system Trigonal | **Composition** MnCO$_3$ | **Color** Rose pink, brown, or gray | **Form/habit** Rhombohedral | **Hardness** 4 | **Cleavage** Perfect rhombohedral | **Fracture** Uneven | **Luster** Vitreous to pearly | **Streak** White | **SG** 3.5–3.7 | **Transparency** Transparent to translucent | **RI** 1.59–1.82

Most examples of rhodochrosite—manganese carbonate—have some calcium and iron substituting for manganese, and some also contain magnesium. Rhodochrosite is classically rose pink in color, although it can be brown or gray. Its crystals tend to be rhombohedrons and scalenohedrons, but it is more often found in masses. When in stalactite form, it develops bands of various pink shades. It is found in hydrothermal veins formed at moderate temperatures, in high-temperature metamorphic deposits, and as an alteration product in sedimentary manganese deposits. It was named in 1800 from the Greek *rhodokhros*, "of rosy color."

▷ CALCITE

Crystal system Hexagonal/trigonal | **Composition** CaCO$_3$ | **Color** Colorless, white | **Form/habit** Scalenohedral, rhombohedral | **Hardness** 3 | **Cleavage** Perfect rhombohedral | **Fracture** Subconchoidal, brittle | **Luster** Vitreous | **Streak** White | **SG** 2.6–2.7 | **Transparency** Transparent to translucent | **RI** 1.48–1.66

Calcite is the most common form of calcium carbonate. Although over 70 carbonate mineral species are known, three of them account for most of the carbonate material in the Earth's crust: calcite, dolomite, and siderite. Calcite is particularly known for the great variety and beautiful development of its crystals. They are frequently twinned and can form dramatic heart-shaped butterfly twins. Those with steep scalenohedrons and rhombohedrons are known as dogtooth spar; and shallow rhombohedrons are called nailhead spar. Calcite is frequently found in highly transparent crystals, sometimes called optical spar, a reference to their use in polarizing filters. Calcite readily cleaves into rhombohedra. Although it forms spectacular crystals, most calcite is found in masses, either as limestone or marble. It is also found as fibers, nodules, stalactites, and as an earthy aggregate. Although chiefly a chemical precipitate, it can be found in metamorphic deposits, hydrothermal veins, and igneous rocks.

Calcite deposits in Carlsbad Cave, Carlsbad Caverns National Park, New Mexico, US

WITHERITE

Crystal system Orthorhombic | **Composition** $BaCO_3$ | **Color** White, colorless, yellow, brown, or green | **Form/habit** Pseudohexagonal | **Hardness** 3–3½ | **Cleavage** Distinct, imperfect | **Fracture** Uneven, brittle | **Luster** Vitreous | **Streak** White | **SG** 4.3 | **Transparency** Transparent to translucent

Witherite is barium carbonate. It is white or colorless, and may be tinged with yellow, brown, or green. Its crystals can be prisms or pyramids, but they are typically twinned. This causes them to appear hexagonal in shape, or as paired pyramids. They can be short to long prisms or blocks, with grooves running across prism faces, too. Witherite can also be fibrous, columnar, granular, or massed. The presence of barium makes specimens feel relatively heavy. Next to baryte, it is the most common barium mineral. Good crystals occur in Illinois and California, US, and commercial quantities are found in Japan, England, France, and Turkmenistan.

Twinned witherite crystals with galena

STRONTIANITE

Crystal system Orthorhombic | **Composition** $SrCO_3$ | **Color** Colorless, gray, green, yellow, or reddish | **Form/habit** Acicular, columnar | **Hardness** 3½–4 | **Cleavage** Good | **Fracture** Uneven, brittle | **Luster** Vitreous | **Streak** White | **SG** 3.7–3.8 | **Transparency** Transparent to translucent

Usually colorless to gray, strontianite—strontium carbonate—can be pale green, yellow, and yellow brown to reddish. It forms crystals that are needlelike or spear shapes, or grows in radiating aggregates. It can also be granular and in masses. Strontianite forms as a low-temperature hydrothermal mineral, and is also found in geodes and concretions. It is the principal source of strontium, used in sugar refining, and to produce the red color in fireworks. It was

named in 1791 for its discovery locality of Strontian in Scotland. Other localities include Germany, Canada, India, and California in the US.

Translucent, acicular, strontianite crystals

DOLOMITE

Crystal system Trigonal | **Composition** $CaMg(CO_3)_2$ | **Color** Colorless, white, cream, pale brown, or pink | **Form/habit** Rhombohedral | **Hardness** 3½–4 | **Cleavage** Perfect rhombohedral | **Fracture** Subconchoidal | **Luster** Vitreous | **Streak** White | **SG** 2.8–2.9 | **Transparency** Transparent to translucent

The chemical name of dolomite is calcium magnesium carbonate. It is a white, pale brown, or pink mineral. It forms rhombohedron crystals that often have curved faces or cluster in saddle-shaped aggregates. Dolomite can also be coarse to fine granular, in masses, or (rarely) in fibers. It is one of the three major carbonate minerals, is important as a rock-forming mineral in carbonate rocks, and is the principal component of the rock of the same name. Aside from chemical precipitation, it occurs as a replacement of limestone by the action of magnesium-bearing solutions, in marbles, talc schists, and other magnesium-rich metamorphic rocks, and in hydrothermal veins associated with lead, zinc, or copper ores. It was named after the French mineralogist D. de Dolomieu.

Dolomite in quartz groundmass

BARYTOCALCITE

Crystal system Monoclinic | **Composition** $BaCa(CO_3)_2$ | **Color** White, gray, green, or yellowish | **Form/habit** Prismatic | **Hardness** 4 | **Cleavage** Perfect, imperfect | **Fracture** Uneven, brittle | **Luster** Vitreous to resinous | **Streak** White | **SG** 3.7 | **Transparency** Transparent to translucent

Barytocalcite is named for its composition (barium calcium carbonate). It is colorless to white, grayish, greenish, or yellowish. Its crystals are short to long prisms, usually grooved. It is also found in masses. Barytocalcite forms in hydrothermal veins, especially where hydrothermal solutions have invaded limestone. It is often found with baryte, witherite, and calcite, and it can be a minor ore of barium. Crystals up to 2 in (5 cm) long are found in Cumbria, England. Other localities include Sweden and Siberia. It was first described in 1824 for its occurrence in Cumbria.

Prismatic crystals of barytocalcite on limestone

CERUSSITE

Crystal system Orthorhombic | **Composition** $PbCO_3$ | **Color** White, gray, blue to green | **Form/habit** Tabular, prismatic | **Hardness** 3–3½ | **Cleavage** Distinct | **Fracture** Conchoidal, brittle | **Luster** Adamantine to vitreous | **Streak** Colorless | **SG** 6.5 | **Transparency** Transparent to translucent | **RI** 1.8–2.1

The crystals of cerussite—lead carbonate—are highly varied, but include blocks, prisms, double-ended pyramids, needlelike, and pseudohexagonal. Blue or green hues indicate copper impurities. Cerussite is formed by the action of carbonated water on other lead minerals, particularly galena and anglesite. It is the most common ore of lead after galena, and is found in Australia, Bolivia, Spain, Namibia, and the US. It takes its name from the Latin cerussa, a white lead pigment. It is also a collector's gemstone.

AZURITE

Crystal system Monoclinic | **Composition** $Cu_3(CO_3)_2(OH)_2$ | **Color** Azure to dark blue | **Form/habit** Tabular, prismatic | **Hardness** 3½–4 | **Cleavage** Perfect | **Fracture** Conchoidal, brittle | **Luster** Vitreous to dull to earthy | **Streak** Blue | **SG** 3.7–3.9 | **Transparency** Transparent to translucent | **RI** 1.73–1.84

Azurite is deep blue copper carbonate hydroxide. It is highly unusual in that its crystals are prisms or blocks with a wide variety of faces—over 45 are commonly seen, and it can have over 100 more rare faces. It is also found in masses, as stalactites, and in botryoidal form (grapelike). Azurite forms in copper deposits through the action of carbonated waters on other copper minerals. Fine specimens come from the US, France, Mexico, Chile, Australia, Russia, Morocco, and especially Namibia. Its name comes from the same Persian word as lapis lazuli—lazhuward, meaning "blue."

ANKERITE

Crystal system Trigonal | **Composition** $Ca(Fe,Mg,Mn)(CO_3)_2$ | **Color** Colorless to pale buff | **Form/habit** Rhombohedral | **Hardness** 3½–4 | **Cleavage** Perfect | **Fracture** Subconchoidal | **Luster** Vitreous to pearly | **Streak** White | **SG** 2.9–3.8 | **Transparency** Translucent

Ankerite is calcium carbonate with varying amounts of iron, magnesium, and manganese in the structure. Its color depends on the amount of the various elements it contains. Its crystals are rhombohedrons that appear similar to those of dolomite. It can also be in masses or coarsely granular. Some ankerite is fluorescent. Ankerite forms from the action of iron-bearing fluids on limestone or dolomite. Specimens can be found in Japan, Canada, Namibia, South Africa, Austria, the US, and elsewhere. Ankerite was named in 1825 after the Austrian mineralogist M. J. Anker.

Perfect cleavage

Rhombohedral crystals of ankerite

▽ MALACHITE

Crystal system Monoclinic | **Composition** $Cu_2CO_3(OH)_2$ | **Color** Bright green | **Form/habit** Massive, botryoidal | **Hardness** 3½–4 | **Cleavage** Perfect | **Fracture** Subconchoidal to uneven, brittle | **Luster** Adamantine to silky | **Streak** Pale green | **SG** 3.6–4.0 | **Transparency** Translucent | **RI** 1.85

Malachite is a green copper carbonate hydroxide. It is usually found as grapelike or encrusting masses, often with a radiating fibrous structure. It is also found as delicate fibrous aggregates and as concentrically banded stalactites. It tends to be banded in various shades of green. Single crystals are not common but when found they are short to long prisms. Malachite occurs in the altered zones of copper deposits, where it is frequently accompanied by azurite. Fine crystals come from the Democratic Republic of Congo, which also provides much of the world's supply of ornamental malachite—the mineral is prized or its decorative use. Cutting material also comes from South Australia, Morocco, Arizona in the US, and Lyons, France. Single masses weighing up to 56 tons (51 tonnes) once came from mines in the Ural Mountains of Russia, but that supply is now exhausted.

AURICHALCITE

Crystal system Monoclinic | **Composition** $(Zn,Cu)_5(CO_3)_2(OH)_6$ | **Color** Sky blue, green blue, or pale green | **Form/habit** Acicular | **Hardness** 2 | **Cleavage** Perfect, brittle | **Fracture** Uneven | **Luster** Silky to pearly | **Streak** Pale blue green | **SG** 3.6–4.2 | **Transparency** Transparent to translucent

Aurichalcite (zinc copper carbonate) has needlelike crystals, and it is often found as radiating, tufted masses or as velvetlike encrustations. Crystals rarely exceed more than a few millimeters in length. It is also found in masses and in columns. Aurichalcite forms in the oxidized zones of copper and zinc deposits. It was named in 1839, after the Latin for golden copper. Localities include the western US; Nagato, Japan; Chihuahua, Mexico; Chessy, France; and Tsumeb, Namibia.

Aurichalcite crystals in limonite groundmass

PHOSGENITE

Crystal system Tetragonal | **Composition** $Pb_2(CO_3)Cl_2$ | **Color** White, yellow, brown, or green | **Form/habit** Prismatic | **Hardness** 2½–3 | **Cleavage** Perfect | **Fracture** Conchoidal, sectile | **Luster** Resinous | **Streak** White | **SG** 6.0–6.3 | **Transparency** Transparent to translucent

Phosgenite is a rare lead chlorocarbonate. Its color varies from white to green or brown. Its crystals are short prisms, or less commonly, thick and blocky. Sometimes, the crystals have an unusual helical twist. Phosphate is also found in masses and in granular form. It occurs as an alteration product of other lead minerals under surface conditions. Localities include Italy, Tasmania, Australia, England, and the US. The name phosgenite was given in 1841, from phosgene (carbon oxychloride), since the mineral contains the elements carbon, oxygen, and chlorine.

Malachite crystals closeup, showing its botryoidal (like bunches of grapes) form

Phosphates, arsenates, vanadates

▽ MONAZITE

Crystal system Monoclinic | **Composition** (Ce,La,Th,Nd)PO$_4$ | **Color** Yellowish brown to brown, greenish, or nearly white | **Form/habit** Prismatic | **Hardness** 5–5½ | **Cleavage** Perfect, good, poor | **Fracture** Conchoidal to uneven, brittle | **Luster** Resinous, waxy, or vitreous | **Streak** White | **SG** 4.6–5.7 | **Transparency** Translucent

The phosphate monazite is three different species, all sharing the same crystal structure. Each is designated by the chemical symbol of the element it combines with its phosphorous and oxygen. The most common is monazite-(Ce), cerium phosphate, which forms prisms or elongated crystals. Its color is yellowish or reddish brown to brown, greenish, or nearly white. The other two are monazite-(La), lanthanum phosphate and monazite-(Nd), neodymium phosphate. Monazite is a common minor mineral in granites and gneisses, pegmatites, and fissure veins. Monazite in the form of monazite sands can accumulate in commercial quantities.

XENOTIME

Crystal system Tetragonal | **Composition** YPO$_4$ | **Color** Yellowish to reddish brown, grayish white, yellow | **Form/habit** Short to long prismatic | **Hardness** 4–5 | **Cleavage** Perfect | **Fracture** Uneven to splintery | **Luster** Vitreous to resinous | **Streak** Pale brown or reddish | **SG** 4.4–5.1 | **Transparency** Transparent to opaque

Xenotime is yttrium phosphate. Erbium commonly replaces large portions of yttrium in the structure. Its colors are usually yellowish. Its crystals are generally found as glassy, short to long prisms, or as rosette-shaped crystal aggregates. Xenotime occurs as a minor mineral in igneous rocks and in their associated pegmatites, where it can form large crystals. Notable occurrences are in Japan, Sweden, Norway, Germany, Brazil, Madagascar, and the US. Its name comes from the Greek for "vain honor," because the yttrium in xenotime was mistakenly believed to be a new element when it was described in 1832.

Pyramidal xenotime crystals on aggregate

Monazite crystals in rock groundmass

CARNOTITE

Crystal system Monoclinic | **Composition** $K_2(UO_2)_2(VO_4)_2.3H_2O$ | **Color** Yellow | **Form/habit** Powdery microcrystalline | **Hardness** 4 | **Cleavage** Perfect | **Fracture** Uneven | **Luster** Pearly to dull | **Streak** Yellow | **SG** 4.5–4.7 | **Transparency** Semitransparent to opaque

Carnotite is a potassium uranyl vanadate hydrate. Its color is bright to lemon yellow, or greenish yellow. Generally, it is found as powdery or microcrystalline masses, as tiny, disseminated grains, or as crusts. Crystals are uncommon, but when they do form, they are platy, rhombohedrons, or lathlike. Pure carnotite contains about 53 percent uranium and is highly radioactive. Carnotite is also about 12 percent vanadium, with traces of radium. It is formed by the alteration of uranium-vanadium minerals and occurs principally in sandstone, either disseminated or in concentrations around petrified wood or other fossilized vegetable matter. Deposits are found in Southwestern US, Uzbekistan, Congo, and Australia.

Crust of powdery carnotite on sandstone

AUTUNITE

Crystal system Tetragonal | **Composition** $Ca(UO_2)_2(PO_4)_2.10–12H_2O$ | **Color** Lemon yellow to pale green | **Form/habit** Tabular | **Hardness** 2–2½ | **Cleavage** Perfect basal | **Fracture** Uneven | **Luster** Vitreous to pearly | **Streak** Pale yellow | **SG** 3.1–3.2 | **Transparency** Translucent to transparent

The crystals of autunite—calcium uranium phosphate—have a rectangular or octagonal outline. Autunite is also found as coarse groups, but scaly coatings and crusts are more common. It is usually formed as an alteration product of uraninite and other uranium-bearing minerals, and also in hydrothermal veins and pegmatites. It is a popular collector's mineral that fluoresces bright yellow green under ultraviolet light. But because it contains uranium and is thus radioactive, it should be stored carefully and handled as little as possible to minimize exposure to radiation. Sealed in a container, radioactive radon gas builds up, so ventilation is needed upon opening. Autunite is named for the place of its discovery, Autun, France.

Vitreous luster

Thin, tabular, twinned crystals of autunite

VIVIANITE

Crystal system Monoclinic | **Composition** $Fe_3(PO_4)_2.8H_2O$ | **Color** Colorless to green or blue | **Form/habit** Elongated to prismatic | **Hardness** 1½–2 | **Cleavage** Perfect | **Fracture** Uneven | **Luster** Vitreous to earthy | **Streak** Bluish white | **SG** 2.6–2.7 | **Transparency** Transparent to translucent

Vivianite—iron phosphate—is colorless on freshly exposed surfaces but becomes pale blue to greenish blue or indigo on oxidation. Crystals occur as elongated forms, in prisms, or as chunky blades. They may be rounded or corroded, in starlike groups, appear as encrustations, and be earthy, powdery, massive, or fibrous. Vivianite is widespread, forming in the weathered zones of iron ore and phosphate deposits, and complex granite pegmatites. It can also be found in recent soils and sediments, and as an alteration coating on fossil bone. Large crystals can be found in pegmatites such as those in the US and Brazil.

Crystallized cluster of vivianite

TORBERNITE

Crystal system Tetragonal | **Composition** $Cu(UO_2)_2(PO_4)_2.8–12H_2O$ | **Color** Bright green | **Form/habit** Thin to thick tabular | **Hardness** 2–2½ | **Cleavage** Perfect, basal | **Fracture** Uneven | **Luster** Vitreous to subadamantine | **Streak** Pale green | **SG** 3.3–3.7 | **Transparency** Transparent to translucent

Torbernite—copper uranyl phosphate hydrate—is bright mid-green, emerald green, leek green, or grass green. It is one of the principal uranium-bearing minerals and a minor ore of uranium and, being radioactive, it needs to be stored and handled with great care. Its crystals are blocky and commonly square in outline. It also forms foliated micalike masses, sheaflike crystal groups, or scaly coatings. Torbernite is formed in the oxidation zone of deposits containing uranium and copper as an alteration product of uraninite or other uranium-bearing minerals. Exceptional crystals of torbenite come from the Democratic Republic of Congo, the Czech Republic, Cornwall in England, and Australia. Torbernite was named in 1793 after the Swedish mineralogist Torbern Olaf Bergmann.

Tabular crystal

Torbernite crystals on an iron-rich groundmass

ERYTHRITE

Crystal system Monoclinic | **Composition** $CO_3(AsO_4)_2.8H_2O$ | **Color** Purple pink | **Form/habit** Prismatic to acicular | **Hardness** 1½–2½ | **Cleavage** Perfect | **Fracture** Uneven, sectile | **Luster** Adamantine to vitreous | **Streak** Pink | **SG** 3.1 | **Transparency** Transparent to translucent

The bright purplish pink color of erythrite (a cobalt arsenate hydrate) in a rock indicates the presence of cobalt. This is known to miners as "cobalt bloom," an important tool for prospectors looking for related silver deposits. The color can vary from crimson red to peach red, with the lighter colors indicating higher nickel content, which substitutes for the cobalt. Well-formed crystals are rare, but when found they often take the form of radiating, globular tufts. Erythrite is also found as powdery coatings. It is found in the oxidized zones of cobalt-nickel-arsenic deposits. Fine specimens come from Canada and Morocco. Other locations are Southwestern US, Mexico, France, the Czech Republic, Germany, and Australia.

Erythrite crystals

Acicular crystals of Moroccan erythrite

VARISCITE

Crystal system Orthorhombic | **Composition** $AlPO_4.2H_2O$ | **Color** Pale to apple green | **Form/habit** Cryptocrystalline aggregates | **Hardness** 4–5 | **Cleavage** Good but rarely visible | **Fracture** Splintery in massive types | **Luster** Vitreous to waxy | **Streak** White | **SG** 2.6 | **Transparency** Opaque | **RI** 1.5–1.6

The colors of variscite (hydrous aluminum phosphate) range from pale to emerald green, to blue green, to colorless. It rarely forms crystals, and even when it does, they are often visible only under a microscope. Variscite is predominantly found as cryptocrystalline or fine-grained masses, in veins, crusts, or nodules. Black spiderwebbing sometimes occurs in the matrix of variscite found in Nevada, US. Variscite forms in cavities in near-surface deposits, produced by the action of phosphate-rich waters on aluminum-rich rocks. It commonly occurs in association with apatite and wavellite, and with chalcedony and various hydrous oxides of iron. This mineral was named for Variscia, the old name for the German district of Voightland, where it was first discovered. It is found in Austria, the Czech Republic, Australia, Venezuela, and in North Carolina, Utah, and Arizona in the US.

SCORODITE

Crystal system Orthorhombic | **Composition** $FeAsO_4 \cdot 2H_2O$ | **Color** Pale leek green, brown, blue, or yellow | **Form/habit** Pyramidal, tabular, prismatic **Hardness** $3\frac{1}{2}-4$ | **Cleavage** Imperfect **Fracture** Subconchoidal | **Luster** Vitreous to resinous or waxy | **Streak** White | **SG** 3.1–3.3 **Transparency** Transparent to translucent

Scorodite is hydrated iron arsenate. It varies in color depending on the lighting—pale leek green, grayish green, liver brown, pale blue, violet, yellow, pale gray, or colorless. It can be blue green in daylight but bluish purple to grayish blue in incandescent light; in transmitted light, colorless to pale shades of green or brown. Crystals may be blocks or short prisms, but they are usually double pyramids, having the appearance of octahedrons, and often with a number of modifying faces. Coatings of minute crystals are common, but scorodite can also be porous and earthy, or grow in masses. Its name is from the Greek *scorodion* meaning "garliclike," alluding to its odor when heated. Well-crystallized material comes from Germany and Minas Gerais in Brazil.

Well-formed crystal

Scorodite crystal aggregate

HERDERITE

Crystal system Monoclinic | **Composition** $CaBePO_4(F,OH)$ | **Color** Colorless to pale yellow or greenish white | **Form/habit** Prismatic, tabular **Hardness** $5-5\frac{1}{2}$ | **Cleavage** Irregular | **Fracture** Subconchoidal | **Luster** Vitreous | **Streak** White **SG** 3.0 | **Transparency** Transparent to translucent

A calcium beryllium phosphate, herderite varies in color from colorless through yellow to green. Its crystals occur as short, blocky prisms. Some can appear orthorhombic or hexagonal in shape or as spheroids with fibrous, radiating crystals. Some specimens fluoresce deep blue when viewed under ultraviolet light. Herderite forms mainly in granite pegmatites, associated with quartz, albite, topaz, and tourmaline. It is found in many places including Germany, Finland, Russia,

and the US. First described in 1828, it was named after Saxon mining official Sigmund August Wolfgang von Herder.

Herderite crystals

CLINOCLASE

Crystal system Monoclinic | **Composition** $Cu_3(AsO_4)(OH)_3$ | **Color** Green blue | **Form/habit** Elongated, tabular | **Hardness** $2\frac{1}{2}-3$ | **Cleavage** Perfect | **Fracture** Uneven, brittle | **Luster** Vitreous to pearly | **Streak** Bluish green | **SG** 4.3 **Transparency** Subtransparent to translucent

Clinoclase is hydrous copper arsenate. Phosphorous may substitute in small amounts for arsenic. Its dark green and blue crystals can be elongated or blocky and occur as single, isolated crystals or they can be found in aggregates, forming rosettes. It is also found as crusts or coatings with a fibrous structure. Clinoclase forms in the oxidized zones of deposits containing copper sulfides. It is found in the US, Australia, France, Namibia, Germany, Austria, Russia, and Zaire. Its name comes from the Greek "to incline," and "to break," referring to the oblique plane along which it breaks.

Rosette of radiating crystals

Clinoclase crystals growing with olivenite

TRIPLITE

Crystal system Monoclinic | **Composition** $(Mn,Fe,Mg)_2PO_4(F,OH)$ | **Color** Dark brown to chestnut brown | **Form/habit** Rough masses of crystals | **Hardness** $5-5\frac{1}{2}$ | **Cleavage** Good in three directions | **Fracture** Uneven to subconchoidal **Luster** Vitreous to resinous | **Streak** White to brown | **SG** 3.5–3.9 | **Transparency** Translucent

Triplite is a fluoridated manganese phosphate, but in most samples, the manganese is partially replaced by iron. Its colors are chestnut brown, reddish brown, red, salmon pink; brownish black to black if altered; in transmitted light, pale brownish yellow to dark reddish brown. Crystals are usually very rough and poorly developed, but may have many indistinct forms. It is most commonly in nodules or masses. Triplite forms in complex zoned granite pegmatites, and in hydrothermal tin veins, sometimes accompanied by vivianite, apatite, tourmaline, sphalerite, pyrite, and quartz. Its name is from the Greek *triplos* for "triple" because its breakage is at three right angles to each other.

Massive triplite aggregate from Cornwall, England

BRAZLIANITE

Crystal system Monoclinic | **Composition** $NaAl_3(PO_4)_2(OH)_4$ | **Color** Yellow | **Form/habit** Equant to short prismatic | **Hardness** $5\frac{1}{2}$ **Cleavage** Good in one direction | **Fracture** Conchoidal | **Luster** Vitreous | **Streak** Colorless | **SG** 3.0 | **Transparency** Transparent **RI** 1.60–1.62

A relatively rare mineral, brazilianite is sodium aluminum phosphate hydroxide. The majority of brazilianite is chartreuse yellow to pale yellow. Its crystals are often well-formed short to long prisms with grooved faces. Brazilianite is also found with a globular or radiating fibrous structure. It forms in phosphate-rich granitic pegmatites, often accompanied by muscovite, albite, tourmaline, and apatite. It is named for Brazil, its major source and where it was discovered, although it is also found in lesser amounts in the US.

AMBLYGONITE

Crystal system Triclinic | **Composition** (Li,Na)$AlPO_4(F,OH)$ | **Color** White, yellow, or lilac **Form/habit** Prismatic | **Hardness** $5\frac{1}{2}-6$ **Cleavage** Perfect | **Fracture** Uneven to subconchoidal | **Luster** Vitreous to greasy or pearly | **Streak** White | **SG** 3.0 | **Transparency** Transparent to translucent | **RI** 1.57–1.65

Amblygonite is a lithium phosphate mineral. Its crystals tend to be short prisms, often with rough faces. More commonly, the mineral is found in large, white, translucent masses. Gemstone material is usually yellow, greenish yellow, or lilac. Amblygonite often occurs with other lithium-bearing minerals in pegmatite veins. Its rough gemstone material is easily mistaken for albite and other feldspars. It is an important source of lithium and phosphorus, and, to a lesser extent, as a gem mineral. Huge amblygonite crystals occur in granitic pegmatites in Zimbabwe and in the US.

LIBETHENITE

Crystal system Orthorhombic | **Composition** $Cu_2PO_4(OH)$ | **Color** Green | **Form/habit** Short prismatic | **Hardness** 4 | **Cleavage** Indistinct **Fracture** Conchoidal to uneven | **Luster** Vitreous **Streak** Olive green | **SG** 3.8–4.0 | **Transparency** Translucent

Libethenite is a rare copper phosphate hydroxide. It forms striking crystals that are short or slightly elongated prisms and vertically grooved. It can be light to dark or olive green. Libethenite is believed to form a chemical series with olivenite, in which its phosphorus substitutes for the arsenic in olivenite. It forms in the oxidized zone of copper deposits, often with malachite and azurite. Sources include Russia, England, and the US. It was discovered in 1823 in Lubietová, Slovakia, and is named after the German name for that locality.

Libethenite crystals in groundmass

Green olivenite crystals (closeup) growing in rock groundmass

△ OLIVENITE

Crystal system Orthorhombic | **Composition**
$Cu_2AsO_4(OH)$ | **Color** Olive green, brown green,
yellow, or gray white | **Form/habit** Prismatic
Hardness 3 | **Cleavage** Indistinct | **Fracture**
Conchoidal to uneven | **Luster** Adamantine
Streak Green to brown | **SG** 4.3–4.5
Transparency Translucent to opaque

Olivenite is copper arsenate. Aside from
its more common olive-green color,
olivenite can also be brownish green,
straw-yellow, or grayish white. Its crystals
are short to long prisms, sometimes found
in small brilliant crystals with domed ends.
Crystals can additionally be needlelike
or blocky, and it is also found in masses,
fibers, and in grains. Olivenite forms in
the oxidation zone of copper-bearing ore
deposits, often accompanied by
malachite, dioptase, and azurite. The
name alludes to its usual olive-green color
(German, *olivenerz*). Localities include
Namibia, Australia, Germany, Greece,
Chile, and the US.

ADAMITE

Crystal system Orthorhombic | **Composition**
$Zn_2AsO_4(OH)$ | **Color** Yellow, green, pink, or violet
Form/habit Tabular, prismatic | **Hardness** 3½
Cleavage Good | **Fracture** Subconchoidal to
uneven, brittle | **Luster** Vitreous | **Streak** White
SG 4.4 | **Transparency** Transparent to translucent

Adamite is a zinc arsenate hydroxide.
Traces of other elements substituting
for its zinc give it a number of colors:
copper substitutes for zinc to yield
yellow or green crystals depending
on its concentration; and cobalt
substitutes for zinc resulting in pink or
violet crystals. It is rarely colorless or
white. Its crystals are elongated or blocky,
often rounded nearly to the point of
appearing spherical. It also forms rosettes
and spherical masses of radiating
crystals. Adamite forms in the oxidized
zones of zinc and arsenic deposits,
and is often found with azurite,
smithsonite, mimetite, and limonite.
It has no commercial uses, but its
bright and lustrous crystals are highly
sought after by mineral collectors.
Well-crystallized specimens come
from Chile, Namibia, Mexico, Germany,
Italy, France, and the US. Much
adamite material is spectacularly
fluorescent, often glowing green,
depending on its locality.

Fluorescence

Adamite under shortwave ultraviolet light

Turquoise mineral in Albuquerque, New Mexico, US

◁ TURQUOISE

Crystal system Triclinic | **Composition** $CuAl_6(PO_4)_4(OH)_8 \cdot 4H_2O$ | **Color** Blue, green **Form/habit** Massive | **Hardness** 5–6 **Cleavage** Good | **Fracture** Conchoidal **Luster** Waxy to dull | **Streak** White to green **SG** 2.3–2.8 | **Transparency** Usually opaque

Turquoise is hydrous copper aluminum phosphate. It varies in color from sky blue to green, depending on the varying amount of iron and copper it contains. Its crystals are rare, but when found they rarely exceed 0.07 in (2 mm) in length, and occur as short, often transparent prisms. Turquoise usually occurs in masses or in microcrystalline forms as encrustations or nodules, or in veins. It is principally found in arid environments, probably derived from the decomposition of apatite and copper sulfides, and deposited from circulating waters in surrounding rock fractures. Turquoise from several sources was first transported to Europe through Turkey, from which its name is probably derived—*turquoise* is French for "Turkish." Sky-blue turquoise from Iran has been mined for centuries and is regarded as the most desirable. The local source was Neyshabur, in the Khorasan region of Iran. Other localities for turquoise include northern Africa, Australia, Siberia, England, Belgium, France, Poland, Ethiopia, Mexico, Chile, China and the US.

PYROMORPHITE

Crystal system Hexagonal | **Composition** $Pb_5(PO_4)_3Cl$ | **Color** Green, yellow, orange, or brown **Form/habit** Prismatic | **Hardness** 3½–4 **Cleavage** Poor | **Fracture** Uneven to subconchoidal, brittle | **Luster** Resinous **Streak** White | **SG** 6.7–7.1 | **Transparency** Subtransparent to translucent

Pyromorphite is a lead phosphate chloride and forms a continuous chemical series with mimetite, in which the phosphorus in pyromorphite and the arsenic in mimetite replace each other. Its color is dark green to yellow green, shades of brown, a waxy yellow, or yellow orange. Its crystals may be simple hexagonal prisms or rounded and barrel shaped, spindle shaped, or cavernous. It occurs in the oxidized zone of lead deposits with cerussite, smithsonite, vanadinite, galena, and limonite. It is a minor ore of lead but also a very popular collector's mineral. Pyromorphite takes its name from the Greek *pyr*, meaning "fire," and *morphe*,

"form," an allusion to its property of taking up a crystalline form upon cooling after it has been melted to a globule.

Double termination

Pyromorphite crystals

MIMETITE

Crystal system Hexagonal | **Composition** $Pb_5(AsO_4)_3Cl$ | **Color** Pale yellow to yellowish brown, orange, green | **Form/habit** Barrel-shaped crystals and rounded masses | **Hardness** 3½–4 **Cleavage** Poor | **Fracture** Conchoidal to uneven, brittle | **Luster** Resinous | **Streak** White **SG** 7.1 | **Transparency** Subtransparent

The colors of mimetite—lead chloride arsenate—include shades of yellow, orange, brown, and sometimes green or colorless. Mimetite forms heavy, barrel-shaped hexagonal crystals or rounded masses. It is also found in granular and blocky form, and in needlelike aggregates. It is named after the Greek *mimetes*, "imitator," referring to its resemblance to pyromorphite, in which mimetite's arsenic is replaced by phosphorous. Mimetite forms in the oxidized zone of lead deposits and other localities where lead and arsenic occur together. Excellent specimens come from Chihuahua, Mexico; Saxony, Germany; Broken Hill, Australia, and Arizona, US. One single crystal from Tsumeb in Namibia measured 2½ x 1 in (6.4 x 2.5 cm) in length.

Prismatic crystal

Prismatic crystals of mimetite

EUCHROITE

Crystal system Orthorhombic | **Composition** $Cu_2(AsO_4)(OH) \cdot 3H_2O$ | **Color** Emerald green **Form/habit** Prismatic | **Hardness** 3½–4 **Cleavage** Imperfect | **Fracture** Uneven to subconchoidal | **Luster** Vitreous | **Streak** Green **SG** 3.4 | **Transparency** Transparent to translucent

A copper arsenate hydrate, euchroite is a vitreous leek green to brilliant emerald green mineral. Its crystals are short prisms or are blocky, and they may reach up to 1 in (2.5 cm) in length. Crystals that grow in cavities result in numerous small, crystal-tipped surfaces. It is a very rare mineral, formed in the oxidized zone of some copper-bearing hydrothermal mineral deposits, often associated with olivenite, azurite, and malachite. Euchroite was named in 1823 from the Greek for "beautiful color." Among its localities are Slovakia, where it was originally found. There are important deposits in Bulgaria and Greece, and it is also found lining crevices in mica schist in Cramer Creek, a copper mining area near Missoula, Montana, US.

Short, prismatic euchroite crystals from Slovakia

VANADINITE

Crystal system Hexagonal | **Composition** $Pb_5(VO_4)_3Cl$ | **Color** Orange red, yellow **Form/habit** Hexagonal prisms | **Hardness** 3 | **Cleavage** None | **Fracture** Uneven, brittle | **Luster** Adamantine | **Streak** Whitish yellow | **SG** 6.5–7.1 **Transparency** Transparent to translucent

Relatively rare, vanadinite is lead vanadate chloride. Small amounts of calcium, zinc, and copper may substitute for lead, and arsenic can completely substitute for vanadium in the crystal structure to form the mineral mimetite. It is generally bright red or orange red in color, but sometimes brown, red brown, gray, yellow, or colorless. Its crystals are usually short hexagonal prisms, but they can also be found as hexagonal pyramids, or occasionally forming hollow prisms. It is also found as rounded masses or crusts.

It forms in oxidized ore deposits containing lead, often associated with galena, baryte, wulfenite, and limonite. It is an industrial source of vanadium, and its distinctive color makes it popular among mineral collectors. Superb specimens come from many US locations; Minas Gerais, Brazil; Chihuahua, Mexico; Mibladen, Morocco, Leadhills, Scotland; and Tsumeb, Namibia.

Prismatic crystals

Typically smooth-faced crystals of vanadite

APATITE

Crystal system Hexagonal | **Composition** $Ca_5(PO_4)_3(F,OH,Cl)$ | **Color** Green, blue, violet, purple, colorless, yellow, or rose | **Form/habit** Short to long prismatic, tabular | **Hardness** 5 | **Cleavage** Indistinct, variable | **Fracture** Conchoidal to uneven **Luster** Vitreous, waxy | **Streak** White | **SG** 3.1–3.2 | **Transparency** Transparent to translucent **RI** 1.63–1.64

Apatite is the name given to a series of calcium phosphate minerals. Within the apatites, fluorapatite contains fluorine; hydroxylapatite contains a hydroxyl (OH); and carbonate hydroxylapatite is the principal component of human bones and teeth. Apatites are all structurally identical, and are usually found as well-formed and transparent, colored, glassy crystals, and in masses, or nodules. They are often intensely colored. Apatite is a minor mineral in a range of igneous rocks, including pegmatites and high-temperature hydrothermal veins, and in marbles, and other metamorphic deposits. Crystals weighing up to 485 lb (200 kg) have been found in Canada. Other localities include the US, Mexico, Namibia, and Russia.

▽ WAVELLITE

Crystal system Orthorhombic | **Composition**
$Al_3(PO_4)_2(OH,F)_3 \cdot 5H_2O$ | **Color** Green, white,
yellow, brown, turquoise blue, black | **Form/
habit** Radiating aggregates | **Hardness** 3½–4
Cleavage Good | **Fracture** Subconchoidal
to uneven | **Luster** Vitreous to resinous
Streak White | **SG** 2.4 | **Transparency**
Translucent

Wavellite—aluminum phosphate—is a
classic radiating mineral. It is usually green
but it can also range from white, greenish
white, green yellow, yellowish brown,
turquoise blue, brown to black. Its colors
may be zoned. It rarely forms individual
crystals but when they occur they are
short to long prisms, elongated and
grooved parallel to the prism faces. It is
more commonly found as translucent,
greenish, globular aggregates of radiating
crystals up to just over 1 in (3 cm) in
diameter, as crusts, or occasionally
as stalactites. It forms in crevices in
aluminum-rich metamorphic rocks, limonite
and phosphate rock deposits; and, rarely,
hydrothermal veins. Wavellite was named
in 1805 after William Wavell, the English
physician who discovered this mineral.
European localities include France,
Bohemia in the Czech Republic; the
Highdowns quarry in north Devon,
England; the Laharran Quarry, Tracton,
Co. Cork, Ireland; and the Lichtenberg
Mine, Ronneburg, Thuringia, Germany.
Large, cream-colored specimens
come from the Siglo Veinte Mine,
Llallagua, Potosí Department, Bolivia.
In the US, several localities in Arkansas
produce excellent specimens; large
yellow and green spheres are found
in Pennsylvania, and a deep green
form in Slate Mountain, California.

Needlelike crystals of wavellite forming radiating aggregates on slate in Devon, England

LAZULITE

Crystal system Monoclinic | **Composition** $(Mg,Fe)Al_2(PO_4)_2(OH)_2$ | **Color** Various shades of blue | **Form/habit** Pyramidal | **Hardness** 5–6 **Cleavage** Indistinct | **Fracture** Uneven to splintery **Luster** Vitreous | **Streak** White | **SG** 3.0–3.4 **Transparency** Transparent to translucent **RI** 1.61–1.64

Lazulite is a magnesium aluminum hydroxophosphate, not to be confused with the silicate lazurite, a component of lapis lazuli. It is azure blue, sky blue, or bluish white to blue green in color. Its crystals are pyramidal in form, and it can also be found in masses. It forms by high-grade metamorphism of high silica quartz-rich rocks, quartz veins, and in pegmatites. It occurs in association with quartz, andalusite, rutile, kyanite, corundum, muscovite, pyrophyllite, dumortierite, tourmaline, and beryl. It takes its name from the old German *lazurstein*, meaning "blue stone." Significant localities are Switzerland, Austria, and Brazil. In North America, it occus in the White Mountains, California; Newport, New Hampshire; the Chowders Mountain, North Carolina; and Graves Mountain, Georgia, US, and Mount Fitton, Yukon, Canada. Faceting material is rare, but when found it can be carved, tumble polished, or made into beads or other decorative items. It is pleochroic—it shows blue or white depending on the angle of view.

Pyramidal crystals

Lazurite crystals in quartz groundmass

PHARMACOSIDERITE

Crystal system Cubic | **Composition** $KFe_4(AsO_4)_3(OH)_4.6–7(H_2O)$ | **Color** Green, yellow **Form/habit** Cubic | **Hardness** 2½ | **Cleavage** Imperfect to good | **Fracture** Uneven | **Luster** Adamantine to greasy | **Streak** Pale green yellow **SG** 2.8 | **Transparency** Transparent to translucent

Pharmacosiderite—potassium iron arsenate hydrate—has a wide range of greens, browns, and yellows: olive green, honey yellow, yellow brown, dark brown, hyacinth red, brown red, grass green, and emerald green. Its crystals are commonly diagonally grooved cubes, and not many exceed a few millimeters. Pharmacosiderite occurs in hydrothermal deposits, or as an alteration product of arsenic-bearing minerals, such as arsenopyrite. It takes its name from the Greek *pharmakon*, meaning "poison" or "drug," an allusion to its arsenic content, and *sideros*, meaning "iron." It was first discovered in Cornwall, England and it is also found in Germany, Greece, Italy, Algeria, Namibia, Brazil, and Utah, New Jersey, and Arizona in the US.

Cubic crystals of pharmacosiderite

DUFRENITE

Crystal system Monoclinic | **Composition** $CaFeFe_5(PO_4)_4(OH)_6.2H_2O$ | **Color** Dark olive green to black | **Form/habit** Massive | **Hardness** 3½–4½ | **Cleavage** Perfect in two directions **Fracture** Indeterminate, brittle | **Luster** Vitreous to silky or dull | **Streak** Yellow green | **SG** 3.1–3.5 **Transparency** Subtranslucent to opaque

Dufrenite is a hydrous phosphate of iron. It is greenish brown, green black, dark green, or olive green in color, turning brown when oxidized. Its crystals are relatively rare, but when found they are in sheaflike aggregates with rounded ends. Dufrenite is usually found in grapelike masses or in crusts. It occurs as an alteration product in the weathered zone of metallic veins and in iron-ore deposits. It is an uncommon mineral, often associated with other rare minerals, which makes it interesting to collectors. Notable localities for dufrenite include Hirschberg, Westphalia, Germany; Cornwall, England; Hureau, France; New Hampshire and Cherokee County, Alabama, US; and Bushmanland, South Africa. It also occurs in Portugal, Australia, and Argentina. Dufrenite was named in 1833 by French mineralogist Alexandre Brongniart after the Professor of Mineralogy at Ecole des Mines, Paris, Pierre Armand Dufrénoy.

Botryoidal mass of dufrenite on goethite

CHILDRENITE

Crystal system Orthorhombic | **Composition** $FeAl(PO_4)(OH)_2.H_2O$ | **Color** Brown to yellowish brown, white, or brownish black | **Form/habit** Pyramidal to short prismatic | **Hardness** 5 **Cleavage** Poor | **Fracture** Subconchoidal to uneven | **Luster** Vitreous | **Streak** White to yellowish | **SG** 3.2 | **Transparency** Translucent

Childrenite is an iron aluminum phosphate hydrate, and is translucent brown or yellow in color. As increasing amounts of manganese substitute for iron, it grades into the pink mineral eosphorite. Childrenite is denser than eosphorite. The crystals of childrenite are pyramids or short prisms with grooved faces, thick blocks, or plates. The mineral is sometimes found in radiating, grapelike clusters, crusts with a fibrous structure, or in masses. It probably forms as an alteration product of other phosphate minerals. Childrenite is found in Maine and Custer, South Dakota, in the US; Devon and Cornwall in England; Germany; and Minas Gerais in Brazil. It was first discovered in the George and Charlotte mine in Devon, England, and was named in 1823 after the English mineralogist John George Children.

Prismatic childrenite crystals in quartz

CHALCOPHYLLITE

Crystal system Trigonal | **Composition** $Cu_{18}Al_2(AsO_4)_2(SO_4)_3(OH)_{27}.33H_2O$ | **Color** Vivid blue green | **Form/habit** Tabular | **Hardness** 2 **Cleavage** Perfect basal | **Fracture** Uneven to subconchoidal | **Luster** Pearly to vitreous **Streak** Pale green | **SG** 2.4–2.6 | **Transparency** Transparent to translucent

Chalcophyllite is a copper aluminum arsenate sulfate hydrate. It is vivid blue green in color, and takes its name from the Greek *chalco* for "copper" and *phyllon* for "leaf," in reference to its copper content and its common leafy-appearing occurrence. Its crystals are six-sided, flattened plates and it can also be found as rosettes, sheets of tiny crystals, or in masses. It is a widespread mineral that occurs from the alteration zone of hydrothermal copper deposits, often accompanied by azurite, malachite, brochantite, cuprite, and clinoclase. It is related to barrotite and can be confused with tabular spangolite. However, chalcophyllite has a green streak and a distinctive crystal habit, both of which make conclusive identification clear. It has an unusually large amount of hydroxides and water molecules—in any specimen, half of the mineral is made up of water or hydroxide. Chalcophyllite is pleochroic, appearing blue green and almost colorless depending on the angle of view. The mineral was first described from material collected in Germany and named in 1847. It is a prized collectors' mineral because of its high luster, attractively bright color, and six-sided crystals that may be arranged in rosettes. Specimens come from southwest England, notably Wheal Gorland, Cornwall, and occasionally Wales; Alsace and Languedoc-Roussillon in France; Germany (the type material is conserved at the Mining Academy in Freiberg); Austria; Russia; Namibia; Chile; and Utah, Arizona and Majuba Hill Mine, Antelope District, Nevada in the US.

Mass of tabular chalcophyllite crystals

Chalcophyllite crystals on groundmass

Borates and nitrates

ULEXITE

Crystal system Monoclinic | **Composition**
$NaCaB_5O_6(OH)_6 \cdot H_2O$ | **Color** Colorless, white
Form/habit Nodular, acicular | **Hardness** 2½
Cleavage Perfect | **Fracture** Uneven | **Luster**
Vitreous to silky | **Streak** White | **SG** 2.0
Transparency Transparent to translucent

Ulexite—sodium calcium borate hydrate—
is found in nodular, rounded, or lenslike
crystal aggregates (resembling cotton
balls) Less commonly, it is found in dense
veins of parallel fibers which behave like
natural fiber optics, transmitting light from
one end of the crystals to the other. In
this form it is known as "television stone."
It also occurs in radiating or compact
aggregates of crystals. Ulexite is often
found in playa lakes and other evaporite
basins in desert regions—it occurs in
the dry plains of Chile, Argentina, and
Kazakhstan, and in the Kramer District
of Death Valley, California, US. Ulexite
frequently forms in association with
anhydrite, colemanite, and glauberite.
An important economic borate mineral,
ulexite is a major source of boron, since
it is derived from boron-rich fluids. It is
used in glazes for glass and pottery, as a
fertilizer and soap additive, and as a water
softener. It was named after German
chemist George Ludwig Ulex, who
analyzed it in 1850.

Slice of ulexite from boron deposits of California

BORACITE

Crystal system Orthorhombic | **Composition**
$(Mg,Fe)_3B_7O_{13}Cl$ | **Color** White to gray, green
Form/habit Pseudocubic | **Hardness** 7–7½
Cleavage None | **Fracture** Conchoidal to uneven
Luster Vitreous | **Streak** White | **SG** 3.0
Transparency Subtransparent to translucent

This glassy mineral is magnesium iron
borate chlorate. Generally white to gray, it
shades toward light green with increasing

iron content. Its crystals appear cubic,
and it also occurs in massive aggregates.
Boracite is found as crystals embedded
in sedimentary deposits of anhydrite,
gypsum, and halite. It is, as its name
suggests, a minor source of the element
boron. Localities include the Khorat Plateau
of Thailand; Stassfurt, Hannover, and
Lüneberg, Germany; Cleveland, England;
Lunéville, France; Inowroclaw, Poland; and
Missouri and Louisiana in the US.

Pseudocubic
habit

Iron gives
green color

Boracite crystals from Saxony, Germany

COLEMANITE

Crystal system Monoclinic | **Composition**
$CaB_3O_4(OH)_3 \cdot H_2O$ | **Color** Colorless, white
Form/habit Short prismatic | **Hardness** 4–4½
Cleavage Perfect, distinct | **Fracture** Uneven to
subconchoidal | **Luster** Vitreous to adamantine
Streak White | **SG** 2.4 | **Transparency**
Transparent to translucent

Colemanite (hydrous calcium borate) is
colorless, white, yellowish white, or gray
and occurs as short prisms, in nodules, or
as granular or coarse massive aggregates.
In commercial deposits, it is usually in
masses, but individual crystals up to 8 in
(20 cm) long have been found. Colemanite
is found in evaporite basins that were
originally huge inland lakes, where it
replaces other borate minerals, such as
borax and ulexite, and where deposits may
be several meters thick. Heat-resistant
glass has been developed by the inclusion
of boron compounds. These make glass
resistant to not only heat but also

chemicals and electricity. One such
glass is sold under the trademark Pyrex.
Colemanite was named in 1884 for
William Coleman, the owner of the
Californian mine where it was discovered.

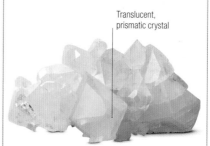

Translucent,
prismatic crystal

Complex crystals of colemanite

KERNITE

Crystal system Monoclinic | **Composition**
$Na_2B_4O_6(OH)_2 \cdot 3H_2O$ | **Color** Colorless, white
Form/habit Fibrous mass | **Hardness** 2½
Cleavage Perfect | **Fracture** Splintery | **Luster**
Vitreous, silky, dull | **Streak** White | **SG** 1.9
Transparency Transparent

Kernite, a sodium borate hydrate, is
colorless or white, but is usually covered
by a surface film of opaque white
tincalconite, a dehydration product of
borate minerals. Its crystals are relatively
rare, but when found they can be large,
often 2–3 ft (60–90 cm) long. The biggest
discovered so far measured 8 x 3 ft (240 x
90 cm). Kernite is associated with other
borate minerals as veins and irregular
masses formed in saline lake deposits,
and as crystals embedded in shale. The
Boron area of Kern County (which gave
kernite its name) and the deposits at Inyo,
California yield a large portion of the boron
used in the US. Catamarca and Salta
Provinces, Argentina, and Kirka, Turkey,
are major world sources.

Fibrous mass of kernite from Kern County

HOWLITE

Crystal system Monoclinic | **Composition**
$Ca_2B_5SiO_9(OH)_5$ | **Color** White | **Form/habit**
Nodular masses | **Hardness** 3½ | **Cleavage**
None | **Fracture** Conchoidal to uneven
Luster Subvitreous | **Streak** White
SG 2.6 | **Transparency** Translucent to opaque

Generally, howlite—calcium borosilicate
hydroxide—forms nodular masses,
sometimes resembling cauliflower.
These masses are white with fine gray or
black veins of other minerals in an erratic,
often weblike pattern, running throughout.
Crystals of howlite are blocky, and
colorless, white, or brown. They are rare,
however. Howlite occurs with other boron
minerals, such as kernite and borax. It was
named in 1868 after Canadian chemist,
geologist, and mineralogist Henry How,
who discovered it. When dyed, it
resembles, and is sometimes sold as,
turquoise. Howlite is found in quantity in
the Kramer district of Death Valley and San
Bernadino County, California, US. It also
occurs in Nova Scotia and Newfoundland,
Canada; Magdalena, Mexico; Saxony
Germany; the southern Urals of Russia;
and Suserlak, Turkey.

▷ BORAX

Crystal system Monoclinic | **Composition**
$Na_2B_4O_5(OH)_4 \cdot 8H_2O$ | **Color** Colorless
Form/habit Short prismatic | **Hardness** 2–2½
Cleavage Perfect, imperfect | **Fracture** Conchoidal
Luster Vitreous to earthy | **Streak** White | **SG**
1.7 | **Transparency** Transparent to translucent

Borax (hydrated sodium borate) can
be white, gray, pale green, or pale blue.
Its colorless crystals dehydrate in air to
become the chalky mineral tincalconite.
Borax crystals are short prisms or blocks,
although in commercial deposits it is
predominantly found in masses. It is an
evaporite mineral formed in dry desert
lake beds, and accompanied by halite,
other borates, and evaporite sulfates and
carbonates. Borax fuses easily to become
a colorless glass. It is the principal source
of boron compounds used in industry.
In metallurgy, it is used as a solvent for
metal-oxide slags in steelmaking and
metal casting, and as a flux in welding and
soldering. About half of the world's supply
comes from the borax crusts and brine of
Searles Lake in Southern California, US.
Borax also occurs in Boron, California;
Turky; Salar Cauchari, Salta Province,
Argentina; Ladakh and Kashmir in the
Indian Himalayas; and Inder, Kazakhstan.

Borax deposits (white) at the Laguna Colorada, Altiplano plateau, in Bolivia, South America

Sulfates, chromates, tungstates, molybdates

BARYTE

Crystal system Orthorhombic | **Composition**
$BaSO_4$ | **Color** Colorless, white, gray, bluish,
greenish, beige | **Form/habit** Tabular to prismatic
Hardness 3–3½ | **Cleavage** Perfect | **Fracture**
Uneven | **Luster** Vitreous, resinous, pearly
Streak White | **SG** 4.5 | **Transparency**
Transparent to translucent | **RI** 1.63–1.65

Baryte—barium sulfate—is the principal
source of barium. Blocky or prismatic
crystals are common. It can be fibrous,
and in masses, stalactites, or concretions.
Its aggregate rosettes of crystals are
known as "desert roses." Transparent blue
crystals can resemble aquamarine. A rare,
rich golden baryte comes from Colorado,
US. Baryte is a common minor mineral in
lead and zinc veins. It is also found in
sedimentary rocks such as limestone, in
clay deposits formed by the weathering of
limestone, and in cavities in igneous rock.

"Desert rose" baryte rosette of crystals

CELESTINE

Crystal system Orthorhombic | **Composition**
$SrSO_4$ | **Color** Colorless, white, red, green, blue, or
brown | **Form/habit** Tabular | **Hardness** 3–3½
Cleavage Perfect | **Fracture** Uneven | **Luster**
Vitreous, pearly on cleavage | **Streak** White
SG 4.0 | **Transparency** Transparent to translucent
RI 1.62–1.63

Celestine, also called celestite, is strontium
sulfate. Barium can substitute freely for
strontium in the structure. It often forms
highly collectable, beautifully transparent
light or medium blue crystals which have
been known to reach more than 30 in
(75 cm) in length. It is also found in

masses, fibers, or nodules. Celestine
forms in sedimentary rock such as
limestones, dolomites, and sandstones,
and it occasionally forms in hydrothermal
deposits. Beds of massive celestine
10–20 ft (3–6 m) thick occur in California,
US. Collectors' specimens come from
Madagascar, Mexico, Italy, Canada,
and the US.

ANGLESITE

Crystal system Orthorhombic | **Composition**
$PbSO_4$ | **Color** Colorless to white, yellow,
green, or blue | **Form/habit** Thin to thick tabular
Hardness 3–3½ | **Cleavage** Good, distinct
Fracture Conchoidal, brittle | **Luster** Adamantine
to resinous, vitreous | **Streak** White | **SG** 6.4
Transparency Transparent to opaque

In anglesite (lead sulfate), barium may
substitute for minor amounts of its lead.
It is colorless to white, grayish, yellow,
green, or blue and often fluoresces yellow
under ultraviolet light. It has a number of
crystal forms: thin to thick blocks, prisms,
pseudorhombohedrons, or pyramids,
and grooved along the length. Its crystals
resemble those of baryte and celestine,
with which it shares a similar structure.
Exceptionally large crystals of up to 32 in
(80 cm) long have been found in Touissit,
Morocco. It is also commonly found in
masses. Anglesite is frequent in the
oxidation zone of lead deposits. It is an
alteration product of galena, formed when
galena is subjected to sulfate solutions
generated from the oxidation of sulfide
minerals and sometimes having a core
of unaltered galena.

Prismatic anglesite
crystal

Anglesite crystals in rock groundmass

GLAUBERITE

Crystal system Monoclinic | **Composition**
$Na_2Ca(SO_4)_2$ | **Color** Gray, yellowish, colorless,
or reddish | **Form/habit** Prismatic, tabular
Hardness 2½–3 | **Cleavage** Perfect, indistinct
Fracture Conchoidal | **Luster** Vitreous to waxy
Streak White | **SG** 2.8 | **Transparency**
Transparent to translucent

Usually gray or yellowish, the surface
of glauberite (sodium calcium sulfate)
may alter to white, powdery sodium
sulfate. Crystals can be prisms,
blocks, and double pyramids, all
with combinations of forms, and
all of which may have rounded edges.
Glauberite forms under a variety of
conditions, although it is primarily an
evaporite, forming in both marine and
salt-lake environments. It is also found
in cavities in basaltic igneous rocks and
in volcanic fumaroles. Glauberite was
named in 1808 for its similarity to another
chemical, Glauber's salt, which in turn
was named after the German alchemist
Johann Glauber. Glauberite crystals are
often replaced by another mineral, such
as calcite or gypsum.

Dipyramidal glauberite crystals

ANHYDRITE

Crystal system Orthorhombic | **Composition**
$CaSO_4$ | **Color** Colorless to white, pink, blue,
violet, brownish, reddish, or grayish | **Form/habit**
Massive | **Hardness** 3½ | **Cleavage** Perfect, good
Fracture Uneven to splintery | **Luster** Vitreous to
pearly | **Streak** White | **SG** 3.0 | **Transparency**
Transparent to translucent

Along with gypsum, anhydrite is a form
of calcium sulfate. Individual crystals are
uncommon, but when found, they are
blocky. Crystals up to 4 in (10 cm) long

come from Swiss deposits. Anhydrite
is usually found in masses, or coarsely
crystalline. It is one of the major minerals
in deposits formed by the evaporation of
inland lakes and seas, and occurs most
often in salt deposits associated with
gypsum and halite. An important rock-
forming mineral, its name comes from
the Greek, *anhydrous*, meaning "without
water." Gypsum, which does contain water
in its structure, is a much more common
mineral. Anhydrite alters to gypsum in
humid conditions.

Anhydrite rough found in Germany

MELANTERITE

Crystal system Monoclinic | **Composition**
$FeSO_4 \cdot 7H_2O$ | **Color** White, greenish, bluish
Form/habit Stalactitic, concretionary
Hardness 2 | **Cleavage** Perfect | **Fracture**
Conchoidal, brittle | **Luster** Vitreous | **Streak**
White | **SG** 1.9 | **Transparency** Translucent

Melanterite—hydrous iron sulfate—is
the iron analogue of the copper sulfate
chalcanthite. Usually colorless to white,
it becomes green to blue with increasing
substitution of copper for iron. Crystals are
rare, but when they occur, they are short
prisms or pseudooctahedrons. Melanterite
is generally found in masses. It is formed
by the oxidation of pyrite, marcasite, and
other iron sulfides, and is often deposited
on the wood structures of old mine
workings. It also occurs in altered zones
of pyrite-bearing rocks, especially in arid
climates, and in coal and lignite deposits.

Melanterite nodule with typical massive habit

Gypsum at White Sands New Mexico, US, an expanse of gypsum sand so large it can be seen from space

CHALCANTHITE

Crystal system Triclinic | **Composition**
$CuSO_4 \cdot 5H_2O$ | **Color** Blue | **Form/habit**
Stalactitic, encrustations | **Hardness** 2½
Cleavage Not distinct | **Fracture** Conchoidal
Luster Vitreous | **Streak** Colorless
SG 2.3 | **Transparency** Transparent

Formerly known as blue vitriol,
chalcanthite is hydrated copper
sulfate. It is commonly peacock
blue, but may tend to greenish.
Natural crystals are relatively rare; it
usually occurs in thin veins, and as
masses and stalactites. Chalcanthite forms
through the oxidation of chalcopyrite and
other copper sulfates, occurring in the
oxidized zone of copper deposits. Because
it is water soluble, it is often found forming
crusts and stalactites on the wood
structure and walls of old mine workings
where it has crystallized from mine waters.
It takes its name from the Greek *khalkos,*
"copper" and *anthos,* "flower."

Chalcanthite crystals

△ GYPSUM

Crystal system Monoclinic | **Composition**
$CaSO_4 \cdot 2H_2O$ | **Color** Colorless, white, light brown,
yellow, pink | **Form/habit** Prismatic to tabular
Hardness 2 | **Cleavage** Perfect | **Fracture**
Splintery | **Luster** Subvitreous to pearly
Streak White | **SG** 2.3 | **Transparency**
Transparent to translucent | **RI** 1.52–1.53

Gypsum—calcium sulfate hydrate—is
colorless or white, but impurities may
tint it light brown, gray, yellow, green, or
orange. It is found in a number of forms.
It often occurs in well-developed crystals:
single crystals can be blocky with a
slanted parallelogram outline, bladed, or
in a long, thin, curving shape like a ram's
horn. Gypsum takes its name from the
Greek *gypsos,* meaning "chalk," "plaster,"
or "cement." Gypsum occurs in extensive
beds formed by the evaporation of ocean
brine, along with other minerals similarly
formed—in particular, anhydrite and halite.
It has several varieties, each of which
has its own name: selenite, which often
forms transparent,swordlike crystals
that can reach lengths of 30ft (10m);
satin spar, which has parallel, fibrous
crystals with a silky luster; alabaster, the
massive, fine-grained variety; and the
rosette-shaped crystals called desert
roses. Twinned crystals are also common,
and frequently form characteristic
"swallowtails" or "fishtails." Gypsum occurs
widely throughout the world, but the US,
Canada, Australia, Spain, France, Italy,
and England are among the leading
commercial producers.

Brochantite formed in acicular (needlelike) crystals

EPSOMITE

Crystal system Orthorhombic | **Composition** $MgSO_4.7H_2O$ | **Color** Colorless or white
Form/habit Fibrous crusts | **Hardness** 2–2½
Cleavage Perfect | **Fracture** Conchoidal
Luster Vitreous to silky | **Streak** White
SG 1.7 | **Transparency** Translucent

Epsom salts is the more familiar name for epsomite, called hydrated magnesium sulfate. Crystals are uncommon, as are prisms or fibers. Usually, it occurs as crusts, powdery or woolly coatings, or in masses. Epsomite occurs in solution in seawater, saline lake water, and spring water. It was first found around springs near the town of Epsom in Surrey, England, and named for that locality in 1805. It is also found as crusts in coal, in the weathered portions of magnesium-rich rocks and in the oxidized zones of sulfide mineral deposits.

Epsomite rough showing fibrous habit

◁ BROCHANTITE

Crystal system Monoclinic | **Composition** $Cu_4SO_4(OH)_6$ | **Color** Emerald or blue green
Form/habit Short prismatic to acicular | **Hardness** 3½–4 | **Cleavage** Perfect | **Fracture** Uneven to subconchoidal | **Luster** Vitreous | **Streak** Pale green | **SG** 4.0 | **Transparency** Translucent

Brochantite—hydrous copper sulfate— usually forms needlelike crystals or prisms, rarely more than a few millimeters long. It is also found in tufts, crusts and fine-grained masses. Brochantite forms in the oxidation zone of copper deposits, especially in arid regions. It is usually associated with azurite, malachite, and other copper minerals. Spectacular specimens come from Tsumeb, Namibia, and Arizona, US. It was named for the French geologist and mineralogist A. J. M. Brochant de Villiers.

ALUNITE

Crystal system Trigonal | **Composition** $KAl_3(SO_4)_2(OH)_6$ | **Color** White, yellowish | **Form/habit** Massive | **Hardness** 3½–4 | **Cleavage** Distinct | **Fracture** Conchoidal to splintery | **Luster** Dull to vitreous to pearly | **Streak** White | **SG** 2.6–2.9 | **Transparency** Translucent

Alunite—hydrated aluminum potassium sulfate—is informally called "alumstone." Its colors are yellow, red to reddish brown, colorless if pure; it may be white, or pale shades of gray. Alunite crystals are rare, but when found, they are rhombohedra with interfacial angles of about 90 degrees, causing them to resemble cubes. Alunite occurs as veins and replacement masses in potassium-rich volcanic rocks. It is also found near volcanic fumaroles. Its white, finely granular masses closely resemble limestone and dolomite. Earthy masses of alunite are commonly mixed with quartz and kaolinite clay. Large deposits are found in Ukraine, Spain, and Australia.

CYANOTRICHITE

Crystal system Orthorhombic | **Composition** $Cu_4Al_2SO_4(OH)_{12}\cdot2H_2O$ | **Color** Blue | **Form/habit** Acicular, fibrous | **Hardness** 2–3 | **Cleavage** None | **Fracture** Uneven | **Luster** Silky | **Streak** Pale blue | **SG** 2.7–2.9 | **Transparency** Transparent to translucent

Cyanotrichite (hydrous copper aluminum sulfate) is sky blue to azure blue in color. It forms needlelike crystals and fibrous aggregates, and it is also found as encrustations. It is an oxidation product of copper mineralization in a weathering environment where there is abundant aluminum and sulfate. It is often accompanied by brochantite, chalcophyllite, olivenite, azurite, and malachite. The name is from the Greek *kyaneos* for "blue" and *triches,* "hair," referring to its color and typical needlelike crystals. It is found principally in France, Romania, and the US.

Acicular (needlelike) cyanotrichite crystals

CROCOITE

Crystal system Monoclinic | **Composition** $PbCrO_4$ | **Color** Orange, red | **Form/habit** Prismatic | **Hardness** 2½–3 | **Cleavage** Distinct in one direction | **Fracture** Conchoidal to uneven, brittle | **Luster** Vitreous | **Streak** Orange yellow | **SG** 5.9–6.1 | **Transparency** Transparent to translucent

Crocoite—lead chromate—is one of the most eye-catching of minerals, always bright orange to red. Crystals are prisms, commonly square sectioned, slender and elongated, and sometimes cavernous or hollow. They usually occur in radiating or randomly intergrown clusters. Crocoite can also form in grains and masses. It is rare because of the conditions required for its formation: an oxidation zone of lead ore and presence of very low-silica igneous rocks serving as its source of chromium. Exceptional crystals 3–4 in (7.5–10 cm) in length are found in Tasmania.

Prismatic crystals of crocoite from Tasmania

HÜBNERITE

Crystal system Monoclinic | **Composition** $MnWO_4$ | **Color** Reddish brown | **Form/habit** Prismatic, bladed | **Hardness** 4–5½ | **Cleavage** Perfect | **Fracture** Uneven | **Luster** Submetallic/adamantine to resinous | **Streak** Yellow to brown | **SG** 7.3 | **Transparency** Transparent to translucent

Hübnerite—manganese tungstate—is generally reddish brown in color. It is found as short to long prisms, or blocky or flattened crystals with grooved faces, and it can also form groups of parallel crystals, or radiating groups. It is the manganese-end member of the manganese-iron series in which the iron and manganese are interchangeable. The iron-end member is ferberite, and together they constitute the mineral formerly known

as wolframite. This now obsolete name is still used in industry to refer to tungsten ores that are principally ferberite-hübnerite. Good specimens of hübnerite come from Peru and Colorado, US.

Prismatic hübnerite crystal

Hübnerite crystals on quartz groundmass

FERBERITE

Crystal system Monoclinic | **Composition** $FeWO_4$ | **Color** Black | **Form/habit** Bladed, prismatic | **Hardness** 4–4½ | **Cleavage** Perfect in one direction | **Fracture** Uneven, brittle | **Luster** Submetallic | **Streak** Black to brown | **SG** 7.5 | **Transparency** Opaque

The black crystals of ferberite, an iron tungstate, are commonly elongated or flattened with a wedge-shaped appearance, and are frequently grooved. Ferberite is found in granitic pegmatites and in high-temperature hydrothermal veins. It is the iron-end member of the manganese-iron series wherein the iron and manganese are fully interchangeable. The manganese-end member is hübnerite, and together they constitute the mineral formerly known as wolframite. The now obsolete mineral name wolframite is still used in industry to refer to tungsten ores that are principally ferberite-hübnerite. Excellent specimens of ferberite come from Japan, South Korea, Rwanda, Portugal, and Romania, and it is also found in Myanmar and the Czech Republic.

Ferberite crystal from Cínovec, Czech Republic

WULFENITE

Crystal system Tetragonal | **Composition** $PbMoO_4$ | **Color** Yellow, orange, red | **Form/habit** Square tabular, prismatic | **Hardness** 2½–3 | **Cleavage** Distinct | **Fracture** Subconchoidal to uneven | **Luster** Subadamantine to greasy | **Streak** White | **SG** 6.7–6.9 | **Transparency** Transparent to translucent

Wulfenite (lead molybdate) is variable in color and ordinarily forms as thin, square plates, or square, beveled, blocky crystals, but it can also be found in masses or granular aggregates. It is formed in the oxidized zone of lead and molybdenum deposits, and occurs with minerals such as cerussite, vanadinite, and pyromorphite. It is the second most common molybdenum mineral after molybdenite. It is often found in superb crystals, occasionally up to 4 in (10 cm) on an edge. Its bright, colorful, and sharply formed crystals are popular with collectors.

Wulfenite crystals from Arizona, US

SCHEELITE

Crystal system Tetragonal | **Composition** $CaWO_4$ | **Color** White, yellow, brown, or green | **Form/habit** Bipyramidal | **Hardness** 4½–5 | **Cleavage** Distinct | **Fracture** Uneven to subconchoidal | **Luster** Vitreous to greasy | **Streak** White | **SG** 5.9–6.1 | **Transparency** Transparent to translucent

The crystals of scheelite (calcium tungstate) are generally bipyramidal. In irregular masses, scheelite can be difficult to spot, but most specimens fluoresce vivid bluish white under a shortwave ultraviolet light. Opaque crystals weighing up to 15 lb (7 kg) come from Arizona, US. It commonly occurs in contact with metamorphic deposits, in high-temperature hydrothermal veins, and less commonly in granitic pegmatites. Scheelite is sometimes associated with native gold, and its fluorescence is used by geologists to search for gold deposits. Scheelite is a major source of tungsten.

Silicates: tectosilicates

▽ QUARTZ

Crystal system Trigonal | **Composition** SiO_2
Color Various | **Form/habit** Various | **Hardness**
7 | **Cleavage** None | **Fracture** Conchoidal
Luster Vitreous | **Streak** White | **SG** 2.65
Transparency Transparent to opaque | **RI** 1.53–1.55

All quartz is silicon dioxide. It occurs in nearly all silica-rich metamorphic, sedimentary, and igneous rocks, and is found worldwide. Along with feldspar, it is a common mineral in the crust of the Earth. Quartz comes in two essential forms: crystalline (forming distinct crystals) and cryptocrystalline (forming microscopic crystalline particles). Crystalline quartz is usually colorless and transparent (rock crystal) or white and translucent (milky quartz), pink and translucent (rose quartz), transparent to translucent lavender or purple (amethyst), transparent to translucent black to pale smoky brown (smoky quartz), and transparent to translucent yellow or reddish brown (citrine), and blue quartz, yet to be given a variety name. Crystalline varieties all form hexagonal prisms and pyramids, although crystals of rose quartz are rare and always small. Crystalline quartz may have inclusions. Rutilated quartz encloses needles of the titanium mineral rutile. The needles can occur as sprays, or they can be randomly oriented. There can be just a few, to so many that the stone may be nearly opaque. The rutile is usually golden in color, but it can be reddish to deep red, appearing black without intense light. Aventurine is quartz that has a spangled appearance due to sparkling internal reflections from uniformly oriented minute inclusions of other minerals. Cryptocrystalline and microcrystalline varieties of quartz either have crystals that are microscopic (microcrystalline) or too small even for a microscope (cryptocrystalline). The purest variety is chalcedony, which is white but often contains trace elements or microscopic inclusions of other minerals, giving a range of colors. Many of these colored chalcedonies have their own variety names. Chalcedony that shows distinct banding is called agate. Carnelian is red translucent chalcedony. Bloodstone is dark opaque green colored by traces of iron silicates and with patches of bright red jasper. Chrysoprase is a translucent apple green variety, colored by nickel. Sard is brown chalcedony, and sardonyx is color-banded sard. Jasper, chert, and flint are opaque, fine-grained or dense, impure varieties of cryptocrystalline quartz.

Crystals of amethyst, a crystalline quartz variety

OPAL

Crystal system Amorphous | **Composition** $SiO_2.nH_2O$ | **Color** Colorless, white, yellow, orange, rose red, black, or dark blue | **Form/habit** Massive | **Hardness** 5–6 | **Cleavage** None | **Fracture** Conchoidal | **Luster** Vitreous | **Streak** White | **SG** 1.9–2.3 | **Transparency** Transparent to translucent | **RI** 1.40–1.47

Opal is hardened silica (SiO_2) gel, and usually contains 5–10 percent water in submicroscopic pores. The vast majority is common opal in opaque yellows and reds. Its structure varies from essentially structureless to partially crystalline. Precious opal is the least crystalline form of the mineral, and consists of an ordered arrangement of tiny, transparent, silica spheres. Color play occurs when the spheres are precisely arranged and of the correct size, causing the diffraction of light through the spheres. Opal is widespread, and in its pure form it is colorless. It is deposited at low temperatures from silica-bearing, circulating waters, and is found as nodules, stalactitic masses, thin veins, and encrustations in most kinds of rocks. Opal constitutes important parts of many sedimentary accumulations such as diatomaceous earth.

Opal in cavity in ironstone host

ORTHOCLASE

Crystal system Monoclinic | **Composition** $KAlSi_3O_8$ | **Color** Colorless, white, cream, yellow, pink, brown red | **Form/habit** Short prismatic | **Hardness** 6–6½ | **Cleavage** Perfect | **Fracture** Subconchoidal to uneven, brittle | **Luster** Vitreous | **Streak** White | **SG** 2.5–2.6 | **Transparency** Transparent to translucent | **RI** 1.51–1.54

Orthoclase is potassium aluminosilicate. It is the potassium-bearing end-member of the potassium-sodium feldspar series. Pure orthoclase is relatively rare, since some sodium is usually present in the structure. Orthoclase is colorless, white, pink, cream, pale yellow, and brownish red. It appears as well-formed, short prisms, and also in massive aggregates. Orthoclase is an important rock-forming mineral—it is a major component of granite. Its pink crystals give common granite its characteristic pink color. Orthoclase is abundant in igneous rocks rich in potassium or silica, in pegmatites, and in gneisses. It occurs worldwide. The word "orthoclase" is derived from the Greek for "straight" in reference to the manner in which it breaks. Adularia is a colorless or white variety. Moonstone is an opalescent variety of orthoclase and other feldspars, which have a blue or white sheen as a result of the interlayering of orthoclase with albite.

HYALOPHANE

Crystal system Monoclinic | **Composition** $(K,Ba)(Al,Si)_4O_8$ | **Color** Colorless, white, pale yellow, or pale pink | **Form/habit** Prismatic | **Hardness** 6–6½ | **Cleavage** Perfect, good | **Fracture** Conchoidal, brittle | **Luster** Vitreous | **Streak** White | **SG** 2.6–2.8 | **Transparency** Transparent to translucent

Hyalophane (potassium barium aluminosilicate) is one of the less common feldspar minerals. It is colorless, white, pale yellow, or pale to deep pink. Its crystals are frequently transparent and glassy prisms, either similar in appearance to those of adularia or short prisms like those of orthoclase. It is also found as glassy masses. Chemically, its composition is intermediate between the feldspars orthoclase and celsian, a barium aluminosilicate. Hyalophane is found with manganese deposits in contact-metamorphic zones, often in association with rhodonite, spessartine, epidote, and analcime. Localities include Japan, Sweden, Wales, Switzerland, Australia, New Zealand, Bosnia, and the US. It takes its name from the Greek for "glass" and "to appear," a reference to its transparent, glassy crystals.

Hyalophane crystal in Swiss dolomite marble

Hyalophane crystal

SANIDINE

Crystal system Monoclinic | **Composition** $(K,Na)AlSi_3O_8$ | **Color** Colorless, white | **Form/habit** Tabular | **Hardness** 6–6½ | **Cleavage** Perfect, good | **Fracture** Conchoidal to uneven | **Luster** Vitreous | **Streak** White | **SG** 2.6 | **Transparency** Transparent to translucent

Sanidine—potassium sodium aluminosilicate—is the high-temperature form of potassium feldspar. It is a member of the potassium-sodium feldspar chemical interchange series. Its crystals are generally short prisms or blocky, with a square cross section, and have been known to reach 20in (50cm) in length. Sanidine is also found as granular and cleavable masses. It is a widespread mineral, and is found in feldspar and quartz-rich volcanic rocks such as rhyolite, phonolite, and trachyte, as well as in high-temperature and low-pressure metamorphic rocks. A notable occurrence is when sanidine forms spherical masses of needlelike white crystals in black obsidian, called "snowflake" obsidian. Some sanidine falls under the general category of moonstone.

Sanidine crystal

Well-formed prismatic sanidine crystal in trachyte

CELSIAN

Crystal system Monoclinic | **Composition** $BaAl_2Si_2O_8$ | **Color** White, colorless, or yellow | **Form/habit** Short prismatic | **Hardness** 6–6½ | **Cleavage** Perfect, good | **Fracture** Uneven, brittle | **Luster** Vitreous | **Streak** White | **SG** 3.1–3.5 | **Transparency** Transparent to translucent

Celsian—barium aluminosilicate—is one of the less common of the feldspar minerals. Its crystals are commonly twinned and can be colorless, white, or yellow, often with glassy, lustrous faces. They are short prisms that have a similar shape to adularia crystals. The mineral also occurs in easily split masses. Celsian is found in manganese-

and barium-rich metamorphic rocks, often in association with diopside, witherite, and quartz. Localities include Sweden, Japan, Wales, Australia, Namibia, and Fresno and Santa Cruz counties, California, US. It is named after Anders Celsius (1701–1744), the Swedish naturalist who invented the Celsius temperature scale.

Granular habit

Celsian from Big Creek, Fresno Co., California

ADULARIA

Crystal system Monoclinic | **Composition** $KAlSi_3O_8$ | **Color** Colorless, white, yellow, pink, or reddish brown | **Form/habit** Short prismatic | **Hardness** 6–6½ | **Cleavage** Perfect, good | **Fracture** Conchoidal to uneven | **Luster** Vitreous | **Streak** Colorless | **SG** 2.5–2.6 | **Transparency** Transparent to translucent

Adularia (potassium aluminum silicate) is not technically a distinct mineral, but is a low-temperature form of either of the feldspars microcline or orthoclase, and is generally considered to be a more highly ordered form. Its crystals vary in color and are glassy prisms. It can also be in masses, grains, or in cryptocrystalline form. It is found in low-temperature hydrothermal veins in mountainous areas, such as the Adular Mountains of Switzerland, after which it is named. It can also be found in cavities in crystalline schists. Noteworthy localities include the Alps, where it occurs in schists, and Betroka, Madagascar, where large, transparent crystals are found.

Glassy adularia crystal

Adularia crystals in actinolite groundmass

MICROCLINE

Crystal system Triclinic | **Composition** KAlSi$_3$O$_8$
Color White, pale yellow | **Form/habit** Short prismatic | **Hardness** 6–6½ | **Cleavage** Perfect, good | **Fracture** Conchoidal to uneven, brittle
Luster Vitreous, dull | **Streak** White | **SG** 2.6
Transparency Transparent to translucent | **RI** 1.5

Microcline is one of two forms of potassium aluminosilicate, the other being orthoclase. It is one of the most common feldspar minerals, and can be colorless, white, cream to pale yellow, salmon pink to red, and bright green to blue green. Microcline's crystals are either short prisms or are blocky. They can be of considerable size: single crystals from granite pegmatites can weigh several tons, and reach tens of yards in length. Microcline is the most common feldspar in feldspar-rich rocks such as granite, granodiorite, and syenite. It is also found in granite pegmatites and in metamorphic rocks such as schists and gneisses.

Microcline in amazonite

ANORTHOCLASE

Crystal system Triclinic | **Composition** (Na,K)AlSi$_3$O$_8$
Color Colorless, white | **Form/habit** Prismatic
Hardness 2–2½ | **Cleavage** Perfect, good
Fracture Conchoidal to uneven, brittle
Luster Vitreous | **Streak** White | **SG** 2.6
Transparency Transparent to translucent | **RI** 1.5

Anorthoclase is a sodium potassium aluminosilicate and is member of the sodium and potassium-rich feldspar group. Crystals are prismatic or blocky, and are often multiple twinned; it is also found in masses and grains. It shows two sets of fine lines at right angles to each other which are the result of layered crystallization. As a result of this layering, much anorthoclase exhibits a bluish, greenish, or gold schiller, making it one of several feldspars known as moonstone

when cut *en cabochon*. Its body color can be white, colorless, cream, pink, pale yellow, gray, and green. A type of syenite called larvikite has schillerized crystals of anorthoclase enclosed within it, and is highly prized as an ornamental building or facing stone. Anorthoclase is widespread, but particularly fine examples come from Colorado, US; Larvik, Norway; and Fife, Scotland. Other localities include Scotland, Kenya, New Zealand, Australia, Sicily, and Antarctica.

Anorthoclase rough

ALBITE

Crystal system Triclinic | **Composition** NaAlSi$_3$O$_8$
Color White, colorless | **Form/habit** Tabular
Hardness 6–6½ | **Cleavage** Perfect, good
Fracture Conchoidal to uneven, brittle | **Luster** Vitreous to pearly | **Streak** White | **SG** 2.6
Transparency Translucent | **RI** 1.52–1.54

Albite is sodium aluminosilicate. It takes its name from the Latin *albus*, "white," a reference to its usual color; it can also be colorless, yellowish, pink, or green. It is the sodium end member of both the plagioclase and the sodium and potassium-rich feldspars. Its crystals are blocky or platy, glassy, and brittle, and it is also found as masses and grains. Albite occurs widely in pegmatites and feldspar and quartz-rich igneous rocks such as granites, syenites, and rhyolites, and it is also found in low-grade metamorphic rocks. The primary geological importance of albite is as a rock-forming mineral.

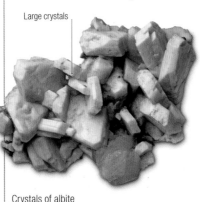

Large crystals

Crystals of albite

OLIGOCLASE

Crystal system Triclinic | **Composition** (Na,Ca)Al$_2$Si$_2$O$_8$ | **Color** Gray, white | **Form/habit** Massive | **Hardness** 6 | **Cleavage** Perfect
Fracture Conchoidal to uneven, brittle | **Luster** Vitreous | **Streak** White | **SG** 2.6
Transparency Translucent | **RI** 1.54–1.55

Oligoclase is sodium potassium aluminosilicate, the most common variety of the plagioclase feldspars. It can be gray, white, red, greenish, yellowish, brown, or colorless, and is usually in masses or granular aggregates, although it also forms blocky crystals. It occurs in granite, granitic pegmatites, diorite, rhyolite, and other feldspar and quartz-rich igneous rocks. In metamorphic rocks, it occurs in highly metamorphosed schists and gneisses. Its name was given by August Breithaupt in 1826 from the Greek *oligos*, "little," and *clasein*, "to break," because it was thought to break less uniformly than albite. The best developed and largest crystals are found in veins in granite at Arendal in Norway.

Rough specimen of oligoclase

ANDESINE

Crystal system Triclinic | **Composition** NaAlSi$_3$O$_8$ – CaAl$_2$Si$_2$O$_8$ | **Color** Gray, white | **Form/habit** Short prismatic | **Hardness** 6–6½ | **Cleavage** Perfect
Fracture Conchoidal to uneven | **Luster** Subvitreous to pearly | **Streak** White | **SG** 2.7 | **Transparency** Transparent to translucent | **RI** 1.54–1.55

Andesine is sodium calcium aluminosilicate, an intermediate member of the plagioclase series. It can be white, gray, green, yellow, or red in color. Andesine tends to be massive, or occur as rockbound grains. Crystals are uncommon, but when found are well formed, although rarely exceeding 1 in (2.5 cm). It is named is for the Andes Mountains in South America due to its abundance in the andesite lavas in those mountains; it is also found in other igneous rocks such as diorite, and syenite. It is commonly associated with quartz, biotite, hornblende, and magnetite.

BYTOWNITE

Crystal system Triclinic | **Composition** NaAlSi$_3$O$_8$—CaAl$_2$Si$_2$O$_8$ | **Color** Gray, white
Form/habit Short prismatic to tabular
Hardness 6–6½ | **Cleavage** Perfect
Fracture Uneven to conchoidal | **Luster** Vitreous to pearly | **Streak** White
SG 2.7 | **Transparency** Transparent to translucent | **RI** 1.56–1.57

Consisting of calcium, sodium aluminum silicate, Bytownite is the rarest member of the plagioclase feldspar group. Its color can be white, gray, yellow, or brown, with gem bytownite varying in color from pale, straw yellow to light brown. Well-developed crystals are relatively uncommon, and when found are short prisms or blocky. Bytownite occurs in iron and magnesium-rich igneous rocks, both intrusive and extrusive, and it is also found in stony meteorites. Bytownite was named for the place of its discovery at Bytown (now Ottawa), Canada. Localities include Canada, Mexico, Scotland, Greenland, and the US.

ANORTHITE

Crystal system Triclinic | **Composition** CaAl$_2$Si$_3$O$_8$
Color White, gray, pink | **Form/habit** Short prismatic | **Hardness** 6–6½ | **Cleavage** Perfect
Fracture Conchoidal to uneven, brittle | **Luster** Vitreous | **Streak** White | **SG** 2.7 | **Transparency** Transparent to translucent | **RI** 1.57–1.59

Anorthite is calcium aluminosilicate and can contain up to 10 percent albite. It is the calcium-rich end member of the plagioclases, with white, grayish, or reddish, brittle, and glassy crystals. It develops well-formed crystals which are short prisms, and it can also be found in masses and granular aggregates. Anorthite is a major rock-forming mineral present in many magnesium and iron-rich igneous rocks, intrusive and extrusive, in contact metamorphic rocks, and in chondroditic meteorites. Anorthosite, a rock composed mainly of anorthite, makes up much of the lunar highlands.

Anorthite crystals in rock groundmass

NEPHELINE

Crystal system Hexagonal | **Composition**
(Na, K)AlSiO$_4$ | **Color** White, gray, yellow, or red brown
Form/habit Massive | **Hardness** 5½–6 | **Cleavage**
Poor | **Fracture** Subconchoidal, brittle | **Luster**
Vitreous to greasy | **Streak** White | **SG** 2.6
Transparency Transparent to opaque

Nepheline is an aluminosilicate of sodium
and potassium. It is the most common
feldspathoid mineral, a group intermediate
between the feldspars and the zeolites.
Usually white, often with a yellowish or grayish
tint, it can also be colorless, white, gray,
yellow, or red brown, depending on
inclusions. It is generally found in masses, but
when crystals are found, they are commonly
hexagonal prisms, and may exhibit a variety
of prism and pyramid shapes. Large crystals
are often rough and pitted, and are rarely
well formed. It is a characteristic mineral of
iron and magnesium-rich igneous rocks,
where it may occur with spinel, perovskite,
and olivine. In other igneous rocks, such
as nepheline syenites, it is often found with
augite, and aegirine.

Nepheline syenite rough

▷ LABRADORITE

Crystal system Triclinic | **Composition**
NaAlSi$_3$O$_8$–CaAl$_2$Si$_2$O$_8$ | **Color** Blue, gray, white
Form/habit Usually massive | **Hardness** 6–6½
Cleavage Perfect | **Fracture** Uneven to conchoidal
Luster Vitreous | **Streak** White | **SG** 2.7
Transparency Transparent to translucent
RI 1.56–1.57

Labradorite is sodium calcium aluminosilicate,
the calcium-rich, middle-range member of the
plagioclase feldspars. Its base color is often
blue or dark gray, but it can be colorless or
white; when transparent, it can be yellow,
orange, red, or green. It rarely forms crystals—
when crystals do occur, they are blocky.
Most often found in crystalline masses that
can be up to a yard or more across, it is
a major or important constituent of certain
medium-silica and silica-poor igneous and
metamorphic rocks, including basalt, gabbro,
diorite, andesite, and amphibolite.

Slice of blue-green labrodorite, close-up

leucite, nepheline, and nosean (a sodium aluminosilicate). The volcanic rocks of Germany, especially Niedermendig, Laacher See, and Eifel, are some of its main sources.

LEUCITE

Crystal system Tetragonal | **Composition** KalSi$_2$O$_6$ | **Color** White, gray, or colorless
Form/habit Trapezohedral | **Hardness** 5½–6
Cleavage Poor | **Fracture** Conchoidal, brittle
Luster Vitreous | **Streak** White | **SG** 2.5
Transparency Transparent to translucent

Leucite (potassium aluminosilicate) is commonly found in good crystals, which can be up to 3½ in (9 cm) across. Leucite is cubic at high temperatures, forming trapezohedral crystals, but at lower temperatures it is tetragonal. The trapezohedral form is preserved as the mineral cools and develops tetragonal structure. It is most often found in masses and granular aggregates, and as disseminated grains. Leucite is found only in igneous rocks, especially those that are potassium rich and silica poor, where it can make up nearly the entire rock composition. It occurs worldwide.

Leucite crystal in tuff groundmass

Natrolite in needlelike crystals on a basalt groundmass

△ NATROLITE

Crystal system Orthorhombic | **Composition** Na$_2$Al$_2$Si$_3$O$_{10}$·2H$_2$O | **Color** Pale pink, colorless, white, gray, red, yellow, or green | **Form/habit** Acicular | **Hardness** 5–5½ | **Cleavage** Perfect
Fracture Uneven, brittle | **Luster** Vitreous to pearly
Streak White | **SG** 2.3 | **Transparency** Transparent to translucent

Natrolitehydrated sodium aluminosilicate is one of the zeolite minerals, a group of more than 50 water-containing silicates that have a particularly open type of crystal structure. Natrolite can be pale pink, colorless, white, red, gray, yellow, or green. Some specimens fluoresce orange to yellow under ultraviolet light. Its crystals are generally long and slender and up to

3 ft (1 m) in length, with vertical grooving and a square cross section. Natrolite also forms in radiating masses of needlelike crystals, and in granular or compact masses. It is found in cavities or fissures in basalt, volcanic ash deposits, and veins in gneiss, granite, and other rock types. It often forms together with other zeolites, or with apophyllite, quartz, or heulandite. Natrolite takes its name from the Greek *natrium*, which means "soda," in reference to its sodium content. Particularly fine specimens come from Golden, British Columbia, Canada; Larne, Northern Ireland; Hegau, Germany; Mumbai, India; Canterbury, New Zealand; Cape Grim, Tasmania; and Bound Brook, New Jersey, and the Dallas Gem Mine, California, US.

HAÜYNE

Crystal system Cubic | **Composition** Na$_3$Ca(Al$_3$Si$_3$O$_{12}$)(SO$_4$) | **Color** Blue, white, gray, yellow, green, or pink | **Form/habit** Dodecahedral, octahedral | **Hardness** 5½–6 | **Cleavage** Distinct
Fracture Uneven to conchoidal, brittle | **Luster** Vitreous to greasy | **Streak** Blue to white
SG 2.4–2.5 | **Transparency** Transparent to translucent | **RI** 1.49–1.51

Haüyne is sodium, calcium aluminosilicate with sulfate. It is blue when found in lapis lazuli, of which it is one of the components. In addition to crystals, it is found in rounded grains. Haüyne is primarily found in silica-poor volcanic rocks, although it has also been found in a few metamorphic rocks. It is commonly associated with

LAZURITE

Crystal system Cubic | **Composition** Na$_3$Ca(Al$_3$Si$_3$O$_{12}$)S | **Color** Blue | **Form/habit** Dodecahedral when crystallized | **Hardness** 5–5½
Cleavage Indistinct | **Fracture** Uneven, brittle
Luster Dull to vitreous | **Streak** Blue | **SG** 2.4
Transparency Translucent to opaque | **RI** 1.5

Lazurite is sodium, calcium, aluminosilicate with sulfide, and is the principal mineral in the rock lapis lazuli, which can also contain sodalite, pyrite, haüyne, and calcite. Lazurite is light to deep blue, violet blue, greenish blue, and colorless. Distinct crystals of lazurite were rare until large numbers came out of the mines of Badakhshan, Afghanistan, in the 1990s. These crystals are dodecahedrons or combinations of cubes and octahedrons. The highest-quality lapis lazuli is an intense dark blue, with minor patches of white calcite and brassy yellow pyrite. Lapis lazuli is relatively rare and commonly forms in crystalline limestones as a product of heat and pressure. The mines in Afghanistan were the principal ancient source, and remain a major locality today. Lighter blue lapis is found in Chile, and lesser amounts are also found in Italy, Argentina, the US, and Tajikistan.

Lazurite crystals in calcite groundmass

HELVITE

Crystal system Cubic | **Composition** MnBe₃(SiO₄)₃S | **Color** Yellow or brown | **Form/habit** Tetrahedral | **Hardness** 6–6½ | **Cleavage** Distinct | **Fracture** Conchoidal to uneven, brittle | **Luster** Vitreous to resinous | **Streak** White | **SG** 3.2–3.4 | **Transparency** Translucent

Helvite (manganese beryllium silicate with sulfide) is yellow to yellow green, red brown to brown, and darkens with weathering. Its crystals are tetrahedrons or pseudooctahedral. It is also found in rounded aggregates. Helvite forms in high- to medium-silica pegmatites, and in hydrothermal deposits. Crystals up to 5 in (12 cm) long come from Argentina. Other localities include Canada, Mexico, and the US. Its name comes from *helios*, Greek for "sun," for its yellow color.

Yellow helvite crystals on rock groundmass

CANCRINITE

Crystal system Hexagonal | **Composition** (NaCa)₈(Al₆Si₆)O₂₄(CO₃)₂.2H₂O | **Color** Pale to dark yellow, orange, violet, pink, or purple | **Form/habit** Massive | **Hardness** 5–6 | **Cleavage** Perfect, poor | **Fracture** Uneven | **Luster** Vitreous | **Streak** White | **SG** 2.5 | **Transparency** Transparent to translucent

Cancrinite is a complex carbonate and silicate of sodium, calcium, and aluminum. It can be pale to dark yellow, pale orange, pale violet, or pink to purple. A feldspathoid mineral, its crystals are relatively rare, but when they do occur, they can be several centimeters across. Cancrinite is usually found in fine-grained or columnar masses. It forms in syenites, in metamorphics, and in some pegmatites. Fine specimens come from Litchfield, Maine, US; the Lovozero Massif, Russia; and Laacher See, Germany.

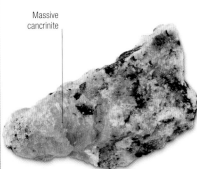

Massive cancrinite

Cancrinite in syenite groundmass from Maine

SODALITE

Crystal system Cubic | **Composition** Na₄Al₃Si₃O₁₂Cl | **Color** Gray, white, blue | **Form/habit** Massive | **Hardness** 5½–6 | **Cleavage** Poor to distinct | **Fracture** Uneven to conchoidal | **Luster** Vitreous to greasy | **Streak** White to light blue | **SG** 2.1–2.3 | **Transparency** Transparent to translucent | **RI** 1.48

Sodalite is sodium aluminum silicate chloride. Blue sodalite is a popular decorative stone, but this mineral can also be colorless, gray, pink, and other pale shades. Crystals are relatively rare, but when found are dodecahedrons or octahedrons. It fluoresces bright orange under ultraviolet light. It almost always forms massive aggregates or disseminated grains. Single pieces of sodalite can weigh many pounds. A feldspathoid mineral, sodalite principally occurs in igneous rocks such as nepheline syenites and their associated pegmatites. It was named in 1811 for its high sodium content. Its principal use is as a gemstone.

HARMOTOME

Crystal system Monoclinic | **Composition** (Ba₀.₅, Ca₀.₅,K,Na)₅Al₅Si₁₁O₃₂.12H₂O | **Color** Colorless, white, gray | **Form/habit** Blocky, multiple twins | **Hardness** 4½–5 | **Cleavage** Good, fair | **Fracture** Uneven to conchoidal, brittle | **Luster** Vitreous | **Streak** White | **SG** 2.5 | **Transparency** Transparent to opaque

Harmotome is hydrated barium potassium sodium calcium silicate. All forms of this zeolite appear glassy, and most vary from colorless to gray. It is one of the more rare zeolites. It commonly forms blocky crystals, with twinning-producing crystals that appear to be tetragonal or orthorhombic. It also forms cross-shaped twins. Harmotome occurs in low-temperature hydrothermal veins, in volcanic rocks, and as an alteration product of barium-containing feldspars. It is widespread, with good specimens from Idar-Oberstein, Germany; Thunder Bay, Canada; and Taimyr, Russia. Harmotome was named in 1801 from the Greek words *harmos*, meaning "joint," and *temseis*, meaning "cut," for the way in which it breaks.

Blocky harmotome crystals

Well-formed harmotome crystals on groundmass

GMELINITE

Crystal system Hexagonal | **Composition** (Na,K,Ca₀.₅)₄Al₈Si₁₆O₄₈.22H₂O | **Color** Colorless to white, yellow, green, orange, or red | **Form/habit** Platy, tabular | **Hardness** 4½ | **Cleavage** Good | **Fracture** Uneven to conchoidal, brittle | **Luster** Vitreous to dull | **Streak** White | **SG** 2.1 | **Transparency** Transparent to opaque

Gmelinite is hydrous sodium potassium calcium aluminosilicate. It can be colorless, white, pale yellow, greenish, orange, pink, and red. It is a member of the gmelinite series of zeolites, which includes gmelinite-Ca, gmelinite-K, and gmelinite-Na. Its crystals are hexagonal plates or

hexagonal prisms. It occurs in silica-poor volcanic rocks, sodium-rich pegmatites, and marine basalts. It is often intergrown with chabazite and associated with other zeolites, quartz, aragonite, and calcite. It is widespread but found in only small amounts. Localities include Canada, Australia, and the US. It is named for German mineralogist C. G. Gmelin.

Tabular gmelinite crystals from Nova Scotia

LAUMONTITE

Crystal system Monoclinic | **Composition** CaAl₂Si₄O₁₂.4H₂O | **Color** Colorless, white, or red | **Form/habit** Prismatic | **Hardness** 3–4 | **Cleavage** Perfect | **Fracture** Uneven | **Luster** Vitreous, dull when dehydrated | **Streak** Colorless | **SG** 2.3 | **Transparency** Translucent

The zeolite laumontite is a hydrous calcium aluminosilicate. Potassium or sodium may substitute for the calcium but only in very small amounts. It is generally colorless, white, pink, or red, but it can become powdery gray, pink, or yellow when partially dehydrated. It occurs as simple prisms or often as "swallowtail" twins, and it also forms masses, columns, fibers, and radiating aggregates. Laumontite is commonly found filling veins and hollows in igneous rocks, and is also found in hydrothermal veins, pegmatites, and metamorphic rocks. It is one of the more abundant zeolites present in sedimentary rocks. It is a common mineral, found worldwide, and used in water softeners.

Mass of prismatic laumontite crystals

HEULANDITE

Crystal system Monoclinic | **Composition** $CaAl_2Si_7O_{18} \cdot 6H_2O$ | **Color** Colorless, white | **Form/habit** Tabular | **Hardness** 3½–4 | **Cleavage** Perfect | **Fracture** Uneven, brittle | **Luster** Vitreous to pearly | **Streak** Colorless | **SG** 2.2 | **Transparency** Transparent to translucent

Heulandite-(Ca) is a calcium-rich aluminosilicate hydrate. Heulandite is the name given to a series of zeolite minerals, all of which are called heulandite and all of which look the same. It is usually colorless or white, but it can also be red, gray, yellow, pink, green, or brown. Almost all heulandite crystals are elongated, blocky, and widest at the center, creating a characteristic coffin shape. They can be up to 5 in (12 cm) long. Heulandite is a low-temperature zeolite found in a wide range of geological environments: with other zeolite minerals, filling cavities in granites, pegmatites, and basalts, in metamorphic rocks, and in weathered andesites and diabases. It was named after British mineral dealer J. H. Heuland.

Heulandite in rock groundmass

PHILLIPSITE

Crystal system Monoclinic | **Composition** $K(Ca_{1.5},Na)_2(Si_5Al_3)O_{16} \cdot 6H_2O$ | **Color** Colorless, white, pink, red, or yellow | **Form/habit** Pseudo-orthorhombic, tetragonal, cubic | **Hardness** 4–4½ | **Cleavage** Distinct, indistinct | **Fracture** Uneven, brittle | **Luster** Vitreous | **Streak** White | **SG** 2.2 | **Transparency** Transparent to translucent

Phillipsite is a hydrated potassium calcium sodium aluminosilicate. It is colorless or white, sometimes pink, red, or light yellow. It is found in prismatic crystals, blocky crystals, or radiating aggregates. Spherical groups with a radially fibrous structure and crystals on the surface are common. It is a widespread, low-temperature zeolite filling cavities and fissures in basalt, and in veins.

Phillipsite was named in 1825 after William Phillips, the founder of the Geological Society of London.

Blocky twinned phillipsite crystals on groundmass

SCOLECITE

Crystal system Monoclinic | **Composition** $CaAl_2Si_3O_{10} \cdot 3H_2O$ | **Color** Colorless, white, pink, red, or green | **Form/habit** Thin prismatic | **Hardness** 5–5½ | **Cleavage** Perfect | **Fracture** Uneven, brittle | **Luster** Vitreous | **Streak** Colorless | **SG** 2.3–2.4 | **Transparency** Transparent to opaque

Scolecite is hydrated calcium silicate, and a member of the zeolite mineral group. Although generally white or colorless, it can be pink, salmon, red, or green. Scolecite commonly occurs as sprays of thin, prismatic needles, grooved parallel to the length of the needles. Despite being monoclinic, its crystals can appear to orthorhombic or tetragonal, and may have a square cross section. Scolecite is a common hydrothermal zeolite, found in basalts, andesites, gabbros, gneisses, and amphibolites, often associated with quartz, other zeolites, calcite, and prehnite. Most of the world's finest scolecite specimens are found in the Deccan Basalt Pune, India.

Thin, prismatic needles

Sprays of scolecite needles

MESOLITE

Crystal system Monoclinic | **Composition** $Na_2Ca_2(Al_6Si_9)O_{30} \cdot 8H_2O$ | **Color** Colorless, white | **Form/habit** Acicular | **Hardness** 5–5½ | **Cleavage** Perfect | **Fracture** Uneven, brittle | **Luster** Vitreous to silky | **Streak** White | **SG** 2.3 | **Transparency** Transparent to translucent | **RI** 1.54–1.58

Mesolite is hydrous sodium calcium aluminosilicate. It belongs to the zeolite group and is named from the Greek *mesos*, meaning "middle," and *lithos*, meaning "stone," being chemically intermediate in composition between the zeolites natrolite and scolecite. Mesolite is colorless, white, pink, red, yellowish, or green. It forms as long, slender needles, radiating masses, prisms, and less commonly, compact masses or fibrous stalactites. Crystals up to 9 in (20 cm) in length occur in Ahmadnagar, India. Washington, Oregon, and Colorado, US, are also good sources. Mesolite is widespread, found in cavities in basalts and andesites and in hydrothermal veins. Its delicate glassy prisms can occur with stilbite, heulandite, and green apophyllite.

Hairlike tufts of mesolite called cotton stone

THOMSONITE

Crystal system Orthorhombic | **Composition** $NaCa_2(Al_5Si_5O_{20}) \cdot 6H_2O$ | **Color** Colorless, white | **Form/habit** Lamellar or radiating aggregates | **Hardness** 5–5½ | **Cleavage** Perfect | **Fracture** Uneven to subconchoidal | **Luster** Vitreous to pearly | **Streak** White | **SG** 2.3–2.4 | **Transparency** Transparent to translucent

Thomsonite is a series of two zeolite minerals, which are hydrous sodium, calcium, or strontium alumininosilicates and are visually indistinguishable. These are named thomsonite-Ca and thomsonite-Sr, depending on which of the two elements is dominant. Thomsonite-Ca is the most common. Thomsonites can be colorless, white, pink, red, or yellow.

Their crystals can be bladed to blocky prisms, or long, coarse needlelike prisms. It is relatively rare, and occurs in cavities in basalts and syenites, and less often in pegmatites. Localities include Italy, Scotland, Russia, Germany, Japan, and New Jersey, Oregon, and Colorado, US.

Radiating, acicular thomsonite crystals on basalt

SCAPOLITE

Crystal system Tetragonal | **Composition** Marialite $Na_4(Al_3Si_9O_{24})Cl$ Meionite $Ca_4(Al_6Si_6O_{24})(CO_3SO_4)$ | **Color** Colorless, white, gray, yellow, orange, or pink | **Form/habit** Prismatic | **Hardness** 5–6 | **Cleavage** Good | **Fracture** Uneven to conchoidal | **Luster** Vitreous | **Streak** White | **SG** 2.5–2.7 | **Transparency** Transparent to opaque | **RI** 1.53–1.60

Scapolite is sodium, calcium aluminosilicate. It was once believed to be a single mineral, but now it is defined as a compositional group, varying from calcium-bearing meionite to sodium-bearing marialite. "Scapolite" is still used in the gem trade to refer to any members of the group cut as gemstones. Crystals are short to medium prisms, often with prominent pyramid and end-face development, Scapolites are found mainly in metamorphic rocks. The largest crystals, up to 10 in (25 cm) in length, are in marble.

NOSEAN

Crystal system Cubic | **Composition** $Na_8Al_6Si_6O_{24}(SO_4) \cdot H_2O$ | **Color** Colorless or grayish | **Form/habit** Massive, granular | **Hardness** 5½ | **Cleavage** Poor | **Fracture** Uneven to conchoidal | **Luster** Vitreous | **Streak** Colorless | **SG** 2.3 | **Transparency** Transparent to translucent

Nosean is sodium aluminum silicate sulfate hydrate, in the same family as sodalite. Its color varies from colorless and white to gray, brown, gray brown, and blue. It is usually found in masses or granular aggregates. When crystals occur, they are dodecahedral, but rarely exceed ¼ in (6 mm) in length. It is found in low-silica igneous rocks such as phonolites and other volcanics where it occurs as crystals in cavities in the rock. Nosean was first described in 1815 from specimens found in the Laacher See district in Germany, and it is named after the German mineralogist K. W. Nose. It is also found in England, France, Italy, and and in Colorado and Utah, US.

CHABAZITE

Crystal system Hexagonal/trigonal | **Composition** $(Na,Ca_{0.5},K)_4(Al_4Si_8O_{24}).12H_2O$ | **Color** Colorless, white, pink, orange, yellow, or brown | **Form/habit** Pseudocubic rhombohedral | **Hardness** 4–5 | **Cleavage** Indistinct | **Fracture** Uneven, brittle | **Luster** Vitreous | **Streak** White | **SG** 2–2.2 | **Transparency** Transparent to translucent

Chabazite (sodium calcium aluminosilicate hydrate) is a common zeolite mineral that varies in color. Its crystals form prisms or distorted cubes which can look like rhombohedrons. Chabazite is found in cavities in basalt or andesite, volcanic ash deposits, pegmatites, and granitic and metamorphic rocks. It is found widely, with splendid 1–2 in (2.5–5 cm) crystals occurring in Northern Ireland, Iceland, Germany, Scotland, the Czech Republic, Hungary, India, Canada, Australia, and the US. It takes its name from the Greek *chabazios* or *chalazios*, "hailstone."

Pseudocubic chabazite crystals from Canada

STILBITE

Crystal system Monoclinic | **Composition** $(Na,Ca_{0.5},K)_9(Al_9Si_{27}O_{72}).28H_2O$ | **Color** Colorless, white, pink | **Form/habit** Tabular | **Hardness** 3½–4 | **Cleavage** Perfect | **Fracture** Conchoidal, brittle | **Luster** Vitreous to pearly | **Streak** White | **SG** 2.2 | **Transparency** Transparent to translucent

Most stilbites—hydrated sodium calcium aluminosilicate—are calcium rich, but

Analcime crystals on Mt. Etna, Sicily, Italy

some are sodium dominant. Stilbite is usually colorless or white, but it can also be yellow, brown, salmon pink, or red, and, rarely, green, blue, or black. Its crystals are blocky, commonly twinned, and also occur in sheaflike, "bow tie"-appearing aggregates. It is often found in fine crystals, some of which exceed 4 in (10 cm). A member of the zeolite group, stilbite occurs in magnesium and iron-rich igneous rocks, granitic pegmatites, gneisses and schists, and in hot-spring deposits. Stilbite derives its name from the Greek "to shine," in reference to its vitreous to pearly luster. It is widespread, found in many countries.

Tabular stilbite crystals from the Faroe Islands

△ ANALCIME

Crystal system Cubic | **Composition** $Na(AlSi_2)O_6.H_2O$ | **Color** White, colorless, yellow, brown red, or orange | **Form/habit** Trapezohedral | **Hardness** 5–5½ | **Cleavage** None | **Fracture** Subconchoidal, brittle | **Luster** Vitreous | **Streak** White | **SG** 2.3 | **Transparency** Transparent to translucent

Most examples of analcime—sodium aluminum silicate—are colorless or white, but can also be more warmly colored. Most crystals form trapezohedrons. Variations in the order of the sodium-aluminum proportion can vary its structure enough for it to be classified in several crystal systems. Analcime occurs in seams and cavities in basalt, diabase, granite, and gneiss associated with prehnite, calcite, and zeolites. It is also found in extensive beds formed by precipitation from alkaline lakes. Its name is derived from the Greek *analkimos*, "weak," a reference to its weak electrical charge when heated or rubbed.

POLLUCITE

Crystal system Cubic | **Composition** $(Cs,Na)(AlSi_2)O_6.H_2O$ | **Color** Colorless, white, pink, blue, or violet | **Form/habit** Massive | **Hardness** 6½–7 | **Cleavage** None | **Fracture** Conchoidal to uneven | **Luster** Vitreous to greasy | **Streak** White | **SG** 2.7–3.0 | **Transparency** Transparent to translucent | **RI** 1.51–1.52

Pollucite, a complex cesium aluminum aluminosilicate, forms a complete chemical interchange series with analcime. It is usually colorless or white, but can be pink, blue, or violet. It rarely forms distinct crystals; it is commonly found in masses. When crystals are found, they are rounded, and appear corroded. A zeolite, pollucite is found worldwide only in rare-earth-bearing granitic pegmatites, where it occurs with spodumene, petalite, quartz, lepidolite, and apatite. Rare crystals up to 24 in (60 cm) across have been found at Kamdeysh, Afghanistan. There is a huge deposit at Bernic Lake, Manitoba, Canada, and gem material is found in the US.

Silicates: phyllosilicates

SERPENTINE

Crystal system Monoclinic | **Composition** $(Mg,Fe,Ni)_3Si_2O_5(OH)_4$ | **Color** White, gray, yellow, green | **Form/habit** Massive or pseudomorphous **Hardness** $3\frac{1}{2}–5\frac{1}{2}$ | **Cleavage** Perfect but not visible | **Fracture** Conchoidal to splintery **Luster** Subvitreous to greasy, resinous, earthy, dull **Streak** White | **SG** 2.5–2.6 | **Transparency** Translucent to opaque | **RI** 1.55–1.56

Serpentine is a group of at least 16 magnesium silicate minerals. They are usually intermixed, but sometimes individual members can be distinguished. There are four major varieties: chrysotile; antigorite, occurring in corrugated plates or fibers; lizardite, very fine-grained and platy; and amesite, which occurs in platy or pseudohexagonal, columnar crystals. Although their chemistry is complex, the varieties look similar in appearance. Serpentine is derived from the chemical alteration of other minerals such as olivine, pyroxenes, and amphiboles.

CHRYSOTILE

Crystal system Monoclinic | **Composition** $Mg_3Si_2O_5(OH)_4$ | **Color** White, green, yellowish, golden | **Form/habit** Fibrous | **Hardness** 2–3 **Cleavage** Perfect | **Fracture** None | **Luster** Subresinous to greasy | **Streak** White | **SG** 2.6 **Transparency** Translucent to opaque

Chrysotile is the fibrous mineral that is a member of the serpentine group. It is the most important asbestos mineral. Individual chrysotile fibers are white and silky, but aggregate fibers in veins are usually green or yellowish. The mineral can even look gold—its name comes from the Greek for "hair of gold." Chrysotile fibers are actually tubes, in which the structural layers of the mineral are rolled into the form of a spiral. Large deposits of chrysotile are found in Quebec, Canada, and the Ural Mountains of Russia. It also occurs in England, Scotland, Switzerland, Serbia, Australia, Mexico, South Africa, Zimbabwe, Swailand, and the US.

Mass of fibers

Chrysotile fibers on a rock groundmass

White kaolin edge to Mt. Stephanos volcano and mountain range on the Lakki Plateau in Greece

TALC

Crystal system Monoclinic | **Composition**
$Mg_3Si_4O_{10}(OH)_2$ | **Color** White, colorless, green,
yellow to brown | **Form/habit** Foliated and fibrous
masses | **Hardness** 1 | **Cleavage** Perfect
Fracture Uneven to subconchoidal | **Luster** Pearly
to greasy | **Streak** White | **SG** 2.6–2.8
Transparency Translucent

Talc—hydrous magnesium silicate—is
often found mixed with other minerals
such as serpentine and calcite. Crystals
are rare; talc is most commonly found in
leaflike, fibrous, or massive aggregates.
Dense, high-purity talc is called steatite.
Compact talc (and other minerals which
have a soapy or greasy feel) is called
soapstone. Talc is a metamorphic mineral
found in veins and in magnesium-rich
rocks, often associated with serpentine,
tremolite, and forsterite, and occurs as
an alteration product of silica-poor igneous
rocks. It is widespread, and found in
most areas of the world where low-grade
metamorphism occurs. Talc is used in
toiletries, paint, paper, roofing materials,
plastic, and rubber.

Stress lines made during
metamorphism

Talc specimen

◁ KAOLINITE

Crystal system Triclinic | **Composition**
$Al_2Si_2O_5(OH)_4$ | **Color** White when pure
Form/habit Massive | **Hardness** 2–2½
Cleavage Perfect | **Fracture** Unobservable
Luster Earthy | **Streak** White | **SG** 2.6
Transparency Opaque

Kaolinite is aluminum silicate hydrate.
Three other minerals—dickite, nacrite,
and halloysite—are chemically identical
to kaolinite but crystallize in the monoclinic
system. They are often found together
with kaolinite, and can be visually
indistinguishable. Kaolinite is white
to gray, and only forms microscopic
pseudohexagonal plates in compact
or granular masses and in micalike piles.
Kaolinite is a natural product of the
alteration of mica and plagioclase and

sodium-potassium feldspars under the
influence of water, dissolved carbon
dioxide, and organic acids. It is used
in agriculture, as a paint extender, as
a strengthener in rubber, as a filler in
paper, and as a dusting agent in foundry
operations. It even has food and medicinal
applications: as a filler in some food
(such as chocolate), and mixed with pectin
as an antidiarrheal. Kaolinite is the main
component of china clay, a key ingredient
in porcelain. When porcelain began to
be manufactured in Europe in the 17th and
18th centuries, it created a huge
china-clay mining industry, for instance,
in Cornwall, England. Kaolinite is also
mined in China, Chile, Germany, Russia,
the Czech Republic, France, and Brazil.
By providing the raw material for brick,
pottery, and tiles, clay minerals like kaolinite
have played a vital part in the progress of
human civilization.

PYROPHYLLITE

Crystal system Triclinic or monoclinic
Composition $Al_2Si_4O_{10}(OH)_2$ | **Color** White,
colorless, brown green, pale blue, gray | **Form/habit**
Compact masses | **Hardness** 1–2 | **Cleavage**
Perfect | **Fracture** Uneven | **Luster** Pearly to dull
Streak White | **SG** 2.7–2.9 | **Transparency**
Transparent to translucent

Pyrophyllite—aluminum silicate
hydroxide—rarely forms distinct crystals,
but it is sometimes found in coarse laths
and radiating aggregates.
It is usually in granular masses of leaflike
plates, frequently so fine-grained as to
appear textureless. Pyrophyllite forms
by the low-grade metamorphism of
aluminum-rich sedimentary rock, such
as bauxite. Its name is from the Greek for
"fire" and "leaf," referring to its tendency
to break down into flakes when heated.
Finely powdered pyrophyllite gives a sheen
to lipsticks, and some talcum powder is
actually pyrophyllite.

Radiating groups of pyrophyllite laths

MUSCOVITE

Crystal system Monoclinic | **Composition**
$KAl_2(Si_3Al)O_{10}(OH,F)_2$ | **Color** Colorless, silvery
white, pale green, rose, brown | **Form/habit**
Tabular | **Hardness** 2–3 | **Cleavage** Perfect
basal | **Fracture** Uneven | **Luster** Vitreous
Streak Colorless | **SG** 2.7–3.1 | **Transparency**
Transparent to translucent

Muscovite (potassium aluminosilicate
with fluorine and hydroxyl) is also called
common mica, potash mica, or isinglass.
It is the most common member of the
mica group. It is usually colorless or silvery
white, but it can also be brown, light gray,
pale green, or rose red. Its crystals are
generally blocky with a pseudohexagonal
outline. Muscovite also commonly occurs
as leaflike masses and fine-grained
aggregates. It is a common rock-forming
mineral, and occurs in metamorphic rocks
such as gneisses and schists, in granites,
in veins and in pegmatites, the usual
source of large crystals. Single crystals
are known up to 10 ft (3 m) in diameter.

Muscovite crystals in rock groundmass

GLAUCONITE

Crystal system Monoclinic | **Composition**
$(K,Na)(Mg,Al,Fe)_2(Si,Al)_4O_{10}(OH)_2$ | **Color** Green
Form/habit Rounded aggregates | **Hardness** 2
Cleavage Perfect basal | **Fracture** Uneven
Luster Dull to earthy | **Streak** N/D | **SG** 2.4–2.9
Transparency Translucent to opaque

A member of the mica group, glauconite
is an iron potassium magnesium silicate.
Although most glauconite is bluish green,
its color ranges from olive green to
blackish green. It usually occurs as
rounded aggregates and pellets of
fine-grained, scaly particles. It weathers
quickly, and easily crumbles to a fine
powder. Glauconite forms in shallow

marine environments and is a diagnostic
mineral, which indicates a continental
shelf marine deposit with a slow rate of
accumulation. It can also be found in sand
or clay formations, or in impure limestone
or chalk. It was named in 1828 from the
Greek *glaukos*, meaning "blue green."

Aggregate of fine glauconite grains

PHLOGOPITE

Crystal system Monoclinic | **Composition**
$KMg_3AlSi_3O_{10}(OH)_2$ | **Color** Colorless, pale yellow
to brown | **Form/habit** Tabular or pseudohexagonal
Hardness 2–2½ | **Cleavage** Perfect basal
Fracture Uneven | **Luster** Pearly to submetallic
Streak Brownish white | **SG** 2.8–3.0 |
Transparency Transparent to translucent

Phlogopite mica is potassium magnesium
aluminosilicate with fluorine and hydroxyl.
Iron substitutes for its magnesium in
variable amounts as it grades into the
iron-rich biotite. Phlogopite can have a
coppery look. Its crystals are blocky and
look hexagonal, and it usually occurs as
platy aggregates. Phlogopite micas are
found primarily in igneous rocks, and it is
a common constituent of diamond-bearing
kimberlites. The largest documented
crystal of phlogopite, found in Canada,
measured 30 x 5 x 5 ft (10 x 4.3 x 4.3 m)
and weighed about 364 tons (330 tonnes).

Pseudohexagonal phlogopite crystals

LEPIDOLITE

Crystal system Monoclinic | **Composition** $K(Li,Al)_3(AlSi_3)O_{10}(OH,F)_2$ | **Color** Pale lilac, colorless, yellow, or gray | **Form/habit** Tabular to short prismatic | **Hardness** 2½–3½ | **Cleavage** Perfect basal | **Fracture** Uneven | **Luster** Vitreous to pearly | **Streak** Colorless | **SG** 2.8–2.9 | **Transparency** Transparent to translucent

Lepidolite is potassium lithium aluminosilicate with fluorine and hydroxyl. It is one of the major sources of the rare alkali metals rubidium and cesium which occur in minor amounts. This mica is the Earth's most common lithium-bearing mineral. Although typically pale lilac, lepidolite can be colorless, pale yellow, or gray. Crystals can be pseudohexagonal, and it is also found as bubbly-appearing masses, and as fine- to coarse-grained interlocking plates. Its perfect cleavage yields thin, flexible sheets. It occurs almost exclusively in granitic pegmatites, where it is associated with other lithium minerals, beryl, topaz, quartz, tourmaline, and spodumene. "Lepidolite" is no longer a mineral name, and is now the name of a series of minerals with varied chemistry.

Lepidolite
crystal

Lepidolite crystals on pegmatite

VERMICULITE

Crystal system Monoclinic | **Composition** (Mg, Fe,Al)$_3$(Al,Si)$_4$O$_{10}$(OH).4H$_2$O | **Color** Gray white, golden brown | **Form/habit** Foliated, expanded crystals | **Hardness** 1–2 | **Cleavage** Perfect | **Fracture** Uneven | **Luster** Oily to earthy | **SG** 2.6 | **Transparency** Translucent

Vermiculite is the name of a group of magnesium iron aluminosilicate mica minerals in which various chemical substitutions occur in the molecular structure. It is greenish, golden yellow, or brown. It forms blocky, pseudohexagonal crystals, platy aggregates, or is found as small particles in soils and ancient sediments. Vermiculite can be interlayered with other micas and claylike minerals. It is found at the interface between feldspar-rich igneous rocks and iron and magnesium-rich igneous rocks, and by the hydrothermal alteration of iron-bearing micas. When heated to nearly 572° F (300° C), vermiculite can expand quickly and strongly to 20 times its original thickness. Expanded vermiculite is very light and is widely used in concrete and plaster, for thermal and acoustic insulation, as a packing medium, and as a growing medium in horticulture.

Vermiculite specimen from Pennsylvania, US

FUCHSITE

Crystal system Monoclinic | **Composition** K(Cr,Al)$_2$(AlSi$_3$)O$_{10}$(OH,F)$_2$ | **Color** Green | **Form/habit** Curved aggregates | **Hardness** 2½ | **Cleavage** Perfect basal | **Fracture** Uneven | **Luster** Vitreous to pearly | **Streak** White | **SG** 2.8–2.9 | **Transparency** Transparent to opaque

Fuchsite—hydrous potassium chromium aluminosilicate—is the chromium-bearing variety of muscovite mica and is also known as chrome mica. Its chromium concentration may be up to 25 percent, giving it a distinctive apple-green hue, although its color can vary from pale to emerald depending on the amount of chromium. Fuchsite forms in concentric, curved aggregates and as flakes, and its perfect cleavage yields thin, flexible plates. It is found where hydrothermal solutions replace carbonates in gold deposits.

Mexican fuchsite

CHAMOSITE

Crystal system Monoclinic | **Composition** Fe$_5$(Fe$_2$Al)(Si$_3$AlO$_{10}$)(OH)$_8$ | **Color** Green to greenish black | **Form/habit** Massive | **Hardness** 2½–3 | **Cleavage** Perfect | **Fracture** Uneven | **Luster** Oily to vitreous | **Streak** White green to gray | **SG** 3.0–3.3 | **Transparency** Translucent

Chamosite—a hydrous aluminum silicate of iron—may be greenish gray or greenish brown. It is usually found in masses or earthy aggregates, but crystals, when found, can be blocky, pseudohexagonal, or pseudorhombohedral. Chamosite is a relatively uncommon mineral. Only about 15 localities around the world are known, all associated with iron deposits. It is produced in an environment of low- to moderate-grade metamorphism of iron deposits. Chamosite is usually found in occurrence with other chlorite minerals. Localities include France, Poland, the Czech Republic, Germany, Western Australia, Japan, England, South Africa, and Pennsylvania, Arkansas, Colorado, Michigan, and Maine in the US.

Specimen of massive chamosite

ZINNWALDITE

Crystal system Monoclinic | **Composition** K(Li,Fe,Al)$_3$(AlSi$_3$)O$_{10}$(F,OH)$_2$ | **Color** Gray brown | **Form/habit** Tabular to short prismatic | **Hardness** 2½–4 | **Cleavage** Perfect | **Fracture** Uneven | **Luster** Vitreous to pearly | **Streak** White | **SG** 2.9–3.2 | **Transparency** Transparent to translucent

Zinnwaldite mica is potassium lithium iron aluminum silicate hydroxide fluoride. It is gray brown, yellow brown, pale violet, and dark green in color, with color zoning common. When found in crystals, they are well-formed short prisms or blocks, hexagonal in appearance, in rosettes or as fan-shaped groups, scaly aggregates, or as disseminated grains. Zinnwaldite is found in hydrothermal veins and granitic pegmatites. It is commonly associated with cassiterite, topaz, wolframite, lepidolite, beryl, spodumene, tourmaline, and fluorite. Specimens of zinnwaldite come from Altenburg, Germany; Zinnwald (after which the mineral is named) and Severocesk, the Czech Republic; Baveno, Italy; many localities in Japan; Virgem de Lapa, Brazil; and Pike's Peak, Colorado, and Amelia Court House, Virginia, in the US.

Vitreous
luster

Zinnwaldite from the Czech Republic

▷ BIOTITE

Crystal system Monoclinic | **Composition** K(Mg,Fe)$_3$(AlSi$_3$)O$_{10}$(OH,F)$_2$ | **Color** Black, brown, pale yellow, tan, or bronze | **Form/habit** Tabular | **Hardness** 2½–3 | **Cleavage** Perfect basal | **Fracture** Uneven | **Luster** Vitreous to submetallic | **Streak** White | **SG** 2.7–3.4 | **Transparency** Transparent to translucent

Biotite mica is a hydrated, fluoridated potassium magnesium iron aluminosilicate. It tends to be black when iron rich, or brown, pale yellow to tan, or bronze in color with increasing magnesium. It frequently forms large crystals, blocks to short prisms, and is often hexagonal appearing in cross section. It also occurs as scaly aggregates, or disseminated grains. Biotite is widespread and common in both igneous and metamorphic rocks. Biotite mica is a key constituent of many igneous and metamorphic rocks, including granites, nepheline syenites, schists, and gneisses. Biotite is used to age-date rocks, by either the potassium-argon or the argon-argon dating method. Notable localities include Scotland; Ontario; Canada; and Pike's Peak, Colorado, the Adirondak Mountains, New York, and King's Mountain, North Carolina in the US. "Biotite" is now not valid mineral name, and is used for the series including phlogopite, siderophyllite, and others.

Biotite monzogranite in the Alabama Hills, Lone Pine, California, US

Clinochlore specimen close-up

△ CLINOCHLORE

Crystal system Monoclinic | **Composition**
$Mg_3(Mg_2Al)(Si_3AlO_{10})(OH)_8$ | **Color** Green
Form/habit Tabular | **Hardness** 2½ | **Cleavage**
Perfect | **Fracture** Uneven | **Luster** Vitreous
to dull | **Streak** Pale yellow | **SG** 2.6–2.9
Transparency Translucent

Clinochlore—hydrous magnesium
aluminosilicate—forms a chemical
interchange series with chamosite wherein
iron swaps for magnesium. The mineral
is various shades of green in color.
Crystals are blocky, with a hexagonal or
orthorhombic appearance. Clinochore
forms widely as a product of low-grade
metamorphism, as a hydrothermal
alteration product, in marine clays and
soils, in filled-in hollows in volcanic rocks,
and in veins. It takes its name from the
Greek *klinen*, meaning "to incline," and
chloros, "green," in reference to its optical
symmetry and color. Notable localities are
Pajsberg, Sweden; Quebec, Canada;
Banff, Scotland; Valais, Switzerland; Bahia,
Brazil; Marienberg, Germany; Russia's Ural
Mountains; and, in the US, Brewster, New
York; Lancaster County, Pennsylvania;
and Lowell, Vermont.

CAVANSITE

Crystal system Orthorhombic | **Composition**
$Ca(VO)Si_4O_{10}.4H_2O$ | **Color** Blue | **Form/habit**
Prismatic | **Hardness** 3–4 | **Cleavage** Good
Fracture Conchoidal | **Luster** Vitreous
Streak White | **SG** 2.2–2.3 | **Transparency**
Transparent to translucent

Cavansite (hydrous calcium vanadium
silicate) is a distinctive deep blue to
greenish blue in color. It forms aggregates
of prismatic crystals, and also occurs
as platelets and in rosettes. It is found
in basalts in association with, and often
perched on top of, zeolites, and it is
also found in tuffs. Cavansite was named
after some of the elements it contains:
calcium (Ca), vanadium (V), and silicon
(Si). Because of its color and rarity, it is a
sought after collector's mineral. Cavansite
was discovered at Lake Owyhee State
Park, Oregon, US. The finest specimens
come from Pune, India.

Cavansite
rosette

Cavansite on heulandite from Pune, India

CHRYSOCOLLA

Crystal system Orthorhombic | **Composition**
$Cu_2H_2(Si_2O_5)(OH)_4.nH_2O$ | **Color** Blue, blue green
Form/habit Massive | **Hardness** 2–4
Cleavage None | **Fracture** Uneven to conchoidal
Luster Vitreous to earthy | **Streak** Pale blue,
tan, gray | **SG** 2.0–2.4 | **Transparency**
Translucent to nearly opaque

Chrysocolla—copper aluminum silicate—
is generally blue green in color. Crystals
are rarely seen, but the mineral can form
bubbly-appearing, radiating aggregates.
Chrysocolla is commonly very fine grained
and in masses, frequently intergrown
with other minerals such as quartz,
chalcedony, or opal to yield a harder,
more resilient gemstone variety. It forms
as a decomposition product of copper
minerals, especially in arid regions. It is
found worldwide. Important localities
include England, Israel, Mexico, the Czech
Republic, Australia, the Congo, and the

US. The name is derived from the Greek *chrysos*, meaning "gold," and *kolla*, meaning "glue."

Chrysocolla grown with turquoise and malachite

APOPHYLLITE

Crystal system Tetragonal | **Composition** $KCa_4Si_8O_{20}(F,OH).8H_2O$ (fluorapophyllite) | **Color** Colorless, pink, green, or yellow | **Form/habit** Tabular, prismatic | **Hardness** 4½–5 | **Cleavage** Perfect | **Fracture** Uneven, brittle | **Luster** Vitreous | **Streak** Colorless | **SG** 2.3–2.4 | **Transparency** Transparent to translucent

Once considered to be a single mineral, apophyllite is now divided into two distinct species—fluorapophyllite and hydroxyapophyllite. Both apophyllites are hydrous potassium calcium silicates with either fluorine or hydroxyl, which form a chemical interchange series in which fluorine can predominate over oxygen and hydrogen, and vice versa. Specimens whose precise chemical composition have not been established are still called apophyllite. It forms an abundance of large crystals which are square-sided grooved prisms with flat ends that may appear cubic, or can have steep pyramidal terminations; some crystals reach 8 in (20 cm) in length. Apophyllite is highly popular with collectors. It frequently occurs with zeolite minerals in basalt.

Blocky green crystals of apophyllite from India

PETALITE

Crystal system Monoclinic | **Composition** $LiAlSi_4O_{10}$ | **Color** Colorless to grayish white, pink, or green | **Form/habit** Usually as aggregates | **Hardness** 6½ | **Cleavage** Perfect basal | **Fracture** Subconchoidal | **Luster** Vitreous | **Streak** White | **SG** 2.4 | **Transparency** Transparent to translucent | **RI** 1.50–1.51

Petalite (lithium aluminum silicate) is rarely found as individual crystals and most commonly occurs as aggregates. When crystals do occur, they are blocky prisms. Petalite occurs in lithium-bearing pegmatites along with albite, quartz, spodumene, lepidolite, and tourmaline. It was named in 1800 for the Greek word for "leaf," a reference to its perfect cleavage, which allows it to peel off in thin, leaflike layers. Lithium was first recognized as a new chemical element in 1817 by Swedish chemist Johan August Arfvedson from its occurrence in this mineral.

Petalite specimen

ALLOPHANE

Crystal system Amorphous | **Composition** $Al_2Si_2O_7.3H_2O$ to $Al_2SiO_5.2H_2O$ | **Color** White to tan, green, blue, or yellow | **Form/habit** Crusts and masses | **Hardness** 3 | **Cleavage** None | **Fracture** Conchoidal | **Luster** Waxy to earthy | **Streak** White | **SG** 2.8 | **Transparency** Translucent

Allophane (hydrous aluminum silicate) has a variable chemical composition, in part due to impurities. It is amorphous (structureless), and it forms waxy, bubbly-appearing or crusty masses. Allophane is almost invariably mixed with other minerals such as opal, limonite, and gibbsite. It is a weathering product

of volcanic ash or a hydrothermal alteration product of feldspars. It was named from the Greek *allos*, "other," and *phanos*, "to appear," since it gave a deceptive reaction in old mineralogical testing

OKENITE

Crystal system Triclinic | **Composition** $Ca_5Si_9O_{23}.9H_2O$ | **Color** Colorless, white, pale yellow, or blue | **Form/habit** Fibrous, bladed | **Hardness** 4½–5 | **Cleavage** Perfect | **Fracture** Splintery | **Luster** Pearly to vitreous | **Streak** White | **SG** 2.3 | **Transparency** Transparent to translucent

The crystals of okenite (calcium silicate hydrate) are fibrous or bladed, and lathlike. It is sometimes found within geodes in spherical masses of radiating fibers that look like cotton balls. Okenite is often associated with zeolites, apophyllite, quartz, prehnite, and calcite in basalts. It was named in 1828 after the German naturalist Lorenz Oken. It is found in many localities, including India, Azerbaijan, Chile, Ireland, and New Zealand.

Spherical mass of okenite crystals

Acicular (needlelike) okenite with laumontite

PREHNITE

Crystal system Orthorhombic | **Composition** $Ca_2Al_2Si_3O_{10}(OH)_2$ | **Color** Green, yellow, tan, or white | **Form/habit** Tabular, botryoidal masses | **Hardness** 6–6½ | **Cleavage** Distinct basal | **Fracture** Uneven, brittle | **Luster** Vitreous | **Streak** White | **SG** 2.8–3.0 | **Transparency** Transparent to translucent | **RI** 1.61–1.67

Prehnite—calcium aluminum hydroxysilicate—is usually pale to mid-green, but it can also be tan, pale yellow, gray, or white. It often has an oily appearance. It is usually found as globular, spherical, or stalactitic aggregates of fine to coarse crystals and, rarely, in individual crystals. When it forms distinct crystals they show a square cross section with rounded faces. Prehnite is found lining cavities in volcanic rocks and in mineral

veins in granite. Crystals measuring several centimeters come from Canada. Prehnite is found in many localities worldwide. Although not a zeolite, it is associated with zeolite minerals such as apophyllite, stilbite, laumontite, and heulandite.

Green prehnite crystals with calcite in groundmass

SEPIOLITE

Crystal system Orthorhombic | **Composition** $Mg_4Si_6O_{15}(OH)_2.6H_2O$ | **Color** White, gray, pinkish | **Form/habit** Massive | **Hardness** 2–2½ | **Cleavage** Good but rarely seen | **Fracture** Uneven | **Luster** Dull to earthy | **Streak** White | **SG** 2.0–2.1 | **Transparency** Opaque

Sepiolite—magnesium silicate hydrate—is usually white or gray, and may be tinted yellow, brown, or green. It is compact, earthy, claylike, and often porous, and is usually found in nodular masses of interlocking fibers, which give it a toughness that belies its mineralogical softness. Sepiolite is an alteration product of rocks such as serpentinite and magnesite. It is best known by its popular name of meerschaum, from the German for "seafoam." It takes the name sepiolite from its resemblance to the light, porous bone of the cuttlefish Sepia. Eskisehir, Turkey, is the key commercial deposit.

Typical earthy luster

Massive sepiolite specimen

Silicates: inosilicates

PIGEONITE

Crystal system Monoclinic | **Composition** (Mg,Fe,Ca)₂(Si₂O₆) | **Color** Brown to black | **Form/habit** Granular | **Hardness** 6 | **Cleavage** Good | **Fracture** Uneven to conchoidal, brittle | **Luster** Vitreous | **Streak** White to pale brown | **SG** 3.2–3.5 | **Transparency** Semitransparent

Pigeonite (magnesium iron and calcium silicate) is brown, purplish brown, or greenish brown to black in color. An iron-rich variety is sometimes called ferropigeonite. Well-formed crystals are relatively rare; it is generally found as rockbound grains. Pigeonite is found in many lavas and smaller intrusive rock bodies, where it is the dominant pyroxene, and it is an important component of andesites and dolerites. It is also found in meteorites and the Mare, the large, dark, relatively flat areas of the Moon once believed to be seas, which are basalts containing pigeonite. Pigeonite is named for the type locality, Pigeon Point, Minnesota, US. It is widespread, with notable localities in Greenland, Scotland, Canada, Tasmania, and the US.

Pigeonite from Kola Peninsula, Russia

ENSTATITE

Crystal system Orthorhombic | **Composition** Mg₂Si₂O₆ | **Color** Variable | **Form/habit** Massive | **Hardness** 5–6 | **Cleavage** Good | **Fracture** Uneven | **Luster** Vitreous | **Streak** Gray | **SG** 3.2–3.3 | **Transparency** Opaque | **RI** 1.65–1.68

Enstatite—magnesium silicate—is colorless, pale yellow, or pale green, and becomes darker with increasing iron content, turning greenish brown to black. Bronzite is an intermediate variety of enstatite, and is brown with a metallic luster. Well-formed crystals are uncommon but when found, they tend to be short prisms, often with complex end faces. Enstatite generally occurs as rockbound grains, or in massive aggregates. It is also sometimes found as fibrous masses of parallel, needlelike crystals. Enstatite is a very widespread mineral, commonly occurring in iron- and magnesium-rich igneous rocks, and in meteorites. Star-enstatite and iridescent enstatite are gemstone varieties, found in Mysore, India, and Canada respectively.

Massive enstatite from Telemark, Norway

AEGIRINE

Crystal system Monoclinic | **Composition** NaFe(Si₂O₆) | **Color** Dark green, red brown, or black | **Form/habit** Prismatic | **Hardness** 6 | **Cleavage** Good to perfect | **Fracture** Uneven | **Luster** Vitreous | **Streak** Yellow green to pale green | **SG** 3.5–3.6 | **Transparency** Translucent to opaque

Aegirine—sodium iron silicate—forms a chemical interchange series with hedenbergite and diopside. Its crystals are prisms, and often grooved along the length, with steep or blunt end faces. The prism faces are often lustrous, while the end faces tend to be etched and dull. Crystals can be needlelike or fibrous, forming attractive radiating sprays. Aegirine is found worldwide in iron- and magnesium-rich igneous rocks, especially in syenites and syenitic pegmatites. It is also found in metamorphosed iron-rich sediments, and in schists and in some metamorphic rocks. Aegirine was first found in Norway, a notable locality.

Prismatic aegirine crystal in rock groundmass

HEDENBERGITE

Crystal system Monoclinic | **Composition** CaFe(Si₂O₆) | **Color** Pale to dark green or brownish green to greenish black | **Form/habit** Equant to prismatic | **Hardness** 6 | **Cleavage** Distinct in two directions at almost right angles | **Fracture** Uneven to conchoidal | **Luster** Vitreous to resinous | **Streak** Pale green to tan | **SG** 3.5–3.6 | **Transparency** Translucent to nearly opaque

Hedenbergite—calcium iron silicate—forms a complete chemical exchange series with diopside, in which magnesium completely replaces iron. Its color varies depending on how much iron is present. Its crystals are prisms or blocky but more commonly it is found as masses, as blades, or as platy aggregates. Hedenbergite is common in metamorphosed silica-rich limestones and dolomites, in heat-metamorphosed iron-rich sediments, and in some igneous rocks. It is also commonly found in chondrodite meteorites—stony meteorites that have experienced very little alteration since the formation of the solar system 4.56 billion years ago. Hedenbergite is found around the world.

Masses of bladed hedenbergite crystals

AUGITE

Crystal system Monoclinic | **Composition** (Ca,Na)(Mg,Fe,Tl,Al)(Al,Si)₂O₆ | **Color** Greenish black to black, dark green, brown | **Form/habit** Short prismatic | **Hardness** 5½–6 | **Cleavage** Distinct in two directions at almost right angles | **Fracture** Uneven to subconchoidal | **Luster** Vitreous to dull | **Streak** Pale brown to greenish gray | **SG** 3.3 | **Transparency** Translucent to nearly opaque

Augite is a silicate of calcium, magnesium, iron, titanium, and aluminum. It occurs usually as short, thick, prisms with a square or octagonal cross section, and as large masses. It is a member of a chemical-exchange group, along with diopside and hedenbergite. Augite is common in basalts, gabbros, andesites, and various other dark-colored igneous rocks as well as in intermediate rocks, such as andesite. It is a constituent of lunar basalts and some meteorites. It is hard to tell augite, diopside, and hedenbergite apart in hand specimens.

Augite in rock groundmass

SPODUMENE

Crystal system Monoclinic | **Composition** LiAl(Si₂O₆) | **Color** Gray, white to green, pink, colorless | **Form/habit** Prismatic | **Hardness** 6½–7 | **Cleavage** Perfect | **Fracture** Subconchoidal to splintery | **Luster** Vitreous | **Streak** White | **SG** 3.0–3.2 | **Transparency** Transparent to translucent | **RI** 1.66–1.68

The usual color of spodumene (lithium aluminosilicate) is ash gray, but its gem variety kunzite is pink or lilac, and hiddenite is emerald green. Crystals are flattened prisms and usually grooved along the length. Crystal faces are often etched and pitted with triangular markings. One of the largest single crystals of any mineral ever found was a spodumene 47 ft (14.3 m) long, weighing 99 tons (90 tonnes), from South Dakota, US. Spodumene is typically found in lithium-bearing granite pegmatites, often with other lithium-bearing minerals such as lepidolite. Common spodumene is an important source of lithium and is used for making ceramics as well as cell-phone and car batteries, medicine, and as a fluxing agent. Spodume is found in Brazil, the US, Canada, Russia, Mexico, Sweden, Western Australia, Afghanistan, and Pakistan. It is named from the Greek for "ash," an allusion to its usual color.

JADEITE

Crystal system Monoclinic | **Composition** $Na(Al,Fe)Si_2O_6$ | **Color** White, green, lilac, pink, brown, orange, yellow, red, blue, or black | **Form/habit** Massive, crystals rare | **Hardness** 6–7 | **Cleavage** Good | **Fracture** Splintery | **Luster** Vitreous to greasy | **Streak** White | **SG** 3.3–3.4 | **Transparency** Transparent to translucent | **RI** 1.64–1.68

There are two different minerals that are called "jade": jadeite and nephrite. Jadeite is sodium iron aluminum silicate, and a member of the pyroxene mineral group. Pure jadeite is white. Its other colors include green, colored by iron; and lilac, from impurities of manganese and iron. Highly valued emerald-green jadeite, colored by chromium, is called imperial jade. Jadeite is made of interlocking, blocky, granular crystals and commonly has a sugary or granular texture. Distinct crystals are rare, and are usually found in hollows within massive material as short prisms. Jadeite most often occurs in high-pressure metamorphic rocks.

Green color from chromium

Rough jadeite

ASTROPHYLLITE

Crystal system Triclinic | **Composition** $(K,Na)_3(Fe,Mn)_7Ti_2(Si_8O_{24})(O,OH)_7$ | **Color** Yellow brown | **Form/habit** Bladed, stellated | **Hardness** 3 | **Cleavage** Perfect | **Fracture** Uneven, brittle | **Luster** Submetallic to pearly | **Streak** Yellow | **SG** 3.3–3.4 | **Transparency** Translucent in thin laminae

Astrophyllite is hydrous potassium iron titanium silicate. It is golden yellow to dark brown. A rare mineral, it is usually found as

Green chrome diopside in a layer of the igneous rock pyroxenite

bladed crystals radiating from a common center (a form known as stellated). When it breaks, it forms brittle, thin leaves. It therefore derives its name from the Greek *astron*, meaning "star," and *phyllon*, "leaf." It forms in cavities in igneous rocks, such as granites, and especially in syenites and syenite pegmatites. Astrophyllite is also found in gneisses. It is usually associated with albite, aegirine, arfvedsonite, zircon, nepheline, titanite, and other minerals. Specimens come from Norway, where it was first discovered, Spain, Russia, Tajikistan, Egypt, Guinea, Canada, South Africa, and the US.

Radiating, bladed astrophyllite crystals

△ DIOPSIDE

Crystal system Monoclinic | **Composition** $CaMg(Si_2O_6)$ | **Color** White, green, violet blue | **Form/habit** Equant to prismatic | **Hardness** 5–6 | **Cleavage** Distinct in two directions at almost right angles | **Fracture** Uneven | **Luster** Vitreous | **Streak** White to green | **SG** 3.2–3.3 | **Transparency** Transparent to translucent | **RI** 1.66–1.73

Diopside (calcium magnesium silicate) can be colorless, but is more often green. When bright green, colored by chromium, it is known as chrome diopside. Violet-blue crystals, colored by manganese, are found in Italy. Crystals are prisms, often nearly square in section, or blocky. Diopside is found in granular or massive aggregates too. It forms in metamorphosed silica-rich limestones and dolomites and in iron-rich metamorphic rocks. More rarely, it occurs in igneous peridotites and kimberlites.

WOLLASTONITE

Crystal system Triclinic | **Composition** $CaSiO_3$ | **Color** White, gray, green | **Form/habit** Tabular, massive | **Hardness** 4½–5 | **Cleavage** Perfect | **Fracture** Uneven to splintery | **Luster** Vitreous to silky | **Streak** White | **SG** 2.9 | **Transparency** Translucent in thin laminae

Wollastonite—calcium silicate—is white, gray, or pale green in color. Crystals are rare and blocky; it is usually found

in crystalline masses, coarse blades, or leafy or fibrous masses. It has six known structural variations, which cannot be distinguished in hand specimens. Wollastonite occurs as a result of the metamorphism of limestones, and is formed in igneous rocks. It appears in metamorphosed rocks, too, in slates, phyllites, and schists, often accompanied by other calcium-containing silicates such as diopside, tremolite, grossular garnet, and epidote. Wollastonite is used in industry for its electrical- and heat-insulating properties. Its many localities include Utah, Michigan, California, New York, and Arizona, US.

Splintery structure

Coarse-bladed, parallel wollastonite crystals

Grooved, prismatic crystals of riebeckite in Alinci, Macedonia

△ RIEBECKITE

Crystal system Monoclinic | **Composition**
$Na_2(Fe^{2+}_3Fe^{3+}_2)Si_8O_{22}(OH)_2$ | **Color** Dark blue, black
Form/habit Prismatic | **Hardness** 5–5½
Cleavage Perfect | **Fracture** Uneven | **Luster**
Vitreous, silky | **Streak** Blue gray | **SG** 3.3–3.4
Transparency Transparent to translucent
RI 1.68–1.70

Riebeckite is a sodium iron silicate. It is
generally grayish blue to dark blue, its color
depending on the concentration of iron in
the structure. It forms prismatic, grooved
crystals and is also found as masses or
fibrous aggregates. It is one of several
minerals called asbestos. An asbestos
variety of riebeckite is named crocidolite,
and is commonly called blue asbestos.
Riebeckite is found in feldspar- and
quartz-rich igneous rocks such as granites
and syenites, and in feldspar and
quartz-rich volcanics, especially sodium-
rich rhyolites. Crocidolite is of metamorphic
origin and is formed under moderate
temperature and pressure from ironstones.
It occurs at Robertstown, South Australia;
Henan Province in China; Cochabamba,
Bolivia; and South Africa. Riebeckite

granite, known locally as ailsite, is found
on the island of Ailsa Craig in western
Scotland, and is prized for its use in the
manufacture of curling stones.

PECTOLITE

Crystal system Triclinic | **Composition**
$NaCa_2(Si_3O_8)(OH)$ | **Color** White, tan, blue
Form/habit Acicular | **Hardness** 4½–5
Cleavage Perfect | **Fracture** Uneven, brittle
Luster Vitreous to silky | **Streak** White
SG 2.8–2.9 | **Transparency** Translucent

Pectolite is sodium calcium silicate
hydroxide. It can be white, pale tan, or
pale blue. Its crystals are elongated
and flattened, but it more often occurs
as needlelike sprays or radially fibrous
masses. Pectolite is a hydrothermal
mineral found in cavities in basalts and
andesites associated with zeolites,
and in mica peridotites. Some specimens
are triboluminescent—they give off light
with friction. It name is derived from the
Greek *pektos*, meaning "congealed" or
"well put together." Pectolite occurs in

Canada, England, Italy, the Dominican
Republic, Greenland, Russia, Japan, and
the US. There is a blue gemstone variety
of pectolite called larimar.

Larimar is the trade name for gem pectolite

RHODONITE

Crystal system Triclinic | **Composition** $(MnCa)_5$
(Si_5O_{15}) | **Color** Pink | **Form** Tabular | **Hardness**
6 | **Cleavage** Perfect | **Fracture** Conchoidal
Luster Vitreous | **Streak** White | **SG** 3.5–3.7
Transparency Translucent | **RI** 1.71–1.73

Rhodonite—manganese silicate—is pink
to red in color. It takes its name from the
Greek *rhodon*, meaning "rose," from its

typical color. Rhodonite is occasionally
found as rounded crystals, some of which
are transparent. It is more commonly
found as masses, or granular aggregates,
and is often coated or veined with black
manganese oxides. Rhodonite is found
in various manganese ores, often with
rhodochrosite or as a metamorphic
product of rhodohrosite. It is a relatively
widespread mineral, but gemstone
sources of rhodonite are in Canada,
the US, Russia, and Japan. Crystals
come from Australia, Peru, and Brazil.
Occasionally used as a manganese ore,
it is more often mined as a gem.

Rhodonite crystals in rock groundmass

HORNBLENDE

Crystal system Monoclinic | **Composition** eg:
$Ca_2(Fe^2,Mg)_4(Al,Fe^3)(Si,Al)O_{22}(OH,F)_2$ | **Color** Green,
black | **Form/habit** Prismatic | **Hardness** 5–6
Cleavage Perfect | **Fracture** Uneven, brittle
Luster Vitreous | **Streak** White to gray | **SG**
2.9–3.4 | **Transparency** Translucent to opaque

Hornblende is calcium iron magnesium
aluminosilicate with fluorine and hydroxyl.
The name hornblende is applied to what
are now recognized to be two minerals,
but only detailed chemical analysis makes
it possible to tell them apart. Both are
rich in calcium and monoclinic in crystal
structure. They are ferrohornblende, rich
in iron, and magnesiohornblende, in which
magnesium substitutes for iron in the
chemical composition. To the collector
or the geologist, they are both simply
known as hornblende. Its colors are green,
dark green, and brownish green to black.
Crystals are most commonly bladed, and
often show a pseudohexagonal cross
section. Well-formed crystals are short
to long prisms. The mineral also occurs
in masses, and as radiating groups.
Hornblende occurs in metamorphic rocks,
and iron and magnesium-rich igneous
rocks. Well-formed crystals occur in
Norway, Canada, and the US.

TREMOLITE

Crystal system Monoclinic | **Composition** $Ca_2(Mg,Fe^2)_5Si_8O_{22}(OH)_2$ | **Color** Colorless, white, gray green to green black | **Form/habit** Bladed **Hardness** 5–6 | **Cleavage** Perfect | **Fracture** Splintery, brittle | **Luster** Vitreous to silky | **Streak** White | **SG** 2.9–3.1 | **Transparency** Transparent to translucent

Tremolite is calcium magnesium silicate, and is known as nephrite jade when massive and fine grained. Its color varies with increasing iron content from colorless to white for pure tremolite, to nearly black, and traces of manganese may tint it pink or violet. Its crystals are short to long prisms when well formed, but are more commonly found in bladed crystals in parallel aggregates with poorly developed end faces. It is also found in radiating groups. This amphibole is the product of heat and area metamorphism and is an indicator of metamorphic grade.

Featherlike aggregates of tremolite crystals

ANTHOPHYLLITE

Crystal system Orthorhombic | **Composition** $(Mg,Fe)_7Si_8O_{22}(OH)_2$ | **Color** Brown, pale green, gray, white | **Form/habit** Massive, fibrous | **Hardness** 5½–6 | **Cleavage** Perfect, imperfect | **Fracture** Uneven | **Luster** Vitreous | **Streak** Colorless to gray | **SG** 2.8–3.2 | **Transparency** Transparent to nearly opaque

Anthophyllite is magnesium, iron, manganese silicate. The iron, sodium, and magnesium content is variable in its structure. Iron-rich anthophyllite is called ferroanthophyllite; when sodium is present, it becomes sodium-anthophyllite, and with magnesium dominating it becomes magnesioanthophyllite. Anthophyllite crystals are uncommon, and are prisms with no end faces. Anthophyllite is commonly produced by metamorphism of iron and magnesium-rich rocks, especially of silica-poor igneous rocks. It is an important component of some

crystalline schists and gneisses, and is found worldwide. Fibrous anthophyllite has been used as asbestos, with no industrial differentiation between the various forms.

Fibrous, radiating crystals of anthophyllite

NEPHRITE

Crystal system Monoclinic | **Composition** $Ca_2(Mg,Fe)_5(Si_8O_{22})(OH)_2$ | **Color** Cream, green **Form/habit** Massive | **Hardness** 6½ | **Cleavage** Perfect | **Fracture** Splintery, brittle | **Luster** Dull to waxy | **Streak** White | **SG** 2.8–3.1 **Transparency** Translucent to nearly opaque **RI** 1.61–1.63

Nephrite is one of two lapidary materials called jade, the other being jadeite. Nephrite is not a mineral name, but the label applied to the tough, compact form of either tremolite or actinolite. Both are calcium magnesium silicate hydroxides, and structurally identical, except that in actinolite some of the magnesium is replaced by iron. The color varies from dark green when iron rich to cream when magnesium rich. White nephrite is pure tremolite, sometimes called "mutton-fat" jade. Nephrite is composed of a mat of tightly interlocking fibers, and is formed in metamorphic environments, especially metamorphosed iron- and magnesium-rich rocks or in areas where dolomites have been intruded by iron- and magnesium-rich igneous rocks. Large deposits of nephrite are found in Alaska, US, among many other localities.

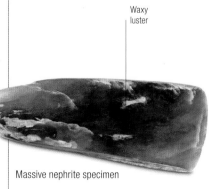

Waxy luster

Massive nephrite specimen

RICHTERITE

Crystal system Monoclinic | **Composition** $Na(Ca,Na)Mg_5Si_8O_{22}(OH)_2$ | **Color** Brown, yellow, red, or green | **Form/habit** Prismatic **Hardness** 5–6 | **Cleavage** Perfect | **Fracture** Uneven, brittle | **Luster** Vitreous | **Streak** Pale yellow | **SG** 3.0–3.5 | **Transparency** Transparent to translucent

Richterite is sodium calcium magnesium silicate. Colors range from brown, grayish brown, yellow, brownish to rose red, or pale to dark green. Its crystals are long prisms, fibrous aggregates, or rockbound crystals. When iron replaces the magnesium in the structure, the mineral is called ferrorichterite; if fluorine replaces the hydroxyl, it is fluorrichterite. Richterite occurs in heat-metamorphosed limestones. It also occurs as a hydrothermal product in magnesium- and iron-rich igneous rocks, and in manganese-rich ore deposits. Localities include Canada, Sweden, Western Australia, Myanmar, and the US. It was named in 1865 after the German mineralogist Theodore Richter.

Brown, prismatic richterite crystals in quartz

GLAUCOPHANE

Crystal system Monoclinic | **Composition** $Na_2(Mg_3Al_2)Si_8O_{22}(OH)_2$ | **Color** Bluish gray, black **Form/habit** Slender prismatic | **Hardness** 5½–6½ **Cleavage** Distinct | **Fracture** Uneven to conchoidal **Luster** Vitreous to pearly | **Streak** Gray blue | **SG** 3–3.3 | **Transparency** Transparent to translucent

Glaucophane is sodium magnesium aluminum silicate. When iron replaces the magnesium it is called ferroglaucophane. It is gray, lavender blue, or bluish black. Its crystals are slender prisms, often lathlike, and grooved lengthwise. It is also found as masses, fibers, and granular aggregates. Glaucophane occurs in schists formed by the low-temperature, high-pressure metamorphism of sodium-rich sediments, or by the introduction of sodium into the metamorphic process, and is an important

indicator of metamorphic grade. Glaucophane and its associated minerals are known as the glaucophane metamorphic facies. It is one of the minerals that fall under the general term of asbestos. Glaucophane is widespread, with Colorado, US, among its localities.

Glaucophane crystals in rock groundmass

ARFVEDSONITE

Crystal system Monoclinic | **Composition** $Na_3(Fe^{2+}_4Fe^{3+})Si_8O_{22}(OH)_2$ | **Color** Blue black **Form/habit** Massive aggregates | **Hardness** 5½–6 | **Cleavage** Perfect | **Fracture** Uneven **Luster** Dull to vitreous | **Streak** Blue gray **SG** 3–3.5 | **Transparency** Translucent

Arfvedsonite—hydrous sodium iron silicate—is black, greenish black to bluish, and deep green on thin edges. It forms well-developed crystals less often than the other amphiboles; it is generally found as massive aggregates or fibrous to radiating masses. When crystals do occur, they are short to long prisms or blocky. It occurs in feldspar-rich igneous rocks and their associated pegmatites, often with aegirine and augite. It is also found in regional metamorphics. Crystals occur in the US, Russia, Norway, and Canada. Arfvedsonite was discovered in 1823 and named for the Swedish chemist Johan Arfvedson.

Arfvedsonite crystal fragments in groundmass

Silicates: cyclosilicates

DIOPTASE

Crystal system Trigonal | **Composition** $CuSiO_2(OH_2)$ | **Color** Emerald to blue green | **Form/habit** Prismatic | **Hardness** 5 | **Cleavage** Perfect | **Fracture** Uneven to conchoidal | **Luster** Vitreous to greasy | **Streak** Pale greenish blue | **SG** 3.3 | **Transparency** Transparent to translucent

The crystals of dioptase (copper silicate) are prisms, often with rhombohedron end faces. Dioptase is also found as masses or granular aggregates. It forms where copper sulfide veins have been altered by oxidation in a complex process, principally in desert environments. Its bright green crystals can superficially resemble emerald, and were originally mistaken for it. Its vibrant color and its tendency to occur as well-formed crystals make dioptase highly popular with mineral collectors. Its crystals can be highly transparent, yielding its name, from the Greek *dia*, "through," and *optazein*, "to see." Superb specimens come from Kazakhstan, Iran, Namibia, Congo, Argentina, Chile, and the US.

Dioptase crystals in groundmass

▽ CORDIERITE

Crystal system Orthorhombic | **Composition** $(Mg,Fe)_2Al_4Si_5O_{18}$ | **Color** Blue, blue green, gray violet | **Form/habit** Short prismatic, granular | **Hardness** 7–7½ | **Cleavage** Moderate to poor | **Fracture** Conchoidal to uneven | **Luster** Vitreous to greasy | **Streak** White | **SG** 2.5–2.6 | **Transparency** Transparent to translucent | **RI** 1.54–1.56

Cordierite is magnesium iron aluminum silicate. Gem-quality blue cordierite is known as iolite, and because of its color is also called "water sapphire." Cordierite can be blue, violet blue, gray, or blue green. Its crystals are prismatic. It occurs in high-grade, heat-metamorphosed, aluminum-rich rocks. It is also found in schists and gneisses, and more rarely in granites, pegmatites and quartz veins. Aside from its use as a gemstone, cordierite is an important industrial mineral in the production of the ceramics used in catalytic converters. Gem material comes from Sri Lanka, Myanmar, Madagascar, Tanzania, and South Africa, and there is a major source of iolite near Madras, India. Flawless crystals are found on Garnet Island, Northwest Territories, Canada.

BENITOITE

Crystal system Hexagonal | **Composition** $BaTiSi_3O_9$ | **Color** Blue, colorless, pink | **Form/habit** Platy | **Hardness** 6½ | **Cleavage** Imperfect | **Fracture** Conchoidal to uneven | **Luster** Vitreous | **Streak** White | **SG** 3.7 | **Transparency** Transparent to translucent | **RI** 1.76–1.80

An extremely rare barium titanium silicate, benitoite is generally blue to dark blue, although it can be colorless, white, or, rarely, pink. Benitoite has a rare five-pointed crystal form, and an even rarer six-pointed "star of David" form. It is found in hydrothermally altered serpentine and in veins in schist. Benitoite typically occurs with an unusual set of minerals such as natrolite, serpentine, and albite. Its principal gem source is the San Benito Mine, California, US, where it was first discovered. In 1985, benitoite was named as the official state gem of California. Gem-quality crystals tend to be small, with cut stones rarely more than 5 carats. Benitoite is also found in Niigata Prefecture, Japan; Esneux, Belgium; and Magnet Cove, Arkansas, US.

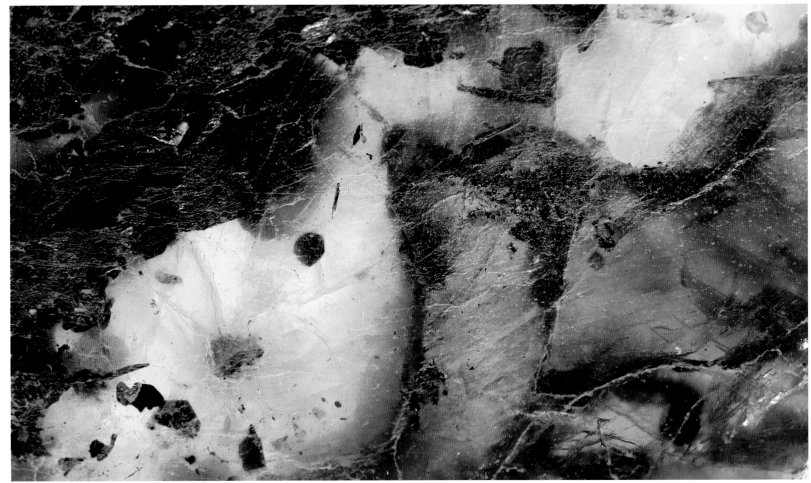

Violet-blue crystals of iolite—gem-quality cordierite

SUGILITE

Crystal system Hexagonal | **Composition** $KNa_2(Fe,Mn,Al)_2Li_3Si_{12}O_{30}.H_2O$ | **Color** Pink, brown yellow, or purple | **Form/habit** Massive **Hardness** $5\frac{1}{2}$–$6\frac{1}{2}$ | **Cleavage** Poor | **Fracture** Subconchoidal | **Luster** Vitreous | **Streak** White **SG** 2.7–2.8 | **Transparency** Translucent to opaque **RI** 1.60–1.61

Sugilite is sodium potassium lithium silicate hydrate. It also contains variable amounts of iron, manganese, and aluminum. It is pale to deep pink, brownish yellow, or purple. The pink-to-purple coloration is caused by manganese; the pink material is aluminum rich, and the purple is iron rich. It is usually in masses or granular aggregates, but when crystals do occur, they are small prisms (less than ¾ in (2 cm) across. Sugilite forms in metamorphosed manganese deposits and as fragments enclosed in marble. It was discovered in 1944 but was only recognized as a mineral in 1976. It is named for Japanese petrologist Ken-ici Sugi, its co-discoverer— one of its localities is Iwagi Island, Japan.

Massive gem-quality sugilite from South Africa

TOURMALINE

Crystal system Trigonal | **Composition** $Na(Mg,Fe,Li,Mn,Al)_3Al_6(BO_3)_3Si_6.O_{18}(OH,F)_4$ **Color** Black, green, brown, red, blue, yellow, or pink **Form/habit** Prismatic, acicular | **Hardness** 7–$7\frac{1}{2}$ **Cleavage** Indistinct | **Fracture** Uneven to conchoidal | **Luster** Vitreous | **Streak** Colorless **SG** 3.0–3.2 | **Transparency** Transparent to translucent | **RI** 1.62–1.65

Tourmaline is the name given to a family of sodium iron lithium magnesium manganese fluoridated hydrated aluminum borosilicate minerals. Their composition is complex and variable but all have the same basic crystal structure. There are over 30 species in the tourmaline group, including elbaite, dravite, schorl, and liddicoatite. Elbaite forms a chemical interchange series with dravite, and dravite forms a chemical interchange series with elbaite and with schorl. Schorl is black and opaque; dravite is usually brown. Elbaite provides the most gemstone material, along with minor amounts of liddicoatite. They are usually green, but they can also be yellow green, pink or red, or blue. Color zoning is common. Tourmaline is abundant, and its best-formed crystals are usually found in pegmatites and in metamorphosed limestones in contact with granitic magmas. Aside from gems, tourmaline is an important industrial mineral, used in pressure devices such as depth-sounding equipment.

BERYL

Crystal system Hexagonal | **Composition** $Be_3Al_2Si_6O_{18}$ | **Color** Green, blue, yellowish | **Form/habit** Prismatic | **Hardness** $7\frac{1}{2}$–8 | **Cleavage** Indistinct | **Fracture** Uneven to conchoidal **Luster** Vitreous | **Streak** White | **SG** 2.6–2.8 **Transparency** Transparent to translucent **RI** 1.56–1.60

Beryl is beryllium aluminum silicate. Pure beryl is colorless; the colors in beryl are caused by minute chemical impurities. Its crystals normally occur as hexagonal prisms. Before 1925, its solitary use was as a gemstone, but from then on many important uses were found for beryllium metal. Since then, common beryl has been widely sought as a source of this rare element. It has several gemstone varieties: emerald (green), aquamarine (blue and blue green), heliodor (golden yellow), morganite (pink), and goshenite (colorless). Beryl of various colors is most commonly found in granitic pegmatites, but it also occurs in mica schists, gneisses, and in limestone. Much beryl production is a by-product of the mining of feldspar and mica, and no large deposits have been found. Localities include Columbia, the US, Brazil, Russia, Mozambique, Italy, Zimbabwe, Pakistan, Zambia, and Madagascar.

Prismatic crystal of heliodor (golden beryl)

Silicates: sorosilicates

HEMIMORPHITE

Crystal system Orthorhombic | **Composition** $Zn_4Si_2O_7(OH)_2.H_2O$ | **Color** Colorless, white, yellow, blue, or green | **Form/habit** Asymmetric, prismatic, tabular, botryoidal | **Hardness** $4\frac{1}{2}$–5 | **Cleavage** Perfect, good, poor | **Fracture** Uneven, brittle **Luster** Vitreous | **Streak** White | **SG** 3.4–3.5 **Transparency** Transparent to translucent

The crystals of hemimorphite (zinc silicate are double-ended prisms with a differently shaped face at each end—pointed at one and flat at the other—a property shown by few other minerals. Crystals are often grouped in fan-shaped clusters. Hemimorphite can also be found in masses, fibers, in granular masses, or in encrustations. It is a mineral formed in the alteration zone of zinc deposits, especially as an alteration product of sphalerite. Well-crystallized specimens come from Algeria, Namibia, Germany, Mexico, Spain, and the US.

Hemimorphite crystals in rounded aggreagates

DANBURITE

Crystal system Orthorhombic | **Composition** $CaB_2Si_2O_8$ | **Color** Colorless, yellow | **Form/habit** Prismatic | **Hardness** 7–$7\frac{1}{2}$ | **Cleavage** Indistinct **Fracture** Subconchoidal to uneven | **Luster** Vitreous to greasy | **Streak** White | **SG** 3.0 **Transparency** Transparent to translucent **RI** 1.63–1.64

Danburite—calcium borosilicate—is colorless, amber, straw yellow, gray, pink, or yellow brown. Its crystals are glassy prisms with wedge-shaped end faces, and they resemble topaz. The Dana classification of minerals categorizes danburite as a sorosilicate, while the Strunz classification system lists it as a tectosilicate, but its structure fits either subgroup. Danburite is generally a moderate- to low-temperature metamorphic mineral, but it is also found in ore deposits formed at relatively high temperatures as well as in pegmatites. It is named for Danbury, Connecticut, US, where it was first discovered in 1839. Localities include Switzerland, Myanmar, Mexico, and Russia.

AXINITE

Crystal system Triclinic | **Composition** $Ca_2FeAl_2(BSi_4O_{15})(OH)$ | **Color** Clove brown, gray to bluish gray, pink, violet, yellow, orange, or red **Form/habit** Axe-shaped crystals | **Hardness** $6\frac{1}{2}$–7 | **Cleavage** Good, poor | **Fracture** Uneven to conchoidal, brittle | **Luster** Vitreous **Streak** Colorless to light brown | **SG** 3.2–3.3 **Transparency** Transparent to translucent **RI** 1.67–1.70

Axinite is a group of four minerals, all hydrous calcium iron borosilicates. Iron-rich ferroaxinite is the most common; in magnesioaxinite, magnesium replaces the iron in ferroaxinite; in manganaxinite, manganese replaces the iron in ferroaxinite; and tinzenite is intermediate in composition between ferroaxinite and manganaxinite. Axinite is usually clove brown. It occurs as flattened, ax-head-shaped crystals, but it is also found as rosettes, and in masses and granular aggregates. This group takes its name from the ax-head shape of its crystals. It is found worldwide; gem axinite comes from the US, Russia, and Australia.

Transparent, tabular axinite crystals

ILVAITE

Crystal system Monoclinic or orthorhombic
Composition $CaFe_3OSi_2O_7(OH)$ | **Color** Black to grayish black | **Form/habit** Short to long prismatic
Hardness 5½–6 | **Cleavage** Distinct, indistinct
Fracture Uneven, brittle | **Luster** Submetallic to dull | **Streak** Greenish to brownish black | **SG** 3.8–4.1 | **Transparency** Opaque

Ilvaite—hydrous iron and calcium silicate—can also have manganese and magnesium in its structure. It is black to brownish black to gray and opaque. Ilvaite frequently occurs in short to long prisms, often vertically grooved. It is also coarsely crystalline, and in masses, or granular aggregates. Ilvaite is found in contact metamorphic zones with zinc, copper, and iron ores. It is occasionally found in syenites. Large crystals come from the island of Elba, Italy, and from Russia, Germany, Japan, and the US. Ilvaite was discovered in 1811 on Elba and is called after the Latin name *Ilva* of the island.

Parallel, prismatic ilvaite crystals

VESUVIANITE

Crystal system Tetragonal or monoclinic
Composition $Ca_{10}(Mg,Fe)_2Al_4(SiO_4)_5(Si_2O_7)_2(OH,F)_4$
Color Green, yellow | **Form/habit** Prismatic
Hardness 6½ | **Cleavage** Poor | **Fracture** Subconchoidal to uneven, brittle | **Luster** Vitreous to resinous | **Streak** White to pale greenish brown
SG 3.4 | **Transparency** Transparent to translucent
RI 1.70–1.72

Formerly called idocrase, vesuvianite is fluoridated hydrated calcium magnesium iron aluminosilicate. Numerous elements may substitute in the structure, including tin, lead, manganese, chromium, zinc, and sulfur. Vesuvianite is usually green or chartreuse, but it can be yellow, brown, red, black, blue, or purple. Its crystals are pyramids or prisms and glassy. Crystals up to 4 in (9 cm) long have been found. Vesuvianite forms by the metamorphism

of impure limestones, and it is found in marbles and granulites, often with garnet, wollastonite, diopside, and calcite. A greenblue, copper-bearing vesuvianite is called cyprine.

High-quality prismatic vesuvianite crystals

ZOISITE

Crystal system Orthorhombic | **Composition** $Ca_2Al_3(SiO_4)_3(OH)$ | **Color** Yellow green, white
Form/habit Prismatic | **Hardness** 6–7
Cleavage Perfect | **Fracture** Conchoidal to uneven, brittle | **Luster** Vitreous | **Streak** White
SG 3.2–3.4 | **Transparency** Transparent to translucent | **RI** 1.69–1.70

Most zoisite (calcium aluminum silicate hydroxide) is pale, but a dark blue variety is called tanzanite. Zoisite forms as deeply grooved prisms, disseminated grains, and as massive aggregates. It is characteristic of area metamorphism and of hydrothermal alteration of igneous rocks. It occurs in medium-grade schists, gneisses, and amphibolites resulting from metamorphism of calcium-rich rocks, and in quartz veins and pegmatites. Tanzania is the source of a zoisite amphibolite favored by collectors containing ruby crystals in vivid green chrome zoisite and black hornblende.

Crystal of tanzanite, a variety of zoisite

CLINOZOISITE

Crystal system Monoclinic | **Composition** $Ca_2Al_3(SiO_4)_3(OH)$ | **Color** Gray, yellow, rose
Form/habit Prismatic | **Hardness** 6½
Cleavage Perfect | **Fracture** Uneven, brittle
Luster Vitreous | **Streak** White | **SG** 3.2–3.4
Transparency Transparent to translucent

Clinozoisite—calcium aluminum silicate hydroxide—is the monoclinic equivalent of zoisite. It is a member of the epidote group. Its crystals are elongated prisms, and it can also be found as granular, massive aggregates, and as fibers. It is colorless, yellowish green, yellowish gray, or rarely, rose to red. It is common in areas of metamorphosed rocks, in feldspar-rich igneous rocks, and as an alteration product of plagioclase feldspar. Localities include Canada, Mexico, Austria, Switzerland, Italy, and the US. Its name reflects its resemblance to zoisite and its monoclinic crystal structure. It was first discovered in 1896 in the East Tyrol region of Austria.

Clinozoisite rock with prismatic crystals

ALLANITE

Crystal system Monoclinic | **Composition** $(Ce,Ca,Y)_2(Al,Fe,Fe)_3(SiO_4)_4(OH)$ | **Color** Light brown to black, yellow, green | **Form/habit** Tabular to long prismatic | **Hardness** 6–6½ | **Cleavage** Imperfect | **Fracture** Conchoidal to uneven
Luster Resinous, greasy, or submetallic | **Streak** Light brown | **SG** 3.1–4.2 | **Transparency** Transparent to translucent

Allanite is a complex group of three hydrous silicate minerals: allanite-(Ce), allanite-(La) and allanite-(Y), depending on the dominant rare-earth element present: cerium, lanthanum or yttrium. These three can also contain iron, aluminum, calcium, manganese, strontium, barium, chromium, uranium, and zirconium. Allanite is often coated with a yellow-brown alteration product. Its crystals are generally blocky to long prisms. It can also be in granular aggregates or occur as embedded grains. Allanite is weakly radioactive.

PIEMONTITE

Crystal system Monoclinic | **Composition** $Ca_2(Al,Mn,Fe)_3(SiO_4)_3(OH)$ | **Color** Reddish brown, black, red | **Form/habit** Prismatic | **Hardness** 6–6½ | **Cleavage** Perfect | **Fracture** N/A
Luster Vitreous | **Streak** Cherry red | **SG** 3.4–3.5 | **Transparency** Translucent to opaque

Piemontite (also spelled piedmontite) is calcium manganese aluminosilicate hydrate. Its redness comes from manganese. Crystals are blocks or slender, elongated prisms. Piemontite is also found in massive aggregates. It is found in low-grade, area metamorphosed rocks, oxidized volcanic rocks, and as a hydrothermal alteration product in manganese deposits. Specimens come from Scotland, Pakistan, Italy, Japan, Egypt, New Zealand, and the US. It was named for the place of its discovery in Piedmont, Italy.

KORNERUPINE

Crystal system Orthorhombic | **Composition** $Mg_3Al_6(Si,Al,B)_5(O_{21}(OH)$ | **Color** Green, white, blue
Form/habit Long prismatic | **Hardness** 6½–7
Cleavage Distinct prismatic | **Fracture** N/A
Luster Vitreous | **Streak** White | **SG** 3.3–3.5
Transparency Transparent to translucent and opaque | **RI** 1.66–1.70

The usual color of kornerupine (hydrous magnesium iron aluminum borosilicate) is dark to sea green, but it can also be white, cream, colorless, blue, pink, or black. Its crystals form grooved prisms of 2 in (5 cm) and more, which can be mistaken for tourmaline. It may be found as radiating or fibrous masses. It occurs in silica-poor, aluminum-rich metamorphic rocks, often with cordierite, sillimanite, and corundum. It is a relatively rare mineral, with less than 60 known sources worldwide. Important localities include Greenland, Madagascar, and Sri Lanka. It is named for Danish geologist A. D. Kornerup.

Kornerupine crystals in rock groundmass

Epidote-rich bands alternate with other rocks in Avacha Bay, Kamchatka, Russia

△ **EPIDOTE**

Crystal system Monoclinic | **Composition** $Ca_2Al_2(Fe,Al)(SiO_4)(Si_2O_7)O(OH)$ | **Color** Light to dark pistachio green; also gray or yellow | **Form/habit** Short to long prismatic | **Hardness** 6–7 | **Cleavage** Good | **Fracture** Uneven to splintery | **Luster** Vitreous | **Streak** Colorless or grayish | **SG** 3.3–3.5 | **Transparency** Translucent | **RI** 1.73–1.77

Epidote is calcium aluminum iron silicate hydroxide. Its characteristic color is light to dark pistachio green, although it can also be fairly gray or yellow. It is strongly pleochroic (it appears to be different colors when viewed from different directions). Epidote is sometimes found in transparent crystals, which are cut for collectors. Rock that is made up primarily of epidote may be polished or tumbled, made into jewelry, and sold as unakite.

Tawmawlite is a bright green, chromium-rich variety. Epidote frequently forms fine crystals, which may be prisms or thick blocks with faces finely grooved parallel to the crystal's length. Crystals are often twinned. Epidote is also found as masses, and granular aggregates. It occurs widely in metamorphosed rocks, and as a product of hydrothermal alteration of plagioclase feldspar. It derives its name from the Greek *epidosis*, meaning "increase," because one side of its crystal prism is always longer than the others. Good crystals are found in Canada, France, Myanmar, Norway, Peru, and, in the US, Colorado, California, Connecticut, and Alaska. Large, dark-green tabular crystals occur in Prince of Wales Island in Alaska. Gem-quality crystals are found near Salzburg in Austria and in Pakistan and Brazil.

Silicates: nesosilicates

MONTICELLITE

Crystal system Orthorhombic | **Composition** CaMgSiO$_4$ | **Color** Colorless, white, brown | **Form/habit** Granular | **Hardness** 5½ | **Cleavage** Poor **Fracture** Conchoidal | **Luster** Vitreous to oily **Streak** White | **SG** 3.0 | **Transparency** Transparent to translucent

Monticellite is calcium magnesium silicate. It is colorless, whitish, pale greenish gray, or yellowish gray. Crystals are rare, but when found they are short prisms up to 2 in (5 cm) and often rounded. It is more often found as granular masses. Monticellite forms in contact-metamorphic zones in limestone, and is sometimes found in magnesium and iron-rich igneous rocks. It is associated with forsterite, magnetite, apatite, biotite, vesuvianite, and wollastonite. It was named in 1831 for the Italian mineralogist Teodoro Monticelli. Localities include Ontario, Canada; Nassau, Germany; Siberia, Russia; Hokkaido Prefecture, Japan; Tasmania, Australia; and several sources in the US.

Rough surface

Brown monticellite

ZIRCON

Crystal system Tetragonal | **Composition** ZrSiO$_4$ **Color** Colorless, brown, red, yellow, orange, blue, green | **Form/habit** Prismatic to dipyramidal **Hardness** 7½ | **Cleavage** Imperfect | **Fracture** Uneven to conchoidal | **Luster** Adamantine to oily **Streak** White | **SG** 4.6–4.7 | **Transparency** Transparent to opaque | **RI** 1.93–1.98

Zircon is zirconium silicate. It can be colorless, yellow, gray, green, brown, blue, and red. It forms prisms to double-pyramid crystals, which can reach a considerable size: examples weighing up to 5½ lb (2 kg) and 10 lb (4 kg) have

been found in Australia and Russia, respectively. Zircon is widespread as a minor constituent of silica-rich igneous rocks, and in metamorphic rocks. It is resistant to physical and chemical weathering, and because of its high specific gravity it concentrates in stream and river gravels, allowing it to survive in many types of rock. Zircons are thus ideal for dating rocks radiometrically.

Crystals in groundmass

Zircon crystals

TOPAZ

Crystal system Orthorhombic | **Composition** Al$_2$SiO$_4$(F,OH)$_2$ | **Color** Colorless, blue, yellow, pink, brown, green | **Form/habit** Prismatic **Hardness** 8 | **Cleavage** Perfect basal | **Fracture** Subconchoidal to uneven | **Luster** Vitreous **Streak** Colorless | **SG** 3.4–3.6 | **Transparency** Transparent to translucent | **RI** 1.62–1.63

Topaz is aluminum silicate fluoride hydroxide. Topaz is found in a wide range of colors, with the sherry yellow stones from Brazil being particularly valuable. Pink topaz is even more valuable, but natural pink stones are rare. Some blue topaz is almost indistinguishable from aquamarine with the naked eye. Its crystals are well-formed prisms and have a characteristic lozenge-shaped cross section and grooving parallel to their length. It is also found in masses and in granular aggregates. Topaz is formed by fluorine-bearing vapors given off during the last stages of the crystallization of various igneous rocks, and typically occurs in cavities in rhyolites, granites, pegmatites, and hydrothermal veins. The world's largest preserved single topaz crystal weighs 596 lb (271 kg). The name topaz is thought to derive from *tapaz*, the

Sanskrit word for "fire." Natural topaz is frequently heat-treated or irradiated to alter its color.

Topaz in pegmatite

TITANITE

Crystal system Monoclinic | **Composition** CaTiSiO$_5$ | **Color** Yellow, green, brown, black, pink, red, blue | **Form/habit** Wedge shaped or prismatic **Hardness** 5–5½ | **Cleavage** Imperfect **Fracture** Conchoidal | **Luster** Vitreous to greasy **Streak** White | **SG** 3.5–3.6 | **Transparency** Transparent to translucent | **RI** 1.84–2.03

Titanite is calcium titanium silicate, and it may also contain significant amounts of iron, and minor amounts of thorium and uranium. Crystals occur in yellow, green, or brown, and can also be black, pink, red, blue, or colorless. Its crystals are classically wedge shaped but are also found in prisms, and in masses or leaflike aggregates. It is one of the few stones with a color dispersion higher than that of diamond. Titanite is widely distributed as a minor component of silica-rich igneous rocks and associated pegmatites, and in metamorphic gneisses, marbles, and schists. Its former name sphene originates from the Greek *sphen*, meaning "wedge," in reference to its crystal shape.

Interpenetrating crystals

Wedge-shaped titanite crystals

STAUROLITE

Crystal system Monoclinic | **Composition** (Fe,Mg)$_4$Al$_{17}$(Si,Al)$_8$O$_{45}$(OH)$_3$$_2$ | **Color** Brown **Form/habit** Pseudo-orthorhombic prismatic **Hardness** 7–7½ | **Cleavage** Distinct | **Fracture** Conchoidal **Luster** Vitreous to resinous | **Streak** Colorless to gray | **SG** 3.7 | **Transparency** Transparent to opaque | **RI** 1.74–1.75

Staurolite is hydrous iron magnesium aluminosilicate. Reddish brown, yellowish brown, or nearly black, it normally occurs as hexagonal or diamond-shaped prisms, which often have rough surfaces. Staurolite is widespread, and occurs with garnet, tourmaline, and kyanite or sillimanite in mica schists and gneisses, and other area-metamorphosed aluminum-rich rocks. It forms under a very specific range of temperatures and pressures, and is useful in determining the conditions under which metamorphic rock formed. It is named from the Greek *stauros* for "cross" and *lithos* "stone," for its crosslike penetration twins.

Staurolite in mica schist

▷ OLIVINE

Crystal system Orthorhombic | **Composition** (Mg,Fe)$_2$SiO$_4$ | **Color** Green, yellow, white, brown, black | **Form/habit** Tabular, massive, granular **Hardness** 6½–7 | **Cleavage** Imperfect **Fracture** Conchoidal | **Luster** Vitreous | **Streak** White | **SG** 3.3–4.3 | **Transparency** Transparent to translucent | **RI** 1.64–1.69

Olivine is magnesium-iron silicate, and includes any mineral belonging to the magnesium-iron silicate chemical exchange series in which iron and magnesium substitute freely for each other. Crystals are blocky, often with wedge-shaped end faces, although well-formed crystals are rare. It is usually found in masses or granular aggregates. Peridot is the name of its gemstone variety, with a composition on the magnesium-rich end of olivine. Olivine is inferred to be a major component of the Earth's upper mantle, and as such is probably one of the most abundant mineral constituents of the planet. It has also been found in lunar rocks and in meteorites.

Olivine-rich sand gives Green Sand Beach on Big Island, Hawaii, its distinctive green coloring

Garnet-encrusted limestone formations on a rockface near Bearmouth Pass, Montana, US

△ GARNET

Crystal system Cubic | **Composition** Various
Color Various | **Form/habit** Dodecahedral,
trapezoidal | **Hardness** 6½–7½ | **Cleavage**
None | **Fracture** Various | **Luster** Vitreous
Streak White | **SG** 3.6–4.3 | **Transparency**
Transparent to translucent | **RI** 1.69–1.89

Garnet is a group of 15 garnet species.
They are all silicates containing various
proportions of calcium, iron, magnesium,
manganese, aluminum, chromium,
titanium, zirconium, or vanadium. There
is much chemical interchange between
many of the garnet species. Several
species make up the majority of all garnets:

spessartine (manganese aluminosilicate);
almandine (iron aluminosilicate); grossular
and hessonite (calcium aluminosilicate);
pyrope (magnesium aluminosilicate);
andradite (calcium iron silicate); and
uvarovite (calcium chromium silicate).
The name an individual garnet specimen
is given is based on the elements that
make up the largest percentage of its
composition. Colors vary from red, pink,
orange, green, emerald green, reddish
brown, white, colorless, honey, brownish
yellow, brownish red, or black. Garnets are
generally found as well-developed crystals
with cubic crystal forms: dodecahedrons,
icosahedrons and trapezohedrons, or
various combinations of these. Garnets

are widespread, and are particularly
abundant in metamorphic rocks, although
some are found in igneous rocks.

EUCLASE

Crystal system Monoclinic | **Composition**
$BeAlSiO_4(OH)$ | **Color** Colorless, white, blue, green
Form/habit Prismatic | **Hardness** 7½ | **Cleavage**
Perfect | **Fracture** Conchoidal, brittle | **Luster**
Vitreous | **Streak** White | **SG** 3.0 | **Transparency**
Transparent to translucent | **RI** 1.65–1.67

Euclase is beryllium aluminum hydroxide
silicate. It is generally white or colorless

but it can also be pale green or pale
to deep blue, a color for which it is
particularly noted. It forms grooved prisms,
often with complex end faces. It can also
be found in masses and fibers. It occurs
in low-temperature hydrothermal veins,
granitic pegmatites, as well as in some
metamorphic schists and phyllites. It
is also found in stream gravels that are
derived from these rocks. Euclase takes
its name from the Greek *eu*, "good," and
klasis, "fracture," in reference to the way
in which it breaks in perfect planes. Gem
euclase is found in Minas Gerais and
several other localities in Brazil, and in
Park County, Colorado, US.

ANDALUSITE

Crystal system Orthorhombic | **Composition** Al_2SiO_5 | **Color** Pink, brown, white, gray, violet, yellow, green, blue | **Form/habit** Prismatic **Hardness** 6½–7½ | **Cleavage** Good to perfect, poor | **Fracture** Conchoidal | **Luster** Vitreous **Streak** White | **SG** 3.2 | **Transparency** Transparent to nearly opaque | **RI** 1.63–1.64

Andalusite is aluminum silicate. It is pink to reddish brown, white, gray, violet, yellow, green, or blue. Its crystals are elongated and tapered prisms with a square cross section. A yellowish gray variety called chiastolite is found as long prisms enclosing symmetrical wedges of carbonaceous material which, in cross section, form a cross. Andalusite is found locally in low-grade metamorphic rocks, and in area metamorphic rocks, where it is associated with corundum, kyanite, cordierite, and sillimanite. It is also found rarely in granites and granitic pegmatites. It is named from the discovery locality in Andalusia, Spain; chiastolite is from the Greek *chiastos*, meaning "cross."

Vitreous luster

Prismatic andalusite crystals

SILLIMANITE

Crystal system Orthorhombic | **Composition** Al_2SiO_5 | **Color** Colorless, white, pale yellow, blue, green, violet | **Form/habit** Prismatic to acicular | **Hardness** 7 | **Cleavage** Perfect **Fracture** Uneven | **Luster** Silky | **Streak** White **SG** 3.2–3.3 | **Transparency** Transparent to translucent | **RI** 1.66–1.68

Sillimanite is aluminum silicate. Commonly colorless to white, it can also be pale yellow to brown, pale blue, green, or violet. It is found as long, slender, glassy crystals, or in blocky prisms with poor end-face development. A common metamorphic mineral, sillimanite is characteristic of high-temperature metamorphosed clay-rich rocks, where it is often found with corundum, kyanite, and cordierite.

It is often found in sillimanite schists and gneisses. Sillimanite is named for American chemist Benjamin Silliman. An important industrial mineral, it is used in heat-resistant ceramics and car spark plugs.

Fibrous sillimanite

KYANITE

Crystal system Triclinic | **Composition** Al_2SiO_5 **Color** Blue, green | **Form/habit** Bladed **Hardness** 4½; 6 | **Cleavage** Perfect | **Fracture** Splintery | **Luster** Vitreous | **Streak** Colorless **SG** 3.6 | **Transparency** Transparent to translucent | **RI** 1.71–1.73

Kyanite is aluminum silicate. It is usually blue and blue gray, generally mixed or zoned within a single crystal, but it can also be green, orange, or colorless. Kyanite occurs principally as elongated, flattened blades that are often bent, and less commonly as radiating, columnar aggregates. Kyanite is formed during the metamorphism of clay-rich sediments. It forms at temperatures between those of andalusite and sillimanite, and occurs in mica schists, gneisses, and associated hydrothermal quartz veins and pegmatites. It is one of the key minerals for estimating the temperature, depth, and pressure at which a rock undergoes metamorphism.

Kyanite blades

WILLEMITE

Crystal system Hexagonal/trigonal | **Composition** Zn_2SiO_4 | **Color** Green, red brown | **Form/habit** Massive | **Hardness** 5–5½ | **Cleavage** Good **Fracture** Conchoidal to uneven | **Luster** Vitreous to resinous | **Streak** Colorless | **SG** 4.0 **Transparency** Transparent to translucent

Willemite is zinc silicate. It is colorless to white, gray, red, dark brown, honey yellow, apple green, blue, yellow brown, and red brown, and fluoresces bright green when exposed to ultraviolet light. It sometimes forms short, prismatic crystals, but is most commonly found in masses or fibrous aggregates. Willemite is found in the oxidized zones of zinc deposits as an alteration of previously existing sphalerite, and in metamorphosed limestones. It is often associated with hemimorphite, smithsonite, franklinite, and zincite. Important localities include Canada, Greece, Sweden, Namibia, Zambia, Australia, and the US. It was discovered in 1830 and named after William I of the Netherlands.

Willemite variety "troostite"

PHENAKITE

Crystal system Hexagonal/trigonal | **Composition** Be_2SiO_4 | **Color** Colorless, white | **Form/habit** Rhombohedral | **Hardness** 7½–8 | **Cleavage** Indistinct | **Fracture** Conchoidal | **Luster** Vitreous **Streak** Colorless | **SG** 3 | **Transparency** Transparent to translucent | **RI** 1.65–1.67

Phenakite is a fairly rare beryllium silicate. It can be colorless and transparent, but

more often it is translucent grayish or yellowish, and occasionally pale rose red. Its crystals are predominantly rhombohedrons, and less commonly short prisms. Phenakite is found in high-temperature pegmatites and in mica schists often accompanied by quartz, chrysoberyl, apatite and topaz. Large crystals are found in Russia, Norway, France, and in the US. Phenakite was named in 1833 from the Greek for "deceiver," alluding to its tendency to be mistaken for quartz. Transparent crystals are faceted for collectors. Its indices of refraction are higher than topaz and its brilliance approaches that of diamond.

HUMITE

Crystal system Orthorhombic | **Composition** $(Mg,Fe)_7(SiO_4)_3(F,OH)_2$ | **Color** Yellow to dark orange **Form/habit** Granular | **Hardness** 6–6½ **Cleavage** Poor | **Fracture** Subconchoidal to uneven **Luster** Vitreous | **Streak** Yellow to orange | **SG** 3.2–3.3 | **Transparency** Transparent to translucent

Humite is a silicate of magnesium and iron. It is yellow to dark orange or reddish orange in color, tending toward brown with increasing manganese content. Manganese substitutes for iron in the structure to form a complete chemical interchange series with manganhumite. It is generally found in granular masses, with well-formed crystals being rare: when they occur they rarely exceed ½ in (10 mm) in length. Humite occurs in metamorphosed limestones and dolomites, where it is often found with cassiterite, hematite, mica, tourmaline, quartz, and pyrite. Named after English mineral collector Sir Abraham Hume in 1813, it is found in Sweden, Scotland, Italy, Switzerland, and the US.

Yellowish-brown humite crystals

Humite crust

NORBERGITE

Crystal system Orthorhombic | **Composition** $Mg_3SiO_4(F,OH)_2$ | **Color** Light yellowish brown, white, or rose | **Form/habit** Granular | **Hardness** 6–6½ | **Cleavage** Distinct | **Fracture** Uneven to subconchoidal, brittle | **Luster** Vitreous | **Streak** Yellowish | **SG** 3.1–3.2 | **Transparency** Transparent to translucent

Norbergite is fluoridated hydrous magnesium silicate. It is light yellowish brown, white, tan, yellow, yellow orange, orange brown, pink with a purplish tint, or rose in color. It rarely forms crystals, and is usually granular. When crystals occur, they are blocky. Norbergite occurs in contact-metamorphic and area-metamorphosed rocks where there is contact between magnesium-rich sedimentary rocks and granites. It is often found with forsterite, diopside, phlogopite, and brucite. Localities include the US, Russia, Italy, Tajikistan, and India. It was discovered in and named after Norberg, Sweden in 1926.

Yellow norbergite encrustation

Norbergite crust

CHONDRODITE

Crystal system Monoclinic | **Composition** $(Mg,Fe,Ti)_5(SiO_4)_2(F,OH,O)_4$ | **Color** Yellow, orange, brown red, or green brown | **Form/habit** Granular | **Hardness** 6–6½ | **Cleavage** Indistinct | **Fracture** Conchoidal to uneven, brittle | **Luster** Vitreous to greasy | **Streak** Yellow | **SG** 3.1–3.2 | **Transparency** Translucent to transparent

Chondrodite is hydrous magnesium fluorosilicate, and a fairly rare mineral. Hydroxide can substitute for the fluorine, and iron and titanium can substitute for the magnesium. Chondrodite is yellow, orange, brownish red, or greenish brown in color. Its crystals are blocky and commonly 1 in (2.5 cm) long, although its usual occurrence is as isolated grains. It forms in metamorphosed limestones and dolomites, kimberlites, rocks rich in iron and magnesium, and in marbles. It takes

its name from the Greek for "granule." Chondrodite should not be confused with chondrite, which is a type of meteorite.

Chondrodite crystals

DATOLITE

Crystal system Monoclinic | **Composition** $CaBSiO_4(OH)$ | **Color** Colorless, white, yellowish, pale pink, or green | **Form/habit** Platy to short prismatic | **Hardness** 5–5½ | **Cleavage** Imperfect | **Fracture** Uneven to subconchoidal, brittle | **Luster** Vitreous to greasy | **Streak** White | **SG** 2.9–3.0 | **Transparency** Transparent to translucent

Datolite is a calcium boron hydroxide silicate. It can be colorless, white, gray, yellowish, pale pink, greenish white, or green. Its crystals are generally platy to short prisms, but may also be blocky or in spherical aggregates. It is also found in masses, and in cryptocrystalline form. Datolite occurs as veins and cavity linings in iron and magnesium-rich igneous rocks and in metallic-ore veins. It is also found in gneisses and diabases. Notable deposits exist in Mexico, Japan, Russia, Germany, the Czech Republic, and in the US. Its name comes from the Greek word for "to divide," a reference to the granular occurrence of some of its varieties.

Datolite crystals

CHLORITOID

Crystal system Monoclinic or triclinic | **Composition** $(Fe,Mg,Mn)_2Al_4Si_2O_{10}(OH)_4$ | **Color** Dark to grayish green | **Form/habit** Foliated, massive | **Hardness** 6½ | **Cleavage** Perfect | **Fracture** Uneven | **Luster** Vitreous to adamantine | **Streak** White | **SG** 3.5–3.8 | **Transparency** Translucent

Chloritoid is iron magnesium manganese aluminosilicate hydroxide. Its color is dark to grayish green, or black, and its crystals are hexagonal appearing or blocky. It is more commonly found in platy, leaflike masses, often with curved plates. Chloritoid occurs in low- to medium-grade, metamorphosed, fine-grained sediments, and is a useful mineral in the interpretation of the metamorphic environment. It is also found in lavas, tuffs, and rhyolites. It was named in 1837 for its similarity to the chlorite group of minerals. Localities include Belgium, Taiwan, and the US.

Chloritoid

Chloritoid crystals

DUMORTIERITE

Crystal system Orthorhombic | **Composition** $(Al,Fe)_7(BO)_3(SiO_4)_3O_3$ | **Color** Pinkish red, bluish violet, brown, or greenish | **Form/habit** Fibrous aggregates | **Hardness** 7–8 | **Cleavage** Distinct, imperfect | **Fracture** Uneven, brittle | **Luster** Vitreous | **Streak** White or bluish white | **SG** 3.2–3.4 | **Transparency** Transparent to translucent | **RI** 1.66–1.72

Dumortierite is aluminum borosilicate. Its usual colors are pinkish red or violet to blue, but it can also be brown or greenish. It typically forms fibrous aggregates of radiating crystals or slender prisms. Dumortierite occurs in pegmatites, in aluminum-rich metamorphic rocks, and in rocks that are metamorphosed by boron-bearing vapor derived from hot, intruding bodies of granite. Sources of

dumortierite include Austria, Brazil, the US, Canada, the Czech Republic, Namibia, Norway, Peru, Poland, Russia, and Sri Lanka. Dumortierite was first described in 1821 and named for the French paleontologist Eugène Dumortier. It is used in the manufacture of high-grade porcelain.

Dumortierite specimen

GADOLINITE

Crystal system Monoclinic | **Composition** $Y_2FeBe_2Si_2O_{10}$ | **Color** Green, blue green | **Form/habit** Compact masses | **Hardness** 6½–7 | **Cleavage** None | **Fracture** Conchoidal, brittle | **Luster** Vitreous to greasy | **Streak** Greenish gray | **SG** 4.4–4.8 | **Transparency** Opaque to nearly transparent

Gadolinite is a complex mineral containing cerium, lanthanum, neodymium, yttrium, beryllium, iron silicate. It exists in two forms: yttrian gadolinite is the most common; and cerian gadolinite. It is green, blue green, or, rarely, pale green when fresh, and brown or black when altered. It is generally found in compact masses, but when crystals occur, they are prisms. It is relatively rare, and is found in granite and granitic pegmatites. Gadolinite was named in 1800 for Johan Gadolin, the Finnish mineralogist-chemist who first isolated yttrium from the mineral in 1792. The mineral gadolinium was also named for him, although gadolinite contains only traces of gadolinium.

Greasy luster

Gadolinite specimen

Organic gems

COPAL

Crystal system None | **Composition** Various
Color Colorless, yellow | **Form/habit** Amorphous
Hardness 2–2½ | **Cleavage** None | **Fracture**
Conchoidal | **Luster** Resinous | **Streak** White
SG About 1.1 | **Transparency** Transparent
to translucent

Copal is resin obtained from various
species of tropical trees. Copal is the
same approximate hardness as amber,
but differs from it in that copal is still wholly
or partially soluble in organic solvents. It
can be collected from living trees or from
accumulations in the soil beneath the
trees, or mined if it is buried. Copal from
different sources may possess similar
physical properties, but different chemical
properties. Buried copal is the closest to
amber in durability, and is in many cases
virtually indistinguishable from it. Deeply
buried copal is sometimes referred to as
"subfossil" copal. The island of Zanzibar is
a major source of buried copal. It is used
in making varnishes, inks, and linoleum.

Copal nugget

PEANUT WOOD

Crystal system Amorphous | **Composition** SiO_2
Color White markings on brown, gray, or green
Form/habit Pseudocubic | **Hardness** 6½–7
Cleavage None | **Fracture** None | **Luster**
Vitreous | **Streak** Black | **SG** 2.58–2.91
Transparency Opaque | **RI** 1.54

Peanut wood is a variety of petrified wood
that is usually dark brown to black in
coloration, with white-to-cream markings
that are ovoid in shape and about the size
of peanuts, hence its name. It was formed
during the Cretaceous period, about
100 million BCE. Driftwood floating
on the oceans was bored into by clam

larvae, leaving the "peanut" holes that were
later replaced by silica—microcrystalline
quartz—during the fossilization process.
It is one of dozens of forms of petrified
(or fossilized) wood with unusual markings
or colors, or retain original cells replaced
by silica, and which are used as
gemstones or for ornaments. Most peanut
wood derives from Western Australia.

▽ AMBER

Crystal system None | **Composition** Hydrocarbon
(C,H,O) | **Color** Yellow, sometimes brown or reddish
Form/habit Amorphous | **Hardness** 2–2½
Cleavage None | **Fracture** Conchoidal | **Luster**
Resinous | **Streak** White | **SG** 1.1 | **Transparency**
Transparent to translucent | **RI** 1.54–1.55

Amber is fossilized resin. It comes
principally from extinct coniferous trees,
although amberlike substances from
earlier trees are known. Amber occurs
in a range of colors: it can be whitish
through a pale lemon yellow, to brown
and almost black. Uncommon colors are
red, green, and blue. Ambers from Europe
and North America derive from at least
three different tree species. There are five
different classes of amber based on their
organic chemical constituents. Amber is
generally found in association with lignite
coal, which is itself the fossilized remains
of trees and other plant material. For
several thousand years, the largest source
of amber has been the substantial
deposits along the Baltic coast, which
extend from Gdánsk in Poland westward
to the coastlines of Denmark and Sweden.
Accounting for 90 percent of the world's
amber, Baltic amber is mined from land
and recovered from the shores after heavy
storms. The oldest amber recovered dates
to the Upper Carboniferous (320 million
years BCE).

Amber inclusions closeup

Coral formations make up the Great Barrier Reef, Australia

◁ CORAL

Crystal system Trigonal, orthorhombic, amorphous **Composition** CaCO₃ / conchiolin | **Color** Red, pink, black, blue, golden | **Form/habit** Coral shaped **Hardness** 3½ | **Cleavage** None | **Fracture** Hackly | **Luster** Dull to vitreous | **Streak** White **SG** 2.6–2.7 | **Transparency** Opaque | **RI** 1.49–1.66

Coral is the skeletal material generated by sea-dwelling coral polyps. For most corals, this material is aragonite, but in the case of black and golden corals, it is a hornlike substance called conchiolin. Precious coral is the name given to corallium rubrum and several related species. It grows in the shape of small leafless bushes, seldom exceeding 3 ft (1 m) in height. Its red color comes from carotenoid pigments within the aragonite skeleton. Red and pink precious corals are found in the warm seas around Japan and Malaysia, in the Mediterranean, and in African coastal waters. Black coral comes from the West Indies, Australia, and around the Pacific Islands. Natural coral has a dull luster, but can take a bright polish.

ANTHRACITE

Crystal system None | **Composition** Various **Color** Black | **Form/habit** Amorphous **Hardness** 2–2½ | **Cleavage** None | **Fracture** Conchoidal | **Luster** Nearly metallic | **Streak** Black | **SG** About 1.1 | **Transparency** Opaque

Anthracite is similar in appearance to the mineraloid jet. It differs from ordinary coal by its greater hardness (2.75–3 on the Mohs scale), its higher density (1.3–1.4), and its luster, which is often semimetallic. All coal is made up of an irregular mixture of different chemical compounds called macerals, which are analogous to minerals in inorganic rocks. Unlike minerals, they have no fixed chemical composition and no crystalline structure. Anthracite has a high percentage of fixed carbon and a low percentage of volatile matter, and may be considered as a transition stage between ordinary bituminous coal and graphite.

Anthracite coal

JET

Crystal system None | **Composition** Various **Color** Black, brown | **Form/habit** Amorphous **Hardness** 2½ | **Cleavage** None | **Fracture** Conchoidal | **Luster** Velvety to waxy | **Streak** Black to dark brown | **SG** About 1.3 | **Transparency** Opaque | **RI** 1.64–1.68

Generally classified as a lignite coal, the mineraloid jet has a high carbon content and a layered structure. It is black to dark brown, and sometimes contains tiny inclusions of pyrite, which have a metallic luster. Jet results from the high-pressure decomposition of wood over millions of years, commonly the wood of trees of the family Araucariaceae. The microstructure of jet, which resembles the original wood, can be seen under high magnification. It tends to occur in rocks of marine origin, perhaps from waterlogged driftwood or other plant material. Jet can occur in distinct beds, such as those at Whitby, England.

Jet specimen

PEARL

Crystal system Orthorhombic | **Composition** Principally CaCO₃ | **Color** White, cream, black, blue, yellow, green, or pink | **Form/habit** Reniform **Hardness** 3 | **Cleavage** None | **Fracture** Uneven, brittle | **Luster** Pearly | **Streak** White **SG** 2.7 | **Transparency** Opaque | **RI** 1.55–1.68

The finest pearls are those produced by certain limited species of saltwater oysters and freshwater clams. Consisting of the same material as the shell—principally the mineral aragonite—pearl is a concretion deposited in concentric layers. In addition to aragonite, the shell contains small amounts of organic substances to produce a composite called nacre. Pearls can be any delicate shade from black to white, cream, gray, blue, yellow, green, lavender, and mauve. Gem-quality pearls are almost always nacreous and iridescent, like the interior of the shell that produces them. The shell-secreting cells are located in the mantle, a layer of the mollusk's body tissue. When a foreign particle enters the mantle, the cells build up concentric layers

of pearl around it to protect the soft mantle. Cultured pearls are created by opening the oyster and inserting an object for the pearl to form around, usually a spherical bead.

SHELL

Crystal system Trigonal, orthorhombic, amorphous **Composition** CaCO₃ | **Color** Red, pink, brown, blue, golden | **Form/habit** Shell shaped | **Hardness** 2½ **Cleavage** None | **Fracture** Conchoidal | **Luster** Dull to vitreous | **Streak** White | **SG** About 1.3 **Transparency** Translucent to opaque

Like coral, shell is mineral matter generated by biological processes. The mineral in shells is either calcite or aragonite. Shell is secreted in layers by cells in the mantle, a skinlike tissue in the mollusk's body wall. Different mollusks are characterized by the number, composition (aragonite or aragonite and calcite), and arrangement of the calcareous layers. The results are distinct microstructures that have differing mechanical properties and, in some shells, differing colors. Shells for carving or other ornamentation can be marine or freshwater. Shells with different colored layers have been carved into cameos since antiquity.

Abalone shell

MOTHER OF PEARL

Crystal system Amorphous | **Composition** CaCO₃ | **Color** All | **Hardness** 3½ **Cleavage** None | **Fracture** None | **Luster** Greasy to pearly | **Streak** None | **SG** 2.70–2.89 | **Transparency** Opaque **RI** 1.530–1.685

Mother of pearl is the substance nacre, an organic-inorganic material produced by some mollusks as an inner shell layer. Nacre is made up of hexagonal platelets of aragonite arranged in continuous layers, which are separated by sheets of organic material such as chitin, lustrin, and silklike proteins. This mixture of brittle platelets and thin layers of organic materials makes mother of pearl tough and resilient. Nacre

is iridescent because the thickness of the aragonite platelets is close to the wavelength of visible light, interfering with different wavelengths at different viewing angles.

Shell with mother of pearl

AMMOLITE

Crystal system Orthorhombic | **Composition** CaCO₃ | **Color** All spectral colors—red, orange, yellow, green, blue, indigo, and violet | **Form/habit** Fractured and nonfractured flat layers | **Hardness** 3½–4 | **Cleavage** None | **Fracture** Uneven **Luster** Vitreous | **Streak** None | **SG** 2.75–2.85 **Transparency** Opaque | **RI** 1.52–1.68

Ammolite is an organic gemstone exhibiting opal-like colors. It is made of the fossilized shells of ammonites, which are themselves composed of aragonite. It retains the microstructure of the original shell, and exhibits iridescence in all the spectral colors due to the microstructure of the aragonite. In addition to aragonite, it may also contain calcite, silica, pyrite, or other minerals involved in the fossilization process. It is found in the eastern Rocky Mountains of North America, especially in the Canadian provinces of Alberta and Saskatchewan, and in Montana in the US.

Green-red iridescence

Ammolite specimen

Rocks

Entries follow the standard geological order. Igneous rocks are divided into intrusive and extrusive, sedimentary rocks into clastic and chemical, and metamorphic rocks are listed by increasing degree of metamorphism.

Igneous rocks

GRANITE

Rock type Felsic, plutonic, igneous | **Major minerals** Potassium feldspar, quartz, mica, sodium **Minor minerals** Sodium plagioclase, hornblende **Color** White, light gray, gray, pink, red | **Texture** Medium to coarse

Granite is familiar as a coarse-grained, mottled pink, white, gray, and black ornamental stone, and is the most common intrusive rock in the Earth's continental crust. Granite's three main minerals are feldspar, quartz, and mica, either as dark biotite and/or as silvery muscovite. Feldspar predominates, and quartz usually accounts for more than 10 percent. Its feldspars are often pink (although they can also be white, buff, or gray), resulting in the pink granite used as an ornamental stone for the facings of buildings and for floors. Because it is formed at depth, the exposure of granite at the surface is evidence that the area has been uplifted, and that the great thickness of rock overlying the granite has been eroded away. Gemstones such as topaz, rock crystal, tourmaline, and aquamarine, and many metal ores—including those of gold, silver, lead, and titanium—are deposited by solutions released by crystallizing granite.

DIORITE

Rock type Intermediate, plutonic, igneous | **Major minerals** Sodium plagiocase, hornblende | **Minor minerals** Biotite | **Color** Mottled black/dark green and gray or white | **Texture** Medium to coarse

Diorite is somewhat difficult to define because it grades imperceptibly into granite. It is a medium- to coarse-grained intrusive igneous rock darker in color

than granite. Its composition is about two-thirds plagioclase feldspar, and one-third dark-colored minerals, such as hornblende or biotite, giving it a mottled appearance. Diorite can be of uniform grain size, with plagioclase or hornblende set into its medium-grained groundmass. Diorite was particularly prized in ancient Egypt where it was used in statuary, for columns, pillars, sarcophagi, and for lining the chambers of some pyramids. Diorite is sometimes sold today as "black granite."

Diorite specimen

GRANODIORITE

Rock type Felsic, plutonic, igneous | **Major minerals** Plagiocase, K-feldspar, quartz, mica **Minor minerals** Hornblende, augite | **Color** Gray, white, or pink | **Texture** Medium to coarse

A medium- to coarse-grained rock that is similar to granite, granodiorite has more plagioclase feldspar than orthoclase feldspar, and is among the most abundant of intrusive igneous rocks. Granodiorite can be pink or white with a grain size and texture similar to granite, but the presence of plagioclase generally makes it appear darker. Hornblende and biotite are often present, which can give it a

speckled appearance. It is sawn and polished to create flooring, facing for buildings, and countertops, due to its toughness and often-attractive speckling. It is one of several stones sold as "black granite," and is also crushed for use as road ballast and curbstones.

Pink granodiorite

▷ BASALT

Rock type Mafic, volcanice, igneous | **Major minerals** Sodium plagioclase, pyroxine, olivine **Minor minerals** Leucite, nepheline | **Color** Dark gray to black | **Texture** Fine-grained to porphyritic

The volcanic rock basalt is the most common igneous rock that occurs at the Earth's surface. Basalt forms on the ocean floors as volcanoes in the ocean basins erupt basalt, and volcanic islands are also made from it. Darkly colored and relatively rich in iron and magnesium, basalts are mainly composed of plagioclase, pyroxene, and olivine. They can also have distinct crystals set in a fine-grained groundmass. On land, basalts have formed enormous plateaus in the Columbia River region of the Northwestern US; the Paraná Basin in South America; and the Deccan Plateau of southern and central India. Basalts form by partial melting of the Earth's mantle. Unmelted fragments—known as xenoliths—of the mantle can be carried upward, providing geologists with their only direct observation of the mantle's composition. Most basalts are very fine grained and compact, even glassy, but many types are porphyritic, with distinct crystals set in a fine-grained base or groundmass.

PEGMATITE

Rock type Felsic, plutonic, igneous | **Major minerals** Quartz, feldspar, mica | **Minor minerals** Tourmaline, topaz | **Color** Light | **Texture** Very coarse

Pegmatites are the source of a large number of gem minerals. They are coarse-grained intrusive igneous rocks, and usually form as sheets, pods, lenses, or cigar-shaped bodies. Most have the same major constituents—quartz and feldspar—as granites or syenites. Although their crystals can be huge—several meters long—the average size is 3–4 in (8–10 cm). Well-formed crystals can occur in hollows in the pegmatite, or the entire pegmatite may be completely filled. Some pegmatites may consist of a single or predominant mineral, from which their name is taken. The silica-rich constituents of granitic and syenitic pegmatites are particularly important because they are the chief commercial source of feldspar and sheet mica, and also a major source of gemstones. Tourmaline, aquamarine, emerald, rock crystal, smoky quartz, rose quartz, topaz, moonstone, amazonite, kunzite, heliotrope, sphene, garnet, and others are all found in pegmatites. Pegmatites are also the source of other important economic minerals, including beryllium, lithium, titanium, molybdenum, tin, tungsten, tantalum, niobium, and other rare elements.

Feldspar   Quartz

Group of crystals in granitic pegmatite

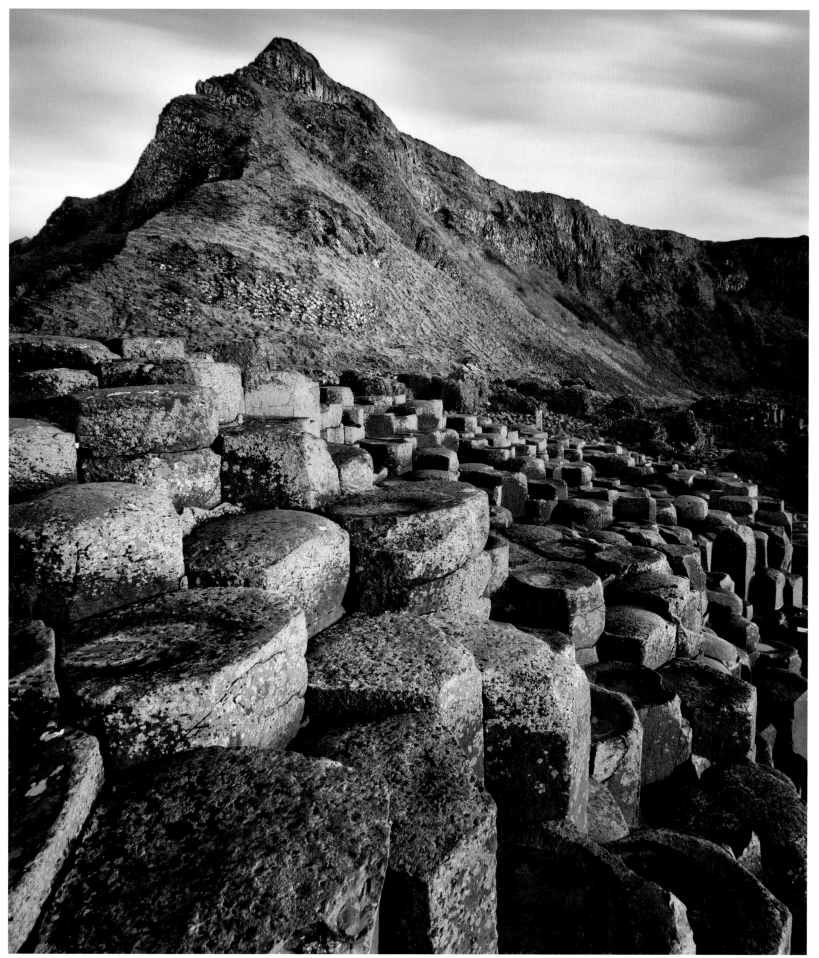

Basalt columns at the Giant's Causeway, Northern Ireland, formed during a volcanic eruption 50 to 60 million years ago

▽ TUFF

Rock type Volcanic, igneous | **Fossils** Generally moulds of nonmarine plants and animals, including humans | **Major minerals** Glassy fragments **Minor minerals** Crystalline fragments | **Color** Light to dark brown | **Texture** Fine

"Tuff" is a catch-all term that describes any relatively soft, porous rock made of solidified materials, such as ash and other small fragments ejected from volcanic vents. Tuffs stem from foaming magma that wells to the surface as a mixture of hot gases and incandescent particles, and is ejected from a volcano. Tuffs differ from pumice, in that they originate as a mass of independent particles that are later cemented or fused, whereas pumice remains, effectively, a sticky mass of magma that is full of bubbles. Most tuff formations include a range of fragment sizes and varieties, ranging from fine-grained dust and ash (ash tuffs), to medium-sized fragments called lapilli (lapilli tuffs), up to large volcanic blocks and bombs (bomb tuffs).

SYENITE

Rock type Intermediate, plutonic, igneous **Major minerals** Potassium feldspar | **Minor minerals** Sodium plagioclase, biotite, amphibole, pyroxene, feldspathoids | **Color** Gray, pink, or red | **Texture** Medium to coarse

Syenite is often so visually similar to granite that it can be confused with it. With close examination, syenite can be distinguished from granite by the absence or scarcity of quartz. Syenites are essentially composed of an alkali feldspar such as orthoclase, albite, or—less commonly—microcline, and sodium-rich plagioclase, along with ferromagnesian minerals—minerals rich in iron and magnesium—such as biotite, hornblende, or pyroxene. There are several types of syenites: quartz syenites, with slightly more quartz, and syenites rich in feldspathoid minerals, especially nepheline, known as nepheline syenites. Syenites can have sphene, apatite, zircon, magnetite, and pyrite in small amounts.

Syenite rough

DOLERITE

Rock type Mafic, plutonic, igneous | **Major minerals** Calcium plagioclase, pyroxene | **Minor minerals** Quartz, magnetite, olivine | **Color** Dark gray to black, often mottled white | **Texture** Fine to medium

A fine- to medium-grained rock, dolerite is from one-third to two-thirds calcium-rich plagioclase feldspar, with the remainder being principally pyroxene. Magnetite and olivine may be present; if the latter, the rock is called olivine dolerite. It is dark gray to black in color, although it sometimes displays white mottling. It is extremely hard and tough, and is usually found intruded into fissures in other rocks. Dolerite's most famous use is at Stonehenge, England, a megalithic stone circle built c.3000–2000 BCE. The dolerite "bluestones" of the inner circle of Stonehenge were transported

Tuff cone rising 148 m (486 ft) above the ocean, Kicker Rock, San Cristobal, Galapagos Islands

240 miles (385 km) from Wales to the site of Stonehenge, and were brought by sea, river, and over land.

Dark gray dolerite

GABBRO

Rock type Mafic, plutonic, igneous | **Major minerals** Calcium plagioclase feldspar, pyroxene **Minor minerals** Olivine, magnetite | **Color** Dark gray to black | **Texture** Medium to coarse

Gabbros are medium to coarse grained, and consist principally of calcium-rich plagioclase feldspar and pyroxene, usually augite. Named by German geologist Christian Leopold von Buch after a town in Tuscany, Italy, gabbros are essentially the intrusive equivalent of basalt, and are dark gray to black in color. The mineralogy and composition of gabbros are both highly variable, since this rock type often has a significant amount of olivine or coarse crystals of plagioclase. Gabbros are low in silica, and quartz is rarely present. They are another rock sold as "black granite," although they owe their major economic significance to the ores that are contained within them, such as nickel, chromium, and platinum. Others containing magnetite and ilmenite are mined for their iron or titanium. Gabbros are widespread, but not common.

Plagioclase feldspar   Ferromagnesium

Layered gabbro

PERIDOTITE

Rock type Ultramafic, plutonic, igneous | **Major minerals** Olivine, pyroxene | **Minor minerals** Garnet, chromite | **Color** Dark green to black **Texture** Coarse

An intrusive igneous rock, peridotite is coarse-grained, dark-colored, and dense. Fresh peridotite contains olivine, pyroxene, and chromite, while peridotite that has been altered by weathering becomes serpentinite, containing chrysotile, other serpentines, and talc. It contains at least 40 percent olivine and a high percentage of pyroxene. Peridotite is a major structural component of the Earth—it is the major constituent of the upper mantle. It is also found in mountain belts as olivine-rich masses and as sheetlike bodies. Peridotite is an important economic rock because of the minerals found within it. It is a key source of chromium ore, garnet is often present as an accessory mineral, weathered peridotites are a source of nickel, and its variety kimberlite is the source of all naturally occurring diamonds.

Green peridotite

KIMBERLITE

Rock type Ultramafic, volcanic, igneous | **Major minerals** Olivine, pyroxene, mica | **Minor minerals** Garent, ilmenite, diamond | **Color** Dark gray **Texture** Fine to coarse/porphyritic

Kimberlite is a form of peridotite, and is the prime source of diamonds. It is dark gray with well-formed crystals of brown mica, pyrope garnet, chrome-bearing diopside, and lesser amounts of ilmenite, serpentine, pyroxene, calcite, rutile, perovskite, and magnetite, and diamonds. Kimberlite occurs in volcanic pipes, intrusive igneous bodies that are roughly circular in cross section with vertical sides, and usually less than ¾ mile (1 km) in diameter. It also occurs in other steep-sided, intrusive igneous bodies. These are often found in the uplifted centers of continental platforms, and appear to be roughly the same age,

having formed during the Late Cretaceous period (100 to 65 million years BCE). Diamond-bearing kimberlite is yellow in color due to weathering, and is hence known as "yellow ground."

Diamond

Kimberlite specimen

OBSIDIAN

Rock type Felsic, volcanic, igneous | **Major minerals** Glass | **Minor minerals** Hematite, feldspar | **Color** Black, brown, red | **Texture** Amorphous

Obsidian is a natural volcanic glass that forms when lava solidifies so quickly that mineral crystals do not have time to grow. The name obsidian refers to the rock's glassy texture and not to its composition, which technically can have any chemical composition. However, most obsidian is similar in composition to rhyolite, and it is commonly found on the outer edges of rhyolite flows. Obsidian is generally jet black, although the presence of other minerals can change its color: hematite (iron oxide) produces red and brown varieties, and the inclusion of tiny gas bubbles can create a golden sheen. In "snowflake" obsidian, scattered spheres of white, radiating, needlelike crystals form after the black obsidian has cooled, appearing like snowflakes when the rock is sawn in cross section. Most obsidian is relatively young, since volcanic glass devitrifies—the process by which minerals crystallize from the glass—over time.

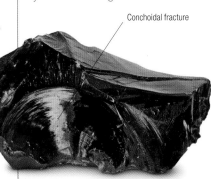

Conchoidal fracture

Black obsidian

RHYOLITE

Rock type Felsic, volcanic, igneous | **Major minerals** Quartz, potassium feldspar | **Minor minerals** Glass, biotite, amphibole, plagioclase **Color** Very light to medium gray, light pink **Texture** Fine or porphyritic

Rhyolite is the volcanic equivalent of granite, and is relatively rare. Because rhyolitic lava is so silica rich, it is also very sticky, or viscous, and thus rarely reaches the Earth's surface. When it does the lava piles up over its vent without flowing away because of its thickness, and tends to form steep-sided domes rather than flows. Rhyolites tend to be light colored and very fine grained, and many are porphyritic, with larger crystals (phenocrysts) in the fine-grained matrix. The phenocrysts are commonly quartz and sanidine, but can be biotite, amphibole, and pyroxene.

Rhyolite specimen

ANORTHOSITE

Rock type Ultramafic, plutonic, igneous | **Major minerals** Calcium plagioclase | **Minor minerals** Olivine, pyroxene, garnet | **Color** Light gray to white | **Texture** Medium to coarse

Anorthosite is an intrusive igneous rock composed of at least 90 percent calcium-rich plagioclase feldspar—principally labradorite and bytownite, as well as olivine, pyroxene, garnet, and iron oxides to make up the remainder. It is light in color, running from white to gray, and its distinct crystals are usually small. It tends to occur as immense masses, or as layers between mafic and ultramafic rocks such as gabbro and peridotite. Anorthosite is not a common rock on Earth, but is extremely common on the surface of the Moon, where it makes up the ancient, rough, light-colored highlands. On Earth there are large anorthosite-bearing rock bodies in New York and Montana, US, in eastern Canada, and in South Africa.

PUMICE

Rock type Volcanic, igneous | **Origin** Extrusive
Major minerals Glass | **Minor minerals** Feldspar,
augite, hornblende, zircon | **Color** White, yellow,
brown, black | **Texture** Fine

Pumice is created when gas-saturated
liquid magma erupts like a carbonated
drink released from a shaken bottle. It
cools so rapidly that the resulting foam
solidifies into a natural volcanic glass, which
is full of gas bubbles. It is very porous, with
hollows that can be rounded, elongated,
or tubular, depending on the flow of the
lava as it solidifies. Pumices formed from
silica-rich lavas are white, while those
from lavas with intermediate silica content
are often yellow or brown, and the rarer
silica-poor pumices (like those in the
Hawaiian Islands) are black. Because of
its hollow pores, pumice has a very low
density and it can easily float in water.
Pumices are frequently accompanied
by obsidian, which forms under similar
conditions but results from greater pressure.

Frothy
texture

Rhyolitic pumice

ANDESITE

Rock type Intermediate, volcanic, igneous | **Major
minerals** Plagioclase feldspars | **Minor minerals**
Pyroxene, amphibole, biotite | **Color** Light to dark
gray, reddish pink | **Texture** Fine, porphyritic

Andesite is the fine-grained or porphyritic
volcanic equivalent of diorite. Unlike
rhyolites, andesites do not contain quartz.
It consists primarily of the plagioclase
feldspar minerals andesine and oligoclase,
and one or more of the dark minerals,
such as pyroxene or biotite. Andesite erupts
from explosive volcanoes and is often found
with volcanic ash and tuff. Amygdaloidal
andesite occurs when the voids left by gas
bubbles in the solidifying magma are later
filled in, and porphyritic andesite occurs
when larger phenocrysts of feldspar and
pyroxene form in a fine-grained matrix.

PHONOLITE

Rock type Intermediate, volcanic, igneous
Major minerals Sanidine, oligoclase | **Minor
minerals** Feldspathoids, hornblende, pyroxene,
biotite | **Color** Medium gray | **Texture** Fine
to medium, porphyritic

Phonolite is a variety of the intrusive
rock trachyte, but contains nepheline or
leucite rather than quartz. Phonolites are
commonly fine grained and compact,
splitting into thin, tough plates, which make
a ringing sound when struck—hence the
rock's name. The principal dark-colored
mineral in phonolite is pyroxene, usually
in the form of aegirine or augite. Phonolites
are common in Europe, especially in
Germany, the Czech Republic, and in
the Mediterranean area, particularly in Italy.
The spectacular Devil's Tower, Wyoming—
the neck of an ancient volcano that
featured in the movie *Close Encounters
of the Third Kind*—is probably phonolite's
most famous occurrence.

TRACHYTE

Rock type Intermediate, volcanic, igneous | **Major
minerals** Sanidine, oligoclase | **Minor minerals**
Feldspathoids, quartz, hornblende, pyroxene, biotite
Color Off-white, gray, pale yellow, pink | **Texture**
Fine to medium, porphyritic

Trachyte is the volcanic, extrusive equivalent
of syenite. Trachyte is similar to rhyolite in
color and occurrence, but contains very
little or no quartz, as with syenite. Its name
comes from the Greek for rough, in that it
has a characteristically rough texture. It is
commonly porphyritic, meaning it has large
crystals set in a fine-grained groundmass.
A common mineral found in trachytes is
the feldspar sanidine, which occurs in well-
formed crystals up to 2 in (5 cm) across.
Dark, iron and magnesium-rich minerals
such as biotite, amphibole, and pyroxene
can also be present in small quantities.
Localities include Ascension Island;
Colorado, US; the Auvergne, France; and
the Rhine region of central Europe.

Trachyte specimen

Sedimentary rocks

LIMESTONE

Rock type Marine, chemical, sedimentary | **Fossils**
Marine and freshwater invertebrates | **Major
minerals** Calcite | **Minor minerals** Aragonite,
dolomite, siderite, quartz, pyrite | **Color** White, gray,
pink | **Texture** Fine to medium, angular to rounded

Limestone is very abundant, and occurs
in thick, extensive, multiple layers. It is
composed mainly of calcite, and it can be
yellow, white, or gray. It is easily identified
by its reaction with dilute hydrochloric acid,
or even the acids in carbonated drinks,
which results in a rapid release of carbon-
dioxide gas and a fizzing sound. Limestone
usually forms in warm, shallow seas, both
from the direct precipitation of calcium
carbonate from sea water and from the
accumulation of the shells and skeletons
of calcareous marine organisms. The
combination of sizes and textures within
the various limestones give important clues
to the environment in which they were
formed. They can be coarse and fossil-rich
or fine and microcrystalline.

Fossiliferous limestone

DOLOMITE

Rock type Marine, chemical, sedimentary | **Fossils**
Invertebrates | **Major minerals** Dolomite | **Minor
minerals** Calcite | **Color** Gray to yellowish gray
Texture Fine to medium, crystalline

Dolomite is the rock formed exclusively
from the mineral dolomite, a calcium
magnesium carbonate. Most dolomites
are thought to be limestones in which
their calcite (calcium carbonate) is in
contact with magnesium-bearing solutions,
causing it to be replaced by dolomite—
a process known as dolomitization.
Dolomites are usually less fossiliferous
than limestones because fossils and other
features are destroyed by dolomitization.

It is distinguished from limestone in that it
does not fizz violently in hydrochloric acid.
Fresh dolomite looks very similar to white-
to light-gray limestone; when weathered,
it is a yellowish gray.

Dolomite specimen

CHALK

Rock type Marine, organic, sedimentary | **Fossils**
Invertebrates, vertebrates | **Major minerals** Calcite
Minor minerals Quartz, glauconite, clays | **Color**
White, gray, buff | **Texture** Very fine, angular
to rounded

Chalk is a white to grayish variety of
limestone, and is soft, fine grained, and
easily pulverized. It is composed almost
entirely of the calcite shells of minute
marine organisms. Other minerals can
also be present in small quantities, such
as glauconite, apatite, and clay minerals.
Silica-based sponge spines, diatom and
radiolarian skeletons, and nodules of chert
and flint are also common. Extensive chalk
deposits were formed in the Cretaceous
period (142 to 65 million years BCE), an era
named from the Latin *creta* for "chalk." Like
other limestones, chalk is used for making
lime, cement, quicklime for mortar, as a
fertilizer, and to lower the acidity of the soil.

Chalk specimen

ROCK SALT

Rock type Marine, evaporite, sedimentary
Fossils None | **Major minerals** Halite | **Minor minerals** Silvite | **Color** White, orange brown, blue | **Texture** Coarse to fine crystalline

Rock salt is the massive rock form of the mineral halite, also known as common table salt. It occurs in beds that range from 3 ft (1 m) to more than 1,000 ft (300 m) in thickness. Rock salt forms from the evaporation of saline water in partially enclosed basins, and is commonly interlayered with beds of limestone, dolomite, and shale. It also occurs with other rocks formed by evaporation, such as gypsum and anhydrite. Beds of rock salt are still mined by traditional underground excavation methods. Rock salt is also important to the petroleum industry, since it often occurs in salt domes—a core of salt and an envelope of surrounding strata—the underside of which can trap oil that has migrated up to the surface from oil-rich shales.

Massive rock salt

▷ TUFA

Rock type Continental, chemical, sedimentary
Fossils Rare | **Major minerals** Calcite or silica
Minor minerals Aragonite | **Color** White
Texture Fine, crystalline

Tufa is the name for two different sedimentary rocks, both of which precipitate from water. The first is calcareous tufa, or calc-tufa, a soft, porous form of limestone deposit composed of calcium carbonate (calcite) from hot springs, lake water, and groundwater. The second is siliceous tufa, also called siliceous sinter, a deposit of amorphous silica that forms through the rapid precipitation of fine-grained silica as an encrustation around hot springs and geysers. The variety of sinter that specifically forms around geysers is called geyserite. It forms as terraces and cones around geyser mouths, where superheated silica-bearing waters cool.

Tufa eroded over millennia into distinctive buttresses, Rose Valley, Cappadocia, Turkey

Navajo sandstone has been eroded by flash flooding to form Antelope Canyon in Arizona, US

◁ SANDSTONE

Rock type Continental, detrital, sedimentary
Fossils Vertebrates, invertebrates, plants | **Major minerals** Quartz, feldspar | **Minor minerals** Silica, calcium carbonate | **Color** Cream to red | **Texture** Fine to medium grained, angular to rounded

Sandstone is the second most abundant sedimentary rock after shale, making up about 10 to 20 percent of all the sedimentary rocks in the Earth's crust. It is the lithified accumulation of sand-sized grains, 0.00025–0.08 in (0.063–2 mm) in diameter. Because quartz is resistant to chemical weathering, quartz sand usually dominates, with other minerals present in varying amounts. Sandstone is abundant, well exposed, and has a wide range of textures and mineralogy. This makes it an important indicator of erosion and deposition processes. Desert sandstone has typically well-rounded grains; river sands are usually angular; and beach sands are somewhere in between. After deposition of the sand, the space between the grains fills with a chemical cement of silica, calcite, or occasionally iron oxide. Sandstones are often bedded, as a series of layers that represent successive deposits of grains, and which can show depositional features, such as ripples, or the cross bedding that is typical of dunes. Sandstones are classified according to texture and mineralogical properties, such as micaceous sandstone, which has a large mica component.

CONGLOMERATE

Rock type Marine, freshwater, and glacial detrital sedimentary | **Fossils** Very rare | **Major minerals** Any hard mineral can be present | **Minor minerals** Any mineral can be present | **Color** Various | **Texture** Very coarse, rounded clasts

Conglomerates are formed by the natural cementation of rounded rock or mineral fragments over 0.08 in (2 mm) in diameter. Conglomerates are further classified by the average size of their constituent materials: pebble conglomerate (fine), cobble conglomerate (medium), and boulder conglomerate (coarse). Conglomerates can also be named to include a description of the rock or mineral fragments of which it is composed; for example, quartz pebble conglomerate. Their depositional environments are indicated by how well their rock fragments are sorted: well-sorted conglomerates— those whose pebbles have a small size variation and are generally of only one rock or mineral type—result from normal

water flow over a long period. Poorly sorted conglomerates—those with pebbles of varying sizes—form from rapid water flow and deposition.

Cobble conglomerate

ROCK GYPSUM

Rock type Marine, evaporite, sedimentary | **Fossils** None | **Major minerals** Gypsum | **Minor minerals** Anhydrite | **Color** White, pinkish, yellowish, gray | **Texture** Medium to fine crystalline

Rock gypsum is the sedimentary rock formed principally from the mineral gypsum. It occurs in extensive beds formed by the evaporation of ocean water, as well as in saline lakes and salt pans. It is commonly granular, but it can also occur in fibrous bands. Beds of rock gypsum occur interbedded with limestones, dolomitic limestones, and shale. It is also found interlayered with rock salt and anhydrite. Rock gypsum is of major economic importance: it is heated to drive off some of its water in order to be used as plaster of paris. When the heated material is mixed with water, it recrystallizes as gypsum and hardens. It is used to make, cast ornamental plasterwork, plasterboard (sheetrock) and in medicine to make plaster casts.

Massive rock gypsum

BRECCIA

Rock type Marine, freshwater, and glacial detrital sedimentary | **Fossils** Very rare | **Major minerals** Any hard mineral can be present | **Minor minerals** Any mineral can be present | **Color** Various | **Texture** Very coarse, angular clasts

Breccias have the same-sized rock fragments as conglomerates—over 0.08 in (2 mm) in diameter—but the fragments are angular or only slightly rounded. The absence of rounding is a sign that little or no transportation has taken place. Breccia can form when a rock body shatters (from the action of frost or earth movements, for example) and then becomes cemented in its original position. Breccias can also form when newly broken rock fragments that accumulate at the base of a cliff become cemented where they fell, or in areas of active faulting, when shattered material along the line of the fault becomes cemented in place.

Breccia specimen

MUDSTONE

Rock type Marine, freshwater, and glacial detrital sedimentary | **Fossils** Invertebrates, vertebrates, plants | **Major minerals** Clays, quartz | **Minor minerals** Calcite | **Color** Gray, brown, or black | **Texture** Fine, microscopic

As its name implies, mudstone is a gray or black rock formed from mud. Made up primarily of a mix of clay- and silt-sized particles like shale, mudstone is deposited and solidified into stone so that it is not easily split into thin layers, like shale. In appearance it looks like hardened clay, and it can show cracks or fissures like sun-baked clay. Mudstones are made up of minerals such as quartz and feldspar, clay minerals, and carbonaceous matter. Types containing a substantial amount of calcite are known as calcareous mudstone, and some are fossiliferous. Mudstones can form deposits several meters thick.

SHALE

Rock type Marine, freshwater, and glacial detrital sedimentary | **Fossils** Invertebrates, vertebrates, plants | **Major minerals** Clays, quartz, calcite | **Minor minerals** Pyrite, iron oxides, feldspar | **Color** Gray | **Texture** Fine

Shale is the most abundant sedimentary rock, making up about 70 percent of those found in the Earth's crust. Deposited by gentle transporting currents, shales are laid down on deep-ocean floors, basins of shallow seas, and river floodplains. They contain a high percentage of clay minerals, significant amounts of quartz, and smaller quantities of feldspars, iron oxides, fossils, carbonates, and organic matter. Unlike mudstones, which comprise particles of similar sizes, shales consist of silt- and clay-sized particles, and are easily split into thin layers. Shales come in a number of colors, depending on their composition: reddish and purple shales due to hematite and goethite; blue, green, and black from ferrous iron; gray or yellowish from calcite.

SILTSTONE

Rock type Marine, freshwater, and glacial detrital sedimentary | **Fossils** Invertebrates, vertebrates, plants | **Major minerals** Quartz, feldspar | **Minor minerals** Mica, chlorite, micaceous clay minerals | **Color** Gray to beige | **Texture** Fine, rounded to angular

Siltstone forms from sediments with grain sizes between sandstone and mudstone. It can form in different environments and have different colors and textures. Reds and grays, and flat bedding planes are typical. Siltstones tend to be hard and durable, and are not easily split into thin layers. In some siltstones, the presence of mica may produce a siltstone that splits into thicker, flagstonelike sheets. They may also contain abundant chlorite. The layering of siltstone is indistinct and tends to weather at oblique angles unrelated to bedding. They are much less common than shales or sandstones, and rarely form thick deposits.

Siltstone specimen

ARKOSE

Rock type Terrestrial, marine, or freshwater detrital, sedimentary | **Fossils** Rare | **Major minerals** Quartz, feldspar | **Minor minerals** Mica | **Color** Pinkish to pale gray | **Texture** Medium, angular

Arkose is a pink sandstone that has been colored by an abundance of feldspar grains, especially pink alkali feldspars. Its high feldspar content—amounting to more than 25 percent of the sand grains—sets it apart from other sandstones. It is a relatively coarse rock, consisting primarily of quartz and feldspar grains with small amounts of mica. Since feldspar is readily decomposed by chemical weathering, arkose is formed by the fast deposition of sand weathered from granites and gneisses. Arkose is thought to develop either under conditions of climatic extreme, or from rapid uplift and high relief of the source area.

GRAYWACKE

Rock type Marine, detrital, sedimentary | **Fossils** Rare | **Major minerals** Quartz, feldspar, mafic minerals | **Minor minerals** Chlorite, biotite, clay, calcite | **Color** Gray, greenish gray | **Texture** Fine to medium, angular

Graywacke is also known as dirty sandstone. It is composed of poorly sorted coarse- to fine-grained quartz, feldspar, and dark-colored iron and magnesium-rich minerals—such as amphibole and pyroxene—set in a fine-grained matrix of clay, quartz, or calcite. It is easily confused visually with igneous basalt. Graywackes are mostly gray, brown, yellow, or black, and hard. Their poor sorting results from the rapid deposition of sediments in a turbulent marine environment—a fast-moving, chaotic mass of water and sediment that travels over the continental shelf into the deeper ocean. A general term for rocks formed in this manner is a "turbidite."

Graywacke specimen

MARL

Rock type Marine or freshwater, detrital, sedimentary | **Fossils** Vertebrates, invertebrates, plants | **Major minerals** Clays, calcite | **Minor minerals** Glauconite, hematite | **Color** Various | **Texture** Fine, angular

Marl, also called calcareous mudstone, is a term applied to a variety of rocks that possess generally earthy mixtures of fine-grained minerals, and a considerable range of compositions. They form in shallow water under both marine and freshwater conditions, and predominantly consist of clay minerals and calcium carbonate. The calcium carbonate component is often made up of shell fragments of marine or freshwater organisms, or it can be precipitated by algae. Marls are usually whitish gray or brownish in color but may also be gray, green, red, or variegated depending on which minerals are present, such as glauconite or iron oxides.

Slice of marl

GREENSAND

Rock type Marine, detrital, sedimentary | **Fossils** Vertebrates, invertebrates, plants | **Major minerals** Quartz, glauconite | **Minor minerals** Feldspar, mica | **Color** Green | **Texture** Medium, angular

Greensand, also called glauconitic sandstone, is a quartz sandstone with a high percentage of glauconite, a green mica mineral. Although green on freshly broken surfaces, it typically weathers to a brown color on surface exposures. It is thought to form in shallow oxygen-poor marine environments that are rich in organic debris, where sediments accumulate slowly. Greensands often contain shell fragments and larger fossils. The potassium found in glauconite is useful in radiometric age dating. Due to its chemical exchange properties, the glauconite component of greensand is used as a water softener, in water treatment systems, and as an organic growing medium.

Greensand specimen

MICACEOUS SANDSTONE

Rock type Marine or freshwater, detrital, sedimentary | **Fossils** Invertebrates, plants, vertebrates | **Major minerals** Quartz, feldspar, mica | **Minor minerals** None | **Color** Buff, green, gray, pink | **Texture** Medium, angular to flattened

Micaceous sandstone is characterized by large amounts of mica and feldspar, in addition to quartz. It is a medium-grained rock, and its grains are well sorted. The majority of the grains are angular with the mica occurring typically as flakes. Its mica is usually muscovite and less commonly biotite, and is particularly visible where the rock is broken along bedding planes, commonly revealing an abundance of sparkling flakes. Micaceous sandstones are likely to have been deposited in water, since the small flakes of mica are very light, and are blown away easily in sediments deposited on the land surface.

Micaceous sandstone

IRONSTONE

Rock type Marine or continental, chemical, sedimentary | **Fossils** None or invertebrates | **Major minerals** Hematite, goethite, chamosite, magnetite, siderite, limonite, jasper | **Minor minerals** Pyrite, pyrrhotite | **Color** Red, black, gray, striped | **Texture** Fine to medium, crystalline to angular, oolitic

Ironstone is a general term for sandstones and limestones that are very rich in hematite, goethite, siderite, chamosite, or other iron minerals, and which contain more than 15 percent iron. These give the rock a dark red, brown, green, or yellow color, depending on which iron minerals are present. The formation of ironstone is something of a mystery since it no longer appears to be forming. Many seem to have formed early in Earth's history when oxygen was not as abundant in the atmosphere as it is now. Precambrian (more than 500 million years BCE) ironstones are common, as well as later ones formed prior to 240 million years ago.

METEORITES

Rock type Igneous, metamorphic, sedimentary | **Fossils** No macroscopic | **Major minerals** Olivine, pyroxene, plagioclase | **Minor minerals** Various | **Color** Gray, greenish, tan, or black | **Texture** Fine to medium

Meteorites are rocks that formed elsewhere in our solar system, were chipped from their parent bodies, and orbited the Sun until colliding with Earth. Friction causes most of these rocks to evaporate completely in the upper atmosphere, forming meteors (shooting stars). Meteorites are classified into three types based on properties visible to the naked eye: iron, stony-iron, and stony meteorites. Iron meteorites are formed mostly of metallic iron, mixed with varying amounts of nickel; stony-iron meteorites are composed of a mixture of iron and silicate minerals; and stony meteorites are composed entirely of stone. Many meteorites appear to be from the crust and mantle of now-disintegrated asteroids. One type of stony meteorites, the chondrites, make up the majority of the meteorites that fall to Earth. These are named for the small igneous, silicate spheres called chondrules, which are the dominant component in most chondritic meteorites. These are believed to be miniature igneous rocks that were melted and then crystallized while floating in space during the birth of the solar system.

Metamorphic rocks

SLATE

Rock type Regional metamorphic | **Temperature** Low | **Pressure** Low | **Structure** Foliated **Major minerals** Quartz, mica, feldspar | **Minor minerals** Pyrite, graphite | **Color** Various **Texture** Fine | **Protolith** Mudstone, siltstone, shale, or felsic volcanics

Slate is known for its characteristic "slaty cleavage," allowing it to be split into relatively thin, flat sheets. This cleavage results from microscopic mica crystals that have all grown oriented in the same plane. These are the planes formed during metamorphism rather than those of the original sedimentary layers. Slate forms when mudstone, siltstone, shale, or felsic volcanic rocks are buried and subjected to low temperatures and pressures. The color of slate depends on the mineralogy and oxidation conditions of the original sedimentary environment: black slate forms in a very oxygen-poor environment, while red slate forms in an oxygen-rich environment. Some slates are spotted, the spots being coarser-grained minerals that are scattered throughout the finer groundmass of the slate. Typical of these are cordierite and andalusite.

Slate specimen

PHYLLITE

Rock type Regional metamorphic | **Temperature** Low to moderate | **Pressure** Low | **Structure** Foliated | **Major minerals** Quartz, feldspar, muscovite mica, graphite, chlorite | **Minor minerals** Tourmaline, analusite, cordierite, pyrite, magnetite **Color** Silvery to greenish gray | **Texture** Fine **Protolith** Mudstone, shale, siltstone, or felsic volcanics

Phyllite is a dark colored, fine-grained rock that is usually gray or dark green with a sheen, because of its mica crystals. It is formed when fine-grained sedimentary rocks, such as mudstones or shales, are buried and then subjected to relatively low temperatures and pressures for a long period of time. Many phyllites have a scattering of large crystals that grow during metamorphism, and may include tourmaline, cordierite, andalusite, staurolite, biotite, and pyrite. Like slate, phyllite has a tendency to split because of the parallel alignment of its mica minerals. Split surfaces are more irregular than slate's, and it splits into slabs rather than sheets.

Phyllite with wavy foliation

▷ GNEISS

Rock type Regional metamorphic | **Temperature** High | **Pressure** High | **Structure** Foliated, crystalline | **Major minerals** Quartz, feldspar **Minor minerals** Biotite, hornblende, garnet, staurolite | **Color** Gray, pink, multicolored **Texture** Coarse | **Protolith** Granite, shale, granodiorite, mudstone, siltstone, or felsic volcanics

Gneiss is characterized by distinct bands of minerals of different colors and grain sizes. Gneiss is medium- to coarse-grained and has little or no tendency to split along planes, unlike schist. Its bands are folded, although the folds may be too large to see in a small specimen. Quartz and feldspar are abundant in most gneisses, but neither is necessary for a rock to be called gneiss, since the term refers to texture and origin rather than mineral content. Gneisses make up the cores of many mountain ranges, and form at very high temperatures and pressures from sedimentary or granitic rocks. They are popular as a stone for building and facing.

SCHIST

Rock type Regional metamorphic | **Temperature** Low to moderate | **Pressure** Low to moderate **Structure** Foliated | **Major minerals** Quartz, feldspar, mica | **Minor minerals** Garnet, actinolite, hornblende, graphite, kyanite | **Color** Silvery, green, blue | **Texture** Medium | **Protolith** Mudstone, siltstone, shale, or felsic volcanics

Schist has a texture composed of wavy, wrinkled, or irregular sheets. Schists have a varied composition, but usually include mica. Most are composed of platy minerals such as muscovite, chlorite, talc, biotite, and graphite, and often show distinct layering of light- and dark-colored minerals. Of the different schists, blueschist is rich in blue glaucophane; greenschist is rich in the green minerals chlorite, actinolite, and epidote; and garnet schist has numerous large crystals of garnet. The minerals present can help to determine the original rock and the schist's metamorphic history.

Schist rough

Lewisian gneiss occurs in the northwest Highlands of Scotland

FULGURITE

Rock type Contact metamorphic | **Temperature** Very high | **Pressure** Low | **Structure** Glassy (amorphous) | **Major minerals** Amorphous silica | **Minor minerals** Various | **Color** Gray, white | **Texture** Glassy to sandy | **Protolith** Usually sand

Fulgurites form when lightning strikes the ground, and are named after the Latin *fulgur*, for "thunderbolt." Fulgurites can form in any type of easily melted rock, but they are most easily discovered in desert regions where sand has been melted, and the surrounding loose material has blown or washed away. They are found as crusts when occuring in rock, or as tubes in sand, some of which branch out like trees. Fulgurite tubes are lined with fused sand, or glass, with partially melted sand stuck to the outside. The largest recorded fulgurite was about 16 ft (5 m) long. Fulgurites have no real use, although they are decorative and collectable.

Fulgurite tube

QUARTZITE

Rock type Regional metamorphic | **Temperature** High | **Pressure** Low to high | **Structure** Crystalline | **Major minerals** Quartz | **Minor minerals** Mica, kyanite, silimanite | **Color** Almost any | **Texture** Medium | **Protolith** Sandstone

This rock is formed by the burial, heating, and squeezing of quartz sandstone, and is thus converted into a solid quartz rock. "Quartzite" can also refer to sedimentary sandstone that has been converted to a much denser form through the quartz grains becoming cemented together with silica cement in pore spaces. Pure quartzite is usually white to gray when formed from relatively pure quartz sand, but can also be various shades of pink, red, yellow and orange, due to mineral impurities in the

parent sandstone. Metamorphic quartzite consists of interlocking crystals of quartz, whereas sedimentary quartzite contains rounded quartz grains.

Quartzite specimen

HORNFELS

Rock type Contact metamorphic | **Temperature** Moderate to high | **Pressure** Low to high | **Structure** Crystalline | **Major minerals** Hornblende, plagioclase, andalusite, cordierite, and others | **Minor minerals** Magnetite, apatite, titanite | **Color** Dark gray, brown, greenish, reddish | **Texture** Microcrystalline to fine | **Protolith** Almost any rock

Hornfels is formed close to igneous intrusions at temperatures as high as 1,300–1,450°F (700–800°C). It can form from almost any parent rock, so its composition depends on the parent rock and the exact temperatures and fluids to which the rock is exposed. Usually dense, hard, hard-to-break, very fine-grained, and sometimes glassy, hornfels is often evenly colored throughout as dark gray, reddish, greenish, or brown. It may also be banded. Garnet hornfels has large crystals of garnet set in a rock matrix, and cordierite hornfels contains large crystals of cordierite. Dark hornfels is easily confused with basalt.

AMPHIBOLITE

Rock type Regional metamorphic | **Temperature** Low to moderate | **Pressure** Low to moderate | **Structure** Foliated, crystalline | **Major minerals** Hornblende, actinolite | **Minor minerals** Feldspar, calcite, pyroxene | **Color** Gray, black, greenish | **Texture** Coarse | **Protolith** Basalt, graywacke, dolomite

Amphibolites are dark-colored, coarse-grained rocks dominated by amphiboles: black or dark-green hornblende, or green tremolite actinolite. Amphibolites form from the metamorphism of iron- and magnesium-rich igneous rocks, and from sedimentary rocks, such as graywacke. Although dominated by amphibole minerals they may

also contain microscopic grains of feldspar, pyroxene, and garnet. Mineral grains are aligned and sometimes the rock is banded. Amphibolites mark one of the major divisions in the classification of metamorphic rocks, and formed under conditions of moderate to high temperatures 950°F (500°C) and pressures.

MIGMATITE

Rock type Regional metamorphic | **Temperature** High | **Pressure** High | **Structure** Foliated, crystalline | **Major minerals** Quartz, feldspar, mica | **Minor minerals** Various | **Color** Banded light and dark gray, pink, white | **Texture** Coarse | **Protolith** Various, including granite and gneiss

Migmatite, meaning "mixed rock," consists of metamorphic schist or gneiss that is interlayered, streaked, or veined with igneous granite rock. These occur at the meeting of igneous and metamorphic rock. Layering may be tightly folded, and is the product of the softening or partial melting of the metamorphic rocks from the heat of the contact with the granite magma. The granitic parts of migmatites consist of granular patches of quartz and feldspar. The schist and gneiss parts are made up of quartz, feldspar, and dark-colored minerals. Migmatites form deep in the crust, and are rocks that represent the base of eroded mountain chains.

Migmatite specimen

GRANULITE

Rock type Regional metamorphic | **Temperature** High | **Pressure** High | **Structure** Crystalline | **Major minerals** Feldspar, quartz, garnet | **Minor minerals** Spinel, corundum | **Color** Gray, pinkish, brownish, mottled | **Texture** Medium to coarse | **Protolith** Felsic igneous and sedimentary rocks

Granulite is a coarse-grained rock that forms at high temperatures and pressures

deep in Earth's crust. Granulite has an even-grained granular texture, from which it takes its name. Tough and massive, granulites usually have a high concentration of pyroxene, with diopside or hypersthene, garnet, calcium plagioclase, and quartz or olivine. They consist of nearly the same minerals as gneisses, but are finer grained, less perfectly laminated, and have more garnet. They are of particular interest to geologists because many granulites represent samples of the deep continental crust, and are characteristic of the highest grade of metamorphism.

Granulite with garnets

▷ MARBLE

Rock type Regional or contact metamorphic | **Temperature** High | **Pressure** Low to high | **Structure** Crystalline | **Major minerals** Calcite | **Minor minerals** Diopside, tremolite, actinolite | **Color** White, pink | **Texture** Fine to coarse | **Protolith** Limestone, dolomite

Marble is a granular rock derived from limestone or dolomite, and is formed under heat and pressure. This occurs through the deep burial of limestone or dolomite in the older layers of the Earth's crust, with consequent heat and pressure from the thick layers of overlying sediments; or due to contact metamorphism near igneous intrusions. Pure limestone marble is nothing but calcite, but impurities in the original limestone can recrystallize during metamorphism to give mineral impurities, such as quartz, mica, graphite, iron oxides, and small pyrite crystals. Because the impurities were originally layers thinly interbedded in the original limestone, they sometimes occur as bands and swirls. Veined and patterned marbles are created when an existing marble was cracked or shattered, and the resulting spaces are filled in with other minerals.

Folded and weathered layers of marble form the Marble Caves, General Carrera Lake, Chile

7

Glossary
and
Index

Glossary

A

Acicular
Needlelike; the crystal habit of some minerals.

Adamantine
A bright, diamondlike luster.

Adularescence
See *Opalescence*.

Allochromatic
Gems colored by impurities, without which they would be colorless.

Asterism
A four- or six-ray star effect displayed by certain gems, including some sapphires and rubies, that have been cut *en cabochon*; the optical effect is caused by the reflection of light on fibrous or rutile inclusions.

B

Bezel
The part of the mounting that surrounds the girdle of a stone with a metal band.

Birefringence
In doubly refractive gems, this is the difference between the highest and lowest refractive indices.
See also *Double refraction*.

Brilliant cut
A round cut featuring mathematically calculated proportions of triangular facets top and bottom, which are designed to maximize a diamond's fire and brilliance.

C

Cabochon
A polished cut with a domed upper surface and a flat or domed under surface; gems cut in this way are described as being cut *en cabochon*.

Cameo
A low-relief design that has been cut into layered stone or shell, with the background material cut away.

Carat
A unit of gemstone weight. One carat equals 0.007oz (0.2g). (Not to be confused with karat, a measure of gold purity.)
See also *Karat*.

Chatoyancy, chatoyant
The cat's-eye effect shown on certain gems that have been cut *en cabochon*.

Clast
A fragment or grain of rock, usually broken off as a result of physical weathering; clastic rocks are a form of sedimentary rock composed of such clasts.

Cleavage
The way that some minerals break along planes determined by their atomic structure.

Crown
The top part of a cut stone, above the girdle.

Cryptocrystalline
An extremely fine-grained crystalline mineral habit, in which individual crystallized components can only be seen under a microscope.

Crystal
A solid with an ordered internal atomic structure that produces a typical external shape, along with characteristic physical and optical properties.

Crystal structure
The internal atomic structure of a crystal. All crystalline gems may be classified according to the symmetry of their structure as cubic; tetragonal; hexagonal; trigonal; orthorhombic; monoclinic; and triclinic.

Culet
The lowest part of a cut stone, either a point or a ridge.

Cushion
A square cut with rounded sides and corners.

Cut
The shaping of a gemstone by grinding and polishing; the shape of the final gem, as, for example, in brilliant cut.

D

Dendritic
Treelike; the crystal habit exhibited by some minerals.

Diffraction
The splitting of white light into its component colors—the colors of the rainbow—when it passes through a hole or grating; the bending of light rays around the edge of an obstacle.

Dispersion
The splitting of white light into its component colors—the colors of the rainbow—as it passes through an inclined surface such as those on a prism or a faceted gem. Dispersion in gems is known as fire.

Double refraction (DR)
The splitting of light into two separate rays as it enters a gem. Each ray travels at a different speed and has its own refractive index.

E

Extrusive
A type of rock formed from lava that has either flowed onto the Earth's surface or was ejected from a volcanic vent.

F

Faces
The flat surfaces that make a crystal's external shape.

Facet, faceting
The cutting and polishing of multiple flat surfaces (facets) on a gem. The cut is named according to the number and shape of the facets.

Fancy
A gem cut with an unconventional shape, such as a heart.

Fire
See *Dispersion*.

Fluorescence, fluorescent
The glow of some gems under ultraviolet light, caused by impurities in their crystal structure.

Fracture
A mineral breakage or chipping unconnected to cleavage planes, which is thus often uneven.

Freeform cut
A fancy cut that does not follow a regular geometric pattern.

G

Geode
A rock cavity, often rounded, that is lined with crystals.

Girdle
A band around the widest part of a cut gem, dividing the crown and the pavilion.

Granular
Having grains, or being in the form of grains.

Groundmass
A fine-grained rock in which larger crystals are set or upon which they rest. See also *Matrix*.

H

Habit
The external shape in which a crystal grows because of its molecular structure.

I

Idiochromatic
A self-colored gem, in which the color comes from its chemical composition, not from impurities.

Igneous
A type of rock formed from solidified molten rock.

Inclusion
A crystal or fragment of another substance occurring within a gem; it is sometimes a way of identifying a species of gem.

Intaglio
A design in which the subject is cut lower than the background; the reverse of a cameo. See also *Cameo*.

Intrusive
An igneous rock that has solidified within other rocks under the Earth's surface.

Iridescence, iridescent
The rainbow array of colors displayed when light reflects off elements within a gem.

K

Karat
A unit describing the purity of gold. It refers to the amount of gold in 24 parts of a gold alloy. 24-karat is pure gold; 18-karat is three-quarters gold; 12-karat is half gold; and so on. See also *Carat*.

L

Lapidary
A person who cuts and polishes gems.

Luster
The shine of a gem, which is caused by reflected light.

M

Massive
A mineral that has no definite shape or consists of small crystals in masses.

Matrix
The rock in which a gem is found. Also known as groundmass, host rock, or parent rock.

Metamorphic
A rock that has been transformed from one type of rock into another, due to the effects of heat or pressure, or a combination of the two.

Microcrystalline
A mineral habit in which crystals are too small to be seen with the naked eye.

Mineral
An inorganic, naturally occurring material that has a fixed chemical composition and a regular internal atomic structure.

Mixed cut
A cut in which the facets above and below the girdle differ. This usually takes the form of a brilliant cut above and a step cut below.

Mohs scale
The measure of a gem's relative hardness based on its resistance to scratching.

Mounting
The jewellery piece that a gem is, or gems are, set into. Also called a setting.

N

Native element
A chemical element that occurs naturally uncombined with other elements.

O

Opalescence
A bluish-white form of iridescence.

Ore
A rock or mineral from which a metal can be commercially extracted.

Organic gem
A gem that is composed of material made by, or from, living organisms.

P

Parti-colored
Single crystals that are made up of different colors.

Pavilion
The lower part of a faceted gem, below the girdle.

Pegmatite
A type of mineral vein that is characterized by the presence of large, often well-formed, crystals.

Phenocryst
A relatively large crystal set into the matrix of an igneous rock, giving it a porphyritic texture.

Placer deposit
A (secondary) deposit of minerals derived by weathering, and concentrated in streams or beaches because of their high specific gravity.

Pleochroic
A gem that exhibits different colors when viewed from different angles.

Polymorph
A substance that can exist in two or more crystalline forms; one crystalline form of such a substance.

Porphyry, porphyritic
An igneous rock that is textured by large crystals set in a finer matrix.

Precipitation
The condensation of a solid from a liquid or gas.

Prismatic
A mineral habit in which parallel rectangular faces form prisms.

Pseudomorph
A crystal with the outward form of another mineral species.

R

Refraction
The bending of light rays as they pass from one medium into another.

Refractive index (RI)
The measure of light slowing down and bending as it enters a gem. It can be used to identify cut gems and some mineral species.

Rhombohedral
A crystal that is shaped like a skewed cube.

Rock
Material made up of one or more minerals.

Rough
An uncut gem crystal.

S

Scalenohedral
A crystal composed of two base-to-base hexagonal pyramids.

Schiller effect, sheen
A brilliant play of bright colors in a crystal that is often due to minute, rodlike inclusions. It is a form of iridescence.

Sedimentary
A rock formed by the consolidation and hardening of fragments of pre-existing rock, organic remains, or other material.

Species
Of gemstones, individual gems that have definite, verifiable characteristics.

Specific gravity (SG)
The ratio of the mass of a mineral to the mass of an equal volume of water. Specific gravity is numerically equivalent to density (mass divided by volume) in grams per cubic centimetre.

Step cut
A type of cut with a rectangular table facet and girdle, with parallel rectangular facets.

Striation
A series of parallel grooves or lines on a crystal.

T

Table facet
The central facet on the crown of a gem.

Tabular
A habit in which crystals take the shape of a cereal box.

Tetrahedral
A crystal made up of four triangular faces in pairs, rotated 90 degrees from each other.

Tumble polishing
The process of rotating gemstones in a barrel with abrasives in order to round and polish them.

Twinned crystals
Crystals that grow together as mirror images with a common face or at angles of up to 90 degrees to each other.

V

Variety
Of gemstones, a subspecies that has characteristics which are similar but not distinct enough to justify being classed as a separate species.

Vein
A thin, sheetlike mass of rock that fills fractures in other rocks.

Vitreous
Possessing a glasslike luster, common in gemstones.

Index

Acknowledgments

The publisher would like to thank the following for their work on the book:
Contributors: Ronald Bonewitz, Iain Zaczek, Alison Sturgeon, Alexandra Black. UK consultant: Andrew Fellows. Indexer: Margaret McCormack. Editorial assistance: Fergus Day, Richard Gilbert, Georgina Palffy, Helen Ridge, Anna Limerick, Kate Taylor, Sam Atkinson, Kathryn Hennessy. Design assistance: Phil Gamble, Saffron Stocker, Phil Fitzgerald, Steve Crozier, Tom Morse, Ray Bryant, Paul Reid at cobalt id, Vanessa Hamilton. DTP Designers: Syed Mohammad Farhan, Vijay Kandwal, Ashok Kumar, Mohammad Rizwan. Additional photography: Gary Ombler, Richard Leeney.

Dorling Kindersley would especially like to thank the following for their assistance with this book:
Robert Acker Holt, Samantha Lloyd, and all at **Holts Gems** for kindly allowing us to photograph their collection; **The Al Thani Collection**; Laura Behaegel and Harriet Mathias at **Cartier**; Judy Colbert at the **GIA** (Gemological Institute of America); Benjamin Macklowe and Antonio Virardi at the **Macklowe Gallery**; Sonya Newell-Smith at the **Tadema Gallery**; Kealy Gordon and Ellen Nanney at the **Smithsonian Institution**; Megan Taylor at **Luped** for picture research assistance.

The publisher would like to thank the following for their kind permission to reproduce their photographs:

(Key: a-above; b-below/bottom; c-centre; f-far; l-left; r-right; t-top)

Smithsonian Institution, Washington, DC: Cooper-Hewitt, National Design Museum 215cr, 215bl, National Museum of Natural History / Chip Clark 5tr, 6tl, 27 (C), 46, 54tr, 54br, 56tl, 70, 71bc, 72fbr, 76fcr, 77tl, 113tl, 116c, 132, 152tl, 152cr, 159br, 161tr, 161fcr, 169bl, 170tl, 171br, 206bl, 209br, 216fcl, 233cr, 240fbr, 243cl, 259bl, 285bl, 332c, 332crb, 333crb, National Museum of Natural History / Ken Larsen 95bl, 98fcr, 105bl, 117bc, 168cl, 168cr, 168br, 168fcr, 197br, 340br, 346crb, National Museum of Natural History / Paula Crevoshay 56cr, 56bc, 73c, 105br, National Museum of Natural History / Sena Dyber 138l, 345clb, Sherris Cottier Shank 245cra

2 Alamy Stock Photo: Private Collection. **4 Fellows Auctioneers. Sothebys Inc. 5 akg-images:** (tl). **Fellows Auctioneers. Getty Images:** Peter Macdiarmid (ftl). **6 Bridgeman Images:** De Agostini Picture Library / E. Lessing (tl). **© Cartier. 7 The Al-Thani Collection:** Servette Overseas Ltd 2012, all rights reserved. Photographs taken by Prudence Cuming Associates Ltd (tr). **Alamy Stock Photo:**Ian Dagnall (tl); Adam Eastland Art + Architecture (ftr). **Dorling Kindersley:** Holts (ftl). **9 © Cartier. 11 Dorling Kindersley:** Holts. **12 Dorling Kindersley:** Natural History Museum, London (clb). **13 Dorling Kindersley:** Tim Parmenter / Natural History Museum, London (cr). **14 Dorling Kindersley:** Natural History Museum, London (clb). **15 Dorling Kindersley:** Holts (cl, ca, fbl, bc/A, fbr, br); Ruth Jenkinson (bl). **Science Photo Library:** Dirk Wiersma (bc/B). **16 Dorling Kindersley:** Natural History Museum, London (clb, bc); Holts (cl, bl, br). **Science Photo Library. 16-17 Dorling Kindersley:**Holts (c, b). **17 Dorling Kindersley:** Natural History Museum, London (clb, br); Oxford University Museum of Natural History (cr). **19 Dorling Kindersley:** Natural History Museum, London (cra). **20 Alamy Stock Photo:** Wildlife GmbH (cl). **Dorling Kindersley:** Natural History Museum, London (bl, fcr, fbr); Holts (c, br). **Gemological Institute of America Reprinted by Permission:** Jian Xin (Jae) Liao (fcrb); Robert Weldon / courtesy Dr. E. J. Gübelin Collection (fbl). **Science Photo Library:** Alfred Pasieka (cr). **21 Alamy Stock Photo:** Martin Baumgaertner (l). **Dorling Kindersley:** Natural History Museum, London (G, J, M, N, P); Holts (A, F, L, O, S Q); Richard Leeney (B); Richard Leeney (D); Richard Leeney (H); Richard Leeney (R). **Gemological Institute of America Reprinted by Permission:** Jian Xin (Jae) Liao (K). **Science Photo Library:** Joel Arem (E); Dorling Kindersley / UIG (C). **22 Dorling Kindersley:** Natural History Museum, London (C); Holts (A, B, D, E, F, H, I, J, K, L, M, N, O, P, Q); Ruth Jenkinson (S). **Science Photo Library:** Vaughan Fleming (R). **23 Dorling Kindersley:** Natural History Museum, London (tr, fbl, bc); Holts (cla, cl, c, cr). **Science Photo Library:**Paul Biddle (fcl). **25 Getty Images:** samvaltenbergs (tr). **26 Alamy Stock Photo:** Alan Curtis (c). **Dorling Kindersley:** Natural History Museum, London (cr, bc); Holts (br). **27 123RF.com:** Ingemar Magnusson (H). **Dorling Kindersley:** Natural History Museum, London (I); Holts (A, D, E, F, G, J, L, M, N, O, P, Q, R, S, U, W, X, Y, Z, ZA); Natural History Museum, London / Tim Parmenter (K); Richard Leeney (V). **Gemological Institute of America Reprinted by Permission:** Robert Weldon / courtesy Minerales y Metales del Oriente, Bolivia, SA (T). **29 Dorling Kindersley:** Natural History Museum, London (tr); Holts (cra, crb, tc, ca, cb, bc). **Fellows Auctioneers. 30 Dorling Kindersley:** Holts (c, cra, bl, br). **31 Gemological Institute of America Reprinted by Permission:** Robert Weldon (br, bl, c, cl, tr); Robert Weldon (bc); Robert Weldon (cr). **32 Bridgeman Images:** Birmingham Museums and Art Gallery (bc); Indianapolis Museum of Art, USA / Gift of Mr. and Mrs. Eli Lilly (cl).**The Art Archive:** Ashmolean Museum (fcla); Musee du Louvre Paris / Kharbine-Tapabor (fcl); Egyptian Museum Cairo / Araldo De Luca (cr). **33 Bridgeman Images:** Christie's Images (ftr); Museo Nazionale del Bargello, Florence, Italy (ftl); State Hermitage Museum, St. Petersburg, Russia (fbl); Weltliche und Geistliche Schatzkammer, Vienna, Austria (cl); De Agostini Picture Library (bl); Fitzwilliam Museum, University of Cambridge, UK (tl); Fitzwilliam Museum, University of Cambridge, UK (tr). **Van Cleef & Arpels. 36 Verdura**. **37 akg-images:** Bildarchiv Steffens (bl). **© Cartier:** (br).**Dorling Kindersley:** Natural History Museum, London (tl). **Getty Images:** Araldo de Luca / Corbis (bc). **38 Bridgeman Images:** De Agostini Picture Library / E. Lessing (bl). **The Trustees of the British Museum. © Cartier. Corbis:** David Lees (ftr). **39 The Trustees of the British Museum:** (bl, bl). **Bulgari:** (fbr). **© Cartier. Dorling Kindersley:** The Trustees of the British Museum (bl). **Fellows Auctioneers. Antonio Virardi of Macklowe Gallery, New York. 39 The Trustees of the British Museum:** (bl, bl). **Bulgari:** (fbr). **© Cartier. Dorling Kindersley:** The Trustees of the British Museum (bl). **Fellows Auctioneers. Antonio Virardi of Macklowe Gallery, New York. 40 akg-images:** Pictures From History. **41 akg-images:** Nimatallah (tl). **Bridgeman Images:** De Agostini Picture Library / Chantilly, Château, Musée Condé (Picture Gallery And Art Museum) (cr); French School, (14th century) / Bibliotheque Municipale, Castres, France (bl). **Muenze Oesterreich AG:** (cl). **42 1stdibs, Inc:**(tl); Macklowe Gallery / Anotonio Virardi (fbr). **Alamy Stock Photo:** David J. Green - technology (bl). **Dorling Kindersley:** Colin Keates / Natural History Museum, London (fcl); Tim Parmenter / Natural History Museum, London (cl). **Fellows Auctioneers. 43 1stdibs, Inc. Bonhams Auctioneers, London. Rijksmuseum Amsterdam:** (tc). **Sothebys Inc. 44 Alamy Stock Photo:** philipus (bl). **Dorling Kindersley:** Holts (fcr); Natural History Museum, London (tl); Natural History Museum, London (fcl); Natural History Museum, London (cr). **Getty Images:** DEA / R.Appiani (cl). **45 1stdibs, Inc. © Cartier. The Goldsmiths' Company:** Leo De Vroomen (tc). **Antonio Virardi of Macklowe Gallery, New York. 47 Bridgeman Images:** (br); Christie's Images (tl). **Corbis:** Leemage (cb). **48 Dorling Kindersley:** Canterbury City Council, Museums and Galleries (br); Ruth Jenkinson (c); University of Pennsylvania Museum of Archaeology and Anthropology (cr). **49 Dorling Kindersley:** Newcastle Great Northern Museum, Hancock (fcr); The Trustees of the British Museum (tl, cl); University of Pennsylvania Museum of Archaeology and Anthropology (cr); University of Pennsylvania Museum of Archaeology and Anthropology (bl). **Dreamstime.com:** Wojpra (br). **50 Photo Scala, Florence:** Marie Mauzy. **51 akg-images:** Erich Lessing (cl). **Getty Images:** Universal Images Group (tl). **Library of Congress, Washington, D.C.:** (cr). **The Art Archive:** Musée du Louvre Paris / Gianni Dagli Orti (bl). **52 Graff Diamonds**. **53 Bridgeman Images:** Christie's Images (br). **© Cartier. The Royal Collection Trust © Her Majesty Queen Elizabeth II:** 2016 (bc). **Photo Scala, Florence:** bpk, Bildagentur fuer Kunst, Kultur und Geschichte, Berlin (bl). **54 Dorling Kindersley:** Natural History Museum, London (tl, cl, bl); Holts (l, ftr, fcr). **55 Dorling Kindersley:** Holts (t, fcl, fbl, fcr, br). **Science Photo Library:** Vaughan Fleming (cl). **56 Bridgeman Images:** Christie's Images (bl). **Fellows Auctioneers:** (tc, tr). **57 Bridgeman Images:** Private Collection / Photo © Christie's Images (br). **© Cartier. Dorling Kindersley:** Holts (fcl). **Fellows Auctioneers. Getty Images:** Peter Macdiarmid (bc). **Tadema Gallery:** (bl). **Van Cleef & Arpels:** (cr). **58 The Royal Collection Trust © Her Majesty Queen Elizabeth II**. **59 Alamy Stock Photo:** V&A Images (cl). **Bibliothèque nationale de France, Paris:** (cr). **The Royal Collection Trust © Her Majesty Queen Elizabeth II:** 2016 (tl, bl). **60-61 The Trustees of the British Museum. 61 Bridgeman Images:** Egyptian National Museum, Cairo, Egypt (br). **62 Corbis:** Smithsonian Institution. **63 Corbis:** Smithsonian Institution (tl); Leemage (cl). **Library of Congress, Washington, D.C.:** (bc). **Museum National d'Histoire Naturelle:** François Farges (cr). **65 Dorling Kindersley:** Holts. **66 Dorling Kindersley:** Ruth Jenkinson / Holts (cl, br). **67 Dorling Kindersley:** Tim Parmenter / Natural History Museum, London (bc). **68 The Royal Collection Trust © Her Majesty Queen Elizabeth II:** 2016. **69 Bridgeman Images:** Her Majesty Queen Elizabeth II, 2016 (bc); Chetham's Library, Manchester, UK (tl); Walker, Robert (1607-60) / Leeds Museums and Galleries (Leeds Art Gallery) U.K. (cr). **71 Bridgeman Images:** Kremlin Museums, Moscow, Russia (bl). **© Cartier. 72 Alamy Stock Photo:** Rhea Eason (cl); ZUMA Press, Inc (bl); Greg C Grace (ftr). **Dorling Kindersley:** Natural History Museum, London (ftl, c, cr); Holts (tl, tr). **Science Photo Library:** Joel Arem (br). **73 Dorling Kindersley:** Holts (br, fbr); Judith Miller / Sloane's (fbl). **Fellows Auctioneers. 74 Getty Images:** STF. **75 Dulong Fine Jewellery:** Sara Lindbaek (tl, bl). **Getty Images:** Print Collector (c). **Rex by Shutterstock:** Tim Rooke (cr). **76 Dorling Kindersley:** Holts (l, cl, cr).**Antonio Virardi of Macklowe Gallery, New York. 77 © Cartier. Dorling Kindersley:** Holts (bc); Tim Parmenter / Natural History Museum, London (br). **Fellows Auctioneers.. Antonio Virardi of Macklowe Gallery, New York. 78 Alamy Stock Photo:** V&A Images. **79 Alamy Stock Photo:** Dinodia Photos (cr). **The Royal Collection Trust © Her Majesty Queen Elizabeth II:** 2016 (tl, cl). **80 Dorling Kindersley:** Colin Keates / Natural History Museum, London (cr); Tim Parmenter / Natural History Museum, London (tl, c); Holts (bl). **The Royal Collection Trust © Her Majesty Queen Elizabeth II:** 2016 (br). **81 1stdibs, Inc. © Cartier. Dorling Kindersley:** Holts (tl, bc); Tim Parmenter / Natural History Museum, London (fcl). **Gemological Institute of America Reprinted by Permission:** Robert Weldon / Ring courtesy of a Private Collector and Mona Lee Nesseth, Custom Estate Jewels (tc). **82 Alamy Stock Photo:** Granger, NYC. **83 Bridgeman Images:** Tretyakov Gallery, Moscow, Russia (bl); Kremlin Museums, Moscow, Russia (c). **Getty Images:** Leemage (tl); Mondadori Portfolio (cr). **84 Dorling Kindersley:** Natural History Museum, London (c, fbl, cl); Holts (cr, bc, br). **85 1stdibs, Inc. © Cartier. Dorling Kindersley:** Natural History Museum, London (c, bl). **Gemological Institute of America Reprinted by Permission. Tadema Gallery. 86 Dorling Kindersley:** Natural History Museum, London (br); Ruth Jenkinson (bc). **87 Bonhams Auctioneers, London:** (cr). **Dorling Kindersley:** Natural History Museum, London (fcl, bc). **Gemological Institute of America Reprinted by Permission. 88 Alamy Stock Photo:** Annie Eagle (bl); Universal Images Group North America LLC / DeAgostini (tl). **Dorling Kindersley:** Tim Parmenter / Natural History Museum, London (bc, br). **89 Dorling Kindersley:** Oxford University Museum of Natural History (br). **Getty Images:** Matteo Chinellato - Chinellato Photo (tl, cr). **90 © Cartier. 91 Bridgeman Images:** Archives Charmet (cr); Christie's Images (cla, bc). **© Cartier. Getty Images:** Universal History Archicve / UIG (fcra). **92 The Al-Thani Collection:** Servette Overseas Ltd 2012, all rights reserved. Photographs taken by Prudence Cuming Associates Ltd (tl, cra, bl, clb). **Van Cleef & Arpels. 92-93 The Al-Thani Collection:** Servette Overseas Ltd 2012, all rights reserved. Photographs taken by Prudence Cuming Associates Ltd (b). **93 The Al-Thani Collection:** Servette Overseas Ltd 2012, all rights reserved. Photographs taken by Prudence Cuming Associates Ltd (c, t). **94 Alamy Stock Photo:** Eddie Gerald (bl). **Bonhams Auctioneers, London:** (fcr, br). **Corbis. Dorling Kindersley:** Holts (fcl, cl). **96 Dorling Kindersley:** Holts (br). **97 Dorling Kindersley:** Colin Keates / Natural History Museum, London (cl); Tim Parmenter / Natural History Museum, London (bc). **98 Alamy Stock Photo:** imageBROKER (bl). **Corbis. Dorling Kindersley:** Natural History Museum, London (tl). **99 Alamy Stock Photo:** Goran Bogicevic (bl). **Dorling Kindersley:** Oxford University Museum of Natural History (cl); Tim Parmenter / Natural History Museum, London (fbl). **Dreamstime.com:** (fbr). **100 Bonhams Auctioneers, London. Dorling Kindersley:** Ruth Jenkinson (cl, fcr). **101 Dorling Kindersley:** Natural History Museum, London (cl, bc). **Getty Images:** Print Collector (br). **102-103 Alamy Stock Photo:** Susana Guzman. **104 Alamy Stock Photo:** The Natural History Museum (bl); Universal Images Group North America LLC / DeAgostini (c); Valery (cr). **105 Alamy Stock Photo:** PjrStudio (cr). **106 Dorling Kindersley:** Natural History Museum, London (bc). **Getty Images:** Universal History Archive (bl). **107 Dorling Kindersley:** Natural History Museum, London (bc); Ruth Jenkinson (cl, br/pendant, r). **108 Alamy Stock Photo:** Heritage Image Partnership Ltd. **109 Alamy Stock Photo:** SilverScreen (crb); Lilyana Vynogradova (ca). **Nordiska Museet:** Mats Landin (tl, bl). **110 Dorling Kindersley:** Holts (br, tl, cl, bc, cr); Tim Parmenter / Natural History Museum, London (bl). **111 Dorling Kindersley:** Holts (tc, tl). **Fellows Auctioneers. Tadema Gallery. 112 akg-images**. **113 Corbis:**Underwood & Underwood (br). **RMN:** Gérard Blot (cr); Jean-Gilles Berizzi (bl). **114 Bridgeman Images:** Natural History Museum, London, UK (G). **Dorling Kindersley:** Natural History Museum, London (F); Holts (H); Holts (C); Holts (E, D, K). **Dreamstime.com:** Nastya81 (A). **115 Alamy Stock Photo:** Jon Helgason (B). **Bridgeman Images:** Natural History Museum, London, UK (N). **Dorling Kindersley:** Natural History Museum, London (H, I, L); Natural History Museum, London (E); Holts (M); Holts (C, tr/D, K, A, F). **Getty Images:** t_kimura (G). **116 Alamy Stock Photo:** SPUTNIK (bl); Universal Images Group North America LLC / DeAgostini (tl, br). **Dorling Kindersley:** Natural History Museum, London (cl); Natural History Museum, London (cr). **117 Dorling Kindersley:** Natural History Museum, London (cr); Natural History Museum, London (br). **118 Dorling Kindersley:** Natural History Museum, London (cr, fbl, bl, br); Holts (fbr). **119 Alamy Stock Photo:** John Cancalosi (bc); Valery Voennyy (bl). **Dorling Kindersley:** Natural History Museum, London (br). **120 Alamy Stock Photo:** blickwinkel (c). **Dorling Kindersley:** Natural History Museum, London (br, fbr). **Science Photo Library:** Larry Berman (bl). **121 Alamy Stock Photo:** Karol Kozlowski (cr); rep0rter (cl); Universal Images Group North America LLC / DeAgostini (fcr); PjrStudio (br). **Dorling Kindersley:** Natural History Museum, London (bl). **122 Alamy Stock Photo:** geoz (cl, fcl); Andrew Holt (bl). **Bonhams Auctioneers, London. Corbis. Dorling Kindersley:** Durham University Oriental Museum (cr); University of Pennsylvania Museum of Archaeology and Anthropology (bc). **123 Alamy Stock Photo:** Fabrizius Troy (br). **Dorling Kindersley:** Natural History Museum, London (tl); Ruth Jenkinson (bc).**124-125 Bridgeman Images:** Lukas - Art in Flanders VZW / Photo: Hugo Maertens. **126 Alamy Stock Photo:** John Cancalosi (cl); RF Company (tl); Corbin17 (bc). **Corbis:** (bl); Visuals Unlimited (cr).**Dorling Kindersley:** Natural History Museum, London (br). **127 1stdibs, Inc. Alamy Stock Photo:** Oleksiy Maksymenko (cr); Steve Sant (cl). **Dorling Kindersley:** Natural History Museum, London (fcl, fcr). **128 Bridgeman Images:** Walker Art Gallery, National Museums Liverpool. **129 Bridgeman Images:** Victoria & Albert Museum, London, UK / The Stapleton Collection (bl); Walker Art Gallery, National Museums Liverpool (tl). **Corbis:** Heritage Images (crb). **Van Cleef & Arpels. 130 123RF.com:** Michał Barański (ca); Laurent Renault (c). **Bridgeman Images:** Christie's Images (tl). **Dorling Kindersley:** Natural History Museum, London (br); Holts (fcl, clb); Holts (bl). **130-131 Dorling Kindersley:** Ruth Jenkinson (cb). **131 123RF.com:** Dipressionist (ftl). **Bridgeman Images:** Natural History Museum, London, UK (tr). **Dorling Kindersley:** Holts (ftr, cr); Natural History Museum, London (cb). **133 Bridgeman Images:** Purchase from the J. H. Wade Fund (bl). **© Cartier. V&A Images / Victoria and Albert Museum, London. 134 Bridgeman Images:** Natural History Museum, London, UK (bl). **Dorling Kindersley:** Natural History Museum, London (cr); Holts (tl, tc). **135 Dorling Kindersley:**Natural History Museum, London (c, fbl, bl); Holts (cr, br, fbr). **Science Photo Library:** Mark A. Schneider (tc). **136 Dorling Kindersley:** Natural History Museum, London (ftl, tr, ftr, c); Holts (bl). **Science Photo Library:** Natural History Museum, London (br). **137 Alamy Stock Photo:** Universal Images Group North America LLC / DeAgostini (cr). **Dorling Kindersley:** Natural History Museum, London (tc, cl, fbr); Holts (tl, c, fbl, bl, br). **138 Bridgeman Images:** Heini Schneebeli (tr). **Fellows Auctioneers:** (br). **Tadema Gallery:** (cr). **139 Bridgeman Images:** Boltin Picture Library (br). **Dorling Kindersley:**Natural History Museum, London (bl); Holts (cr); Judith Miller / Private Collection (bc). **Fellows Auctioneers:** (tl, cl, c). **140 Photo Scala, Florence:** bpk, Bildagentur fuer Kunst, Kultur und Geschichte, Berlin. **141 akg-images:** (cr); Erich Lessing (tl). **Alamy Stock Photo:** Prisma Bildagentur AG (bl). **Photo Scala, Florence:** bpk, Bildagentur fuer Kunst, Kultur und Geschichte, Berlin (cl). **142 Dorling Kindersley:** Holts (bl, fcr, fbl); Natural History Museum, London (fbr). **RMN:** Droits réservés (cl). **Science Photo Library:** Natural History Museum, London (br, clb). **142-143 Dorling Kindersley:** Harry Taylor (t). **143 Alamy Stock Photo:** Mykola Davydenko (br). **Dorling Kindersley:** Natural History Museum, London (ca); Ruth Jenkinson (tr); Harry Taylor (cr). **Dreamstime.com:** Ismael Tato Rodriguez (cb). **144 Bayerische Schlösserverwaltung:** Maria Scherf / Rainer Herrmann, München. **145 Getty Images:** DEA / A. DeGregorio (cr); Imagno (tl); Heritage Images (cl); Print Collector (bl). **146 Bonhams Auctioneers, London**. **147 Alamy Stock Photo:** The Art Archive (bl); World History Archive (bc). **148 Alamy Stock Photo:** bilwissedition Ltd. & Co. KG (bl); Universal Images Group North America LLC / DeAgostini (br). **Bonhams Auctioneers, London:** (cr). **Dorling Kindersley:** Holts (bc); Natural History Museum, London / Tim Parmenter (cl). **149 1stdibs, Inc. Bonhams Auctioneers, London. Dorling Kindersley:** Holts (tl, ftl). **150 V&A Images / Victoria and Albert Museum, London:** (tl, tc, tr). **151 Bridgeman Images:** (crb); Look and Learn (bc). **V&A Images / Victoria and Albert Museum, London. 152 Dorling Kindersley:** Holts (bl, c, br). **153 Dorling Kindersley:** Natural History Museum, London (fcl, cl, fbr); University of Pennsylvania Museum of Archaeology and Anthropology (ftl); Oxford University Museum of Natural History (tl); Holts (tr, cr). **Fellows Auctioneers. 154 Dorling Kindersley:** Natural History Museum, London (tl, cl, c, br); Holts (cr, fbr). **155 © Cartier. Dorling Kindersley:** Natural History Museum, London (bl, fcr); Holts (bc). **Fellows Auctioneers. 156-157 Bridgeman Images**. **158 © Cartier. 159 Bridgeman Images:** (bl); Christie's Images (bc). **Gemological Institute of America Reprinted by Permission:** Thomas Cenki (tl). **160 Alamy Stock Photo:** Zoonar GmbH (ftr). **Bonhams Auctioneers, London. Dorling Kindersley:** Natural History Museum, London (ftl); Holts (tl, tr, fbr). **161 1stdibs, Inc. © Cartier. Fellows Auctioneers. Tadema Gallery. Van Cleef & Arpels. 162 Bonhams Auctioneers, London. 163 Alamy Stock Photo:** AF Fotografie (cr). **Gayle Beveridge:** (bl). **Bonhams Auctioneers, London. 164 Dorling Kindersley:** Colin Keates / Natural History Museum, London (fbl); Tim Parmenter / Natural History Museum, London (tl, fcl); Holts (cr, fcr, br, fbr). **Gemological Institute of America Reprinted by Permission. 165 © Cartier. Dorling Kindersley:** Holts (tl). **Fellows Auctioneers. Tadema Gallery. 166 Corbis:** Marc Dozier (cla). **Dorling Kindersley:** Holts (clb); Holts (br). **Science Photo Library:** (bc). **167 Alamy Stock Photo:** John Cancalosi (br). **Dorling Kindersley:** Natural History Museum, London (bl). **Gemological Institute of America Reprinted by Permission:** Jeff Scovil (ca). **Science Photo Library. 168 Dorling Kindersley:** Natural History Museum, London (fcl); Ruth Jenkinson (bl). **169 Dorling Kindersley:** Holts (cl, bc); Ruth Jenkinson (cr). **170 Alamy Stock Photo:** The Natural History Museum (bc). **Dorling Kindersley:** Natural History Museum, London (cl, br, fcr); Holts (cr/Cabochon). **171 Dorling Kindersley:** Natural History Museum, London (tl, bl, bc). **172 Dorling Kindersley:** Natural History Museum, London (bl, br). **173 Alamy Stock Photo:** Susan E. Degginger (fbl); Siim Sepp (cl). **Getty Images:** Ron Evans (cr). **iRocks.com/Rob Lavinsky Photos:** (fcr). **174 David Webb**. **175 Bridgeman Images:** Ashmolean Museum, University of Oxford, UK (bc); Egyptian National Museum, Cairo, Egypt (bl). **© Cartier. Dorling Kindersley:** Holts (tl). **176 Alamy Stock Photo:** The Art Archive (fbr); Interfoto (fcr). **Dorling Kindersley:** Natural History Museum, London (cl, cr/Cabochon); Holts (tl, tc, tr, fcl). **Science Photo Library:** Joel Arem (br). **177 1stdibs, Inc. Bulgari. Fellows Auctioneers. Getty Images:** DEA / A. Dagli Orti (tr); Mark Moffet / Minden Pictures (cl).**178 Bridgeman Images:** Boltin Picture Library (cl). **Dorling Kindersley. Getty Images:** DEA / S. Vannini (br). **179 Bridgeman Images. The Trustees of the British Museum. Dorling Kindersley:**Durham University Oriental Museum (fcr); University of Pennsylvania Museum of Archaeology and Anthropology (crb); University of Pennsylvania Museum of Archaeology and Anthropology (cr). **180 Dorling Kindersley:** Natural History Museum, London (bl); The Science Museum, London (fcr); Ruth Jenkinson (fcl, br); Natural History Museum, London / Tim Parmenter (tl). **181 Bonhams Auctioneers, London. Bridgeman Images:** De Agostini Picture Library (bc). **Dorling Kindersley:** Natural History Museum, London (tl, fcl). **iRocks.com/Rob Lavinsky Photos. 182 Bridgeman Images:** Louvre-Lens, France. **183 Alamy Stock Photo:** Everett Collection Historical (cla). **Bridgeman Images:** Christie's Images (tl); Tallandier (crb). **TopFoto.co.uk:** Woodmansterne (bl). **184 Bonhams Auctioneers, London. Dorling Kindersley:** Natural History Museum, London (fcr, bl). **Gemological Institute of America Reprinted by Permission:** Robert Weldon (fbl). **Science Photo Library:** Science Stock Photography (cr). **185 Alamy Stock Photo:** Lanmas (br); Universal Images Group North America LLC / DeAgostini (cl). **Gemological Institute of America Reprinted by Permission. iRocks.com/Rob Lavinsky Photos. 186 Science Photo Library:** Mark a. Schneider (tc); Natural History Museum, London (cl). **187 Science Photo Library:** Mark a. Schneider (cr). **188-189 Bridgeman Images:** Werner Forman Archive. **190 Corbis:** Scientifica (tl). **Dorling Kindersley:** Natural History Museum, London (fbl, cr, fbr); The University of Aberdeen (fcr); Ruth Jenkinson (bl).**191 Corbis:** Scientifica (cl). **Dorling Kindersley:** Durham University Oriental Museum (tl); Pennsylvania Museum of Archaeology and Anthropology (cr, fcr); Natural History Museum / Colin Keates (fcl).**192 Bonhams Auctioneers, London. Gemological Institute of America Reprinted by Permission:** Nathan Renfro (bl); Robert Weldon (fbr); Kevin Schumacher (tl). **Getty Images:** Matteo Chinellato - ChinellatoPhoto (cr). **193 Alamy Stock Photo:** Universal Images Group North America LLC / DeAgostini (tl). **Dorling Kindersley:** Natural History Museum, London (fbl, cl). **194 Bridgeman Images:**Christie's Images. **195 Alamy Stock Photo:** GL Archive (cl). **Bridgeman Images:** Christie's Images (tl). **Corbis:** Chris Wallberg / dpa (crb). **Getty Images:** Universal History Archive (bl). **196 Bonhams Auctioneers, London. Bridgeman Images:** Private Collection / Ken Welsh (bc). **Dorling Kindersley:** Ruth Jenkinson (cl). **197 Bonhams Auctioneers, London. Dorling Kindersley:** Natural History Museum, London (cl); Tim Parmenter / Natural History Museum, London (tl). **Getty Images:** Ron Evans (cr). **198 Bridgeman Images:** Yale Center for British Art, Paul Mellon Collection, USA (clb). **Dorling Kindersley:** Natural History Museum, London (cr); Tim Parmenter / Natural History Museum, London (tl); Ruth Jenkinson (cl, br); Harry Taylor (fcl). **199 Alamy Stock Photo:** Corbin17 (cr).**Dorling Kindersley:** Natural History Museum, London (cl, bc); Tim Parmenter / Natural History Museum, London (tl, c). **Gemological Institute of America Reprinted by Permission:** Robert Weldon (bl). **200-201 Getty Images:** Handout. **202 123RF.com:** Valentin Kosilov (cr). **Alamy Stock Photo:** Alan Curtis / LGPL (c); Susan E. Degginger (bl). **Dorling Kindersley:** (cl); Tim Parmenter / Natural History Museum, London (br, bc); Harry Taylor (tl). **203 Dorling Kindersley:** Natural History Museum, London (cr); Holts (fcr). **Fellows Auctioneers. 204 Alamy Stock Photo:** Blend Images (bl).**Corbis. Dorling Kindersley:** Natural History Museum, London (tl, br, fbr). **Getty Images:** Ron Evans (fcr). **205 123RF.com:** vvoennyy (tl, cl). **Dorling Kindersley:** Oxford University Museum of Natural History (fcl). **Getty Images:** Arpad Benedek (br); Ron Evans (bl). **206 Dorling Kindersley:** Natural History Museum, London (tl, cla). **207 Dorling Kindersley:** Holts (cla); Ruth Jenkinson (tl); Ruth Jenkinson (tc, cra); Holts (c). **Science Photo Library:** Millard H. Sharp (bl). **208 Alamy Stock Photo:** Universal Images Group North America LLC / DeAgostini (br). **Bonhams Auctioneers, London. Dorling Kindersley:** Natural History Museum, London (cl); Ruth Jenkinson (bl). **209 Bonhams Auctioneers, London. Dorling Kindersley:** Holts (tl, cl, cr). **210 © Cartier. 211 © Cartier. Getty Images:**Cecil Beaton (bl); Alfred Eisenstaedt (cr). **Van Cleef & Arpels. 212 Dorling Kindersley:** Holts (tl, fcr, fcl); Natural History Museum, London (br); Holts (fbr). **Getty Images:** UniversalImagesGroup (bl).**213 Dorling Kindersley:** Holts (cl); Holts (bc). **Fellows Auctioneers. Kent Raible:** kentraible.com (c). **Tadema Gallery. 214 Bridgeman Images:** Pictures from History. **215 Bridgeman Images:**Pictures from History (tl). **images reproduced courtesy of Powerhouse Museum:** Gift of Mr Alastair Morrison, 1992 (ca). **216 Alamy Stock Photo:** Valery Voennyy (tl). **Bonhams Auctioneers, London. Dorling Kindersley:** Natural History Museum, London (cr, cl). **217 Alamy Stock Photo:** Nika Lerman (tl). **Bonhams Auctioneers, London. Bridgeman Images:** Vorontsov Palace, Crimea, Ukraine (br). **Corbis. Dorling Kindersley:** Ruth Jenkinson (bl). **218-219 Corbis. 220 Alamy Stock Photo:** Alan Curtis (tl); Greg C Grace (fcr). **Bonhams Auctioneers, London. 221 Alamy Stock Photo:** Universal Images Group North America LLC / DeAgostini (cl). **Bonhams Auctioneers, London. Corbis. 222 Bonhams Auctioneers, London. Bridgeman Images:** Pictures from History (bl).**Dorling Kindersley:** Natural History Museum, London (tl, fcr); Holts (cl). **223 Dorling Kindersley:** Natural History Museum, London (cl, c). **Gemological Institute of America Reprinted by Permission. Science Photo Library:** Natural History Museum, London (bl). **224 Alamy Stock Photo:** Private Collection. **225 Bridgeman Images:** British Royal Family (cr). **Corbis:** Hulton-Deutsch Collection (tl). **Getty Images:** Peter Macdiarmid (tc); Ben Stansall / AFP (bl). **226 Bulgari. 227 Bonhams Auctioneers, London. © Cartier. Dorling Kindersley:** Holts (tl). **228 Alamy Stock Photo:**Arco Images GmbH (fbr). **Dorling Kindersley:** Natural History Museum, London (cl, cr, fbl); Holts (tc, tc/Tourmaline Cut, fcl, bl, br). **229 Bulgari. © Cartier. Dorling Kindersley:** Holts (c). **Lang Antiques:** (cl). **Tadema Gallery. 230 Photo Scala, Florence:** The Metropolitan Museum of Art / Art Resource. **231 Bridgeman Images:** Bolivar Museum, Caracas, Venezuela (cr). **Getty Images:**Universal History Archive (cla). **Photo Scala, Florence:** (bl); The Metropolitan Museum of Art / Art Resource (tl). **232 Alamy Stock Photo:** Zoonar GmbH (tl). **Dorling Kindersley:** Natural History Museum, London (bc); Holts (c, bl). **233 Bonhams Auctioneers, London. © Cartier. Van Cleef & Arpels. 234 akg-images:** The British Library Board. **235 akg-images:** Album / Oronoz (bl). **Alamy Stock Photo:** Moviestore collection Ltd (fcr). **Bridgeman Images:** Topkapi Palace Museum, Istanbul, Turkey (tl). **Corbis:** R. Hackenberg (crb). **photographersdirect.com:** (c). **236 The Trustees of the British Museum**. **237 Bridgeman Images:** Christie's Images (br); Kremlin Museums, Moscow, Russia (bl). **The Trustees of the British Museum. Dorling Kindersley:** Natural History Museum, London (tl). **238 Dorling Kindersley:** Colin Keates (tr); Holts (tl, c); Natural History Museum, London (tc, cl). **239 Dorling Kindersley:** Natural History Museum, London (cr); Holts (tl, c, fbl, bl, br/Morganite).**iRocks.com/Rob Lavinsky Photos. 240 1stdibs, Inc. Bonhams Auctioneers, London:** (tl, bl). **Bridgeman Images:** Christie's Images (fcr). **© Cartier. Dorling Kindersley:** Holts (fbl). **241 1stdibs, Inc. Bonhams Auctioneers, London:** (cl). **© Cartier. 242 Getty Images:** Brendan Smialowski / AFP. **243 Atelier Munsteiner:** (tl). **Bridgeman Images:** Christie's Images (cr). **244 Atelier Munsteiner. Alice Cicolini:** (bc). **Michael M. Dyber:** Sena Dyber (br). **245 © Cartier. Michael M. Dyber:** Sena Dyber (l, c, br). **246 Bonhams Auctioneers, London:** (fcl, cl, cr, br). **Dorling Kindersley:** Natural History Museum, London (bc). **Getty Images:** DEA / R. APPIANI (bl). **247 Alamy Stock Photo:** Universal Images Group North America LLC / DeAgostini (fbr). **Bonhams Auctioneers, London:** (fcl).**Dorling Kindersley:** Natural History Museum, London (cr, fbl, bl). **Gemological Institute of America Reprinted by Permission:** Robert Weldon (tl, br). **248-249 Alamy Stock Photo:** Hemis. **249 Photo Scala, Florence:** The Metropolitan Museum of Art / Art Resource (bc). **Courtesy of Sotheby's Picture Library, London:** Private Collection (br). **250 Bonhams Auctioneers, London. Dorling Kindersley:** Natural History Museum, London (fcl); Ruth Jenkinson (bl). **Getty Images:** Corbis / Ron Evans / Ocean (cl). **251 123RF.com:** Chatchai Chattranusorn (bl). **Alamy Stock Photo:** Aysegul Muhcu (br). **Dorling Kindersley:** Natural History Museum, London (tl, bc). **Getty Images:** (cl); Mark Schneider (cr). **252 Bonhams Auctioneers, London. Dorling Kindersley:** Natural History Museum, London (tl, bc). **Gemological Institute of America Reprinted by Permission:** Robert Weldon (cr); Robert Weldon (br); Robert Weldon (fcr). **253 Alamy Stock Photo:** Nika Lerman (br). **Dorling Kindersley:** Natural History Museum, London (fcl, cl, cr); Holts (fcr). **Fellows Auctioneers. 254 Dorling Kindersley:** Natural History Museum, London (c); Holts (tl, br). **255 Alamy Stock Photo:**INTERFOTO (br). **Bonhams Auctioneers, London. Dorling Kindersley:** Holts (fcl, fcr). **256 Getty Images:** JTB Photo. **257 Alamy Stock Photo:** Kumar Sriskandan (bl). **Bridgeman Images:** Pictures from History (c). **Corbis:** Melvyn Longhurst (tl). **Getty Images:** Keystone (cr). **258 Bridgeman Images:** Walters Art Museum, Baltimore, USA. **259 Fabergé:** (br). **Tadema Gallery:** (bc). **260 Dorling Kindersley:** Natural History Museum, London (tl, ftr); Oxford University Museum of Natural History (fcl); Holts (tr, cr, fcr, bc, br). **261 Dorling Kindersley:** Natural History Museum, London (fbl); Holts (cl, tr, fbr). **Science Photo Library:** Joel Arem (br). **262 1stdibs, Inc. Bridgeman Images:** Christie's Images (br); Private Collection / Photo © Christie's Images (fbl). **The Trustees of the British Museum. © Cartier. Dorling Kindersley:** Holts (tl, cl). **Fellows Auctioneers:** (bl). **263 1stdibs, Inc. Bridgeman Images:** Christie's Images (tr). **Bulgari. © Cartier. Dorling Kindersley:** Holts (ftl, fcl, fbl). **264 Universal News And Sport:** Birmingham Museums Trust. **265 Alamy Stock Photo:** World History Archive (tl, c, bl). **The Trustees of the British Museum. Getty Images:** George Munday / Design Pics / Corbis (cr). **266 Dorling Kindersley:** Natural History Museum, London (c); Natural History Museum, London (clb); Natural History Museum, London (bl); Holts (tl, bc). **266-267 Dorling Kindersley:** Holts (c). **267 123RF.com:** Vvoennyy (bc). **Dorling Kindersley:** Holts (bl, tl); Ruth Jenkinson (tc, cra); Holts (crb). **268 Dorling Kindersley:** Natural History Museum, London (cl, c); Holts (tl, bl, r). **269 1stdibs, Inc. Bonhams Auctioneers, London. Dorling Kindersley:** Holts (tl, fcl, bl, fcr). **270 Bridgeman Images:** J. Paul Getty Museum, Los Angeles, USA. **271 Alamy Stock Photo:**Oldtime (cr); Universal Art Archive (tl). **Corbis:** Stapleton Collection (bc). **Rex by Shutterstock:** Nils Jorgensen (cl). **272 Dorling Kindersley:** Natural History Museum, London (tl, r); Holts (bc). **Gemological Institute of America Reprinted by Permission:** Eric Welch (cl). **273 Bonhams Auctioneers, London.**

Dorling Kindersley: Holts (tl, cl, bl, bc). **274 1stdibs, Inc. Alamy Stock Photo:**Stela Knezevic (cl). **Dorling Kindersley:** Natural History Museum, London (tl, cr, bl); Oxford University Museum of Natural History (fcl); Holts (fcr). **275 Gemological Institute of America Reprinted by Permission:** Robert Weldon (br). **Science Photo Library:** Dorling Kindersley / UIG (tl). **276 Bonhams Auctioneers, London. Dorling Kindersley:** Natural History Museum, London (cl, bl, br).**Gemological Institute of America Reprinted by Permission. Science Photo Library:** Natural History Museum, London (cr). **277 123RF.com:** Vvoennyy (bl, tl). **Bonhams Auctioneers, London. Dorling Kindersley:** Natural History Museum, London (c, bc). **278 The Walters Art Museum, Baltimore:** Acquired by Henry Walters, 1930 / Photographer Susan Tobin. **279 Corbis:** Historical Picture Archive (bl). **Getty Images:** DEA / G. Dagli Orti (crb); Print Collector (cr). **Press Association Images:** ABACA Press (tr, ca). **280 Bonhams Auctioneers, London:** (cr, br). **Dorling Kindersley:** Ruth Jenkinson (tl, c). **Dreamstime.com:** David Porter (bl). **281 Bonhams Auctioneers, London:** (tl, br). **Dorling Kindersley:** Natural History Museum, London (bl). **282 Bonhams Auctioneers, London:**(c). **Dorling Kindersley:** Natural History Museum, London (cl, bc, r). **Science Photo Library:** Natural History Museum, London (tl). **283 Bonhams Auctioneers, London:** (tl, c). **Dorling Kindersley:**Natural History Museum, London (bl, cr). **Gemological Institute of America Reprinted by Permission:** Robert Weldon (bc); Robert Weldon (br). **284 Bridgeman Images:** De Agostini Picture Library / G. Cigolini. **285 Alamy Stock Photo:** AF Fotografie (c). **Corbis:** Marc Dozier (tl). **Getty Images:** Jean-Claude Deutsch / Paris Match (cr). **286 Bridgeman Images:** Natural History Museum, London, UK (tr). **Dorling Kindersley:** Holts (cla, cr, tr/Emerald); Holts (ca). **Science Photo Library:** (tc); Dorling Kindersley / UIG (clb). **286-287 Dorling Kindersley:** Holts (c). **288 Corbis:** Michele Falzone / JAI.**289 Alamy Stock Photo:** PRISMA ARCHIVO (cl). **Bridgeman Images:** Abate, Niccolo dell' (c.1509-71) / Galleria Estense, Modena, Italy / Ghigo Roli (bc). **Corbis:** (tl). **Getty Images:** Bettmann / Corbis (cr). **291 Dorling Kindersley:** Ruth Jenkinson (c). **292 V&A Images / Victoria and Albert Museum, London. 293 Bridgeman Images:** Christie's Images (bc, br, bl). **Gemological Institute of America Reprinted by Permission. 294 Bonhams Auctioneers, London. Bridgeman Images:** Christie's Images (fcr, fbr). **Dorling Kindersley:** Natural History Museum, London (cl); Holts (c). **Fellows Auctioneers:** (tr). **295 AB JEWELS Ltd.:** (cl). **© Cartier. Mikimoto:** (bl). **Antonio Virardi of Macklowe Gallery, New York. YOKO London:** (br). **296 Alamy Stock Photo:** Ian Dagnall. **297 Bridgeman Images. Rex by Shutterstock:** SNAP (bl). **298 Mary Evans Picture Library:** Alinari Archives, Florence - Reproduced with the permission of Ministero per i Beni e le Attivit… Cu (c). **The Art Archive:**DeA Picture Library (cr/r). **299 Alamy Stock Photo:** Graham Clarke (cl). **Dorling Kindersley:** Jewellery design Maya Brenner (cr); Ruth Jenkinson (bl). **Fellows Auctioneers. V&A Images / Victoria and Albert Museum, London. 300 The Trustees of the British Museum. 301 Bridgeman Images. Corbis:** 145 / Burazin / Ocean (tl). **Getty Images:** Time Life Pictures (cr). **302-303 Harry Winston.304 Dorling Kindersley:** CONACULTA-INAH-MEX. **305 Alamy Stock Photo:** Aurora Photos (c); Granger, NYC. (tl). **Corbis:** Robert Harding Productions (crb). **Dorling Kindersley:** CONACULTA-INAH-MEX (bl). **Getty Images:** DEA / G. Dagli Orti (cr). **306 Dorling Kindersley:** Natural History Museum, London (cr, fbr); Holts (cl). **Getty Images:** Print Collector (bl). **307 1stdibs, Inc. Dorling Kindersley:** Natural History Museum, London (cl, cr, fcr); Holts (tc). **308 Alamy Stock Photo:** The Natural History Museum (br); PjrStudio (bl). **Dorling Kindersley:** Ruth Jenkinson (cl, bc). **309 Alamy Stock Photo:** Siim Sepp (cl). **Dorling Kindersley:** Natural History Museum, London (tl). **Getty Images:** Don Emmert (br); Ron Evans (bc). **310 Alamy Stock Photo:** Evgeny Parushin (fcr); David Sanger photography (bl). **Dorling Kindersley:** Holts (br); Oxford University Museum of Natural History (cl). **311 Alamy Stock Photo:** Editorial (br). **Dorling Kindersley:** Holts (cr). **iStockphoto.com:**desnik (tc). **Tadema Gallery. 312 akg-images:** Ruhrgas AG. **313 akg-images:** Ruhrgas AG (bl); Universal Images Group / Sovfoto (tl). **Photo Scala, Florence:** Photo Josse (cr). **TopFoto.co.uk:** RIA Novosti (cl). **314 Dorling Kindersley:** Natural History Museum, London (c); Holts (bl, cr, br). **315 Bridgeman Images:** (br). **Dorling Kindersley:** Holts (tl, tc, c); Judith Miller / Christbal (cl). **Fellows Auctioneers. Getty Images:** Christie's Images (cr). **316 Bayerische Schlösserverwaltung:** Maria Scherf / Rainer Herrmann, München. **317 Alamy Stock Photo:** DBI Studio (fcr). **Getty Images:** NY Daily News Archive (bl). **Photo Scala, Florence:** 2016. Image copyright The Metropolitan (tl). **Courtesy of Sotheby's Picture Library, London. 318 Bonhams Auctioneers, London. Geology. com:**(tl). **Roland Smithies / luped.com. 319 Alamy Stock Photo:** Age Fotostock (br). **Dorling Kindersley:** Natural History Museum, London (bc); Holts (bl). **321 Dreamstime.com:** Milahelp S.r.o.. **322 Alamy Stock Photo:** Universal Images Group North America LLC / DeAgostini (bl); WILDLIFE GmbH (fcr). **Dorling Kindersley:** Natural History Museum, London (tl, c). **Dreamstime.com:** Milahelp S.r.o. (cr, bc, br). **323 Dorling Kindersley:** Pitt Rivers Museum, Oxford (fbr, br); Natural History Museum, London (tl, fcl, cl); Ruth Jenkinson (fcr, bl); Holts (cr/obsidian). **324 Dorling Kindersley:**Aberdeen (bc); Natural History Museum, London (tl, br); Pennsylvania Museum of Archaeology and Anthropology (bl). **325 Dorling Kindersley:** Aberdeen (cl); Pennsylvania Museum of Archaeology and Anthropology (tl, fcr); Ruth Jenkinson (cr/sandstone). **Dreamstime.com:** Srinakorn Tangwai (bl). **326 Corbis:** Christian Handl. **327 123RF. com:** Patrick Guenette (cl). **Alamy Stock Photo:** PRISMA ARCHIVO (cr). **Corbis:** Free Agents Limited (tl). **iStockphoto. com:** &#169 Bj°rn Gjelsten (bc). **328 1stdibs, Inc. Alamy Stock Photo:** Wladimir Bulgar (cl). **Bridgeman Images. 329 Dorling Kindersley:** Natural History Museum, London (tl, fcl, cl). **330 Alamy Stock Photo:** Adam Eastland Art + Architecture. **331 Bridgeman Images:** Bonhams, London, UK (cla). **Corbis:** (tl, bl). **ICCD – Fondo Ministero della Pubblica Istruzione Gabinetto fotografico della Regia Soprintendenza alle Gallerie:** (crb). **332 Bridgeman Images:** Christie's Images (cl). **Getty Images:** Ishara S. Kodikara (bl). **Rex by Shutterstock:** Universal History Archive / UIG (tr). **Science Photo Library:** Tom McHugh (br). **333 Crater of Diamonds, State Park:** (fcrb). **Kaufmann de Suisse:** (l). **Press Association Images:** AP Photo / Keystone, Laurent Gillieron (cra). **334-335 © Cartier. 337 Dorling Kindersley:** Holts (c). **338 Alamy Stock Photo:** Fabrizius Troy (cla). **Bonhams Auctioneers, London:** (fcra).**Dorling Kindersley:** Natural History Museum, London (clb, fbl, bl, br); Holts (tc, tr, fcla); Natural History Museum, London (crb); Natural History Museum, London (fbr). **iRocks.com/Rob Lavinsky Photos. 339 Alamy Stock Photo:** Arco Images GmbH (br); Steve Sant (fcla). **Bonhams Auctioneers, London. Dorling Kindersley:** Natural History Museum, London (cra, fcra, crb/Alexandrite, fcrb, fbl, fbr); Ruth Jenkinson / Holts (tl); Holts (clb). **Getty Images:** Corbis / ION / amanaimages (ftr). **iRocks.com/Rob Lavinsky Photos:** (fclb). **340 Alamy Stock Photo:** Goran Bogicevic (fbr). **Dorling Kindersley:** Natural History Museum, London (ftr, fcla, cra, crb); Ruth Jenkinson (fcra); Holts (clb). **Getty Images:** Arpad Benedek (cla); Ron Evans (ftl). **Roland Smithies / luped.com:** (tl). **341 Bonhams Auctioneers, London. Dorling Kindersley:** Natural History Museum, London (ftl, tl, tr, cla, fcra, crb, br); Tim Parmenter / Natural History Museum, London (ftr); Ruth Jenkinson (cra); Holts (fcla, fcrb, bc/Heliodor); Natural History Museum, London (br). **V&A Images / Victoria and Albert Museum, London. 342 Dorling Kindersley:** Natural History Museum, London (tr, br, fbr); Holts (tl, cla, cra, fcrb); Ruth Jenkinson (fcla). **343 Alamy Stock Photo:** Zoonar GmbH (fcra). **Dorling Kindersley:** Natural History Museum, London (fcla/Moldavite, cla/Kornerupine, fbl, bl/Microcline); Holts (tl, cra/Ammolite, fclb, clb, fbr). **344 Dorling Kindersley:** Natural History Museum, London (ftr, fcla, cra); Holts (tl, tr, fclb, cb, crb, br); Holts (fcra). **Getty Images:** Arpad Benedek (clb). **345 Dorling Kindersley:** Natural History Museum, London (ftr, cra, fcra); Holts (ftl, tr, fcla, fclb, fcrb, br, fbr). **346 Bonhams Auctioneers, London. Dorling Kindersley:** Natural History Museum, London (fcra, fcrb); Holts (ftl, tl, cra, fbr); Ruth Jenkinson (fclb, clb). **347 Bonhams Auctioneers, London:** (clb). **Dorling Kindersley:** Natural History Museum, London (tr, fclb); Holts (ftl); Ruth Jenkinson (cla). **Science Photo Library:** Natural History Museum, London (tl). **349 Dorling Kindersley:** (br); Natural History Museum, London / Colin Keates (fclb); Natural History Museum, London / Tim Parmenter (bl). **350 Alamy Stock Photo:** assistant (b). **Dorling Kindersley:** Oxford University Museum of Natural History / Gary Ombler (ca). **Science Photo Library:** Natural History Museum, London (tr). **351 Alamy Stock Photo:** Alan Curtis / LGPL (bc). **Dorling Kindersley:** Linda Burgess (fcl); Natural History Museum, London / Colin Keates (cla); Harry Taylor (fbr); Harry Taylor (ftr); Harry Taylor (cr). **352 Dorling Kindersley:** Oxford University Museum of Natural History / Gary Ombler (bc); Harry Taylor (fcr); Harry Taylor (tr); Harry Taylor (cla); Harry Taylor (fcl). **353 Alamy Stock Photo:** YAY Media AS (t).**Dorling Kindersley:** Harry Taylor (bl). **354 Alamy Stock Photo:** Fabrizio Troiani (tl). **Dorling Kindersley:** Harry Taylor (br); Harry Taylor (fcr). **355 Alamy Stock Photo:** Phil Degginger (cr). **Dorling Kindersley:** Linda Burgess (ftr); Harry Taylor (fbr); Harry Taylor (bl); Harry Taylor (fcla); Harry Taylor (tl). **356 Alamy Stock Photo:** Universal Images Group North America LLC / DeAgostini (br). **Dorling Kindersley:** Harry Taylor (cl); Harry Taylor (fcr); Harry Taylor (cr); Harry Taylor (fclb). **357 Dorling Kindersley:** Natural History Museum, London / Colin Keates (cr); Gary Ombler (fbr); Harry Taylor (bl); Harry Taylor (ftr); Harry Taylor (tl); Harry Taylor (fcl). **358 Alamy Stock Photo:** RGB Ventures / SuperStock (bl). **Dorling Kindersley:** Harry Taylor (cla); Harry Taylor (ftr). **359 Alamy Stock Photo:** The Natural History Museum (bl); The Natural History Museum (fcl). **Dorling Kindersley:** Alamy / D. Hurst (cla); Natural History Museum, London / Colin Keates (fcra); Harry Taylor (fbr). **360 Alamy Stock Photo:** Pat Behnke (tr). **Dorling Kindersley:** Linda Burgess (cr); Linda Burgess (fbr); Harry Taylor (fclb); Courtesy of Holts / Ruth Jenkinson (cl). **361 Alamy Stock Photo:** blickwinkel (r). **Dorling Kindersley:** Harry Taylor (fcl). **362 Alamy Stock Photo:** Universal Images Group North America LLC / DeAgostini. **363 Alamy Stock Photo:** John Cancalosi (cra). **Dorling Kindersley:** Natural History Museum, London / Tim Parmenter (fbr). **364 Alamy Stock Photo:** Hemis (b). **Dorling Kindersley:** Harry Taylor (ftr). **365 Dorling Kindersley:** Harry Taylor (tr); Harry Taylor (fbr); Harry Taylor (br); Harry Taylor (fbl); Harry Taylor (clb). **366 Alamy Stock Photo:** Siim Sepp (cl). **Dorling Kindersley:** Gary Ombler (tr); Gary Ombler (br); Gary Ombler (fclb); Harry Taylor (fcr). **367 Alamy Stock Photo:** Julie Thompson (b). **Dorling Kindersley:** Harry Taylor (tl); Harry Taylor (ftr). **368 Dorling Kindersley:** Harry Taylor (tr); Harry Taylor (fcl); Harry Taylor (br). **369 Alamy Stock Photo:** Galyna Andrushko. **370 Dorling Kindersley:** Harry Taylor (tl); Harry Taylor (fbr); Harry Taylor (fcl); Harry Taylor (cr); Harry Taylor (bl). **371 Alamy Stock Photo:** Diego Barucco (b). **Dorling Kindersley:** Harry Taylor (cra). **372 Alamy Stock Photo:** Universal Images Group North America LLC (b). **Dorling Kindersley:** Harry Taylor (ftr). **373 Dorling Kindersley:** Linda Burgess (fcra); Oxford University Museum of Natural History / Gary Ombler (bl); Harry Taylor (fcl); Harry Taylor (crb); Harry Taylor (cla). **374 Dorling Kindersley:** Gary Ombler (cr); Harry Taylor (cla); Gary Ombler (br, fcl); Harry Taylor (bl). **375 Alamy Stock Photo:** Universal Images Group North America LLC / DeAgostini (t). **Getty Images:** Corbis (br). **376 Alamy Stock Photo:** Antony Souter. **377 Corbis:** (cla). **Dorling Kindersley:** Linda Burgess (cr); Harry Taylor (clb); Harry Taylor (fcra); Harry Taylor (bl). **378 Alamy Stock Photo:** GC Minerals (b). **379 Dorling Kindersley:** Linda Burgess (cra); Linda Burgess (br); Gary Ombler (cl); Harry Taylor (fclb).**380 Dorling Kindersley:** Gary Ombler (fclb); Harry Taylor (br); Harry Taylor (tr). **381 Alamy Stock Photo:** Friedrich von Hörsten. **382 Dorling Kindersley:** Colin Keates / Natural History Museum (fcl); Gary Ombler (fbr); Harry Taylor (bl); Gary Ombler (cr); Gary Ombler (fcra). **383 Alamy Stock Photo:** Jamie Pham (t); PjrStudio (bl). **384 Alamy Stock Photo:** Universal Images Group North America LLC / DeAgostini (l). **Dorling Kindersley:** Harry Taylor (cr). **385 Dorling Kindersley:** Colin Keates / Natural History Museum, London (fcr); Gary Ombler (fbl); Gary Ombler (br); Gary Ombler (cl). **386 Alamy Stock Photo:** David Chapman (b). **387 Dorling Kindersley:** Gary Ombler (tr); Gary Ombler (br); Gary Ombler (bl); Harry Taylor / Courtesy of the Natural History Museum, London (fcl); Harry Taylor (cr). **388 Dorling Kindersley:** Harry Taylor (cla); Harry Taylor (fbr); Harry Taylor (bl); Harry Taylor (fcl); Harry Taylor (cr). **389 Alamy Stock Photo:** repOrter (r). **Dorling Kindersley:** Harry Taylor (cl). **390 Alamy Stock Photo:** Universal Images Group North America LLC / DeAgostini (tl). **Dorling Kindersley:** Harry Taylor (cr). **391 Dorling Kindersley:** Linda Burgess (tr); Gary Ombler (bl); Harry Taylor (br); Harry Taylor (cr); Gary Ombler (cl); Harry Taylor (fcla). **393 Alamy Stock Photo:** Borislav Dopudja (tr). **Dorling Kindersley:** Gary Ombler (bl); Gary Ombler (fclb). **394 Alamy Stock Photo:**Doug Houghton (b). **Dorling Kindersley:** Harry Taylor (tr). **395 Dorling Kindersley:** Harry Taylor (tr); Harry Taylor (fcl); Harry Taylor (cr); Harry Taylor (fbr); Harry Taylor (bl). **396 Dorling Kindersley:**Linda Burgess (bl); Linda Burgess (fcra); Gary Ombler (cla); Harry Taylor (fcl); Harry Taylor (cr). **397 Alamy Stock Photo:** age fotostock. **398 Alamy Stock Photo:** Sergey Skleznev (t). **Dorling Kindersley:** Gary Ombler (br). **399 Dorling Kindersley:** Gary Ombler (cr); Harry Taylor (tr); Gary Ombler (bl); Harry Taylor (br); Tim Parmenter / Natural History Museum, London (cl). **400 Dorling Kindersley:** Gary Ombler (cl); Harry Taylor (br); Harry Taylor (tr); Harry Taylor (fcr); Gary Ombler (fclb). **401 Alamy Stock Photo:** Siim Sepp (tr). **Dorling Kindersley:** Tim Parmenter / Natural History Museum, London (fclb); Harry Taylor (fbr); Harry Taylor (bl). **402 Alamy Stock Photo:** Borislav Dopudja (tl). **Dorling Kindersley:** Linda Burgess (crb); Harry Taylor (fcr). **403 Dorling Kindersley:** Linda Burgess (bl); Harry Taylor (tl); Harry Taylor (ftr); Harry Taylor (fbr); Harry Taylor (fcl); Harry Taylor (cr). **404 Alamy Stock Photo:** PjrStudio (b). **Dorling Kindersley:** Harry Taylor (cl). **405 Dorling Kindersley:** Gary Ombler (fcl); Harry Taylor (fbr); Harry Taylor (bl); Harry Taylor (cr). **406 Dorling Kindersley:** Gary Ombler (fbr); Tim Parmenter / Natural History Museum, London (bl); Harry Taylor (fcl); Harry Taylor (cr); Tim Parmenter / Natural History Museum, London (tl). **407 Alamy Stock Photo:** (t). **408 Dorling Kindersley:** Gary Ombler (br); Harry Taylor (tr); Harry Taylor (fcr); Tim Parmenter / Natural History Museum, London (cl); Gary Ombler (fclb). **409 Alamy Stock Photo:** Bruce Beck. **410 Alamy Stock Photo:** Don Johnston_WU (t). **411 Dorling Kindersley:** Gary Ombler (tl); Gary Ombler (bl); Tim Parmenter / Natural History Museum, London (fcl); Gary Ombler (cr); Harry Taylor (fbr). **413 Alamy Stock Photo:** Oleksiy Maksymenko (br). **Dorling Kindersley:** Gary Ombler (fcl). **414 Alamy Stock Photo:** Paul Kingsley. **415 Dorling Kindersley:** Colin Keates / Natural History Museum, London (ftr); Gary Ombler (fbl); Richard Leeney (fbr); Harry Taylor (cl); Gary Ombler (cr). **416 Dorling Kindersley:** Colin Keates / Natural History Museum, London (fbr); Harry Taylor (cl); Harry Taylor (cra). **417 Alamy Stock Photo:** Stephen Emerson. **418 Alamy Stock Photo:** imageBROKER (b). **Dorling Kindersley:** Gary Ombler (cra). **419 Dorling Kindersley:** Gary Ombler / Oxford University Museum of Natural History (br); Harry Taylor (ftl); Harry Taylor (fbl); Tim Parmenter / Natural History Museum, London (tr). **Getty Images:** John Cancalosi (fcr). **420 Dorling Kindersley:** Linda Burgess (cr); Tim Ridley (fbr); Gary Ombler (fcl); Harry Taylor (fcra); Harry Taylor (bl). **421 Alamy Stock Photo:** Gavin Hellier (r). **Dorling Kindersley:** Linda Burgess (cl). **422 Alamy Stock Photo:** Andres Rodriguez. **423 Alamy Stock Photo:** Sabena Jane Blackbird (br). **Dorling Kindersley:**Andreas Einsiedel (tl); Tim Parmenter (b); Andreas Einsiedel (cr). **424 Alamy Stock Photo:** Siim Sepp (tr). **Dorling Kindersley:** Andreas Einsiedel (cl); Harry Taylor (fbl); Harry Taylor (br). **425 Alamy Stock Photo:** nick mclaren (fbr). **Dorling Kindersley:** Colin Keates / Natural History Museum, London (ftr); Harry Taylor (fcl). **426 Dorling Kindersley:** Linda Burgess (fcl); Colin Keates / Natural History Museum, London (crb); Harry Taylor (tl); Tim Parmenter / Natural History Museum, London (fcr). **427 Alamy Stock Photo:** imageBROKER

All other images © Dorling Kindersley
For further information see: **www.dkimages.com**